# DICTIONARY OF COMMUNICATIONS TECHNOLOGY

## Second Edition

# DICTIONARY OF COMMUNICATIONS TECHNOLOGY

## Terms, Definitions and Abbreviations
## Second Edition

**Gilbert Held**

4-Degree Consulting
Macon, Georgia, USA

JOHN WILEY & SONS
Chichester · New York · Brisbane · Toronto · Singapore

*Other Wiley Editorial Offices*

John Wiley & Sons, Inc., 605 Third Avenue,
New York, NY 10158-0012, USA

Jacaranda Wiley Ltd, 33 Park Road, Milton,
Queensland 4064, Australia

John Wiley & Sons (Canada) Ltd, 22 Worcester Road,
Rexdale, Ontario M9W 1L1, Canada

John Wiley & Sons (SEA) Pte Ltd, 37 Jalan Pemimpin #05-04,
Block B, Union Industrial Building, Singapore 2057

*British Library Cataloguing in Publication Data*

A catalogue record for this book is available from the British Library

ISBN 0 471 95554 2      ISBN 0 471 95126 9 (pbk)

Typeset in 9.5/11.5pt Times from author's disks by Production Technology
Department, John Wiley & Sons Ltd, Chichester
Printed and bound in Great Britain by Bookcraft Ltd

# PREFACE

The telecommunications field is perhaps the most rapidly evolving area of all disciplines. In just a few years the use of microprocessors, fiber optics and satellites has dramatically changed the structure of the telecommunications industry, the product offerings of vendors and the facilities provided by common carriers. Accompanying this rapid evolution are a wealth of terms, definitions and abbreviations which are probably beyond the capability of any one individual to remember. Thus, my goal in developing this handbook was to provide the reader with a comprehensive up-to-date reference that could be used to refresh one's memory or to determine the specific meaning of a term or what an abbreviation might represent.

In developing this handbook I attempted to make its use as easy as possible by including both terms and abbreviations together instead of separating the two. As you thumb through this handbook you will note that all entries are arranged in a structured, alphabetical order, with letters followed by numerics. Thus, remembering that numerics follow letters will enable you to easily locate entries in this book.

In addition to noting the arrangement of entries as you browse through this handbook, you will also discover almost 300 illustrations and tables that contain important information probably never previously compiled into one, easy-to-use reference source. Our handbook was developed as an extension to a conventional dictionary, because in addition to defining terms and abbreviations it provides the reader with a comprehensive reference to probably the single largest collection of telecommunications information published within one binding.

Although we have attempted to develop a most comprehensive handbook of telecommunications terms, definitions and abbreviations, it is most probable that we have missed some vendor-specific entries. In addition, new telecommunications terms and abbreviations are constantly being added to reflect the introduction of new products and technologies. For this reason I welcome any information readers may care to provide; it can be sent to me directly or through my publisher.

**Gilbert Held**
*Macon, Georgia*

# ACKNOWLEDGEMENTS

The author is most appreciative of the cooperation and assistance of many organizations that resulted in the comprehensiveness of this book. Specifically, I would like to acknowledge the following organizations that granted permission to extract glossary information from previously published manuals, books and trade literature.

Atlantic Research Corporation
AT&T Network Systems
Auerback Publishers, Inc.
Codex Corporation
COM DEV
Department of the Navy
Digilog, Inc.
Digital Equipment Corporation
Dynatech Communications
General Datacomm Industries, Inc.
International Business Machines Corporation
Micom Systems, Inc.
Navtel
Premier Telecom Products, Inc.
Satellite Communications
Shure Brothers, Inc.
Siemens Information Systems, Inc.
Telefile, Inc.
Telenet
Tymnet

In addition to the previously mentioned organizations two individuals deserve special mention for their efforts. I would like to thank Ms Patricia Peel for her fine effort that was indispensable in the development of this book as well as for doublechecking the efforts of a young typist who provided the data entry effort for the electronic draft of the first edition of this book. Concerning the young typist, I would like to thank my son,

Jonathan, for spending a portion of his summer vacation using his computer to convert several shoeboxes of 3-by-5 index cards into the first edition of this book, which through an evolutionary process grew in size and scope of coverage into the handbook you are now reading.

**A**   1. Ampere. 2. Angstrom. 3. Attention.

**A Block**   In cellular communications one of the two sets of channels assigned by the U.S. Government to one of two telephone companies operating in a metropolitan area.

**AB Signaling**   A technique for carrying "on-hook/off-hook" telephony signaling information over T1 spans that use the D4 framing (12-frame superframe) format. In AB Signaling, the circuit provider (telephone company or PTT) "robs" the least significant bit of each channel octet in the 6th and 12th frames of each superframe to convey signaling information. Bits robbed from the 6th frame are called "A" bits and correspond to signals on the E-wire of an analog telephone. Bits robbed from the 12th frame are called "B" bits, and correspond to signals on the M-wire of an analog telephone.

**abandoned call**   A call in which the caller cancels the call after a connection has been made, but before conversation takes place.

**ABAnet**   American Bar Association Network.

**abbreviated addressing**   In packet switched networks, addressing in which a simple mnemonic code is used in lieu of the complete addressing information; the cross reference to complete address is stored in the packet assembler/disassembler (PAD).

**abbreviated and delayed ringing**   A private branch exchange (PBX) function which allows the ringing associated with a call on a station line to be transferred to another station that has an appearance of that line. The transfer occurs automatically after two ringing cycles.

**abbreviated dialing**   The feature of some PABX systems and other switches whereby frequently called numbers can be dialed by means of a brief dial code which the switch translates into a conventional telephone number with appropriate access and area codes.

**ABCD Signaling**   A technique for carrying "on-hook/off-hook" telephony signaling information over T1 spans that use the Extended Superframe (24-frame superframe) format. In ABCD Signaling, the circuit provider (telephone company or PTT) "robs" the least significant bit of each channel octet in the 6th, 12th, 18th, and 24th frames of each superframe to convey signaling information. Bits robbed from the 6th and 18th frames are called "A" bits and correspond to signals on the E-wire of an analog telephone. Bits robbed from the 12th and 24th frames are called "B" bits, and correspond to signals on the M-wire of an analog telephone. ABCD Signaling is used only as an interim technique during conversions from D4 framing to Extended Superframe format.

**ABM**   Accunet Bandwidth Manager.

**ABM**   Asynchronous Balanced Mode.

**abort**   A predefined software-controlled or operator-initiated action which results in the immediate cessation of an activity.

**ABR**   Automatic Baud Rate.

**absolute address**   A reference to a storage location that has a fixed displacement from absolute memory location zero.

**absolute calling**   Coding in which instruction are written in machine language (i.e. coding using absolute operators and addresses); coding that does not require processing before it can be understood by the computer.

**absorption**   The removal of energy from a radiated field by objects that retain the energy or conduct it to ground. Loss by absorption reduces the strength of a radiated signal.

**Abstract Syntax Notation 1 (ANS.1)**   An Open System Interconnection standard method for representing and encoding binary data to be transmitted over a network.

**ABT**   Abort Timer or Answer Back Tone.

**AC**   Alternating current.

**AC Signaling**   The use of alternating current signals or tones to accomplish transmission of information and/or control signals.

**Academic Operating System (AOS)**   IBM's version of the Berkeley Unix operating system.

**Academnet**   A network within Russia which connects universities.

**ACB**   1. In IBM's VTAM, Access Method Control Block. 2. In IBM's NCP, Adapter Control Block.

**ACB address space**   In IBM's VTAM, the address space in which the ACB is opened.

**ACB-based macro instruction**   In IBM's VTAM, a macro instruction whose parameters are specified by the user in an access method control block.

**ACB name**   In IBM's VTAM: 1. The name of

anACB macro instruction. 2. A name specified in the ACBNAME parameter of an APPL statement. Contrast with network name. *Note.* This name allows an ACF/VTAM application program that is used in more than one domain to specify the same application program identification (pointed to by the APPLID parameter of the program's ACB statement) in each copy. ACF/VTAM knows the program by both its ACB name and its network name (the name of the APPL statement). Program users within the domain can request a session using the ACB name or the network name; program users in other domains must use the network name (which must be unique in the network).

**ACC**   AlloCatable Channel.

**ACC**   Army Communications Command.

**Accent**   An AT&T telephone trademark.

**accept**   In an ACF/VTAM application program, to accept a CINIT request from an SSCP to establish a session with a logical unit; the application program acts as the primary end of the session. *Note.* The accept process causes a BIND request to be sent from the primary end of the session to the logical unit that will act as the secondary end of the session, requesting that the session be established and passing session parameters. For example, the session-initiation request that originally caused the SSCP to send the CINIT request may have resulted from a logon by the terminal operator, from a macro instruction issued by an ACF/VTAM application program, or from an ACF/VTAM operator command.

**acceptor of data (acceptor)**   A term used to describe any device capable of accepting data in a controlled manner; it is used in British Standard Interface Specifications to refer to devices which take data from a source.

**access**   In a local area network, the ability to work with files. Usually used in connection with files stored on network disks. Various access rights may be assigned to users.

**access charges**   Charges levied by local telephone

operating companies for providing their customers with access to long distance (interchange) carrier services.

**access code** 1. A series of digits which must be dialed to link your telephone line to a specific type of telephone service. 2. A series of letters and/or numbers which must be entered into a computer terminal or personal computer to provide access to specific data bases or mainframe computer software. 3. The five-digit code you must dial to use a long distance service other than your primary long distance company. 4. With some PBX equipment, the code you must dial to access special services such as WATS lines or private lines.

**access control** The management of rights to use resources.

**access group** All stations which have identical rights to make use of computer, network, or data PABX resources.

**access line** The connection between a subscriber's facility and a public network—either a PDN, public switched network, or public telephone network. Also called local line or local loop.

**access method** 1. In IBM environments, a host program managing the movement of data between the main storage and an input/output device of a computer system; BTAM, TCAM, VTAM are common data communications access methods. 2. In LAN technology, a means to allow stations to gain access to—to make use of—the network's transmission medium; classified as shared access (which is further divided into explicit access or contended access) or discrete access method.

**Access Method Control Block (ACB)** In IBM's VTAM, a control block that links an application program to VSAM or ACF/VTAM.

**Access Process Parameter (APP)** In packet switching, a list of parameters governing the operation of a specified port. (This list is communicated to the remote Packet Assembler/Disassembler (PAD) or Data Terminal Equipment (DTE) by the Data Qualifier (DQ) packet (X.29 protocol)).

**Access Request** The event that notifies the system of a user's desire to initiate a data communications session. It begins the access function and starts the counting of access time. Two specific examples of Access Request events are the off-hook event in the public switched telephone network, and the completion of a Connect Request by a terminal operator in the ARPANET.

**access rights** Privileges which are granted or not granted in order to control how users may work with files. For example, you must have the proper rights before you may read a file, delete a file, or modify a file.

**Access Tandem (AT)** An AT&T ESS switch used to provide carrier access to end offices (and possibly collocated stations).

**access time** A measurement of performance normally used to define the time interval it takes a disk drive to retrieve a byte of data and deliver it to a computer, or to receive data from a computer and record to disk. Access time is expressed in milliseconds (ms).

**access unit (AU)** 1. A wiring concentrator that allows multiple attaching devices access to a token-ring network at a central point such as a wiring closet or in an open work area. 2. In the CCITT Message Handling System (MHS) an access unit provides a gateway between the MHS and external communications services.

**accessibility** The number of network elements, such as ports or trunks, that can be reached by a user group.

**Access/one** A local area network marketed by Ungernn-Bass of Santa Clara, CA, that operates at 10 Mbps over unshielded twisted pair wiring.

**ACCOR** Army COMSEC Central Office of Record.

**accounting exit routine** In IBM's ACF/VTAM, an optional, installation exit routine that collects statistics about session initiation and termination.

**Acculink** An AT&T trademark for a series of multiplexers.

**ACCUMASTER Management Services** A contractual service offered by AT&T for multivendor, network management activities in a customer's Network Management Center (NMC).

**Accunet** Digital service from AT&T, including Accunet T1.5, terrestrial wideband at 1.544 Mbps, Accunet Reserved T1.5, satellite-based channels at 1.544 Mbps primarily for video teleconferencing applications; Accunet Packet Services, packet switching services; Accunet Dataphone digital service (DDS), private-line digital circuits at 2400, 4800, 9600, and 56 Kbps; Accunet switched service providing 56 Kbps dial digital transmission.

**Accunet Bandwidth Manager (ABM)** A service of AT&T which enables customers using an on-site personal computer to order, reconfigure, reroute and report problems with digital private lines.

**Accunet Packet Service** A data communications service offered by AT&T that is based on packet switching technology. It permits customers to efficiently transmit bursts of data through a switched network.

**Accunet Reserved Digital Service (ARDS)** An AT&T digital transmission offering which allows customers to temporarily lease a digital circuit between two locations.

**Accunet Switched 56 Service** A data communications service offered by AT&T which provides full-duplex, digital data transmission at 56 kilobits per second (Kbps) via terrestrial digital facilities, which can be accessed through dedicated lines.

**Accunet T1.5 Information Manager (AIM)** A network management tool marketed by AT&T to customers of that vendor's Accunet T1.5 service which integrates end-to-end circuit performance, sectionalized alarms, and network configuration in graphic format for display on a personal computer.

**Accunet T1.5 Service** A communication service offered by AT&T which provides full-time, full-duplex, dedicated, point-to-point, transmission of digital information at 1.544 Mbps. The service supports applications that require transmissions of voice, data, video, or any signal that can be digitally encoded—in any combination—entirely over dedicated terrestrial channels.

**Accupulse** The name of the switched digital transmission service marketed by BellSouth.

**accuracy** A general performance criterion expressing the correctness with which a specific communication function is accomplished.

**ACD** Automatic Call Distributor.

**AC/DC ringing** A widely used technique for causing a telephone to ring, in which an AC voltage is used to power the bell and the DC voltage is used to power a relay which cuts off the ring when the receiver is taken off-hook.

**ACF** Advanced Communications Function.

**ACF/NCP** In IBM's VTAM, Advanced Communications Function for the Network Control Program.

**ACF/SSP** In IBM's VTAM, Advanced Communications Function for the System Support Programs. Synonym for SSP.

**ACF/TAP** In IBM's VTAM, Advanced Communications Function for the Trace Analysis Program. Synonym for TAP.

**ACF/TCAM** In IBM's VTAM, Advanced Communications Function for the Telecommunications Access Method.

**ACF/VTAM** In IBM's VTAM, Advanced Communications Function for the Virtual Telecommunications Access Method.

**ACF/VTAM application program** In IBM's VTAM, a program that has opened an ACB to identify itself to ACF/VTAM and can now issue ACF/VTAM macro instructions.

**ACF/VTAM definition** In IBM's VTAM, the process of defining the user application network to ACF/VTAM and modifying IBM-defined characteristics to suit the needs of the user.

**ACF/VTAM definition library** In IBM's VTAM, the operating system files or data sets that contain the definition statements and start options filed during ACF/VTAM definition.

**ACF/VTAM operator**  A person or program authorized to issue ACF/VTAM operator commands.

**ACF/VTAM operator command**  A command used to monitor or control an ACF/VTAM domain.

**ACF/VTAME**  In IBM's VTAM, Advanced Communications Function for the Virtual Telecommunications Access Method Entry.

**acknowledgment (ACK)**  A control character used (with NAK) in BSC communications protocol to indicate that the previous transmission block was correctly received and that the receiver is ready to accept the next block. Also used as a ready reply in other communications protocols, such as Hewlett-Packard's ENQ/ACK protocol (see following diagram) and the ETX/ACK method of flow control.

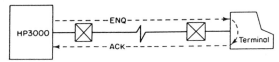

ENQ/ACK protocol diagram

**ACM**  Association for Computing Machinery.

**ACOC**  Area Communications Operations Center.

**ACOnet**  Akademisches COmputer NEtz, a research network in Austria.

**Acorn**  A trademark of AT&T for a network control system.

**acoustic coupler**  A device that converts electrical signals into audio signals, enabling data to be transmitted over the public telephone network via a conventional telephone handset; it also converts the audio signals back into electrical signals at the receiving end. A kind of modem.

**ACP**  Advanced Communications Package.

**ACP**  Allied Communication Publication.

**acquire**  1. In IBM's VTAM, the operation in which

an authorized ACF/VTAM application program initiates and establishes a session with another logical unit; the application program acts as the primary end of the session. *Note.* The acquire process causes an Initiate request to be sent to the SSCP which causes the SSCP to return a CINIT request to the application program (the PLU); this in turn causes the PLU to send a BIND request to the SLU. Contrast with Accept. 2. In relation to ACF/VTAM resource control, to take over resources (communication controllers or other physical units) that were formerly controlled by a data communication access method in another domain, or to assume control of resources that were controlled by this domain but released.

**ACR**  Abandon Call and Retry.

**ACS**  Advanced Communications Services.

**ACSC-E**  Assistant Chief of Staff for Communications—Electronics.

**ACSE**  Association Control Service Element(s).

**ACSNET**  Australian Computer Science Network.

**activate**  An operator action resulting in the resumption of a previously suspended task.

**activate/passive device**  In current loop applications, a device capable of supplying the current for the loop (active) and a device that must draw its current from connected equipment (passive).

**active**  1. Able to communicate. 2. An adapter is active if it is able to communicate. 3. In IBM's VTAM, pertaining to a major or minor node for which a VARY NET, ACT command has been issued. Also, a major or minor node in a list of major nodes to be activated when ACF/VTAM is started. Contrast with Inactive. *Note.* For a major node, this makes the node and its minor nodes known to ACF/VTAM. For a minor node, this generally results in the execution of an SNA protocol to make the minor node usable by the network. For an LU minor node, this indicates that the ACF/VTAM operator has given permission for the LU to participate in an LU–LU session.

**active logical terminal** An active logical terminal is the currently displayed logical terminal. Synonymous with foreground logical terminal.

**active monitor** The computer on a token-ring network that is responsible for initiating a token, maintaining ring timing and delay, and recovering from certain network errors. Any machine on the network has the ability to assume this function if the current active monitor fails.

**active token monitor** In a token-ring network the active token monitor is a node which periodically issues an "active monitor present" control frame. This frame makes other nodes aware that there is an active monitor present on the ring and is used to detect the loss of a token. If a token loss is detected the monitor then issues a free token.

**ACU** Automatic Calling Unit.

**A/D** Analog/Digital (conversion).

**adapter** A device that (1) enables different sizes or types of plugs to mate with one another or to fit into an information outlet; (2) provides for the rearrangement of leads; (3) allows large cables with numerous wires to fan out into smaller groups of wires; or (4) makes interconnections between cables.

**adapter board** A removable board which is inserted into a system expansion slot in a computer to add functionality to the computer. As an example, installing an asynchronous adapter board would provide a computer with the ability to transmit data asynchronously.

**Adapter Control Block (ACB)** In IBM's NCP, a control block that contains line control information and the states of I/O operations for BSC lines, start–stop lines, or SDLC links.

**Adapter Support Interface** The software used to operate IBM Token-Ring Network adapters in an IBM Personal Computer and provide a common interface to application programs.

**Adaptive Differential Pulse Code Modulation (ADPCM)** An encoding technique (CCITT) that allows an analog voice conversation to be carried within a 32 KB digital channel; 3 or 4 bits are used to describe the difference between two adjacent samples at 8000 times a second.

**adaptive equalizer** An equalizer that adjusts to meet varying line conditions; most operate automatically.

**adaptive routing** Message routing which is automatically adjusted to compensate for changes in network traffic patterns and channel availability.

**ADCCP** Advanced Data Communications Control Procedure.

**ADCU** Association of Data Communications Users.

**added channel framing** A frame alignment signal in which the signal elements occupy consecutive digit time slots. Also called bunched frame alignment signal.

**added digit framing** A frame alignment signal in which the signal elements occupy non-consecutive digit time slots. Also called distributed frame alignment signal.

**added main line carrier** An analog carrier system which enables two telephone services to be provided over one physical telephone.

**additional facilities** In packet switched networks, standard network facilities which are selected for a given network but which may or may not be selected for other networks. Contrast with essential facilities.

**ADDRESS** 1. (noun) A unique designation for the location of data or the identity of an intelligent device. Multiple devices on a single communications line must have unique addresses to allow each to respond to its own messages (see polling). 2. (verb) To add or include the coded representation of the desired receiving device (as in to address a message).

**address prefix** In Digital Equipment Corporation Network Architecture (DECnet), Any leading portion of an NSAP address.

**address resolution** In Digital Equipment Corporation Network Architecture (DECnet), the Session Control function which maps from a Naming Service object name to the identifiers of protocols and corresponding addresses which are mutually supported by the local system and the remote system(s) on which the named object resides.

**Address Resolution Protocol (ARP)** The Internet protocol used to dynamically bind a high-level Internet address to a low-level physical hardware address. ARP is only used across a single physical network and is limited to networks that support hardware broadcast.

**address selection** In Digital Equipment Corporation Network Architecture (DECnet), the Session Control function which provides transparent selection of protocols and addresses for Transport Connection establishment based upon the destination object name.

**address space** The complete range of addresses that is available to a programmer.

**addressability** The capability of Direct Broadcast Satellite (DBS) or cable communications system to allow computer control of program distribution from a central location.

**addressing** 1. In data communications, the way that the sending or control station selects the station to which it is sending data. 2. A means of identifying storage locations. 3. Specifying an address or location within a file.

**addressing authority** In Digital Equipment Corporation Network Architecture (DECnet), the authority responsible for the unique assignment of Network layer addresses within an addressing domain.

**addressing domain** In Digital Equipment Corporation Network Architecture (DECnet), a level in the hierarchy of Network layer addresses. Every NSAP address is part of an addressing domain that is administered directly by one and only one addressing authority. If that addressing domain is part of a hierarchically higher addressing domain (which must wholly contain it), the authority for

the lower domain is authorized by the authority for the higher domain to assign NSAP addresses from the lower domain.

**A-disk** IBM's name for a user's PROFS or OFFICE VISION personal storage area.

**adjacent** Network devices or programs that are directly connected by a data link.

**adjacent NCPs** In IBM's VTAM, network control programs (NCPs) that are connected by subarea links with no intervening NCPs.

**adjacent networks** Two SNA networks joined by a common gateway NCP.

**adjacent nodes** In IBM's VTAM, two nodes that are connected by one or more data links with no intervening nodes.

**adjacent SSCP table** A list of SSCPs that can be used to determine the next SSCP on the session-initiation path to a same-network destination SSCP or to a destination network for an LU–LU session. The table is filed in the VTAM definition library.

**adjacent subareas** In IBM's VTAM, two subareas connected by one or more links with no intervening subareas.

**adjusted ring length** In a multiple-wiring-closet ring used to form a token-ring network, the adjusted ring length is the sum of all wiring- closet-to-wiring-closet cables in the main ring path less the length of the shortest of these cables.

**ADLC** Advanced Data Link Controller.

**ADMD** Administrative Management Domain.

**administration subsystem** That part of a premises' distribution system that includes the distribution hardware components where you can add or rearrange circuits. The components include cross-connects, interconnects, information outlets, and the associated patch cords and plugs. Also called "administration points."

**administrative domain** In Digital Equipment Corporation Network Architecture (DECnet), a

7

collection of End Systems, Intermediate Systems, and Subnetworks operated by a single organization or administrative authority. It may be subdivided into a number of routing domains.

**Administrative Management Domain (ADMD)** Under X.400 addressing the ADMD represents a public messaging serivce, such as CompuServe or MIC Mail.

**Administrative Module (AM)** The AM is part of the AT&T 5ESS switch which performs the part of call processing, administration, and maintenance which cannot be economically distributed to switching modules. The AM consists of the processor, disk storage, and tape backup units. The AM processor performs the centralized processing functions, high-speed tape backup, and controls the flow of data between the other dedicated processors distributed throughout the remaining units. The processor functions are fully duplicated (except for the port switch) in order to assure continued processing capability.

**ADMSC** Automatic Digital Message Switching Center.

**Adonis** A network operated by the Institute for Automated Systems in Moscow which connects compter centers located in the former Soviet Union.

**ADP** Answering Detection Pattern.

**ADP** Automatic Data Processing.

**ADPCM** Adaptive Differential Pulse Code Modulation.

**ADPE** Automatic Data Processing Equipment.

**ADPS** Automatic Data Processing System.

**ADU** Automatic Dialing Unit.

**Advanced Communications Function (ACF)** IBM communications software for mainframes and front end communications controllers, which implements various aspects of IBM's Systems Network Architecture.

**Advanced Communications Package (ACP)** An AT&T 3B-based set of Centrex features.

**Advanced Communications Service (ACS)** A shared data communications network service formerly proposed by the Bell System.

**Advanced Communications Function for the Network Control Program (ACF/NCP)** In IBM's VTAM, a program product that provides communication controller support for single-domain and multiple-domain data communication.

**Advanced Communications Function for the Network Control Program (ACF/NCP)** In IBM's VTAM, a program product that provides communication controller support for single-domain and multiple-domain data communication.

**Advanced Communications Function for the Telecommunications Access Method (ACF/TCAM)** In IBM's VTAM, a program product that provides single-domain data communication capability, and, optionally, multiple- domain capability.

**Advanced Communications Function for the Virtual Telecommunications Access Method (ACF/VTAM)** In IBM's VTAM, a program product that provides single-domain data communication capability and, optionally, multiple-domain capability.

**Advanced Communications Function for the Virtual Telecommunications Access Method Entry (ACF/VTAME)** In IBM's VTAM, a program product that provides single-domain and multiple-domain data communication capability for IBM 4300 systems that may include communication adapters.

**Advanced Data Communications Control Procedures (ADCCP)** The USA Federal Standard communications protocol. A standard which specifies a bit-oriented protocol similar to SDLC and HDLC.

**Advanced Mobile Phone Service (AMPS)** The name given to AT&T's first cellular telephone system.

**Advanced Program to Program Communication (APPC)** IBM SNA facility for communicating between programs rather than between a human operator and a program. APPC uses SNA Logical Unit Type 6.2.

**Advanced Program-to-Program Communications/ Personal Computer (APPC/PC)** An IBM product that runs on PCs on the Token-Ring Network; an implementation of the LU6.2 protocol.

**Advanced Radio Data Information Service (ARDIS)** A radio-based communications network managed by a company called ARDIS of Lincolnshire, IL, that was formed by IBM and Motorola. ARDIS enables personal computers or terminals equipped with radio frequency modems to access corporate data bases.

**Advanced Research Projects Agency (ARPA)** Agency that developed the first major packet switched network, ARPANET.

**Advanced Speech Processor (ASP)** A proprietary device of General DataComm, Inc. which compresses a 64K, PCM derived bandwidth into 16K, thereby using 1/4 of the TDM bandwidth to carry a voice circuit while retaining the voice quality.

**Advantage** A trademark of Omniphone, Inc., of Mobile, AL, as well as the name for a series of Automated Coin Telephone Systems (ACTS) from that vendor.

**AEA** American Electronics Association.

**AEA** Asynchronous Emulation Adapter.

**AEA port** A communication connector on the Asynchronous Emulation Adapter.

**AEA port set** 1. One or more 3174 ports that support individual AEA station sets; they must have the same port (connection) type and modem type, but different station types. 2. One or more 3174 station sets that have different station types, but the same port type, modem type, and amount of default destinations.

**AEA station** A 3270 or ASCII display station, printer, or host that communicates through the Asynchronous Emulation Adapter.

**AEA station set** 1. One or more AEA stations that have the same attributes, for example, line speed and parity. 2. One or more AEA stations that share the same characteristics of station type, port type, modem type, and default destination.

**Aeronautical Mobile Satellite Service (AMSS)** An air-to-ground communications system for crew and passengers of commercial airlines. Global space capacity in the L-band for AMSS is provided by the International Maritime Satellite Organization.

**AESC** Automatic Electronic Switching Center.

**AF** Audio Frequency.

**AFC** Air Frequency Coordination.

**AFC** Automatic Frequency Control.

**AFCS** Air Force Communications Service.

**AFI** Authority and Format Indicator.

**AFIPS** American Federation of Information Processing Societies.

**Afrimail** An electronic mail network in Tunisia which has UUCP connections to the rest of the world.

**AFLC** Air Force Logistics Command.

**AFS** Andrew File System.

**AFSC** Air Force Systems Command.

**AFTRCC** Aerospace and Flight Test Radio Coordinating Council.

**AFUTT** Association Française des Utilisateurs du Telephone et des Telecommunications.

**AGC** Automatic Gain Control.

**agent** In Digital Equipment Corporation Network Architecture (DECnet), that part of an entity which provides the interface to network management.

**AGFNET** A network consisting of research centers and universities located in Germany.

**aggregate** The total bandwidth (expressed in bits per second) of a multiplexed bit stream, or the collection of all channels within that bit stream.

9

The aggregate bandwidth of a North American T1 stream is 1.544 Mbps.

**aggregate input rate**   The sum of all data rates of the terminals or computer ports connected to a multiplexer or concentrator; burst aggregate input rate refers to the instantaneous maximum.

**aggregate user**   A collection of entities outside a defined subsystem, comprising one or more end users and the data communication system elements that connect those users with the subsystem.

**AIAG**   Automotive Industry Action Group.

**AIG**   Address Indicating Group.

**AIM**   Accunet T1.5 Information Manager.

**AIOD**   Automatic Identification of Outward Dialing.

**AIPS**   Army Information Processing Standards.

**Air Call**   A partially owned subsidiary of BellSouth Enterprises, the holding company for all unregulated BellSouth companies which market cellular, paging and telephone answering services in the United Kingdom.

**AIRCOM**   AIR COMmunications.

**Airline mileage**   The distance between two points in the USA as determined by a standard set of vertical and horizontal (V–H) coordinates for the major cities; the basis for distance-sensitive circuit service rates.

**AIRS**   Alarm Identification Reporting System.

**Airtime**   In cellular communications the amount of billable time a user has accumulated during a billing period. Airtime is usually measured in minutes.

**AIS**   Alarm Indication Signal.

**AIS**   Automatic Intercept System.

**AL**   Analog Loop.

**Alarm**   An asynchronous event that implies abnormal operation.

**Alarm Identification Reporting System (AIRS)**   A network management tool developed by Harris Corporation. AIRS gathers alarms from asynchronously attached network devices or element management systems and displays alarm information on a central console.

**Alarm Indication Signal (AIS)**   An all-ones pattern on a T1 circuit which is transmitted out when the incoming signal has failed.

**A-LAW**   An algorithm used in Europe for the digitization of voice signals by PCM encoding of PAM samples.

**ALB**   Analog LoopBack.

**Alcatel One**   A private branch exchange marketed by Alcatel PABX Systems Corporation of Alexandria, VA, which uses voice prompts to assist users in programming its features to include call-forwarding.

**alert**   1. In IBM's VTAM NPDA, a high priority event that warrants immediate attention. The NPDA data base record is generated for certain event types that are defined by user-constructed filters. 2. In IBM LAN Manager, a notification appearing on the bottom line of any panel to indicate an interruption or a potential interruption in the flow of data around the ring, or loss of LAN Manager function.

**ALGOL**   Algorithmic Language.

**Algorithm**   A logical or mathematical model that incorporates a specific set of rules that tell how information is to be manipulated to give a desired result.

**Algorithmic Language (ALGOL)**   A computer language used to precisely present complex mathematical procedures and algorithms.

**ALI**   Automatic Link Intelligence.

**ALI–ABA**   American Law Institute–American Bar Association.

**alias name**   In IBM's VTAM, a name defined in a host used to represent a logical unit name, logon mode table name, or class of service name in another network. This name is defined to a name translation program when the alias name does not match the real name. The name translation program is used to associate the real and alias names.

**alias name translation facility**  In IBM's VTAM, a function of the Network Communications Control Facility (NCCF) program product for converting logical unit names, logon mode table names, and class of service names used in one network into equivalent names to be used in another network.

**alias network address**  In IBM's VTAM, an address used by a gateway NCP and a gateway SSCP in one network to represent an LU or SSCP in another network.

**ALIT**  Automatic Line Insulation Test.

**All ones**  In T1 transmission, 1024 or more consecutive ones. When a device loses synchronization, it will send all ones to keep the network up and, at the same time, indicate that there is a problem in transmission.

**All Trunks Busy (ATB)**  A condition in which all trunks in a given trunk group are busy.

**all-dielectric**  A non-metallic insulating material that will store but which will not conduct electricity.

**Alliance**  A teleconferencing service offered by AT&T. Operator setup is obtained by dialing 1-800-544-6363 or you can dial 0-700-456-1000 to do it yourself.

**All-In-1**  An electronic mail system marketed by Digital Equipment Corporation of Maynard, MA for use on that vendor's VAX/VMS computer systems.

**Allocate**  To assign a resource for use in performing a specific task.

**all-rings broadcast**  A function in a token-ring network used to cause bridges to forward the frame to other rings.

**all-stations broadcast**  A feature in a token-ring network used to send a frame of information to all stations through the use of a global address. Bridges may forward the frame on to other rings if appropriate routing information is contained in the frame.

**Aloha**  An experimental packet-switched network implemented on radio by the University of Hawaii in the mid-1970s.

**alphabet**  A table of correspondence between an agreed set of characters and the signals which represent them. The best-known, standard alphabet is International Alphabet No. 5 (IA5) or CCITT or ISO 7-bit code.

**alphageometric**  A scheme for displaying letters, numbers, and various graphic elements by means of combining small geometric building block patterns in Videotex and similar information display systems.

**alphamosaic**  A scheme for displaying letters, numbers, and various graphic elements by means of assembling arrays of individual "tiles" of identical shapes in Videotex and similar information display systems.

**alphanumeric**  Describing a character set that contains letters, numerals (digits), and other characters such as punctuation marks.

**Alphapage**  A paging network in France which enables page subscribers to be contacted in France, Germany (Cityruf), the United Kingdom (Europage) and Italy (Teldrin). Each network operates in the UHF 466 frequency band.

**Alt**  Alternate.

**alternate field**  A partition in IBM PC 3270 Emulation that is accessed via the Alt Task Key (Alt+Esc) while IBM PC 3270 Emulation is running. Compatible programs may be loaded while in this field.

**alternate buffer**  A section of memory in a communications device set aside for the transmission or reception of data. The alternate buffer is used in conjunction with a primary buffer, e.g., when the alternate buffer is empty (transmission) or full (reception), data transmission continues using the primary buffer, while the associated computer or terminal device transfers data to or from the alternate buffer in anticipation of its use when the primary buffer is empty or full.

11

**Alternate Mark Inversion (AMI)**   A physical technique for bipolar transmission of digital signal. In AMI, the logical value of zero is represented by bit spaces with neutral polarity, and the logical value of one is represented alternately by pulses of positive and negative polarity.

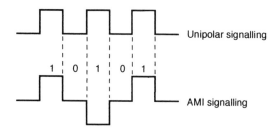

**Alternate Mark Inversion (AMI) Violation**   A "mark" which has the same polarity as the previous "mark" in the transmission of an AMI signal. Also called a bipolar violation.

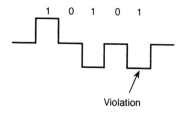

**Alternate mode**   This is a mode of using a virtual terminal by which each of two interacting systems or users has access to its data structure in turn. The associated protocols include facilities to allow the orderly transfer of control from one user to the other. This is in contrast to free running mode.

**alternate/no answer option**   A PBX and switch feature in which calls are automatically transferred to another line within the system after a preset number of rings.

**alternate path**   In IBM's VTAM: 1. Another channel an operation can use after a failure. 2. In CCP, one of two paths that can be defined for information flowing to and from physical units attached to the network by means of an IBM 3710 Network Controller.

**alternate path retry**   In IBM's VTAM, a facility that allows an I/O operation that has failed to be retried on another channel assigned to the device performing the I/O operation. It also provides the capability to establish other paths to an on-line or off-line device.

**alternate route**   A secondary communications path used to reach a destination if the primary path is occupied or otherwise unavailable.

**Alternating Current (AC)**   Electrical current which is used for analog signaling.

**alternative access code**   An 800 number, a 950 number or a five-digit number beginning with 10 which allows callers to reach their preferred long-distance telephone company from telephones that are serviced by other long distance companies.

**Alternative Operator Services (AOS)**   A service for O-plus calls provided by non-communications carriers for large institutions to include hotels, motels, hospitals, universities, and other establishments with non-Bell pay telephones.

**ALU**   Arithmetic Logic Unit.

**Alvyn**   Aluminum PVC sheath used on cable for building risers or other similar areas where a flame-resistant sheath is required to meet the National Electrical Code standards.

**AM**   Administrative Module.

**AM**   Amplitude Modulation.

**AMA**   Automatic Message Accounting.

**AMACC**   Automatic Message Accounting Collection Center.

**AMARS**   Automatic Message Address Routing System.

**AMATPS**   Automatic Message Accounting Teleprocessing System.

**ambient noise**   Communications interference which is present in a communications line at all times. Also known as background, Gaussian, or white noise, distinct from impulse noise.

**AMC**   Administrative Maintenance Center.

**AM/DSB**   Amplitude Modulation/Double SideBand.

**American Bar Association Network (ABAnet)**   A communications network that links ABA members via electronic messaging.

**American Law Institute – American Bar Association (ALI–ABA)**   A network used to provide on-site legal training via satellite-based business television services.

**American Mobile Satellite Corporation (AMSC)**   A U.S. communications satellite industry consortium formed by 12 vendors to provide mobile satellite services. Original members of AMSC include Globesat Express, Global Land Mobile Satellite, Inc., Mobile Satellite Service, Inc., Satellite Mobile Telephone Co., Omninet Corp., Hughes Communications, MCCA American Satellite Service Corp., McCraw Space Technologies, Inc., Mobile Satellite Corp., North American Mobile Satellite, Inc., Skylink Corp., and Wisner & Becker Transit Communications.

**American Cellular Communications**   A partially owned subsidiary of BellSouth Enterprises, the holding company for all unregulated BellSouth companies which provides cellular mobile telephone service primarily outside the southeastern United States.

**American National Standards Institute (ANSI)**   The American organization, founded in 1918, made up of carriers, manufacturers, and users, that addresses a wide variety of information processing standards topics. ANSI, is a voluntary organization that represents the USA in the International Standards Organization (ISO). ANSI also produces Federal Information Processing Standards (FIPS) for the Department of Defense (DoD).

**American Standard Code for Information Interchange (ASCII)**   A 7-bit-plus-parity character set or code established by ANSI to achieve compatibility between data services; sometimes called USASCII, the USA Standard Code for Information Interchange; normally used for asynchronous transmission. Equivalent to the ISO 7-bit code. The ASCII code is listed under the ASCII entry.

**American Telephone and Telegraph (AT&T)**   The USA's major common carrier for long distance telephone lines.

**American Wire Gage (AWG)**   A numbering system used to express the size (diameter) of electrical wires. The AWG number is inversely proportional to the diameter of the wire. Heavy industrial wiring, as an example, may be AWG#0 or 2, whereas, telephone systems use AWG#22, 24, or 26.

**Ameritech**   One of seven regional Bell operating companies (BROCs), covering the mid-Western United States, based in Chicago, IL.

**AMI**   Alternate Mark Inversion.

**AMIS**   Audio Messaging Interchange Specification.

**AM/ISB**   Amplitude Modulation/Independent Side-Band.

**AmLink3**   A trademark of Advanced Micro Devices as well as a chip set which provides complete ISDN voice and data connection capabilities for both B channels and the D channel through OSI layer three.

**Amos**   Israel's first communications satellite which was launched in 1993. Amos weighed 1200 kilograms and has two 12-meter solar panels.

**AMPL**   Amplifier.

**amplification**   The process of increasing the electrical strength of a signal.

**amplifier**   Electronic component used to boost (amplify) signals. Performance (called gain) measured in decibels (dB).

**amplitude**   The maximum departure of a waveform from its average value.

**Amplitude and Phase Shift Keying (APSK)**   A form of modulation that uses a combination of amplitude and phase change to encode a signal.

**amplitude distortion**   An unwanted change in signal

amplitude, usually caused by non-linear elements in the communications path.

**Amplitude Modulation (AM)**  One of three basic ways (see also FM and phase modulation) to add information to a sine wave signal: the magnitude of the sine wave, or carrier, is modified in accordance with the information to be transmitted.

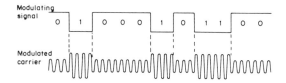

**Amplitude Modulation Vestigial Sideband (AM-VSB)**  A television channel transmission technique that uses 6 MHz as only one complete sideband and only a vestige of the other is sent.

**Amplitude (peak)**  The maximum departure of the value of signal from its reference point. It can be referred to as the strength of the signal.

**amplitude variation (ripple)**  Unwanted variation of signal voltage at different frequencies on a communications line.

**AMPRNET**  Amateur Packet Radio Network.

**AMPS**  Advanced Mobile Phone Service.

**AMSC**  American Mobile Satellite Corporation.

**AMSS**  Aeronautical Mobile Satellite Service.

**AM/SSB**  Amplitude Modulation/Single Side-Band.

**AM-VSB**  Amplitude Modulation Vestigial Sideband.

**AN (Army Navy)**  A system indicator prefix that is part of the U.S. Joint Electronics Type Designation System which identifies U.S. military electronics equipment. The reader is referred to the Joint Electronics Type Designation System entries for the structure of the designation system and its code meanings.

**AN/ASC-15B(V)1**  A U.S. military radio used to provide Army ground and helicopter forces with the ability to communicate with other services. The AN/ASC-15B(V)1 can communicate with radios operating on any ground or aircraft military frequencies between 225 and 400 MHz in either amplitude or frequency modulation mode.

**AN/FRC-93**  A U.S. military radio set that provides amplitude modulation/single sideband (AM/SSB) (upper or lower sideband) or continuous wave (CW) transmission and reception.

**AN/GGC-3**  A U.S. military lightweight transportable teletypewriter set that can be used in both fixed and tactical teletypewriter stations.

**AN/GRA-39B**  A U.S. military transistorized, battery operated remote control system used to provide local and remote control facilities for FM radio sets. The control system consists of a C-2329/GR for local control at the radio site and a C-2328/GR at the remote site.

**AN/GRC-19**  A U.S. military medium power, voice and continuous wave radio set designed for vehicular installations. The frequency range of the transmitter is 1.5 to 20 MHz while the frequency range of its receiver is 0.5 to 32 MHz. The AN/GRC-19 has an 80-kilometer planning range.

**AN/GRC-50**  A U.S. military medium capacity, transportable FM radio set designed to provide multichannel, two-way communications in the UHF range. The AN/GRC-50 can be used with either 12- or 26-channel TDM/PCM multiplex equipment to provide multichannel communications.

**AN/GRC-103**  A U.S. military radio set that can be used with low capacity TDM multichannel systems. The AN/GRC-103 accommodates up to 12 voice frequency channels of communications.

**AN/GRC-106**  A U.S. military high frequency, amplitude modulated/single sideband (AM/SSB) radio receiving–transmitting set used primarily as a mobile link in a communications network. The AN/GRC-106 has a frequency range of 2.0 to 29.999 MHz.

**AN/GRC-144** A U.S. military general purpose tactical FM radio set which provides full duplex operation. The AN/GRC-144 is intended for use in multichannel line-of-sight communications applications.

**AN/GSA-7** A U.S. military small, lightweight electronic switching device used to integrate frequency modulation radio equipment with local battery, push-to-talk telephone circuits.

**AN/GSQ-80** A U.S. military transportable assemblage that is used as a message center distribution facility in a mobile communications system.

**AN/MGC-17** A U.S. military teletypewriter central office which contains facilities for three teletypewriter circuits or two secure half-duplex circuits.

**AN/MGC-19** A U.S. military teletypewriter operations central office that can support seven voice frequency or direct-current teletypewriter circuits or 14 direct-current, full-duplex teletypewriter circuits.

**AN/MRC-54** A U.S. military radio repeater set that can be used as either an FDM multichannel radio relay set or as an FDM multichannel terminal set where the multiplexing equipment is in a separate assemblage.

**AN/MARC-69** A U.S. military FDM multichannel radio terminal set which provides 24 voice frequency channels.

**AN/MRC-73A** A U.S. military FDM multichannel radio terminal configuration that provides 12 channels of voice frequency communications over radio or cable.

**AN/MRC-127** A U.S. military combined radio and multiplex terminal used in airborne and airmobile divisions.

**AN/MSC-29** A U.S. military telegraph terminal that contains facilities for 8 full-duplex or 12 half-duplex voice frequency teletypewriter circuits when communications security equipment is not in use. Facilities are provided for two asynchronous or two synchronous full-duplex on-line communications security equipment circuits, or four asynchronous half-duplex circuits.

**AN/MSC-31** A U.S. military communications operations center that houses a telephone switchboard, intercommunication, local telephone circuits, drafting and display board.

**AN/PGC-1** A U.S. military lightweight, transportable teletypewriter.

**AN/PRC-25** A U.S. military short range manpack portable, frequency modulated receiver–transmitter used to provide two-way voice communications. Receiver–Transmitters RT-505/PRC-25 and RT-505A/PRC-25, part of the AN/PRC-25, are also used as part of Radio Set AN/VRC-53 in vehicular operation, and as part of AN/GRC-125 in vehicular or manpack operation.

**AN/PRC-77** A U.S. military portable FM transmitter–receiver identical to the AN/PRC-25 except that it is fully transistorized, has improved retransmission capability and can provide secure voice transmission.

**AN/PRR-9** A U.S. military battery operated, single channel frequency modulation radio receiver. The AN/PRR-9 is designed to be worn on an operator's helmet or webbing and is used at squad and platoon level in conjunction with the AN/PRT-4 Radio Transmitting Set.

**AN/PRT-4** A U.S. military handheld, low power, battery operated frequency modulation radio transmitter. The AN/PRT-4 is used at the squad and platoon level in conjunction with the AN/PRR-9 Radio Receiving Set.

**AN/PSC-3** A portable man pack tactical satellite communications system used by the U.S. military.

**AN/SLQ-17** A U.S. Navy deception jammer which creates a ghost image of a protected ship to lure away incoming enemy anti-ship missiles.

**AN/SLQ-32** A U.S. Navy deception jammer designed as a replacement for the AN/WLR-1 receiver and the AN/ULQ-6 deception jammer.

**AN/TCC-3**   A U.S. military four-channel tactical FDM telephone terminal. The AN/TCC-3 has a normal range of 25 miles and can provide either four channels of telephone communications plus an orderwire, or one wideband (16 KHz) channel plus an orderwire.

**AN/TCC-4**   A U.S. military telegraph terminal designed to provide frequency shifted, carrier telegraph signals within the voice band with a maximum data rate of 100 words per minute.

**AN/TCC-5**   A U.S. military telephone repeater. The AN/TCC-5 is an attended, 4-wire repeater that equalizes and amplifies signals transmitted from an AN/TCC-3.

**AN/TCC-7**   A U.S. military telephone terminal that supports 12 channels of voice frequency communications over either spiral-four cable or multichannel radio systems, or both in tandem.

**AN/TCC-8**   A U.S. military telephone repeater. The AN/TCC-8 is an attended repeater used in conjunction with the AN/TCC-7 in a 12-channel telephone carrier system.

**AN/TCC-11**   A U.S. military unattended repeater used in the AN/TCC-7 cable carrier system.

**AN/TCC-14**   A U.S. military telegraph/telephone terminal capable of simultaneous transmission of speech and half-duplex frequency-shift telegraph signals.

**AN/TCC-50**   A U.S. military AN/TCC-7 less its power supply.

**AN/TCC-65**   A U.S. military telephone terminal assemblage that is air or vehicular transportable and which provides secure or non-secure TDM multichannel terminal facilities for forward area TDM/PCM.

**AN/TRC-24**   A U.S. military transportable, multi-channel, VHF/UHF-FM radio set designed to operate with FDM telephone carrier equipment. The AN/TRC-24 can support 4 or 12 voice channels.

**AN/TRC-29**   A U.S. military transportable, tactical microwave FM radio set used in a rear area multichannel radio system.

**AN/TRC-110 (V)**   A U.S. military air or vehicular transportable assemblage used to provide extended TDM multichannel facilities for corps or army headquarters.

**AN/TRC-113**   A U.S. military radio repeater set which provides radio relay, radio to cable, or cable to repeater facilities for forward area TDM/PCM multichannel communications systems.

**AN/TRC-117(V)**   A U.S. military air or vehicular transportable assemblage used to provide TDM multichannel radio and cable communications facilities for corps and army headquarters.

**AN/TRC-121**   A U.S. military air or vehicular transportable assemblage used to provide radio terminal or relay facilities for army area PCM communications systems.

**AN/TRC-138**   A U.S. military air or vehicular transportable radio repeater which provides line-of-sight capability in the corps communications system. The AN/TRC-138 can terminate three PCM systems with up to 96 voice channels.

**AN/TRC-145**   A U.S. military radio terminal set which provides secure or non-secure multiplex radio or cable terminal facilities for TDM/PCM multichannel communications at the division level.

**AN/TRC-173**   A U.S. military radio terminal set that operates as a radio or cable terminal and is capable of terminating up to two 18/36 digital multichannel line-of-sight systems.

**AN/TRC-174**   A U.S. military radio repeater that operates as a radio repeater or as a split terminal. The AN/TRC-174 is capable of terminating up to three 18/36 digital multichannel line-of-sight systems.

**AN/TRC-175**   A U.S. military radio terminal set which provides multichannel short range wideband radio or high speed cable links between a bottom-of-the-hill switching node and transmission facilities at the top-of-the-hill communications site.

**AN/TSC-58**  A U.S. military shelter-mounted assemblage on a two-and-a-half-ton truck that serves as a voice-frequency, on-line cryptographic telegraph terminal.

**AN/TSC-85A**  A U.S. military satellite communications terminal which is redundant in the RF, modem, and tactical signal processor sections. The AN/TSC-85A can be used as a nodal (hub) and non-nodal (spoke) terminal in tactical trunking networks.

**AN/TSC-86**  A U.S. military satellite communications terminal capable of providing simultaneous communications with up to four other terminals.

**AN/TSC-93A**  A U.S. military satellite communications terminal which is non-redundant in both the RF and the baseband sections and is used as a non-nodal (spoke) or point-to-point terminal in tactical trunking networks.

**AN/TTC-7**  A U.S. military transportable telephone central office capable of terminating 200 local lines and 40 trunk circuits.

**AN/TTC-29**  A U.S. military telegraph–telephone terminal capable of simultaneous transmission of telegraph signals and speech.

**AN/VRC-12**  A U.S. military family of short range vehicular and fixed radio sets designed for general tactical use.

**AN/VRC-43**  A U.S. military radio set which includes one RT-246 receiver and one antenna.

**AN/VRC-44**  A U.S. military radio set which includes two R-442/VRC receivers; a RT-246/VRC receiver–transmitter and two antennas.

**AN/VRC-45**  A U.S. military radio set which includes two RT-246/VRC receiver–transmitters and one antenna for each RT-246.

**AN/VRC-46**  A U.S. military radio set which includes an RT-524/VRC receiver–transmitter and one antenna.

**AN/VRC-47**  A U.S. military radio set which includes an RT-442/VRC receiver, a RT-524/VRC receiver–transmitter and two antennas.

**AN/VRC-48**  A U.S. military radio set which includes two R-442/VRC receivers, RT-524/VRC receiver–transmitters and two antennas.

**AN/VRC-49**  A U.S. military radio set which includes two RT-524/VRC receiver–transmitters and one antenna for each RT-524.

**AN/VSC-7**  A vehicle-mounted tactical satellite communications system used by the U.S. military.

**ANAC**  Automatic Announcement Circuit.

**analog**  Continuously variable as opposed to discretely variable. Physical quantities such as temperatures are continuously variable and so are described as analog; analog signals vary in accordance with the physical quantities they represent. The public telephone network was designed to transmit voice in analog form.

**analog data**  Data in the form of continuously variable physical quantities.

**analog–digital converter**  A device that converts a signal that is a function of a continuous waveform into a representative number sequence.

**analog extension**  Use of analog transmission facilities (lines and modems) to connect a station not on a digital network (DDS).

**analog loopback**  A diagnostic test that forms the loop at the modem's telephone line interface.

**analog loopback testing**  In data communications, analog loopback testing is a technique whereby a local modem's transmitter is connected to the same modem's receiver input (for full-duplex modems, the transmitter and receiver must be switched into the same channel since they are normally in opposite channels). In the local modem, the

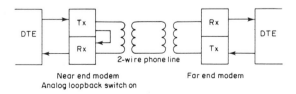

Near end modem          Far end modem
Analog loopback switch on

test pattern is passed through most of the modem circuits and then transmitted back to the test device. The received signal in the test device is compared with the original transmitted signal and any errors induced by the modem are detected.

**analog signaling** An analog signal is one that varies in a continuous manner, such as voice or music.

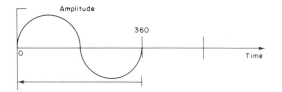

**analog to digital conversion (A/D)** Conversion of an analog signal to a digital signal.

**analog switch** Any of a variety of switching devices which operate without converting the analog signal into a digital signal. Most newer switches operate digitally, converting the analog signal into a digital signal, switching the signal and reconverting to analog for further transmission.

**analog transmission** Transmission of a continuously variable signal as opposed to a discretely variable signal. Physical quantities such as temperature are continuously variable and so are described as "analog."

**Anamix** A full-motion, color desktop videoconferencing system manufactured by Harris Corporation. Anamix includes a coder/decoder integrated into a personal computer.

**ancillary equipment** 1. Devices not required for the provision of basic telephone service, such as a telephone answering machine. 2. In IBM's VTAM, equipment not under direct control of the processing unit.

**Andrew File System (AFS)** A distributing computing system developed at Carnegie-Mellon University with IBM funding. AFS was developed to integrate thousands of workstations in one common environment and operates by connecting

Ethernets and token-ring networks by bridges using the Internet Protocol (IP).

**ANI** Automatic Number Identification.

**ANIK** A series of satellites used for communications throughout Canada as well as for cross-border services to the United States.

**anisochronous signal** A signal which is not related to any clock, and in which transitions could occur at any time.

**Anonymous FTP** A service provided on some computers which enables files to be downloaded by the general Internet community, usually requiring the use of the password GUEST or ANONYMOUS.

**ANSI** American National Standards Institute.

*ANSI Standards*

| | |
|---|---|
| X3.1 | Synchronous Signaling Rates for Data Transmission |
| X3.2 | Print Specifications for Magnetic Ink Character Recognition |
| X3.3 | Bank Check Specifications for Magnetic Ink Character |
| X3.4 | American Standard Code for Information Interchange (ASCII) |
| X3.5 | Flowchart Symbols and Their Usage in Information Processing |
| X3.6 | Perforated Tape Code |
| X3.9 | FORTRAN |
| X3.11 | Specifications for General Purpose Paper Cards for Information Processing |
| X3.14 | Recorded Magnetic Tape (200 CPI,NRZI) |
| X3.15 | Bit Sequencing of ASCII in Serial-by-Bit Data Transmission |
| X3.16 | Character Structure and Character Parity for Serial-by-Bit Data Communications in ASCII |
| X3.17 | Character Set for Optical Character Recognition (OCR-A) |
| X3.18 | One-Inch Perforated Paper Tape for Information Interchange |
| X3.19 | Eleven-Sixteenths Inch Perforated Paper Tape for Information Interchange |

18

Outer Diameter and 3.937 inch (100 mm) Inner Diameter

X3.120 Contact Start/Stop Storage Disk, (95840 Flux Transitions per Track) 7.874 inch (200 mm) Outer Diameter and 2.500 inch (63.5 mm) Inner Diameter

X3.121 Two-Sided Double Density Unformatted 8-inch (200 mm) Flexible Disk Cartridge, for 13262 FTPR Two-Headed Application—General, Physical and Magnetic Requirements

X3.122 Computer Graphics Metafile for the Storage and Transfer of Picture Description Information

X3.124 Graphic Kernel System

X3.124-1 Computer Graphics—Graphical Kernel System (GKS) FORTRAN Binding

X3.124-2 PASCAL Language Binding of the Graphical Kernel System (GKS)

X3.125 Two-Sided Double Density Unformatted 5.25 inch (130 mm) 48 Tracks per inch (1.9 Tracks per mm) Flexible Disk Cartridge for 7958 BPR Use—General, Physical and Magnetic Requirements

X3.126 One- or Two-Sided Double Density Unformatted 5.25 inch (130 mm) 96 Tracks per inch Flexible Disk Cartridge

X3.127 Unrecorded Magnetic Tape Cartridge for Information Interchange 0.250 in (6.30 mm), 6400–10000 ftpi (252–394 ftpmm)

X3.128 Contact Start/Stop Storage Disk, 83000 Flux Transitions per Track, 130-mm (5.118-inch) Outer Diameter 40-mm (1.575-inch) Inner Diameter

X3.129 Intelligent Peripheral Interface, Physical Level

X3.130 Intelligent Peripheral Interface—Logical Device Specific Command Sets for Magnetic Disk Drives

X3.131 Small Computer Systems Interface

X3.132 Intelligent Peripheral Interface—Logical Device Generic Command Set for Optical and Magnetic Disks

X3.133 Network Database Language

X3.135 Database Language SQL

X3.136 Serial Recorded Magnetic Tape Cartridge for Information Interchange, Four and Nine Track

X3.137 Unformatted Flexible Disk Cartridge for Information Interchange, 90mm (3.5 in) 5.3 Tracks per millimeter (135 Tracks per inch), General, Physical and Magnetic Requirements for 7958 BPR Use

X3.139 Fiber Distributed Data Interchange (FDDI) Token Ring Media Access Control (MAC)

X3.140 Information Processing Systems Open Systems—Interconnection—Connection Oriented Transport Layer Protocol Specification

X3.141 Data Communication Systems and Services—Measurement Methods for User-Oriented Performance Evaluation

X3.145 Codes for Identification of Hydrologic Units in the United States and the Caribbean Outlying Area

X3.146 Device Level Interface for Streaming Cartridge and Cassette Tape Drives

X3.147 Intelligent Peripheral Interface Logical Device Generic Command Set for Magnetic Tapes

X3.149 Location of Imprinted Information on a Credit Card Charge Form

X3.150 Office Machines and Business Forms Character and Line Spacing

X3.151 Basic Sheet Sizes and Standard Stock Sizes for Bond Paper and Index Bristols

X3.152 Specifications for Single Ply Non-Carbonized Adding Machine Paper Rolls

X3.153 Open Systems Interconnection, Basic Connection Oriented Session Protocol Specification (1987)

X3.154 Open Systems Interconnection, Basic Connection Oriented Session Protocol Specification (1988)

X3.155 5.25-inch Rigid Disk Removable Cartridge

X3.156 Nominal 8-inch Rigid Disk Removable Cartridge (200/63.5 mm)

X3.157 Recorded Magnetic Tape for Information Interchange, 3100 CPI (126 cpmm) P.E.

X3.158 Serial Recorded Magnetic Tape Cassette for Information Interchange, 0.150 inch (3.81 mm) 8000 BPI (3.5 BPMM) Group Code Recording

| | |
|---|---|
| X3.162 | Unformatted Flexible Disk Cartridge for Information Interchange, 5.25 in (130 mm), 96 Tracks per inch (3.8 Tracks per millimeter), General, Physical, and Magnetic Requirements (for 13262 FTPR Use) |
| X3.163 | Contact Start/Stop Metallic Thin Film Storage Disk, 83,333 Flux |
| | Transitions Per Track, 130 mm Outer Diameter and 40 mm Inner Diameter |
| X4.6 | 10-Key Keyboard for Adding and Calculating Machines |
| X4.9 | Office-Type Dictating Equipment |
| X4.10 | Remote Dictation through an Intercommunication Switching System |
| X4.13 | Financial Transaction Cards |
| X4.16 | Magnetic Stripe Encoding for Financial Transaction Cards |
| X4.19 | Markings on Containers for Printing Ribbons |
| X4.20 | Office Machines and Printing Machines Used for Information Processing Widths of Fabric Ribbons on Spools from 0.1875 inch (4.8 mm) to 0.750 inch (19.0 mm) |
| X4.21 | Data Exchange for Industry Financial Transaction Cards |
| X4.22 | Alternative Keyboard Arrangement for Alphanumeric Machines |
| X9.9 | Financial Institution Message Authentication Standard (FIMAS) |

**answer only**   A data terminal or data set which can accept, but not originate calls on the switched telephone network.

**answer signal**   The signal sent in a backward direction when a call is answered. It is used to begin the computation of the call charge to the calling party.

**answerback**   A response from a data transmission device in response to a request from another transmitting data processing device that is ready to accept or has accepted data.

**Answering Detection Pattern (ADP)**   A pattern sent by a V.42 error correcting protocol modem in response to receiving an originator detection pattern (ODP). The ODP and ADP are used by the originating and receiving modems when they first connect to determine if they have similar error-correcting protocols.

**answering machine**   Equipment which automatically answers a telephone after a preset number of rings. The caller hears the message previously prerecorded and is able to leave a recorded message.

**answering tone**   A signal sent by the called modem (the "answer" modem) to the calling modem (the "originate" modem) on public telephone networks that indicate the called modem's readiness to accept data.

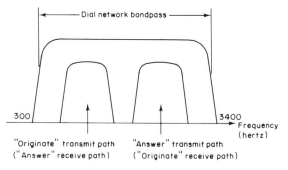

Typical dial modem signaling frequencies

**Ant**   Antenna.

**antenna**   A circuit element designed for radiating or receiving electromagnetic waves, such as radio waves.

**antenna dish**   A device which sends and/or receives signals to or from a satellite. Also called an earth station.

**antipodal signaling**   A technique of encoding binary signals so that the symbol for a 1 is the exact negative of the symbol for a 0. Antipodal signaling provides optimum error performance in terms of the signal to noise ratio.

**anti-streaming**   An anti-streaming feature in a modem enables it to ignore a RTS signal from a DTE if it is held on for longer than a specified amount of time.

**any-mode** In IBM's VTAM: 1. The form of a receive request that obtains input from any one (unspecified) session. 2. The form of an accept request that completes the establishment of a session by accepting any one (unspecified) queued CINIT request.

**AOS** Academic Operating System.

**AOS** Alternative Operative Services.

**APD** Avalance PhotoDiode.

**aperture** The cross-sectional area of an antenna that is exposed to a signal.

**API** Application Program Interface.

**APL** A programming language.

**apogee** The highest point in a satellite's orbit (km); the point in the orbit of a satellite where it is farthest from the object about which it revolves.

**apogee kick motor** A rocket stage used to convert the elliptical transfer orbit of a satellite into a circular geosynchronous orbit.

**APP** Access Process Parameter.

**apparatus boxes** Metallic structures which provide mechanical protection for equipment on a customer's premise.

**apparatus closet** Traditionally, in commercial buildings, the wiring closet, or other enclosed space, on each floor where backbone cables are connected to the switching and power devices associated with key telephone systems. This term is largely outdated, since the original distinction between "apparatus" and "satellite" closets no longer holds true: "apparatus" closets contained key switching and power devices and "satellite" closets did not. Today, backbone and satellite closets are constructed to hold modern power and multiplexing equipment, which is smaller, lighter, and more sophisticated than earlier types of switching and power equipment.

**apparent power** The power in a circuit when the load is reactive, where reactance due to inductance and capacitance in the load causes a shift in the phase angle between current and voltage. The reactance changes with each change of frequency, causing the power in the load to vary with frequency. This causes the measured power output level of a telephone line to vary as the signal frequency is varied.

**APPC** Advanced Program-to-Program Communications.

**APPC/PC** Advanced Program-to-Program Communications/Personal Computer.

**append** To add data to the end of existing data.

**appendage** An application program routine provided to assist in handling a specific occurrence.

**Applelink** Apple Computer Company's electronic mail system.

**Appleshare** Apple Computer Company's file server software (AppleTalk) for local area networking. It lets a Macintosh use a file server (usually another Mac with a hard disk drive) to store files for a network of other Macs.

**Appleshare PC** A desktop communications package marketed by Apple Computer Company for use on MS-DOS personal computers.

**Appletalk** Apple Computer's local area network.

**application** A definable set of tasks to be accomplished as part of the work of an enterprise. An application may be accomplished through manual or computerized procedures including communications control, or both.

**Application layer** Highest (seventh) layer in OSI model, containing all user or application programs.

**applications program** 1. In general, a program that is designed to perform a specific user function. 2. In data communications, a program (that frequently resides in data communications equipment) used to connect and communicate with terminals that performs a set of specified activities for terminal users.

23

**application program exit routine**  In IBM's VTAM, a user-written exit routine that performs functions for a particular application program and is run as part of the application program. Examples are RPL exit routines, EXLST exit routines, and the TESTCB exit routine.

**application program identification**  The symbolic name by which an application program is identified to IBM's ACF/VTAM. *Note.* It is specified in the APPLID parameter of the ACB macro instruction. It corresponds to the ACBNAME parameter in the APPL statement or, if ACBNAME is defaulted, to the name of the APPL statement.

**Application Program Interface (API)**  A set of software calls and routines that can be referenced by an application program to access predefined network services.

**Application program major node**  In IBM's VTAM, a member or book of the ACF/VTAM definition library that contains one or more APPL statements, each representing an application program.

**approved circuit**  In the U.S. military a circuit carrying classified information in plain text, in which the maximum precautions have been taken against radiation, cross modulation, and electrical coupling. An approved circuit is designated by appropriate authority in accordance with the policy stated in ACP-122(-), which states explicitly who approved circuits, under what conditions, in what areas, and at what time.

**APR**  In IBM's VTAM, Alternate Path Retry.

**APS**  Attached Processor System.

**APS**  Automatic Protection Switching.

**APSK**  Amplitude and Phase Shift Keying.

**APT Espana**  A joint venture between the U.S.–Dutch AT&T–Philips Telecommunications BV and the Amper subsidiary of Spanish Telefonica to produce central office switching equipment.

**Arabsat**  A communications satellite consortium which serves Arab countries.

**ARB**  Arbitrator.

**arbitrator (ARB)**  A circuit board that resides in the chassis of a Telenet Processor 4000 (TP4). Its specific function is to allocate main memory addresses among the microprocessor boards.

**architecture**  The manner in which a system (such as a network or a computer) or program is structured. See also closed architecture, distributed architecture, and open architecture.

**Archie**  A service which enables the data bases of many anonymous FTP sites to be searched. Archie searches can be performed by either complete or partial file names and search results include a listing of anonymous FTP site addresses, directory path, file name and the date Archie last visited the site. Archie was developed at McGill University in Montreal. Archie can be accessed via Telnet, a local Archie client or E-mail.

**ARCHIVE**  To back up, or store, data.

**arcing**  A luminous passage of current through ionized gas or air.

**ARCnet**  Attached Resources Computing Network.

**ARCTS**  Automatic Reporting Circuit Test Set.

**ARDIS**  Advanced Radio Data Information Service.

**ARDS**  Accunet Reserved Digital Service.

**area**  In Digital Equipment Corporation Network Architecture (DECnet), the group of systems which constitute a single Level 1 routing subdomain.

**area address**  In Digital Equipment Corporation Network Architecture (DECnet), the concatenation of the IDP and LOC-AREA fields of an NSAP address. A system may have more than one area address, but the systems making up an area must have at least one area address in common with each of their neighbors.

**area code**  A 3-digit number that identifies a particular geographic location. The area code is used in many types of telecommunications for access to hardware devices physically or logically situated at a particular geographic location.

**ARFA**   Allied Radio Frequency Agency.

**ARFAP/S**   Allied Radio Frequency Agency Permanent Staff.

**argument**   User selected items of data that are passed to a subroutine, program, or procedure.

**Ariadne**   A research network located in Greece which uses the X.25 protocol.

**Ariel**   A trademark and interactive video system service of AT&T.

**ARISTOTE**   Association de Réseaux Informatiques en Système Totalement Ouvert et Très Elaboré, a research network located in France.

**ARP**   Addressing Resolution Protocol.

**ARPA**   Advanced Research Projects Agency.

**ARPANET**   A large scale wideband data network interconnecting a large population of terminals and computers at U.S. and European universities and commercial institutions. Sponsored by the Advanced Research Projects Agency; the first packet switched network used to connect different types of computer systems. ARPANET was the predecessor to the Internet.

**ARQ**   1. Automatic Request for Retransmission. 2. Automatic Retransmission Queue.

**array connector**   A connector for use with ribbon fiber cable that joins 12 fibers simultaneously. A fan-out array design can be used to connect ribbon fiber cables to non-ribbon cables.

**arrester**   1. Device which diverts high voltage to ground and away from equipment. 2. The voltage limiting portion of a protector. 3. A lightning arrester.

**ARS**   Automatic Route Selection.

**artificial intelligence**   The quality attributed to programs and computers that gives the illusion of human intellectual activity.

**ARTL**   Analog Responding Test Line.

**ARU**   Audio Response Unit.

**ASA**   Army Security Agency.

**ASCII**   American Standard Code for Information Interchange (7 level).

**ASCII emulation**   In an IBM 3270 network, the ability of a 3270 display station or printer to communicate with an ASCII host using the DEC VT100 or IBM 3101 data stream.

**ASCII pass-through**   In an IBM 3270 network, the transmission of unmodified data between ASCII display stations or printers and an ASCII host or public data network.

**ASCII terminal**   A terminal that uses ASCII; usually synonymous with asynchronous terminal and with dumb terminal.

**ASCNET**   Australian Computer Science Network.

**ASD**   Auto-Speed Detect.

**ASD(T)**   Assistant Secretary of Defense (Telecommunications).

**askERIC**   A question-answering service for teachers, library media specialists and administrators involved with K-12 education.

**Asiasat**   A Hong Kong based consortium, which, in 1989, launched the first privately owned domestic telecommunications satellite to cover the Asian region. Members of Asiasat include Hutchison Telecommunications, Cable & Wireless, and China International Trust and Investment Corporation.

**ASMDR**   Advanced Station Message Detail Reporting.

**ASMMP**   Army Spectrum Management Master Plan.

**ASN.1**   Abstract Syntax Notation 1.

**ASP**   Advanced Speech Processor.

**ASP**   Aluminum, Steel, and Polyethylene, the preferred sheath for waterproof (filled) cable.

**ASP**   Auto-Speed Port.

**ASR**   Automatic Send/Receive.

| HIGH \ LOW B3B2B1B0 | | 0000 | 0001 | 0010 | 0011 | 0100 | 0101 | 0110 | 0111 | 1000 | 1001 | 1010 | 1011 | 1100 | 1101 | 1110 | 1111 |
|---|---|---|---|---|---|---|---|---|---|---|---|---|---|---|---|---|---|
| B7B6B5B4 | | 0 | 1 | 2 | 3 | 4 | 5 | 6 | 7 | 8 | 9 | A | B | C | D | E | F |
| 0000 | 0 | NUL | SOH | STX | ETX | EOT | ENQ | ACK | BEL | BS | HT | LF | VT | FF | CR | SO | SI |
| 0001 | 1 | DLE | DC1 | DC2 | DC3 | DC4 | NAK | SYN | ETB | CAN | EM | SUB | ESC | FS | GS | RS | US |
| 0010 | 2 | SP | ! | " | # | $ | % | & | ' | ( | ) | * | + | , | — | . | / |
| 0011 | 3 | 0 | 1 | 2 | 3 | 4 | 5 | 6 | 7 | 8 | 9 | : | ; | < | = | > | ? |
| 0100 | 4 | @ | A | B | C | D | E | F | G | H | I | J | K | L | M | N | O |
| 0101 | 5 | P | Q | R | S | T | U | V | W | X | Y | Z | [ | \ | ] | ^ | — |
| 0110 | 6 | \ | a | b | c | d | e | f | g | h | i | j | k | l | m | n | o |
| 0111 | 7 | p | q | r | s | t | u | v | w | x | y | z | { | \| | } | ~ | DEL |

BINARY — HEX — ASCII

*ASCII character set*

**Assembler**   A program that translates an assembly programming language into the code of zeros and ones used by computers.

**assembly level code**   A low-level language for computer programming that uses mnemonic commands to represent machine language operation codes. This language allows the programmer to write programs for a specific machine with greater ease than using binary-number machine language instructions.

**associated address space**   In IBM's VTAM, the address space in which RPL-based requests are issued that specify an ACB opened in another address space.

**associated signaling**   Signaling system in which signaling bits are included in the data channels, i.e. bit 8 of frames 6 and 12 in North America (D4 format).

**Association Control Service Elements (ACSE)**   An International Standards Organization (ISO) standard Application layer protocol for handling associations between applications on an Open Systems Interconnection (OSI) based network.

**Association Française des Utilisateurs du Téléphone des Télécommunications (AFUTT)**   The French telecommunications users group which is involved in drafting policies which affect the operation and offerings of France Telecom.

**Astra-Phacs**   A telecommunications network management system marketed by NEC Information Systems that processes call records from a PBX or Centrex system.

**Astroline**   The name for a family of hand-held Pulse Code Modulation (PCM) monitors manufactured by Alcatel STR.

**Astrotec** AT&T trademark for a laser transmitter system.

**ASU** Application Start-Up.

**Asymmetrical transmission** The process by which a modem obtains full duplex transmission by transmitting data in two directions at different speeds. The modem monitors the quantity of data being transmitted in order to assign the higher-speed channel to the major data transmission flow direction.

**ASYNC** Short for asynchronous or for asynchronous transmission.

**asynchronous** A digital network which is not synchronized to a common frequency.

**asynchronous (character framed)** Having a variable or random time interval between successive characters, operations, or events. Transmission in which each character, word, or small block, is individually synchronized (timed), usually by the use of start and stop bits. This is in contrast to "synchronous operation." Also called Start/Stop (S/S).

**asynchronous character** A binary character used in asynchronous transmission which contains equal-length bits, including a start bit and one or more stop bits which define the beginning and end of the character.

**asynchronous communication** A method of communication where the time synchronization of the transmission of data between the sending and receiving stations is set by start and stop bits and the baud rate.

**asynchronous data** Data transmitted character by character where each character has a start bit and one or more stop bits. The transition from stop to start triggers the receiver's internal bit rate clock. The receiver then accepts and discards the start

bit, accepts the number of bits appropriate to the code set, and is conditioned by the stop bit for the next character. Start–stop synchronization enables characters to be sent at random, since each character carries the necessary synchronizing information with itself.

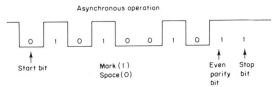

The ASCII character R in start-stop transmission

**asynchronous data channel** A communications channel capable of transmitting data but not timing information. Sometimes called an anisochronous data channel.

**Asynchronous Emulation Adapter (AEA)** An adapter in the IBM 3174 Subsystem Control Unit that enables an ASCII terminal to communicate with a 3270 host using the 3270 data stream, an ASCII terminal to communicate with an ASCII host through the 3174, and a 3270 terminal to communicate with an ASCII host using the DEC VT100 data stream or the IBM 3101 data stream.

**asynchronous exit routine** In IBM's VTAM, an RPL exit routine or an EXLST exit routine other than LERAD or SYNAD.

**asynchronous link** A communications link using asynchronous transmission, in which each transmitted character is framed with start and stop bits. The time interval between characters may be of unequal length.

**asynchronous modem** A modem that uses asynchronous transmission, and, therefore, does not require timing synchronization with its attached DTE or remote modem; also used to describe a modem which converts asynchronous inputs from the DTE to synchronous signals for modem-to-modem transmission.

**asynchronous operation** In IBM's VTAM, an operation, such as a request for session establishment or data transfer, in which the application program is allowed to continue execution while VTAM performs the operation. VTAM informs the program after the operation is completed.

**asynchronous request**   In IBM's VTAM, a request for an asynchronous operation.

**asynchronous routine**   A set of computer instructions that a computer program calls into service on an as-needed basis.

**asynchronous terminal**   A terminal that uses asynchronous transmission; usually synonymous with ASCII terminal and with dumb terminal.

**Asynchronous Time-Division Multiplexer**)   A TDM that multiplexes asynchronous signals by oversampling; also, infrequently used to mean concentrator.

**asynchronous transmission**   Method of sending data in which the interval between characters may be of unequal length; since asynchronous characters are used, no additional synchronizing or timing information need be sent. Also called start–stop transmission.

**AT**   Access Tandem.

**AT command set**   A set of commands used by personal computers to direct dialing operations in a modem. Although the intelligent modem patent is held by BIZcom, it was Hayes Microcomputer Products which popularized the use of intelligent modems. Thus, commonly referred to as the "Hayes command set."

**AT commands**   The "AT" commands are subdivided into three major groups: configuration, immediate action, and diagnostic commands.

The "AT" prefix begins every command line with the exception of the +++ (escape) and the A/ (repeat) commands. "AT," often referred to as attention code, delivers to the modem information about the data rate and the parity setting of the local DTE. Multiple commands can be placed on a single line. Space characters are allowed between commands to improve readability. A command line must be terminated with the ASCII carriage return character. A line feed character following the carriage return character is optional. The backspace or delete key can be used to delete any character entered from the keyboard, except the "AT" prefix. Upon execution of the command the result code is returned by the modem.

Example:  Terminal:   AT DT 998 1234
　　　　　　Modem:　　OK

This example demonstrates the dial command. A string of ASCII characters (the "AT" command) followed by the digits (a number to dial) is entered from the keyboard. Upon successful execution of the dial command, an OK message is returned to the user.

RESULT CODES

| Short | Long Form | Description |
|---|---|---|
| 0 | OK | Command line executed without errors |
| 1 | CONNECT | Connected at 300 bps |
| 2 | RING | Local telephone line ringing |
| 3 | NO CARRIER | Carrier lost, or never received |
| 4 | ERROR | Error in command line, invalid command line, command line exceeds command buffer, invalid character format |
| 5 | CONNECT 1200 | Connected at 1200 bps data rate |
| 6 | NO DIALTONE | No dial tone received within time-out period |
| 7 | BUSY | Called line busy |
| 8 | NO ANSWER | Called line not answered within time-out period |
| 9 | CONNECT 600 | Connection established at 600 |
| 10 | CONNECT 2400 | Connection established at 2400 |

DIAL MODIFIERS

| | |
|---|---|
| P | Pulse dial |
| R | Originate call in Answer mode |
| T | Tone dial |

| | |
|---|---|
| S | Dial stored number |
| W | Wait for dial tone |
| , | Pause |
| ; | Return to command state |
| ! | Flash |
| @ | Wait for silence |

## "AT" COMMANDS SUMMARY

| | |
|---|---|
| AT | Attention code preceded command lines except +++ (escape) and A/ (repeat) |
| A | Go off-hook in answer mode |
| A/ | Repeat previous command line |
| B | CCITT V.22 operation at 1200 bps |
| B1 | Bell 212A operation at 1200 bps (default) |
| D | Dial a number (0–9 ABCD*#) |
| E | Turn Echo off |
| E1 | Turn Echo on (default) |
| H | Go on-hook (hang up) (default) |
| H1 | Go off-hook and switch the auxiliary relay |
| I | Request product code |
| I1 | Compute and return checksum (firmware ROM) |
| I2 | Compute and return checksum with OK or ERROR message |
| L,L1 | Low speaker volume |
| L2 | Medium speaker volume (default) |
| L3 | High speaker volume |
| M | Speaker is off |
| M1 | Speaker is off while carrier is present (default) |
| M2 | Speaker is always on |
| M3 | Speaker is disabled while dialing or receiving carrier |
| O | Return to On-line mode |
| O1 | Return to On-line mode and initiate retrain sequence (in 2400 bps only) |
| Q | Return result codes (default) |
| Q1 | Do not return result codes |
| Sn=x | Write x in S-register n |
| Sn? | Read S-register n |

| | |
|---|---|
| V | Enable short form result codes |
| V1 | Enable full word result codes (default) |
| X | CONNECT result code enabled (300 bps operation) |
| X1 | All CONNECT result codes enabled; dial blind; busy signal is not recognized |
| X2 | All CONNECT result codes enabled; wait for dial tone before dialing; busy signal is not recognized |
| X3 | All CONNECT result codes enabled; dial blind; busy signal is recognized |
| X4 | All CONNECT result codes enabled; wait for dial tone before dialing; busy signal is recognized (default) |
| Y | Enable long space disconnect (default) |
| Y1 | Disable long space disconnect |
| Z | Fetch configuration profile contained in external nonvolatile memory |
| +++ | The default escape code |
| &C | DCD always ON (default) |
| &C1 | DCD tracks the state of the carrier signal |
| &D | DTR ignored (default) |
| &D1 | Assume command state on ON-to-OFF transition of DTR |
| &D2 | Go ON-hook, disable auto-answer, assume command state on ON-to-OFF transition of DTR |
| &D3 | Assume initialization state on ON-to-OFF transition of DTR |
| &F | Fetch factory configuration profile from internal ROM |
| &G | No guard tone (default) |
| &G1 | 550 Hz guard tone |
| &G2 | 1800 Hz guard tone |
| &M | Asynchronous mode (default) |
| &M1 | Synchronous mode 1 (sync/async modes are supported) |
| &M2 | Synchronous mode 2 (dial stored number mode) |
| &M3 | Synchronous mode 3 (manual dial with DTR off) |

&P    Make/Break pulse ratio = 39/61 (USA/Canada) (default)

&P1    Make/Break pulse ratio = 33/67 (CCITT)

&R    CTS OFF-to-ON transition follows RTS OFF-to-ON transition (default)

&R1    CTS always ON; RTS ignored

&S    DSR always ON (default)

&S1    DSR operates accordingly to CCITT V.22 bis/V.22 recommendation

&T    Terminate any test currently in process

&T1    Initiate local analog loopback

&T3    Initiate local digital loopback

&T4    Grant request from remote modem for remote digital loopback

&T5    Deny request from remote modem for remote digital loopback

&T6    Initiate remote digital loopback

&T7    Initiate remote digital loopback with self test

&T8    Initiate local analog loopback with self test

&W    Write configuration to nonvolatile memory

&X    Transmit clock source is modem (default)

&X1    Transmit clock source is DTE

&X2    Transmit clock source is derived from received carrier signal

&Z    Store telephone number

**ATACS**    Army Tactical Communications System.

**ATB**    All Trunks Busy.

**ATC**    Automatic Technical Control.

**ATCAP**    Army Telecommunications Automation Program.

**ATDM**    Asynchronous Time Division Multiplexer.

**ATLAS 400**    An electronic messaging system operated by Transpac in France.

**ATM**    Automated Teller Machine.

**ATT**    Attenuator.

**attach**    1. The process of connecting a device to a communications facility so it can communicate.

2. To participate in the data passing protocol of the ring.

**Attached Resources Computing Network (ARC-net)**    A networking architecture marketed by Datapoint Corporation and other vendors which uses a token-passing bus architecture.

**attaching device**    Any device that is physically connected to a network and can communicate over the network.

**attachment**    1. A resource that connects the control program, the logical link, and the physical link between two or more computers. 2. The process of attaching a file to an electronic message.

**attack dialing**    The process of immediately repeating a dialed call in which a busy signal is encountered, to minimize the time required to eventually connect the call.

**attendant**    A term used to refer to a local switchboard operator, such as the person controlling a PBX.

**attendant conference**    A PBX feature that allows an attendant to establish a conference connection between central office trunks and internal stations.

**attendant monitor**    A special PBX attendant circuit that allows listening in on circuits with the console handset/headset transmitter deactivated.

**attended**    An operation undertaken by, or under the control of, a human operator.

**attention (ATTN)**    1. In the context of the virtual terminal, the attention or interrupt facility allows a user operating in alternate mode to regain access to the virtual terminal data structure when it is under the control of a remote computer. The receipt of an attention signal is acknowledged by a mark. 2. An occurrence external to an operation that could cause an interruption of the operation.

**attention code**    A request for service signal.

**attenuation**    Deterioration of signals as they pass through a transmission medium; generally, attenuation increases (signal level decreases) with both frequency and cable length. Measured in terms of levels or decibels (dB).

**attenuation distortion**  The relative loss at discrete frequencies across the bandwidth when compared to a 1004 Hz reference signal. For example, if a circuit had 6 dB more loss at 2804 Hz than it did at 1004 Hz, it is said to have an attenuation distortion of 6 dB. Attenuation distortion is also referred to as "frequency response" or "slope."

**attenuator**  In fiber optic transmission, an instrument which decreases the amount of light which passes through it.

**ATTIS**  AT&T Information Systems; division of AT&T Technologies that supplies and manufactures customer premises equipment.

**attitude control**  Maintenance of a satellite's orientation with respect to the earth and the sun.

**ATTN**  Attention.

**attribute**  1. Columns in a data table having unique names. 2. A characteristic. 3. A terminal display language that specifies a particular quality for the object with which it is associated.

**attribute value**  A code immediately following the attribute type that specifies a particular property from the set defined by the attribute type.

**AT&T**  American Telephone and Telegraph.

**AT&T ALL PRO America**  A calling plan which integrates AT&T PRO America I and the AT&T PRO State plan in a customer's own state. AT&T ALL PRO offers a discount on direct-dialed AT&T Long Distance calls to locations within your state and across the country, 24 hours a day, seven days a week.

**AT&T Alliance Teleconferencing Service**  An AT&T service which allows customers to dial directly conference calls with as few as three or as many as 59 locations.

**AT&T Call Me card**  A card which permits toll-free calling to your business number, and no other. Suitable for low-volume calling, the card holder

controls how it is publicized and distributed.

**AT&T card**  A free calling card which enables the user to charge all AT&T Long Distance calls from virtually any U.S. location to his/her home or office number. It also works from most foreign countries for calls back to the U.S.

**AT&T Communications**  The regulated subsidiary of post-divestiture AT&T responsible for long distance operations; the successor to AT&T Long Lines.

**AT&T Information Systems**  The unregulated subsidiary of post-divestiture AT&T responsible for equipment sales and leasing.

**AT&T International 800 Service**  An AT&T service which permits customers in more than 20 countries to call a U.S. location toll free. The 800 Service provider specifies the service by the country in which calls originate.

**AT&T Mail**  An electronic mail service marketed by AT&T of Morristown, NJ. Dial 1-800-MAIL-123 to register electronically or 1-800-367-7225, extension 600 to register by voice.

**AT&T Mail Access I**  A software package marketed by AT&T for operation on personal computers which is designed to enhance the use of that vendor's electronic mail service.

**AT&T Mail PMX/X.400**  A software product from AT&T's Data Systems Group which converts AT&T PMX (Private Message Exchange) mail messages to X.400 format.

**AT&T Mail Talk**  A voice response system service marketed by AT&T which permits subscribers of that firm's electronic mail system to read, delete, and save messages via a touch tone telephone. The following table lists AT&T Mail Talk Commands.

AT&T MAIL TALK COMMANDS

| Key | Command |
|---|---|
| 1 | Read first message |
| 2 | Read previous message |
| 3 | Delete/Restore current message |

4   Help
5   Repeat current message
6   Save current message
7   Stop/Start
8   Change rate of speed
9   Read last message
0   Exit system
#   Read next message/terminate login
*   Stop current message, await next command

**AT&T Merlin Communications System**   A family of flexible business communications systems. The smallest Merlin system starts at 4 telephone lines and 10 phones; the largest can handle up to 32 lines and 72 phones.

**AT&T Network Services**   AT&T Technologies' subdivision that manufactures and supplies equipment primarily to telephone organizations.

**AT&T PRO America Family**   A series of calling plans for out-of-state direct-dialed AT&T Long Distance calls. AT&T PRO America includes calls to 39 countries. Businesses which spend more than $120 a month on a combination of AT&T Long Distance interstate calls and AT&T International Long Distance calls to eligible countries generally qualify for PRO America.

**AT&T PRO State**   A calling plan for AT&T Long Distance calls within a state. Only calls outside the subscriber's local calling area are included.

**AT&T Pub 54106**   "Requirements for Interfacing Digital Terminal Equipment to Services Employing the Extended Superframe Format": a basic reference source on ESF in the United States.

**AT&T Pub 62411**   "ACCUNET T1.5 Service Description and Interface Specifications": a basic reference source on AT&T T1 implementations.

**AT&T Reach Uut America Bonus Plan**   A special plan marketed by AT&T for people who do most of their calling in the evenings and weekends. Under this plan, AT&T customers pay a flat hourly rate for any calls made after 10 p.m. on weekdays or during weekend hours.

**AT&T Spirit Communications System**   A basic, easy-to-use phone system with many built-in features. SPIRIT supports from 3 to 24 lines and from 8 to 48 telephones.

**AT&T System 25**   A state-of-the-art digital PBX designed to bring sophisticated telephone system benefits to small and growing businesses with 20 to 150 telephone stations.

**AT&T Technologies**   AT&T equipment manufacturing and supply division; includes AT&T Information Systems and AT&T Network Services (formerly Western Electric).

**AT&T Transaction Services**   An enhanced, distributed, intelligent shared network designed for the delivery of non-cash (check, debit, credit card) transactions.

**AT&T WATS Service**   Low cost, high-quality long distance service for outgoing calls. AT&T WATS requires special access lines, but can provide significant savings for companies with heavy outgoing call volume. Coverage can be tailored from single-state to nationwide in a series of increasing bands.

**AT&T 800 Readyline**   A toll-free service for moderate call volume. It looks like regular AT&T 800 Service to callers, but there are no special lines or equipment to install, making it easy to adjust for changes in geographic range or seasonal calling patterns.

**AT&T 800 Service**   Also known as WATS, this toll-free service is for high-customer call volume. AT&T 800 Service can be designed for in- state callers only or in widening bands for regional or national coverage and can begin with a single line.

**AT-984A/G**   A U.S. military multiple wavelength, long wire antenna used to extend the range of radio sets equipped with receiver–transmitters RT-505/PRC-25 and RT-841/PRC-77.

**AU**   Access Unit.

**audible**   Capable of being heard.

**audible alarm**   1. An alarm that is sounded when

designated events occur that require operator attention or intervention. 2. A special feature that sounds a short, audible tone automatically when a character is entered from the keyboard into the next-to-last character position on the screen. The tone can also be sounded under program control.

**audible ringing tone**  Tone transmitted to the calling telephone indicating that the called telephone is being rung.

**audio frequencies**  The frequencies that can be heard by the human ear (usually between 30 and 20 000 hertz) when transmitted as sound waves.

**audio subcarrier**  In satellite communications the carrier between 5 MHz and 8 MHz containing audio (or voice) information.

**audiographics**  The technology which allows sound and visual images to be transmitted simultaneously. Generally refers to single frame or slow frame visual images as opposed to continuous frame images such as television. Audiographic transmission is often used to teach or train people located remotely from an educational institution or business training center, saving travel and housing expense.

**audiometer**  A calibrated audio oscillator that can be used to vary the amplitude and frequency of sound. The audiometer is normally used for conducting hearing tests.

**audio response unit**  An output device which provides synthesized or pre-recorded speech as the response to inquiries. The response may be transmitted over communications facilities to a remote location from which the inquiry originated.

**Audiotex**  A service which permits a data base host to pass data to a voice-mail computer, where it is interpreted and delivered over the telephone as a natural voice spoken message.

**audit trail**  A printed record of system commands and events identified by date and time of occurrence. The audit trail consists of operator-initiated commands and system-generated events.

**Audix**  An AT&T voice-mail system.

**AUP**  Acceptable Use Policy.

**Australia Overseas Telecommunications Commission**  The communications carrier which provides all of Australia's international satellite communications services.

**authentication**  The process of identifying a user to a network or computer used as a security measure designed to protect a communications system against fraudulent messages.

**authentication of messages**  Addition of a check field to a block of data so that any change to the data will be detected. A secret key enters into the calculation and is known to the intended receiver of the data. In a different form of authentication, the whole block is transformed and this can be a public key cryptosystem. One common method employed for authentication is the challenge and reply technique. This technique employs a predefined rectangular grid formed by columns and rows of characters. The challenging party randomly chooses two letters and transmits them as the challenge. The party being challenged locates the first letter of the challenge in the left-hand column, then proceeds horizontally across this line to the second letter of the challenge. The letter immediately below this letter is the reply. (If the second letter of the challenge is the last letter in the mixed alphabet column, the top letter of the same column will be the reply.) The called party will always make the first challenge. When the caller desires authentication, he must first invite a challenge from the called party by stating that he is prepared to authenticate. If an incorrect reply is received, the called party should rechallenge, with another randomly selected challenge. The following example illustrates the challenge and reply authentication method. (See opposite.)

**Authority and Format Indicator (AFI)**  In Digital Equipment Corporation Networks Architecture (DECnet), the part of a NSAP address which indicates the addressing authority responsible for

33

EXAMPLE:         CHALLENGE:        <u>C</u>harlie <u>X</u>ray         REPLY:         <u>U</u>niform

| A | M | S | C | N | U | P | H | I | B | V | T | J | Y | A | D | E | Z | Q | R | G | L | X | F | W | K | O |
| B | B | J | D | E | P | Y | L | S | X | Q | O | G | U | W | R | C | F | K | A | H | Z | T | I | V | N | M |
| <u>C</u> | U | L | V | H | B | W | Q | E | J | K | P | T | Y | C | D | <u>X</u> | M | R | F | A | S | N | I | Z | O | G |
| D | F | R | V | Y | E | C | Z | A | J | G | S | M | L | I | H | <u>U</u> | Q | P | B | K | O | T | W | N | D | X |
| E | J | Q | L | K | Z | Y | V | S | O | B | F | I | R | N | P | C | U | E | M | T | H | D | A | X | W | G |

authentication of messages

the assignment of the IDP and its format. It also indicates the format (binary or decimal) of the DSP.

**authorization exit routine**   In IBM's VTAM, an optional installation exit routine that approves or disapproves requests for session initiation.

**authorized path**   In IBM's VTAM for MVS, a facility that enables an application program to specify that a data transfer or related operation be carried out in a privileged and more efficient manner.

**authorized user**   A person or organization authorized to use a network or system.

**autoanswer**   A machine feature which allows a station to automatically respond to call over the public switched network.

**autobaud**   See automatic baud rate detection.

**autobaud rate detection**   A devices's ability to adjust itself to the incoming baud rate from the customer's equipment.

**autoconnect**   A feature in which the connecting processor automatically places the call to a predefined address, generally to a host, thus eliminating the need for the terminal user to enter a call command manually each time the terminal is powered on. Autoconnect is implemented on a terminal that accesses one particular host and is usually effected as soon as the terminal is turned on.

**autodelivery**   A type of electronic mail delivery where a message is sent to a user's catalog and a copy is also automatically generated and sent to a station (hardcopy device).

**auto-dial**   A type of modem used on the public telephone network that automatically originates calls (dials the desired number).

**AUTODIN**   AUTOmatic DIgital Network.

**Automated Coin Toll Service (ACTS)**   The feature that automatically computes charges on coin toll calls, announces charges to the customer, counts coin deposits, and sets up coin calls—all without need of an operator.

**automatic activation**   In IBM's VTAM, the activation of links and link stations in adjacent subarea nodes as a result of channel device name or RNAME specifications related to an activation command naming a subarea node.

**automatic answering modem**   A modem that will answer a phone call in an unattended mode and prepare for receipt of data.

**Automatic Answering Unit (AAU)**   A hardware device which performs the auto-answer function.

**Automatic Baud Rate Detection (ABR)**   A process by which a receiving device determines the speed, code level, and stop bits of incoming data by examining the first character—usually a preselected sign-on character. ABR allows the receiving device to accept data from a variety of transmitting devices operating at different data rates without needing to configure the receiver for each specific data rate in advance.

**Automatic Call Director (ACD)** Device used for telemarketing applications to direct incoming calls to the next free agent position. This function is also a feature of many PBX products.

**Automatic call distribution** A PABX feature that allows incoming calls to be automatically connected to an agent's telephone without the agent taking any action.

**Automatic Call Distributor (ACD)** A hardware device which performs the auto-answer function and switches the calls to the first available station or communications port.

**automatic callback** A feature that enables a user with a busy line to be connected automatically to the called line when it becomes idle.

**automatic calling** Calling in which the elements of the selection signal are entered into the data network contiguously at the full data signaling rate.

**Automatic Calling Unit (ACU)** A hardware device which connects to a telephone line and allows an attached computer to dial phone numbers. The ACU is used in conjunction with a modem. When the Auto Call function is included with the modem, the modem is said to have "automatic calling" capability.

**automatic deactivation** In IBM's VTAM, the deactivation of links and link stations in adjacent subarea nodes as a result of a deactivation request naming a subarea node. *Note.* Automatic deactivation occurs only for automatically activated links and link stations that have not also been directly or indirectly activated.

**automatic dialer** 1. A device which automatically dials preset telephone numbers, usually those dialed frequently. 2. A device which automatically dials a series of random numbers; most often used in computerized telephone solicitations.

**Automatic Dialing Unit (ADU)** A hardware device

which is capable of automatically generating dialing pulses, or signals.

**AUTOmatic DIgital Network (AUTODIN)** A data communications network developed and operated by contractors for the U.S. Department of Defense.

**automatic equalization** Equalization of a transmission channel which is adjusted automatically while sending special signals.

**automatic exchange** A telephone company central office through which communication between subscribers is effected by the use of an electronic switching device without the intervention of an operator.

**automatic exclusion** The process whereby the first station accessing a line automatically prevents any other station from gaining access to that line.

**Automatic Identification of Outward Dialing (AIOD)** A PBX service feature that identifies the calling extension, permitting cost allocation.

**Automatic Link Intelligence (ALI)** A feature of the Symplex Communications Corporation Datamizer in which data transmission is routed to alternate lines of throughput degrades below an acceptable level on a line being used.

**automatic logon** In IBM's VTAM, a process by which VTAM creates a session-initiation request (logon) for a session between a secondary logical unit (other than a secondary application program) and a designated primary logical unit whenever the secondary logical unit is not in session with, or queued for a session with, another primary logical unit. *Note.* Specifications for the automatic logon can be made when the secondary logical unit is defined or can be made using the VARY NET,LOGON command.

**Automatic Message Accounting (AMA)** The automatic collection, recording, and processing of call-related information for billing by a telecommunications provider.

**Automatic Message Accounting Teleprocessing System (AMATPS)** A billing system feature where the AT&T 5ESS switch forwards billing information over a data link to a centralized AMA data collection system. The AMA data collection system interfaces with an RAO.

**Automatic Modem selection** A PABX feature that allows a subscriber to be connected to a modem from a modem pool so an internal or external data call can be completed.

**automatic number identification** Means provided at a central office switch in which the switch is able to determine the identity of a calling party and forward that information to the called party.

**Atomatic Potection Switching (APS)** Switch equipment designed to reroute communications automatically to a secondary circuit should the primary fail.

**automatic repeat request** Same as automatic request for retransmission.

**Automatic Request for Repetition (ARQ)** Same as automatic request for retransmission.

**Automatic Request for Retransmission (ARQ)** An error control method in which the receiving device informs the transmitting device which transmission blocks were received successfully; the transmitting device retransmits any blocks not successfully received.

**Automatic Route Selection (ARS)** The capability of a PABX and other network devices to select the optimum route for a call (in terms of cost, reliability, quality, etc.) without operator intervention.

**Automatic Send/Receive Teleprinter (ASR)** Teleprinter equipped with paper tape, magnetic tape, or a solid state buffer that allows it to transmit and receive data unattended.

**Automatic Send/Receive (ASR) terminal** A terminal that consists of a printer, keyboard and auxiliary storage which permits messages to be composed "off-line" and then transmitted at the operating rate of the terminal.

**automatic switching system** A communications

system in which the switching operations are performed by electrically controlled devices without human intervention.

**Automatic Technical Control (ATC)** A computer system used to maintain operational control of a data communications network.

**Automatic Teller Machines (ATM)** The remote terminal used by banks to allow customers to perform banking transactions.

**AUTOmatic VOice Network (AUTOVON)** A voice network developed and operated by contractors for the U.S. Department of Defense.

**Automode** The term which describes the functions required to negotiate between CCITT V.22bis and V.32 modem modulation standards when a connection is made between those two types of modems.

**Automotive Industry Action Group (AIAG)** The developer of Electronic Data Interchange (DEI) standards for the automotive industry.

**Autoplex** A trademark of AT&T as well as the name of a third generation cellular telecommunications system service.

**Autoplex System 100** The name of AT&T's first commercial cellular telephone system which was introduced in Chicago in 1983.

**AUTOSEVOCOM** AUTOmatic SEcure VOice COMmunications.

**Auto-Speed Port (ASP)** A network access port which automatically detects the speed of the terminal and matches operation at that speed. Also called Auto-Speed Detect (ASD).

**autoswitch** A series of matrix switches manufactured by Bytex Corporation of Southborough, MA.

**AutoSync** A firmware feature of Hayes Microcomputer Products 2400 bps and higher speed Smartmodems that enables the transmission of synchronous data directly out of the asynchronous serial port in a personal computer.

**AUTOVON**   AUTOmatic VOice Network.

**auxiliary equipment**   Equipment, such as a printer, not under direct control of the processing unit.

**auxiliary network address**   In IBM's ACF/ VTAM, any network address, except the main network address, assigned to a logical unit capable of having parallel sessions.

**AVA**   Automatic Voice Answering.

**AVAIL**   An automated bank teller network servicing the state of Georgia.

**availability**   A measure of equipment, system, or network performance—usually expressed in percent (a value between 0 and 1 or 0 and 100 percent) that relates to a user obtaining connectivity on a communications link. Availability consists of reliability expressed as the mean time between failures (MTBF) and serviceability expressed as the mean time to repair (MTTR) as follows: Availability = MTBF/ (MTBF + MTTR).

**available**   In IBM's VTAM, pertaining to a logical unit that is active, connected, enabled, and not at its session limit.

**available line**   A circuit between two points that is ready for service but idle.

**Avalance PhotoDiode (APD)**   A light detector that has inherent gain through and electron multiplication process known as avalanche. The APD is a diode which increases its electrical conductivity by a multiplication effect when struck by light. Used in receivers for lightwave transmission.

**average busy-hour traffic count**   The average number of calls received during the busy hour.

**AVD**   Alternate Voice Data.

**AWG**   American Wire Gage.

**AXE**   Ericsson's digital switching system.

**AXESS**   The name of the computer reservation system of Japan Air Lines.

**AZ-EL mount**   Antenna mount that requires two separate adjustments—of azimuth and elevation—to move from one satellite to another.

**azimuth**   In satellite transmission the angle between an antenna beam and the meridian plane which is measured along a horizontal plane.

# B

**B Block** In cellular communications one of two sets of channels assigned by the U.S. Government to one of two telephone companies operating in a metropolitan area.

**B channel "bearer channel"** In ISDN, the 64-Kbps channel of a Basic Rate Interface (BRI) where there are two or a Primary Rate Interface (PRI) (where there are 23) that is circuit switched and can carry either voice or data transmission.

**babbling tributary** In a local area network, a station that continuously transmits meaningless messages.

**BABT** British Approvals Board for Telecommunications.

**back reflection** In a fibre optic system the portion of transmitted light which is deflected from the original direction of travel and reflected back on the fibre towards the optical transmitter. Back reflections are normally caused by inadequate connectors and poorly aligned splices.

**backboard** Wooden or metal panels used for mounting miscellaneous apparatus. May be equipped with premounted connecting blocks, mounting brackets, and/or predrilled holes.

**backbone** 1. In packet switched networks, the major transmission path for a PDN. 2. In a local area network multiple-bridge ring configuration, high-speed link to which the rings are connected by means of bridges. 3. A transmission facility that interconnects bridges in hierarchical networks. 4. In a bulletin board system network the series of echo conferences carried in a zone.

**backbone closet** The closet where backbone cable is terminated and cross-connected to either horizontal distribution cable or other backbone cable. The backbone closet houses cross-connect facilities, and may contain auxiliary power supplies for terminal equipment located at the user work location. Sometimes called a "riser closet."

**backbone network** A transmission facility designed to interconnect generally lower-speed distribution networks or clusters of dispersed user devices.

**backbone subsystem** The part of a premises' distribution system that includes a main cable route and facilities for supporting the cable from an equipment room (often in the building basement) to the upper floors, or along the same floor, where it is terminated on a cross-connect in a backbone closet, at the network interface, or distribution components of the campus subsystem. The subsystem can also extend out on a floor to connect a backbone and satellite closet or other satellite location.

**background** In a 3270 PC environment, the color of the window "writing surface."

**back-level host** In SNA network interconnection, a host processor containing TCAM, VTAME, or a release of VTAM prior to the current release.

**backlog** In a message transmission system, those messages awaiting or being processed for delivery, refile, or transmission.

39

**back-off** 1. The process of reducing the power levels (input and output) of a traveling wave tube to obtain a more linear operation. 2. In a local area network, the process of delaying an attempt to transmit. Usually results when an earlier attempt to transmit encountered some difficulty, such as a collision on a CSMA/CD network.

**backplane bus** A collection of electrical wiring and connectors that interconnect modules inserted into a computer system or other electronic device.

**back-to-back gateways** Two gateways separated by one intervening network that contains no gateway SSCP function involved with either of the two gateway NCPs.

**backup** 1. The hardware and software resources available to recover after a degradation or failure of one or more system components. 2. To copy files onto a second storage device so that they may be retrieved if the data on the original source is accidentally destroyed.

**backup diskette** A diskette containing information copied from a fixed disk or from another diskette. It is used in case the original information becomes unusable.

**backup path** An alternative path for signal flow through access units and their main ring path cabling in a token-ring network. The backup path allows recovery of the operable portion of the network while problem determination procedures are being performed.

**backward channel** Also known as a reverse channel; this type of channel is used to send data in the direction opposite of the primary (forward) channel. The backward channel is often used for low-speed data as may be found in keyboarded or supervisory controls.

**Backward Explicit Congestion Notification (BECN)** A bit position in the frame relay header which when set to a value of 1 indicates that the path through the network in the direction opposite the frame's path is congested.

**Backward Indicator Bit (BIB)** In CCITT Signaling System 7, a bit position whose value is inverted by a receiver if the last signal unit received is corrupt.

**Backward Sequence Number (BSN)** In CCITT #7 terminology the BSN is the value of the forward sequence number of the last signal unit correctly received.

**backwards learning** A method of routing in which nodes learn the topology of the network by observing the packets passing through, noting their source and the distance they have traveled.

**BAFO** Best and Final Offer.

**balanced circuit** 1. A circuit terminated by a network whose impedance balances the impedance of the line so the return losses are negligible. 2. A circuit with an electrical interface that uses differential signaling to transmit information.

**balanced code** A line code whose digital sum variation is finite due to the absence of a DC component in its frequency spectrum.

**balanced differential signaling** Differential signaling in which the sum of the currents in the differential signaling wires equals zero. As an example of differential signaling consider the following illustration. Note that unipolar signals are divided by a driver into equal and opposite polarities. This type of signaling extends the range of transmission.

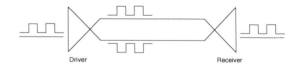

Driver  Receiver

**balanced mode** An arrangement in which two stations are able to transmit to each other without requiring that either be fixed as a "master" or "secondary" station.

**balanced-to-ground** A two-wire circuit, in which the impedance-to-ground on one wire equals the impedance-to-ground on the other wire (compare with unbalanced-to-ground); a favorable condition for data transmission.

**balancing network** Electronic circuitry used to match two-wire facilities to four-wire toll facilities. This balancing is necessary to maximize power transfer and minimize echo. Sometimes called hybrid.

**balun (balanced/unbalanced)** Impedance matching devices used to connect balanced twisted pair cable to unbalanced coaxial cable. They are required on each end of a twisted pair cable when converting a coax to twisted pair medium.

**band** 1. The frequencies lying between two defined limits. 2. The specific geographic area which the customer is entitled to call or be called from in a Wide Area Telephone System (WATS).

**band splitter** A multiplexer (commonly an FDM or TDM) designed to divide the composite bandwidth into several independent, narrower bandwidth channels, each suitable for data transmission at a fraction of the total composite data rate.

**band-limited** A signal or channel of restricted frequency range.

**bandpass** The portion of a band, expressed in frequency differences (bandwidth), in which the signal loss (attenuation) of any frequency when compared with the strength of a reference frequency is less than the value specified in the measurement.

**bandpass filter** A device or circuit which passes a specific group of frequencies between cutoff frequencies.

**Bandsplitting** A technique for transmitting synchronous data in a statistical multiplexer in which a fixed allocation of bandwidth is given to each synchronous channel, regardless of its activity.

**bandwidth** The range of signal frequencies which can be carried by a communications channel subject to specified conditions of signal loss or distortion. Measured in hertz (cycles per second). Bandwidth is an analog term which provides a measure of a circuit's information capacity. There are three general ranges of frequencies for transmission:

Narrow band—2 to 300 Hz
Voice band—300 to 3,000 Hz
Wide band—over 3,000 Hz

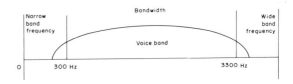

**Bandwidth Management Service (BMS)** An AT&T Accunet T1.5 service option which enables customers to use an on-premises terminal or personal computer to reconfigure channels within T1 trunks. The subscriber may switch any of the twenty-four 64 Kbps channels within a T1 circuit to previously defined alternate paths or switch complete T1 circuits between two BMS nodes.

**Bandwidth on Demand Interoperability Group (BONDING)** A consortium, originally consisting of Digital Access Corporation, General DataComm, Larse Corp., Newbridge Networks, Promptus Communications and Teleos Communications, formed to promote interoperability among customer premises equipment supporting bandwidth-on-demand transmission, such as the use of inverse multiplexers.

**bang** An exclamation point (!) used in a UUCP electronic mail address.

**bang path** A series of UUCP nodes mail will pass through to reach a remote user. Node names are separated by exclamation marks nicknames "bangs." The first node in the path must be on the local system, the second node must be linked to the first, and so on. To reach user1 on sys3 through sys2 if your computer's address is sys1 you would use the following address:

sys1!sys2!sys3!user1

**banner** 1. Predefined identifier which is displayed on terminal screen under certain conditions. 2. A one-page information sheet, printed as the first page of a printout. A banner identifies the printout's creator.

**bantam connector** A miniature tip-and-ring jack connector used in digital telephony; equivalent to WECO 310, but physically smaller.

**bar antenna** An antenna in which the sides give maximum reception of a signal while the ends give minimum reception.

**barrier code** A number of digits that, when dialed prior to an authorization code allow entry to a PBX.

**BARRNet** Bay Area Regional Research Network (San Francisco).

**base address** The first address in a series of addresses in memory. In a computer attached to a local area network, the base address is often used to describe the beginning of its network interface card's I/O space.

**base group** Twelve communications VF paths. A unit of frequency division multiplexing systems bandwidth allocation.

**base style** The style an IBM PROFS or Office Vision user selects to copy or modify when creating a new document style.

**baseband** 1. Direct transmission method used for short distances (less than 10 miles); uses a bandwidth whose lowest frequency is zero (DC level)—that is, transmission of raw (carrier-less) binary data. The transmission medium carries only one signal at a time. 2. Information signal that starts near DC and goes to a finite value. Video baseband is 30 Hz to 4.2 MHz, audio baseband is 30 Hz to 15 KHz.

**baseband (digital)** A digital stream that carries just one information channel.

**baseband or basic signal** The original signal from which a transmission waveform is produced by modulation. In telephony it is the speech waveform. In data transmission many forms are used, and the basic signal is usually made of successive signal elements.

**baseband local area network** A local area network in which information is encoded, multiplexed, and transmitted without modulation of a carrier. The IBM Token-Ring Network is an example.

**baseband modem** A modem which does not apply a complex modulation scheme to the data before transmission, but which applies the digital input (or a simple transformation of it) to the transmission channel. This technique can only be used when a vary wide bandwidth is available and only operates over short distances where signal amplification is not necessary. Sometimes called a limited distance or short-haul modem.

**baseband signaling** Transmission of a signal at its original frequencies, without modulation.

**baseband system** A system that transmits at only one frequency and speed.

**Baseline** A registered mark of AT&T for voice transport service, group transport service, supergroup transport service and T1.5 transport service.

**BASIC** Beginners All-purpose Symbolic Instruction Code.

**Basic Call Management System (BCMS)** A feature of the AT&T Definity Communications System which lets automatic call distribution call data be collected and processed within the PBX.

**Basic Exchange Telecommunications Radio Service (BETRS)** A service offered by telephone companies to remote users in which digital signals are transmitted from the telephone company office to a small antenna next to the customer's home. A drop wire runs from the antenna to the residence and a converter attached to the antenna converts the transmission from analog to digital, permitting the use of existing wiring and phones in the home.

**Basic Information Unit (BIU)** In SNA, the unit of data and control information that is passed between half-sessions. It consists of a request/response header (RH) followed by a request/response unit (RU).

**Basic Input/Output System (BIOS)** A low-level interface set of instructions between application software and hardware. BIOS is normally contained in Read Only Memory (ROM) and consists of a set of functions that are performed when software calls (invokes) a specific BIOS routine.

**basic mode** In ACF/VTAM, Release 1 and in VTAM, a mode of data transfer in which the application program can communicate with non-SNA terminals without using SNA protocols.

**Basic Rate Interface (BRI)** The end-user ISDN service containing two Bearer channels (B-channels) at 64 Kbps each, and 1 Delta channel (D-channel) at 16 Kbps; also called 2B+D Service; transmitted at 144 Kbps.

**Basic Service Elements (BSE)** Telephone company special features.

**Basic Telecommunications Access Method (BTAM)** IBM teleprocessing access method. Used to control the transfer of data between main storage and local or remote terminals. BTAM provides the applications program with macro instructions for using the capabilities of the devices supported. BTAM supports binary synchronous (BSC) as well as start/stop communication.

**Basic Transmission Unit (BTU)** In SNA, the unit of data and control information passed between path control components. A BTU can consist of one or more path information units (PIUs).

**basic services** Refers to transport level services provided by Bell operating companies (BOCs).

**batch** A term applying to any specific group of electronic accounting machine (EAM) cards.

**batch accounting** A facility implemented in the Sprint Telenet Processor (TP) accounting system that reduces overhead, enabling accounting packets to be transferred in groups (batches) rather than individually.

**batch processing** A data processing technique in which input data is accumulated and prepared off-line and processed in batches.

**batched communications** Transmission of large blocks of information without interactive responses from the receiver.

**BATELCO** Bahrain Telecommunications Company.

**baud** Unit of signaling speed. The speed in baud is the number of discrete conditions or signal events per second. If each signal event represents only one bit, the baud rate is the same as bps; if each signal event represents more than one bit (such as in a dibit), the baud rate is smaller than bps.

**Baudot Code** A code (named after Emile Baudot, a pioneer in printing telegraphy) for asynchronous transmission of data in which 5 bits represents a single character. Use of Letters Shift and Figures Shift enables 64 alphanumeric characters to be represented. Used mainly in teleprinter systems which add one start bit and 1.5 stop bits. Although 32 characters could be represented normally with the Baudot code, the necessity of transmitting digits, letters of the alphabet, and punctuation marks made it necessary to devise a mechanism to extend the capacity of the code to include additional character representations. The extension mechanism was accomplished by the use of two "shift" characters: "letters shift" and "figures shift."

The transmission of a shift character informs the receiver that the characters which follow the shift character should be interpreted as characters from a symbol and numeric set or from the alphabetic set of characters. The 5-level Baudot code is illustrated in the following table for one particular terminal pallet arrangement. A transmission of all ones in bit positions 1 through 5 indicates a letter shift, and the characters following the transmission of that character are interpreted as letters. Similarly, the transmission of ones in bit positions 1, 2, 4, and 5 would indicate a figures shift, and the following characters would be interpreted as numerals or symbols based upon their code structure.

**BBS** Bulletin Board System. BBS software enables a computer to be used for message posting and retrieval, file transfer and similar activity.

**bearer channel (B-channel)** In ISDN, a 64 Kbps digital channel that carries end-user voice, data, or image information.

**BCC** Block Check Character.

43

| Letters | Figures | Bit selection 1 | 2 | 3 | 4 | 5 |
|---|---|---|---|---|---|---|
| *Characters* | | | | | | |
| A | — | 1 | 1 | | | |
| B | ? | 1 | | | 1 | 1 |
| C | : | | | 1 | 1 | 1 |
| D | $ | 1 | | | 1 | |
| E | 3 | 1 | | | | |
| F | ! | 1 | | 1 | 1 | |
| G | & | | 1 | | 1 | 1 |
| H | | | | 1 | | 1 |
| I | 8 | | 1 | 1 | | |
| J | ' | 1 | 1 | | 1 | |
| K | ( | 1 | 1 | 1 | 1 | |
| L | ) | | 1 | | | 1 |
| M | . | | | 1 | 1 | 1 |
| N | , | | | 1 | 1 | |
| O | 9 | | | | 1 | 1 |
| P | Ø | | 1 | 1 | | 1 |
| Q | 1 | 1 | 1 | 1 | | 1 |
| R | 4 | | 1 | | 1 | |
| S | ' | 1 | | 1 | | |
| T | 5 | | | | 1 | |
| U | 7 | 1 | 1 | 1 | | |
| V | ; | | 1 | 1 | 1 | 1 |
| W | 2 | 1 | 1 | | | 1 |
| X | / | 1 | | 1 | 1 | 1 |
| Y | 6 | 1 | | 1 | | 1 |
| Z | " | 1 | | | | 1 |
| *Functions* | | | | | | |
| Carriage return | | | | | 1 | |
| Line feed | | | 1 | | | |
| Space | | | | 1 | | |
| Letters shift | < | 1 | 1 | 1 | 1 | 1 |
| Figures shift | = | 1 | 1 | | 1 | 1 |

*Baudot Code*

**BCC**   Blocked Calls Cleared.

**BCD**   Binary Coded Decimal.

**BCH**   An error detecting and correcting technique used at the communications receiver to correct errors as opposed to retry attempts as in ARQ. Named for Messrs Bose, Chaudhuri, and Hocquencgham who developed it.

**BCH**   Blocked Calls Held.

**B-channel (bearer channel)**   In ISDN, a 64 Kbps information carrying channel.

**BCM**   Bit Compression Multiplexer.

**BCMS**   Basic Call Management System.

**BCnet**   British Columbia network.

**BCS**   Block Check Sequence.

**BDCS**   Below Desk Communications System.

**BDLC**   Burroughs Data Link Control.

**beacon**   In a token-ring network, a frame sent by an adapter indicating a serious network problem, such as a broken cable.

**beam**   Microwave radio systems which use ultra/super high frequencies (UHF, SHF) to carry communications, where the signal is a narrow beam rather than a broadcast signal.

**beam splitter**   A device used to divide an optical beam into two or more separate beams; often a partially reflecting mirror.

**beamwidth**   The angular coverage of an antenna beam. Earth station beams are usually specified at the halfpower (or $-3$ dB) point. Satellite beams are based on the area to be covered.

**bearer channel (B-channel)**   A 64 Kbps information channel, time-division-multiplexed with other bearer channels and a delta channel to form an Integrated Services Digital Network interface.

**Bearer Identification Code (BIC)**   A portion of the data user part label in the CCITT Signaling System 7.

**beat frequency**   A frequency resulting from combining two different frequencies. It is usually equal to the sum of or the difference between the two original frequencies.

**Beat Frequency Oscillator (BFO)**   An oscillator used to generate a local signal, which is combined with an incoming unmodulated continuous wave

(CW) signal to produce a beat frequency that is audible. It is used for continuous wave reception.

**BECN**   Backward Explicit Congestion Notification.

**beeper**   A slang term for a radio pager.

**begin bracket**   In SNA, the value (binary 1) of the begin-bracket indicator in the request header (RH) of the first request in the first chain of a bracket; the value denotes the start of a bracket.

**beginning tab**   Any of six index tags that mark the beginning of index information in Revisable-Form Text (RFT) document in IBM's PROFS and Office Vision.

**bel**   Equal to 10 decibels (dB).

**BEL**   A control character that is used when there is a need to call for attention; it may control alarm or attention devices.

**Bell Atlantic**   One of the seven regional Bell operating companies formed as a result of divestiture, whose service area covers the middle-Atlantic states.

**Bell Operating Company (BOC)**   One of the 22 local telephone companies divested from AT&T, and grouped under seven regional Bell holding companies.

**Bellcore (Bell Communications Research)**   Bellcore is the standards organization of the Bell Operating Companies. Because of its market position within the industry, most manufacturers attempt to emulate Bell standards to produce Bell-compatible equipment.

**BellSouth Corporation**   One of the seven Regional Holding Companies (ROC) formed by the divestiture of AT&T. BellSouth consists of Southern Bell, BellSouth, and South Central Bell.

**BellSouth Enterprises, Inc.**   The holding company for all unregulated BellSouth companies.

**BellSouth International**   A subsidiary of BellSouth Enterprises, the holding company for all unregulated BellSouth companies which manages BellSouth activities outside the United States.

**Bellsouth Mobility Inc.**   A subsidiary of BellSouth Enterprises, the holding company for all unregulated BellSouth companies which provides cellular mobile telephone and paging service in the southeastern United States.

**Bell 103**   An AT&T, 0–300 bps modem providing asynchronous transmission with originate/answer capability; also often used to describe any Bell 103-compatible modem.

**Bell 113**   An AT&T, 0–300 bps modem providing asynchronous transmission with originate or answer capability (but not both); also often used to describe any Bell 113-compatible modem.

**Bell 201**   An AT&T, 2400 bps modem providing synchronous transmission; Bell 201 B was designed for leased line applications (the original Bell 201 B was designed for public telephone network applications); Bell 201 C was designed for public telephone network applications; also often used to describe any Bell-201 compatible modem.

**Bell 202**   An AT&T, 1800 bps modem providing asynchronous transmission that requires 4-wire circuit or full-duplex operation; also an AT&T 1200 bps modem providing asynchronous transmission over 2-wire, full-duplex, leased line, or public telephone network applications; often used to describe any Bell 202-compatible modem.

**BELL 208**   An AT&T, 4800 bps modem providing synchronous transmission; Bell 208 A was designed for leased line applications; Bell 208 B was designed for public telephone network applications; also often used to describe any Bell 208-compatible modem.

**BELL 209**   An AT&T, 9600 bps modem providing synchronous transmission over 4-wire leased lines; also often used to describe any Bell 209-compatible modem.

**BELL 212, BELL 212A**   An AT&T, 1200 bps full-duplex modem providing asynchronous transmission or synchronous transmission for use of public telephone network; also often used to describe any Bell 212-compatible modem.

**BELL 310** A large tip-and-ring jack connector used in digital telephony. Also called WECO 310.

**BELL 2296A** A V.32 modem capable of operating at 9.6 Kbps over the switched telephone network.

**BELL 43401** Bell Publication which defines requirements for transmission over telco-supplied circuits that have DC continuity (that are metallic).

**Bellman–Ford** A dynamic routing algorithm in which each router in a network broadcasts its routing table to all neighbouring routers. Packets are then routed based upon the lowest number of hops to a detination.

**Bemer–Ross code** One of the early versions of the ASCII code which was primarily used in Europe.

**bend loss** In fiber optic transmission attenuation caused by having the fiber curved around a restrictive radius of curvature or by microbends caused by minute distortions in the fiber imposed by external perturbations.

**bend radius** The radius of curvature that a fiber optic cable can sustain without any adverse effects.

**BER** Bit Error Rate.

**BERNET I,II** Berlin Network.

**BERT** Bit Error Rate Test (set).

**BEST-1** A network design tool of BGS Systems of Waltham, MA.

**beta test site** The testing of a product prototype and/or early release at client locations prior to general public marketing.

**BETRS** Basic Exchange Telecommunications Radio Service.

**BEX** Broadband exchange.

**Bezek** A telecommunications company in Israel.

**BFO** Beat Frequency Oscillator.

**BGP** Border Gateway Protocol.

**bias** A type of telegraph that occurs when the significant intervals of the modulation do not all have their exact theoretical durations.

**bias distortion** Communications signal distortion with respect to unequal mark and space durations in a digital transmission. The formula to calculate bias distortion is:

Bias distortion = [(mark pulse width − space pulse width)/total pulse width] * 100%

If mark>space, the result is positive
If space>mark, the result is negative

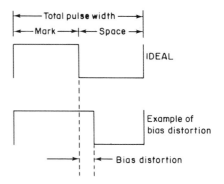

**BIB** Backward Indicator Bit.

**BIC** Bearer Identification Code.

**BIDDER** In SNA, the LU–LU half-session defined at session activation as having to request and receive permission from the other LU–LU half-session to begin a bracket.

**bidirectional printing** Printing in which a printer's print head goes from left to right on every other line. On the other lines it goes from right to left. Since the print head moves and prints in both directions the speed of printing is increased.

**BIDS** Broadband Integrated Distributed Star.

**Bildschirmtext (BTX)** A videotext service introduced by Deutsche Bundespost in l986.

**Billdats** A registered mark of AT&T as well as a data collection apparatus from that vendor.

**Billed Telephone Number (BTN)**  The primary telephone number used for billing, regardless of the number of telephone lines associated with that number.

**billing increment**  The segment of telephone line connection time measured for billing purposes. For example, some services are measured and billed in 1-minute increments; some are measured and billed in 6- or 10-second increments.

**binary**  Digital system with two states, 1 and 0.

**binary code**  Computer language made up of ones and zeros arranged to represent words of computer instruction.

**Binary Coded Decimal (BCD)**  A digital system that uses binary codes to represent decimal digits, where the numbers zero through nine have a unique 4-bit representation.

**binary signal characteristics**  In a digital communication signal, either one state or the other appears on the line but never both at the same time. Further, the signals may be either normal or inverted depending on the circuit specifications. A comparison of circuit conditions necessary for binary signals in normal or inverted form is given in the chart below.

|  |  | NORMAL | INVERTED |  |
|---|---|---|---|---|
| Binary digits | 1 | 0 | 1 | 0 |
| Applied voltage | on | off | off | on |
| Code element | mark | space | space | mark |
| Circuit contacts | closed | open | open | closed |
| Current identification | current | no current | no current | current |

**Binary Synchronous Communications (BSC or BiSync)**  A set of rules for the synchronous transmission of binary-coded data. All data in BSC is transmitted as a serial stream of binary digits. Synchronous communication means that the active receiving station on a communications channel operates in step with the transmitting station through the recognition of SYN bits.

One of the more common modes of employment of BSC is in a multipoint environment where the polling or selection of tributary stations is done. Polling is an "invitation to send" transmitted from the control station to a specific tributary station. Selection is a "request to receive" notification from the control station to one of the tributary stations instructing it to receive the following message(s).

From the illustration below:

SYN SYN is used to establish and maintain synchronization and as a time fill in the absence of any data or other control character.

SOH: Start of Heading precedes a block of heading characters. A heading consists of auxiliary information, such as routing and priority, necessary for the system to process the text portion of the message.

STX: Start of Text precedes a block of text characters. Text is that portion of a message treated as an entity to be transmitted through to the ultimate destination without change. STX also terminates a heading.

ETX: End of Text character terminates a block of characters started with STX or SOH and transmitted as an entity. The BCC (Block Check Character) is sent immediately following ETX. ETX requires a reply indicating the receiving station's status.

ETB: End of Transmission Block indicates the end of a block of characters started with SOH or STX. The BCC is sent immediately after ETB. ETB requires a reply indicating the receiving station's status.

| SYN | SYN | SOH | Header | STX | TEXT | ETX or ETB | BCC |
|---|---|---|---|---|---|---|---|

IBM Bisync (BSC) message format

**binary sychronous transmission**  Data transmission in which synchronization of characters is controlled by timing signals generated at the sending and receiving stations.

**Binary 8-Zero Suppression (B8ZS)**  A technique, used widely in North America, for maintaining one's density on digital circuits. In B8ZS, any string of eight consecutive "zero" bits is replaced by a special bit pattern that contains two bipolar violations in defined positions. Receiving hardware recognizes this pattern by the bipolar violations and substitutes a string of eight "zero" bits to restore the original bit stream. If the pulse preceding an all-zero byte is positive, the inserted code is 000+−0−+. If the pulse preceding an all-zero byte is negative, the inserted code is 000−+0+−. Both inserted codes result in bipolar violations occurring in the fourth and seventh bit positions. Both carrier and customer equipment must recognize these codes as legitimate signals and not as bipolar violations or errors for B8ZS to work to enable a receiver to recognize the code and restore the original eight zeros.

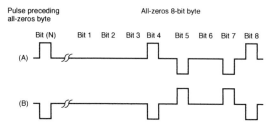

**bind**  In SNA, a request to activate a session between two logical units (LUs).

**binder group**  A group of 25 twisted pairs that are bound together. Two or more group (4-wires or more) are used to form a cable which delivers telephone service.

**Binding posts**  A small screw-type terminal used to make electrical connections to wires. Usually it is part of a terminal or connecting block.

**BIOS**  Basic Input/Output System.

**bipolar coding**  A method of transmitting a binary stream in which binary 0 is sent as no pulse and binary 1 is sent as a pulse which alternates in sign for each 1 that is sent. The signal is therefore ternary.

**bipolar modulation**  A coding technique for binary signals consisting of three states (zero, positive and negative signal levels). Binary 0 in the input is represented by a 0 in the output, and binary 1 is represented by either of the other two levels. Each successive 1 that occurs is coded with the opposite polarity.

**bipolar non-return to zero signaling**  In bipolar non-return to zero signaling, alternating polarity pulses are used to represent marks while a zero pulse is used to represent a space.

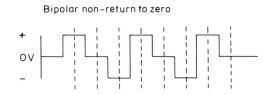

**bipolar return to zero signaling**  In bipolar return to zero signaling, alternate polarity pulses are used to represent marks while a zero pulse is used to represent a space. Here the bipolar signal returns to zero after each mark. As this signal ensures that there is no DC voltage buildup on the line, repeaters can be placed relatively far apart in comparison to other signaling techniques and this signaling technique is employed in modified form on digital networks. Since two or more spaces cause no change from a normal to zero voltage, sampling is required to determine the value of bits.

**bipolar transistor**  A transistor whose operation depends on the flow of both negatively charged electrons and positively charged holes.

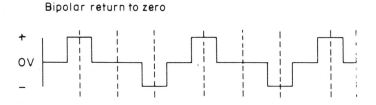

*bipolar return to zero signaling*

**bipolar transmission** Method of sending binary data in which negative and positive states alternate; used in digital transmission facilities such as DDS and T1. Bipolar is a three-level signal format consisting of High, Zero, and Low states where the High and Low states are of opposite voltage potential (i.e. + and − voltages). Sometimes known as polar transmission.

**Bipolar Violation (BPV)** In bipolar violation, the occurrence of consecutive "one" bits with pulses of the same polarity. A bipolar violation is usually regarded as a transmission error, but in some formats, such as B8ZS, specific bipolar violations are used for low-level signaling.

Bipolar encoding: Alternate mark inversion. Shaded bit pulses represent a Bipolar Violation.

**bird** The nickname for any satellite.

**bis** Appended to a CCITT network interface standard, it identifies a second version of the standard.

**BISDN** Broadland Integrated Services Digital Network.

**BiSync** Binary Synchronous Protocol (also BSC).

**BISYNCHRONOUS CONTROL CHARACTERS**

| BISYNCHRONOUS CHARACTER | HEX VALUE | CHARACTER DESCRIPTION |
|---|---|---|
| SYN | 32 | Synchronous idle |
| PAD | 55 | Start of frame pad |
| PAD | FF | End of frame pad |
| DLE | 10 | Data line escape |
| ENQ | 2D | Enquiry |
| SOH | 01 | Start of heading |
| STX | 02 | Start of text |
| ITB | 1F | End of intermediate block |
| ETB | 26 | End of transmission (block) |
| ETX | 03 | End of text |

**bit** The smallest unit of the information that the computer recognizes. A bit (short for binary digit) is represented by the presence or absence of an electronic pulse, 0 or 1. The term was originated at AT&T Bell Laboratories by John Tukey.

**bit bucket** A slang term used to describe the discarding of binary information by relating it to the practice of throwing paper or other information into a wastebucket.

**bit duration** 1. The time it takes one encoded bit to pass a point on a transmission medium. 2. In serial communications, a relative unit of time measurement used for comparing delay times due to propagation.

**Bit Error Rate (BER)** The ratio of bits received in error to the total bits received. If one error occurs in every 100 000 bits, the bit error rate is said to be $10^{-5}$.

**Bit Error Rate Test (BERT)** A data communications measurement in which the total number of received errors are divided by the total number of received data bits. The smaller the resultant number, the higher is the quality of the communications path.

BERT DEFINITIONS

**bits received** The total number of bits received while the test set is in sync with the received data.

**Bit errors** Any received data bit in error while the test set is in sync with the incoming data.

**Blocks received** The total number of blocks received while the device is in sync with the received data.

**Block errors** A block error is any block containing one or more bit errors while in sync. The block size is user definable.

**Error free sec** Number of seconds during which the receiver is in sync with the data and no bit errors are encountered.

**Asynchronous errored sec** The test interval is divided into uniform 1-second intervals which are independent of the incoming data. If an error occurs during an interval while in sync, the second is counted as an errored second.

**Synchronous errored sec** If an error occurs during an interval in which the test set is in sync with the incoming data, a 1-second interval clock is started at the beginning of the detected transmission error. With this method, the errored second is considered to be synchronized with the incoming data errors.

**Time outs** Applies only to half-duplex circuits. When running a half-duplex test, a user-defined period of time is entered to prevent the tester from locking up if a reply is not completely received (i.e. the line must be turned around or SYNC may never be re-established).

**BCC errors** The total number of Block Check Character errors resulting from the receiver checking for transmission errors.

**Frame errors** Applies to asynchronous protocols. The total number of start and or stop bits received in error.

**Parity errors** The total number of parity errors resulting during the transmission and reception of data.

**In sync with data** In BERT applications this will occur when the incoming bit pattern matches the test pattern.

**No data condition** Applies only to the beginning of the BERT and indicates that the receiver is unable to get into SYNC with the incoming data.

**Sync loss sec** The number of seconds that elapse during a Sync loss. Any loss of sync will be counted as at least one Sync loss second.

**Elapsed sec** Refers to the amount of time that the receiver is enabled. It is comprised of the addition of Sync loss secs + Errored secs + Error free secs. For FDX testing this is equivalent to the actual duration of the test after sync has been established. For HDX the receiver is enabled at most 50% of the time, so the Total elapsed time is less than 50% of the actual duration of the test.

**%Error sec** Refers to the percentage of errored seconds. It is calculated by the following division: (Errored sec/Total test Duration) * 100.

**BER** The Bit Error Rate is the ratio of the total number of bit errors divided by the total number of bits received. The user can decide whether or not to include the errors that occur during a Sync loss.

**BLER** The Block Error Rate is the ratio of the total number of block errors divided by the total number of blocks received. The block length can be defined by the user.

**Sync loss** Synchronization loss occurs when the test set, which was previously in sync with the received data, detects eight consecutive characters in error.

**bit interleaving** A multiplexing technique in which data from the low-speed channels is integrated into the high-speed bit stream one bit at a time (i.e. one bit from channel 1, one bit from channel 2 . . . one bit from channel n). Usually used for digital transmission at rates above T1.

**bit masking** Senses specific binary conditions and ignores others. Don't Care characters are placed in bit positions of *not* interest, and zeros and ones in positions to be sensed.

**bit stream** A binary signal without regard to groupings by character. The stream is also used to identify synchronous transmission.

**bit stripping** Removing the start/stop bits on each ASYNC character and transmitting the data using synchronous techniques. Common with statistical multiplexers.

**bit stuffing** A process in bit-oriented protocols, where a zero bit is inserted (stuffed) into a string of six contiguous one bits. This is to prevent user data being confused with frame flag sequences. Also called zero insertion.

**bit time** The time it takes to send one bit of information. A bit time is inversely proportional to bit rate (bps).

**BITCOM** A menu driven PC communications program from BIT Software, Inc. of Milpitas, CA.

**BITNET** Because It's Time Network. A network which interconnects approximately 3000 computers located at educational institutions throughout the world.

**BITNIC** BITNET Network Information Center.

**BIT-ORIENTED** Used to describe communications protocols (such as SDLC) in which control information may be coded in fields as small as a single bit in length.

**Bit-Oriented Line Discipline (BOLD)** NCR's first

bit-oriented protocol which is now obsolete. BOLD does not follow the international HDLC recommendations.

**Bit-Oriented Protocol (BOP)** The most widely known BOP is IBM's Synchronous Data Link Control (SDLC). In SDLC, information is sent in frames. Within each frame are fields which have specific functions.

The flag field is a unique combination of bits (01111110) which lets the receiving device know that a field in SDLC format is about to follow.

The address field specifies which secondary station is to get the information.

The control field is used by the primary station (which maintains control of the data link at all times) to tell the addressed (or secondary) station what it is to do: to poll devices; transfer data; or retransmit faulty data among other things. The addressed station can use the control field to respond to the primary station: to tell it which frames it has received or which it has sent. The control field is also used by both the primary and secondary stations to keep track of the sequence number of the frames transmitted and received at any one time. SDLC frames which contain information fields and supervisory fields have sequence numbers.

The information field may be any length and may comprise any code structure.

The frame check sequence (FCS) is 16 bits long and contains a CRC. All data transmitted between the start and stop flags is checked by the FCS.

The flag field also marks the end of a standard SDLC frame.

| Flag 8 bits | Address 8 bits | Control 8 bits | Information field | Variable length | Frame check sequence 16 bits | Flag 8 bits |

BOP (SDLC) frame format

**Bit-Oriented Signaling (BOS)** A common signaling protocol used in AT&T's Digital Multiplexed Interface (DMI). In Bit-Oriented Signaling, values of bits in the signaling channel represent physical signals on the E- and M-wires of an analog telephone. BOS is used only as an interim provision on systems migrating between D4 framing (which uses AB signaling) and DMI (which uses Message-Oriented Signaling according to CCITT Common Channel Signaling System No. 7).

**bit-rate** The speed (or rate) at which the individual bits are transmitted across a digital transmission system or circuit, usually expressed in bits per second.

**bit-robbing** A technique for carrying "on-hook/off-hook" telephony signaling information over T1 spans. With bit-robbing, the circuit provider (telephone company or PTT) "robs" the least significant bit of each channel octet in the 6th and 12th frames of each superframe (6th, 12th, 18th, and 24th frames in Extended Superframe Format) to convey signaling information. Bits robbed from the 6th (and 18th) frame are called "A" bits, and correspond to signals on the E-wire of the analog telephone. Bits robbed from the 12th (and 24th) frame are called "B" bits, and correspond to signals on the M-wire of an analog telephone. Bit-robbing reduces the channel bandwidth available for digital data from 64 Kbps to 56 Kbps.

**bits per second (bps)** A measure of the rate of information transfer. Often combined with metric prefixes as in Kbps for thousands of bits per second (K for kilo-) and Mbps for millions of bits per second (M for mega-).

**bits/s (bps)** Bits per second.

**biTS-1** A hand-held ISDN test set marketed by Harris Corporation. The biTS-1 can be used to test ISDN S/T interfaces and 2B+D lines.

**BIU** Basic Information Unit.

**BIU segment** In SNA, the portion of a basic information unit (BIU) that is contained within a path information unit (PIU). It consists of either a request/response header (RH) followed by all or a portion of a request/response unit (RU), or only a portion of an RU.

**black box** 1. A device that performs certain functions on the input(s) before outputting the result(s). 2. An illegal device used to allow long

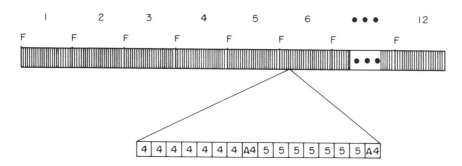

Bit-robbing:
A 12-frame superframe—Two octets from Frame 6 have been enlarged to illustrate robbed-bit signaling—the least significant bit of each octet in Frames 6 and 12 is "robbed" to provide signaling: robbed bits in Frame 6 are "A" bits, those in Frame 12 are "B" bits.

*bit-robbing*

distance calls to be received without being billed. The box allows the called party to flip a switch which prevents enough current to return to the originating city when the phone rings to trip a billing computer. The computer would then assume the phone was ringing and the call was not answered.

**Black Box Corporation** A leading data communications and computer device mail order company; publishes and distributes the *Black Box Catalog*.

**blank** A condition of "no information" in a data recording medium or storage locations. This can be represented by all spaces or all zeros.

**BLAST** 1. BLocked ASynchronous Transmission. 2. A communications software program designed for use on personal computers and mainframes. Marketed by Communications Research Group.

**BLDG/PWR** Building/Power.

**BLERT** BLock-Error-Rate-Test.

**BLF** Busy Lamp Field.

**blink** An extended highlighting attribute value used to emphasize a field or character.

**BLK** Blank.

**BLOCK** A group of characters transmitted as a unit,

over which a coding procedure is usually applied for synchronization and/or error control purposes.

**BLock ASynchronous Transmission (BLAST)** A means of sending asynchronous information synchronously.

**block chaining** Linking blocks of message data in main or secondary storage through pointer arrangements to provide queueing and flexible memory use.

**block check** That part of the error control procedure used for determining that a data block is structured according to a given set of rules.

**Block Check Character (BCC)** A character added to the end of a transmission block for the purpose of error detection—such as a CRC or LRC.

Direction of data flow ⟶

IBM Bisync message format

**Block Check Sequence (BCS)** At least two characters added to the end of a data block for error checking. The BCS is often a cyclic redundancy check (CRC) but can be a checksum result.

**block downconverting** In satellite communications

the multi-conversion process of converting the entire band to an intermediate frequency (4 GHz to 1 GHz) for transmission to multiple receivers, where the next conversion takes place.

**block encryption**  Encryption of a block of data or text as a unit. Successive blocks are transformed independently but dependency may then be introduced by block chaining.

**BLock Error Rate Test (BLERT)**  A data communications testing measurement in which the total number of received block errors are divided by the total number of received blocks. The smaller the resultant number, the higher is the quality of the communications path.

**block mode**  A method of data transmission where data is accumulated in a buffer within the terminal. This accumulation block of data is transmitted from the buffer to the network or the terminal screen.

**block multiplexer channel**  In IBM systems, a multiplexer channel that interleaves bytes of data; also called byte-interleaved channel.

**Block Oriented Network Simulator (BONeS)**  A trademark of Comdisco Systems of Foster City, CA, as well as a software tool for modeling, simulating and analyzing communications networks.

**block overhead**  The bits remaining in a cell after discarding the block payload.

**block payload**  The user information bits within a cell.

**blockage**  A condition which exists when all telephone circuits are busy so that no other telephone call can be carried. Blockage rates—that is, the frequency with which blockage occurs—is one of several standards used to measure performance of local telephone companies and long distance companies.

**Blocked Calls Cleared (BCC)**  A service discipline in which unserviceable requests are rejected by the system without service. Also called lost calls cleared (LCC).

**Blocked Calls Held (BCH)**  A service discipline in which unserviceable requests remain in the system without being serviced but having a portion of their desired service time elapse until service begins. Also called lost calls held (LCH).

**blocking**  1. The process of grouping data into transmission blocks. 2. In LAN technology, the inability of a PABX to service connection requests, usually because its switching matrix can only handle a limited number of connections simultaneously.

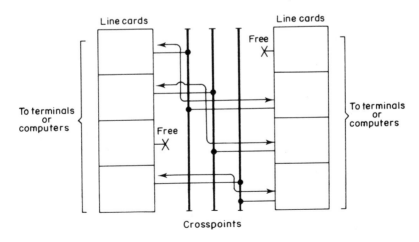

*blocking*

Blocking occurs if a call request from a user cannot be handled due to an insufficient number of paths through the switching matrix; blocking thus prevents free stations from communicating.

**blocking of PIUs**  In SNA, an optional function of path control that combines multiple path information units (PIUs) into a single basic transmission unit (BTU). *Note.* When blocking is not done, a BTU consists of one PIU.

**block-multiplexer channel**  In IBM systems, a multiplexer that interleaves bytes of data; also called byte-interleaved channel.

**blue alarm**  An alarm generated by the receiving device on a DS1 circuit when a Loss of Signal (LOS) condition lasts for more than 150 milliseconds. The receiving device signals a blue alarm by transmitting an unbroken stream of logical "one" bits until the original transmitting device clears the alarm (by restoring normal transmission).

**blue book**  The CCITT edition of colored books in which all of its recommendations on standards for the 1985 through the 1988 period are gathered. The "blue book" contains approximately 18 000 pages.

**blue box**  A box that contains a transmitter that has 13 to 15 buttons which are used to simulate tones. When acoustically coupled to a telephone or wired to a line the blue box is illegally used to obtain free long distance telephone calls.

**B-MAC**  In satellite transmission B-MAC is a method of transmitting and scrambling television signals where the multiplexed analog component (MAC) signals are time-multiplexed with a digital burst which contains digitized sound, video synchronization, authorization and information.

**B-M-P**  The name of a fiber optic cable route connecting Brunei in Malaysia to the Philippines.

**BMS**  Bandwidth Management Service.

**BNA**  Burroughs Network Architecture.

**BNC**  A bayonet-locking connector for miniature coax; BNC is said to be short for bayonet–Neill–Concelman.

**BNC connector**  A small coaxial connector with a half-twist locking shell.

**BNN**  In IBM's SNA, Boundary Network Node.

**BNS**  Billed Number Screening.

**BOB**  Breakout Box.

**BOC**  Bell Operating Company.

**body stabilized**  A satellite altitude control system using the gyroscopic stiffness of one (or more) spinning wheel.

**bogon**  Something that is stupid or nonfunctional.

**BOLD**  Bit-Oriented Line Discipline.

**BOND**  A low-resistance electrical connection between two cable sheaths, two pieces of apparatus, or two ground connections, or between similar parts of two electrical circuits.

**BONDING**  Bandwidth-ON-Demand INteroperability Group.

**BONeS**  Block Oriented Network Simulator.

**book message**  In U.S. military communications, a message that is destined for two or more addressees and is of such nature that the originator considers that no addressee need be informed of any other addressee. Each addressee must be indicated as action or information.

**BOOT**  The process of loading a computer's operating system into the random memory of the computer.

**Boot ROM**  A read-only memory chip allowing a workstation on a local area network to communicate with a file server and read a boot program from the server. A Boot ROM enables workstations to operate on the network without having a disk drive.

**Bootstrap**  A technique or device designed to bring itself into a desired state by means of its own action; for example, a program whose first few instructions are sufficient to load an entire program into the computer from an input device.

**bootstrap routine**   A routine contained in a single record which is read into memory by a bootstrap loader located in ROM. The bootstrap routine then reads the operating system into memory.

**BOP**   Bit-Oriented Protocol.

**BOPI**   Bit-Oriented Protocol Inverted (NRZI).

**Border Gateway Protocol (BGP)**   A routing protocol that uses TCP exchange and passes along full routing path information, allowing a router to build a complete map of the network. The BGP is standardized as RFC 1105.

**BORS(C)HT**   A mnemonic describing the functions of a subscriber's exchange line interface. Battery, Over-voltage, Ringing, Supervision, (Coding), Hybrid, Test. The C is omitted in analog environments.

**BOS**   Bit-Oriented Signaling.

**BOT**   Beginning of Transmission.

**bounced mail**   Mail that is returned to the originator due to an incorrect E-mail address.

**boundary function**   In IBM's SNA: 1. A capability of a subarea node to provide protocol support for adjacent peripheral nodes, such as: (a) transforming network addresses to local addresses, and vice versa; (b) performing session sequence numbering for low-function peripheral nodes; and (c) providing session-level pacing support. 2. The component that provides these capabilities.

**Boundary Network Node (BNN)**   In IBM's TCAM, the programming component that performs FID2 (format identification type 2) conversion, channel data link control, pacing and channel/device error recovery procedures for a locally attached station. These functions are similar to those performed by a network control program for an NCP-attached station.

**boundary node**   In IBM's SNA, subarea node that can provide certain protocol support for adjacent subarea nodes.

**BPDU**   Bridge Protocol Data Unit.

**bps**   Bits per second.

**BPSK**   Binary Phase Shift Keying.

**BPV**   BiPolar Violation.

**BPV errors**   Bipolar Pulse Violations which are not part of a valid B8ZS or other code substitution technique.

**BPV-free second**   A second of time in which no Bipolar Pulse Violations occur.

**BPV-free seconds rate**   A measure of the quality of a bipolar signal over time, obtained by dividing the number of BPV-free seconds by the duration of the test in seconds.

**BPV rate**   A measurement of accuracy in bipolar transmission, derived by dividing the number of bipolar violations by the duration of the test in seconds.

**bracket**   In IBM's SNA, one or more chains of request units (RUs) and their responses that are exchanged between the two LU–LU half-sessions and that represent transaction between them. A bracket must be completed before another bracket can be started. Examples of brackets are data base inquiries/replies, update transactions, and remote job entry output sequences to work stations.

**bracket protocol**   In IBM's SNA, a data flow control protocol in which exchanges between the two LU–LU half-sessions are achieved through the use of brackets, with one LU designated at session activation as the first speaker and the other as the bidder. The bracket protocol involves bracket initiation and termination rules.

**branch**   A transfer of control from one line of code to one other than the next logical instruction. Branches may be conditional, based on some defined circumstance, or unconditional.

**BRCS**   Business and Residence Customer Service.

**breadth of bulletin board access code**   Defines the organizational bounds within which a Telemail user may access bulletin boards.

**breadth of inquiry code**  Defines the organizational bounds within which information can be made available about a Telemail user.

**breadth of posting code**  Defines the organizational bounds within which a Telemail user may send messages.

**breadth of receipt code**  Defines the organizational bounds from which a Telemail user may receive messages.

**break**  A space (or spacing) condition that exists longer than one character time (typical length is 110 milliseconds). Often used by a receiving terminal to interrupt (break) the sending device's transmission, to request disconnection, or to terminate computer output.

**Breakout Box (BOB) (EIA monitor)**  Digital test equipment that monitors the status of signals on the pins of an RS-232C connector and allows signals to be broken, patched, or cross-connected.

**BRI**  Basic Rate Interface.

**bridge**  1. The interconnection between two networks using the same communications method, the same kind of transmission medium, and the same addressing structure; also the equipment used in such an interconnection. Bridges function at the data link layer of the OSI model. 2. The connection of one circuit or component to another. 3. An attaching device connected to two rings simultaneously to allow the transfer of information from one ring to the other. Rings joined together by bridges form multiple-ring networks.

**bridge clip**  Clips that electrically interconnect two adjacent terminals for the purpose of providing a multiplying or testing point.

**bridge lifter**  A device that removes, either electrically or physically, bridged telephone pairs.

**bridge number**  In a local area network, the identifier that distinguishes parallel bridges that is, bridges spanning the same two rings.

**Bridge Protocol Data Unit (BPDU)**  Packets periodically transmitted by bridges to determine the state of the network they are attached to. If a loop is encountered, one of the bridges causing the loop will stop transmission on the port causing the loop until it becomes necessary to reevaluate the state of the network.

**bridge tap**  Is made when a technician bridges across the cable pair to bring it into a customer location. If the service is disconnected, the bridge tap may be left in place. Excessive bridge taps on a cable may be the cause of significant attenuation distortion.

**bridged ringing**  A system where ringers on a line are connected across the line.

**bridged tap**  An extra pair of wires connected in shunt to a main cable pair. The extra pair is normally open circuited but may be used at a future time to connect the main pair to a new customer. Short bridged taps do not effect voice frequency signals but can be extremely detrimental to high frequency digital signals.

**Bridgemaster**  A local area network bridge marketed by Applitek Corporation of Wakefield, MA.

**Bridgeport**  A trademark of NCR Comten (now AT&T) as well as the name for a series of token-ring bridges and related peripheral products from that vendor.

**Bridge+Fiber**  A local area network bridge marketed by Raycom Systems, Inc., of Boulder, CO.

**British Standards Institution (BSI)**  The organization responsible for the development of national standards in the United Kingdom.

**British Telecom International (BTI)**  The major full-service international telecommunications provider in the UK.

**BRN**  Business Radio Network.

**broadband**  1. In general, communications channel having a bandwidth greater than a voice-grade channel and potentially capable of much higher transmission rates; also called wideband. 2. In LAN technology, a system in which multiple channels access a medium (usually coaxial cable) that has a large bandwidth (50 Mbps is typical) using radio-frequency modems.

**Broadband EXchange (BEX)** Public switched communication system of Western Union, featuring various bandwidth, full-duplex connections.

**broadband local area network** A local area network in which information is encoded, multiplexed, and transmitted without modulation of a carrier. The IBM PC Network is an example.

**broadband system** A system that can transmit data at more than one frequency. making it possible for many transmissions to take place concurrently.

**broadcast** 1. Transmission of a message intended for general reception rather than for a specific station. 2. In LAN technology, a transmission method used in bus topology networks that sends all messages to all stations even though the messages are addressed to specific stations.

**broadcast medium** A transmission system in which all messages are heard by all stations.

**broadcast message** A message addressed to all users of a computer network.

**broadcast subnetwork** A multiaccess subnetwork that supports the capability of addressing a group of attached systems with a single message.

**broadcast topology** A network topology in which all stations are capable of simultaneously receiving a signal transmitted by any other station on the network.

**broadcasting** The transmission through space, utilizing preassigned radio frequencies, which are capable of being received aurally or visually.

**Brooklyn Bridge** A file transfer utility program developed by White Crane Systems Inc. of Norcross, GA, for transferring data between personal computers.

**brouter** A hybrid bridge and router product.

**BSA** Basic Servicing Arrangements.

**BSC** Binary Synchronous Communications.

**BSE** Basic Service Elements.

**BSI** British Standards Institution.

**BST** Basic Services Terminal.

**BSY** BuSY (control signal).

**BT** British Telecom plc.

**BTAM** Basic Telecommunications Access Method.

**BTN** Billed Telephone Number.

**Btoa** A UNIX program which translates Binary files into ASCII.

**BTRON** A version of the TRON operating system designed for business use.

**BTU** Basic Transmission Unit.

**BUF** BUFfer.

**buffer** 1. A temporary storage device used to compensate for a difference in either the rate of data flow or the time of occurrence of events in transmissions from one device to another. 2. The protective coating placed on a fiber.

**buffer address** The address of a location in a buffer.

**buffer group** In IBM's VTAM, a group of buffers associated with one or more contiguous, related entries in a buffer list. The buffers may be located in discontiguous areas of storage and may be combined into one or more request units.

**buffer list** In IBM's VTAM, a contiguous set of control blocks (buffer list entries) that allow an application program to send function management data (FMD) from a number of discontiguous buffers with a single SEND macro instruction.

**buffer list entry** In IBM's VTAM, a control block within a buffer list that points to a buffer containing function management (FM) data to be sent.

**buffer storage** Electronic circuitry where data is kept during a buffering operation.

**buffered repeater**  A device which both amplifies and regenerates digital signals to enable them to travel farther along a cable. In a local area network, this type of repeater also controls the flow of messages to prevent collisions.

**buffering**  Process of temporarily storing data in a register or in RAM, allowing transmission devices to accommodate differences in data rates and perform error checking and retransmission of data that was received in error.

**bug**  An error in a program or a malfunction in a piece of equipment.

**building entrance area**  The area inside a building where cables enter the building and are connected to backbone cables and where electrical protection is provided. The network interface may be located here, as well as the protectors and other distribution components for the campus subsystem.

**bulk redundancy**  A method of coding an anisochronous channel on a synchronous stream of bits in which a 1 state is represented by a string of ones while it lasts, and an 0 state by a string of zeros. It is a redundant method because the strings must be long ones to reduce telegraph distortion when the anisochronous signal is reconstructed.

**Bulletin Board System (BBS)**  Software that normally operates on a personal computer and which lets persons leave and retrieve messages and upload and download files.

**bunched frame alignment signal**  A frame alignment signal in which the signal elements occupy consecutive digit time slots. Also called added channel framing.

**Bureaufax Intelpost**  An international facsimile service marketed by Iceland PTT.

**Burroughs Network Architecture (BNA)**  The network architecture developed by Burroughs Corporation for use with its mainframe computers and distributed processing products.

**burst**  1. A group of events occurring together in time.
2. A sequence of signals counted as one unit in accordance with some specific criterion or measure.
3. To separate continuous-form paper into discrete sheets.

**burst error**  A group of bits in which two consecutive erroneous bits are always separated by less than a given number of correct bits. In other words, the erroneous bits are more closely spaced than is statistically expected.

**burst isochronous**  A burst isochronous signal consists of bursts of digits synchronized to a clock, interspersed by "silent" periods when no bits are presented. To indicate the bursts and silence a special clock may be provided which operates only when bits are present. This is called a "stuttering clock."

**burst mode**  The transmission of bulk data in large, continuous blocks, in which the channel is exclusively used for the extent of the transmission.

**burst noise**  Noise which appears on a communication line as a burst of noise.

**burst rate**  A maximum rate of continual signal transmission.

**bursty**  Not uniformly distributed in time.

**BUS, BUSS**  1. In general, a data path shared by many devices such as the input/output bus in a computer.
2. In LAN technology, a linear network topology.

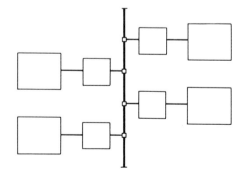

**bus arbitration**  A bus design that allows multiple microprocessors on the system board and adapter cards to share the bus in an organized manner.

Bus arbitration permits multitasking and parallel processing.

**bus hog** A method of operating direct memory access (DMA) devices such that the device obtains control of the backplane bus and remains in control until it has completed as many data transfers as it requires at a point in time.

**bus network** A network in which LAN nodes are all connected to a single length of cable.

**bus segment** A section of a bus that is electrically continuous, with no intervening components, such as repeaters.

**Business and Residence Custom Services (BRCS)** An approach that the 5ESS switch employs to provision revenue-generating services. BRCS features can be assigned to virtually any line type (or trunks) in almost any combination, removing the unnecessary restrictions found in other approaches.

**Business Radio Network (BRN)** A Colorado Springs, CO, based satellite-delivered news service which began broadcast transmission on July 4, 1988.

**business television** A form of satellite video-conferencing used by corporations for meetings, training, and product introductions.

**BUSY** A signal used to indicate the state of the modem. A modem is in the BUSY state during call connection or call processing.

**busy hour** The 60-minute period of a day in which the largest number of transactions occur. Also peak hour or peak period.

**busy lamp** A light on telephone equipment which indicates a busy condition for a certain line or phone.

**busy signal** An audible or visual signal which indicates that the called number or transmission path is unavailable.

**busy verification** A feature in the switched telephone network that permits an attendant to verify the busy or idle state of station lines and/or to break into the conversation.

**BWM** Broadcast Warning Message.

**BX.25** AT&T's rules for establishing the sequence of events and the specific types and forms of signals required to transfer data between computers. BX.25 includes the international rules known as X.25, and more.

**bypass** 1. Refers in general to any private networking scheme used to access long distance transmission facilities without being routed through the local exchange carrier. 2. A T1 multiplexing technique similar to drop-and-insert, used when some channel in a DS1 stream must be demultiplexed at an intermediate node. With bypass, only those channels destined for the intermediate node are demultiplexed –ongoing traffic remains in the T1 signal, and new traffic bound from the intermediate node to the final destination takes the place of the dropped traffic. With drop-and-insert, all channels are demultiplexed, those destined for the intermediate node are dropped, traffic from the intermediate node is added, and a new T1 stream is created to carry all channels to the final destination. 3. To eliminate a component from a network by allowing the path to go past it.

**bypass channels** A term given to channels that are routed through a node without being demultiplexed.

**bypass relay** A relay in a ring network which permits message traffic to travel between two nodes that are not normally adjacent.

**byte** A collection of bits operated upon as a unit; most are 8 bits long; and most character sets use one byte per character. The capacity of storage devices is frequently given in bytes or in Kbytes (K meaning 1024).

**byte multiplexer channel** A mainframe input/output channel that provides multiplexing, of data in bytes.

**byte multiplexing** A byte (or character) for one channel is sent as a unit, and bytes from different channels follow in successive time slots.

**byte stuffing** Insertion into a byte stream of some "dummy" bytes so that the mean data rate is less than the rate of the channel. The qualifying bit, if used, can distinguish the dummy bytes, which then appear as a species of control signal.

**Byte-Oriented Protocol (BOP)** A protocol technique using defined characters (bytes) from a code set for communications control.

**B1** A U.S. Department of Defense computer security level which requires mandatory access controls and the labeling of data to reflect its security classification.

**B7 stuffing** A simple technique for maintaining one's density on digital circuits. With B7 stuffing, each string of eight consecutive zero bits is replaced by a string of seven zeros and a 1.

**B7 zero code suppression** A technique used to keep telephone company repeaters and channel service units in synchronization by providing compliance with digital transmission one's (1's) density requirements. In this technique if all eight bit positions in a T carrier DS-O time slot are zero, B7 zero code suppression will substitute a 1 in bit position 7.

**B8ZS** Binary 8 zero suppression.

B7 zero code suppression

(a) B7 zero code suppression example

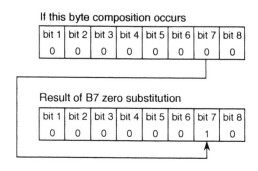

(b) Worst-case scenario

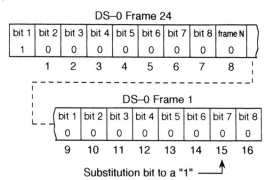

*B7 zero code suppression*

# C

C   Celsius.

C   A programming language developed at Bell Labs in conjunction the UNIX operating system. C is a highly transportable language which permits programs to be run on any machine supporting a C compiler.

C band   A portion of the electromagnetic spectrum used for satellite and microwave transmission; frequencies of approximately 3.9 to 6.2 GHz. The reader is referred to the satellite entry in this book for a table containing C band satellite information to include their orbit, name, operator, year of launch, use and international designation.

C bit   A network control bit originally used in Dataphone Digital Service (DDS) to indicate the subscriber's channel activity and to insure a minimum number of timing-pulse transitions. Today DDS with secondary channel (SC) is derived by the use of the C bit once every third octet.

C conditioning   C conditioning applies attenuation equalization and delay equalization to the line so that line quality falls within certain limits.

C connector   A bayonet-locking connector for coax; C is for Carl Concelman.

cabinet   A physical enclosure for rack-mount equipment; standard cabinets have $1\frac{3}{4}$-inch vertical spacing between mounting holes and 19-inch wide horizontal spacing between mounting rails.

cable   A group of conductive elements (wires) or other media (fiber optics), packaged as a single line to interconnect communications systems.

cable chart   In a local area network environment, a chart, prepared by the network planner, that indicates the location of patch cables connected to components such as the IBM 8228 Multistation Access Unit.

Cable Link   A cable television company which serves over 250 000 subscribers in the Irish Republic.

cable loss   The amount of radio frequency (RF) signal attenuated by coaxial cable transmission. The cable attenuation is a function of frequency, media type, and cable distance.

cable system, cabling system   In LAN technology the medium used to interconnect stations; often called the premises network.

cable-based LAN   A shared medium LAN that uses a cable for its transmission medium.

cache memory   A high-speed computer memory which contains the next most likely instruction or sequence of instructions to be executed.

CAD   Computer Aided Design.

CADN   Cellular Access Digital Network.

CAE   Computer Aided Engineering.

CAI   Computer Aided Instruction.

CALC   Customer Access Line Charge.

61

**call** A request for connection or the connection resulting from such a request.

**call accounting, call accounting record** In packet switched networks, the process of accumulating data on individual calls or of reporting such data; usually includes start and end times, NTN or NUI, and number of data segments and packets transmitted for each individual call.

**call accounting system** Hardware and/or software that tracks outgoing calls and records data for reporting. Also called Station Message Detail Recording (SMDR).

**call add-on** A PABX or central office function that permits adding another party to a conversation already under way.

**call allocator** A communications carrier feature which enables customers to allocate calls between two or more locations by determining the percentage of calls each location should receive.

**call block** A telephone feature offered by many communication carriers and supported by most PBX's which shunts incoming calls from specified phone numbers to a recorded rebuff or which prevents calls being made to a telephone number or group of numbers identified by an area code or prefix.

**call capacity** The ability of a telephone system to provide a specific grade of service.

**call clearing** The orderly termination of a user's call on a network.

**call collision** A condition arising when a trunk or channel is seized at both ends simultaneously, thereby blocking a call.

**call detail recording** The capability of a telephone branch exchange to log calls for accounting purposes.

**call diversion** The automatic switching of a call from the number to which it was directed to another predetermined number.

**call duration** The interval of time from the moment a connection is established to the instant the calling party terminates the call.

**call establishment** The routing of a user's call to its destination and the set up of data transfer.

**call forwarding** A PABX or central office feature that permits a telephone subscriber to reroute incoming calls intended for the subscriber to a different telephone number, either all the time or only when the subscriber's number is busy or doesn't answer.

**call forwarding when busy** Similar to call forwarding, this programming mode routes calls to another designated extension when the primary extension is busy.

**call gapping** A control which regulates the maximum rate at which calls are released towards a destination code. Call gapping control performs the function of limiting access attempts during mass-call situations to specific destination codes.

**call hold** A PABX or central office feature that permits a telephone subscriber to place an existing call on "hold," e.g., in a waiting state, while handling an incoming call or otherwise attending to some other matter.

**call hour (CH or ch)** A unit of traffic intensity. One call hour is the quantity represented by one or more calls with an aggregate duration of 1 hour.

1 ch = 36 CCS = 60 call minutes = 3600 call seconds

**Call Interactive** A joint venture between American Express Information Services Company and AT&T which enables customers to use AT&T's 800 and 900 calling services for market research, telemarketing and interactive audience participation.

**call minute (Cmin)** A unit of traffic intensity. The quantity represented by one or more calls with an aggregate duration of 1 minute.

**call packet** In the CCITT X.25 recommendation a packet which carries addressing and other pertinent information required to establish a switched virtual call (SVC).

**call pickup**  A PABX or central office feature that permits a subscriber to answer the telephone of another subscriber on his/her own instrument, by "picking up" the incoming call. Normally, several phone extensions can be programmed as a group to allow subscribers to answer any calls coming in on each telephone in the group.

**call processing**  The sequence of operations performed by a switching system from the acceptance of the incoming call to its termination.

**call progress tones**  Audible signals returned to the station user by the switching equipment to indicate the status of a call. Common examples of audible signals include dial tones and busy signals.

**call record**  Recorded data pertaining to a single call, usually including the time of the call, its duration and destination number.

**call redirection**  A packet network service which enables subscribers to nominate a second computer or terminal to which calls can be redirected when the first destination is busy or off-line.

**call reference**  A unique value used locally to identify a Modem Connect or X.21 call.

**call request**  A data packet that requests the establishment of a circuit between two network resources.

**call request packet**  In packet switched networks, the packet sent by the originating DTE showing requested NTN or NUI, network facilities, and call user data.

**call restriction**  A PBX or central office feature that prevents selected extension stations from dialing toll calls.

**call second**  A basic unit of measurement in telephone operations (actually, 100 call seconds equals one CCS, the equivalent of one call held for 100 seconds).

**call setup time**  The time required to set up communications between equipment on a switched network or line.

**call sharing**  A form of switched line sharing in which many clients have access to the same call on that line.

**call sign**  A combination of characters (letters and numbers), which identify a communications facility, command, authority, activity or unit.

**call splitting**  A PBX or central office feature that allows an attendant to speak privately to either party in a connection without the other party hearing.

**call termination**  See call clearing.

**call trace**  1. A PABX feature which can be used to identify a trunk or trunk group number on incoming calls and the called number and station on calls going out of the system. 2. A telephone feature offered by communication carriers which allows customers to track down the number of the last call.

**call transfer**  A PABX or central office feature that permits a telephone subscriber or operator to redirect a call to another number.

**call user data (CUD)**  In packet switched networks, user information transmitted in a call request packet to the destination DTE. CUD is a field in an X.25 level 3 Incoming call/Call request packet.

**call waiting**  A signal generated by a PABX or central office which notifies a subscriber currently engaged in an existing call that another call has arrived.

**callback**  The capability for a caller to program a phone to retry another busy extension as soon as it is free.

**callback modem**  A modem which functions as a limited security device, preprogrammed with identification codes, passwords, and telephone numbers. When a caller accesses the callback number and enters his or her identification code and password correctly, the modem will then hang up the connection and use the stored telephone number to dial the caller. Thus, security is effected by the telephone numbers callers are assigned to use. Also called a security modem.

**call-by-call service selection** An AT&T ISDN feature which allows T1 users to change the services to which the 23 B channels are dedicated, such as Megacom Service, Megacom 800 Service or Accunet Switched Digital Service.

**Callcenter** A trademark of Aspect Telecommunications of San Jose, CA, as well as a stand-alone, automatic call distributor from that company that can service up to 7000 calls per hour.

**called, calling, or called/calling channel** In LAN technology and packet switched networks, a called channel is a channel that can receive but not originate calls; a calling channel can originate but not receive calls; and finally, a called/calling channel can both originate and receive calls.

**called-party CAMP-on** A communication system feature that enables the system to complete an access attempt in spite of issuance of a user blocking signal.

**called station** In the U.S. military, a station to which a message is routed (in tape relay) or to which a transmission is directed.

**Call ID** A telephone feature offered by many communication carriers which displays the number of the telephone of the originating call.

**caller identification feature** Customers and users, as specified by the customer, who access a network through public dial-in access ports may be identified to a packet network through the use of the caller identification feature. Each such user, upon verifying his identity by means of valid identification code and password, may establish virtual connections through the network with all associated public port and traffic charges billed to the responsible customer.

**Caller*ID Service** A custom local access signaling service marketed by Bell Atlantic Corporation which identifies the telephone number of the calling party to the called party.

**calling rate** The number of calls per telephone; determined by dividing the count of busy-hour calls by the number of telephones.

**calling signal** A call control signal transmitted over a circuit to indicate that a connection is desired.

**calling station** 1. In general terms the station initiating a transmission. 2. In a U.S. military tape relay system, the station preparing a tape for transmission.

**calling (CNG) tone** A 400 ms tone transmitted at 1100 Hz every 2.5 seconds that identifies the calling device as a fax.

**CALLMASTER** A trademark of AT&T as well as a digital voice terminal marketed by that vendor.

**Callpath** A communications software system marketed by IBM which integrates System/370 CICS applications with the firm's 9751 PBX. Under Callpath information about incoming telephone calls are sent to System/370 computers while the computer transfers corresponding customer data to 3270-type terminals.

**callquest** Call accounting software developed by COMDEV of Sarasota, FL, which operates on PC-DOS/MS-DOS personal computers. Callquest provides businesses with up to 2500 stations the ability to manage their telephone usage.

**CAM** Computer Aided Manufacturing.

**CAMA** Centralized Automatic Message Accounting.

**Cambridge Monitor System (CMS)** An IBM operating system for mainframe computers named after the University of Cambridge where a lot of the original development work was conducted.

**Cambridge Ring** In LAN technology, an empty slot ring LAN. The ring network is connected with a single cable to form a loop to which computers are attached via nodes. Up to 255 nodes can be on a Cambridge Ring, each with a unique 8-bit address. Data on this ring flows at 10 Mbps in "mini-packets" containing two or more bytes of user data. It has not yet achieved a great deal of popularity outside its country of origin, England—where several near-Cambridge Ring systems are being marketed.

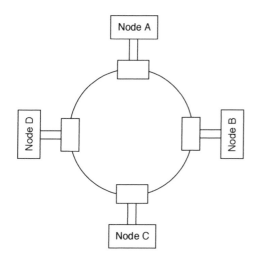

*Cambridge Ring*

**CAMP**   Computer Aided Message Processing.

**CAMP-on**   A mechanism that places a subscriber calling a busy number into a waiting position (queue).

**CAMP-on, CAMP-on-busy**   A PABX or cable-based LAN facility that allows users to wait on line (in queue) if the requested resource is busy and that connects the users in queue—on a first-come, first-served basis—when the requested resource becomes available.

**campus**   In local area networks, a group of buildings or similar installations served by the same network.

**campus subsystem**   The part of a premises distribution system which connects buildings together. The cable, interbuilding distribution facilities, protectors, and connectors that enable communication among multiple buildings on a premises.

**CAN**   Channel character.

**Canadian National/Canadian Pacific Telecommunications (CNCP)**   A Canadian company offering data services which compete against Telecom Canada. CNCP offers services in Canada similar to services offered in the United States by Western Union.

**Canadian Standards Association (CSA)**   A private, non-profit organization which produces standards and certifies products for compliance with their standards. CSA functions similar to a combined U.S. ANSI and underwriters' laboratory. The following table lists several CSA communications related standards.

CSA STANDARDS

| | |
|---|---|
| Z243.4 | ASCII |
| Z243.10 | modified ISO 1155 |
| Z243.11 | modified ISO 1177 |
| Z243.12.2 | modified ISO 4903 |
| Z243.13 | modified ISO 1745 |
| Z243.13.1 | modified ISO 2111 |
| Z243.13.2 | modified ISO 2628 |
| Z243.13.3 | modified ISO 2629 |
| Z243.28.1 | electrical characteristics for balanced interchange circuits |
| Z243.33.1 | HDLC frame structure |
| Z243.35 | modified ISO 2022 |
| Z243.38 | CCITT X.25 |
| Z243.43.1 | modified CCITT X.28 |
| Z243.43.2 | modified CCITT X.3 |
| Z243.43.3 | CCITT X.29 |

**cancel (CAN)**   A character indicating that the data preceding it is in error and should be ignored.

**cancel closedown**   In IBM's VTAM, a closedown in which VTAM is abnormally terminated either because of an unexpected situation or as the result of an operator command.

**CANTV**   Compania Anonima Nacional de Telefonos.

**capacitance**   The ratio of electrical charge between two conductors and their potential difference.

**capacitor**   An electronic circuit component that temporarily stores electric charges. The capacitor's ability to block direct current flow and pass alternating current flow makes the capacitor useful for power supply filters.

**capacity**   An abbreviation of "channel capacity" which is the rate at which data are carried on the channel, measured in bit/s.

**carbon block**  Surge limiting device that will be grounded by arcing across the air gap when the voltage of a conductor exceeds a predetermined level. If the current flow across the gap is large or persists for an appreciable time, the protector mechanism will operate and the protector will become permanently grounded.

**Carbon Copy Mac**  A trademark of Microcom, Inc., of Danbury, CT, as well as a software program from that vendor which allows users to remotely manage an AppleTalk network and control another Macintosh computer.

**Carbon Copy Plus**  A software program from Meridian Technology which enables personal computers and microcomputer local networks to dial in to remote networks to access and share resources as well as to communicate with other non-networked PC's and mainframes.

**card module**  A printed-circuit board that plugs into an equipment chassis.

**CAROT**  Centralized Automatic Reporting On Trunks.

**Carriage Return (CR)**  An ASCII or EBCDIC control character used to position the print mechanism at the left margin on a printer—or the cursor at the left margin on a display terminal.

**carriage width**  The width of a printer expressed in terms of the number of character positions that can be printed in the printer's default mode. Two standard printer carriage widths are 80-column and 132-column. An 80-column printer can print a maximum of 80 characters horizontally on one line.

**carrier**  A continuous signal which is modulated with a second, information-carrying signal.

**carrier, communications common**  A company which furnishes communications services to the general public, and which is regulated by appropriate local, state, or federal agencies.

**carrier band**  A band of continuous frequencies that can be modulated with a signal.

**Carrier Detect (CD)**  An RS-232 control signal (on pin 8) which indicates that the local modem is receiving a signal from the remote modem. Also called Received Line Signal Detector (RLSD) and Data Carrier Detect (DCD).

**carrier failure alarm**  A red or yellow alarm on a T1 circuit, or the end-to-end combination of a red and a yellow alarm.

**carrier frequency**  The frequency of the carrier wave which is modulated to transmit signals.

**carrier heterodyne**  Interference between conventional amplitude modulated (AM) signals caused when the carriers of two adjacent AM signals heterodyne at radio frequency, resulting in an audio beat after modulation falling within the bandpass of the radio receiver output.

**carrier modulation**  A signal at some fixed amplitude and frequency which is combined with an information-bearing signal in the modulation process to produce an output signal suitable for transmission.

**carrier selection**  As a result of the 1982 Federal Court Modified Final Judgment, local telephone companies must offer residence and business customers the opportunity to select which long distance company they wish to use on a primary basis. The company selected is the one which will be used when dialing 1 + area code + telephone number. This opportunity is being phased in over time, as local telephone companies modify their central office equipment to accommodate access to more than one long distance company.

**Carrier Sense Multiple Access (CSMA)**  In LAN technology, a contended access method in which stations listen before transmission, send a packet, and then free the line for other stations. With CSMA, although stations do not transmit until the medium is clear, collisions still occur; two alternative versions (CSMA/CA and CSMA/CD) attempt to reduce both the number of collisions and the severity of their impact.

**Carrier Sense Multiple Access with Collision Avoidance (CSMA/CA)** In LAN technology, CSMA that combines slotted TDM (to avoid having collisions occur a second time) with CSMA/CD; CSMA/CA works best if the time slot is short compared to packet length and if the number of stations is small.

**Carrier-Sense Multiple Access with Collision Detection (CSMA/CD)** A local area network technique, where all devices attached to a LAN listen for transmissions before attempting to transmit and, if two or more begin transmitting at the same time, each backs off for a random period of time (determined by a complex algorithm) before attempting to retransmit.

**carrier signaling** Any signaling technique used in multi channel carrier transmission.

**carrier systems** A system used to maximize the utilization of cable pairs used in the telephone network. Rather than having each call dedicated in individual cables throughout the nation, through the use of electronics, carrier systems allow as many as 24 individual voice/data customers to utilize the same cable pair.

1. Analog carrier systems are an older technology and allow either 12 or 24 customers to use two cable pairs. The channelization of these systems is accomplished in the frequency domain.

2. Digital carrier systems are the latest technology and allow 24 to 96 customers on two cable pairs. The channelization of these systems is accomplished in the time domain. The following figure describes the process of the digital carrier system.

**carrier telegraphy, carrier current telegraphy** A method of transmission in which the signals from a telegraph transmitter modulate an alternating current.

**Carrier to Noise Ratio (C/N)** In satellite transmission the ratio of the satellite's carrier or signal to noise level in a given channel. The carrier to noise ratio is normally measured in dB at the low noise amplifier (LNA).

**carrier wave** Wave upon which a signal is superimposed.

**Carterphone decision** A 1968 FCC decision which held that telephone company tariffs containing blanket prohibition against the attachment of customer-provided terminal equipment to the telecommunications network were unreasonable, discriminatory, and unlawful. The FCC declared the telephone companies could set up reasonable standards for interconnection to insure the technical integrity of the network. Following Carterphone, the telephone companies filed tariffs for protective connecting arrangements to facilitate the interconnection of customer provided terminal equipment.

**CAS** Centralized Attendant Service.

**cascaded network** A network divided into a number of linearly connected, contiguous segments.

**CASE** Common Applications Service Elements.

**CASE** Computer Aided Software Engineering.

**case sensitive** Ability to distinguish between uppercase and lowercase letters.

*carrier systems*

**cassegrain** An antenna comprising two reflectors, the parabolic reflector and an hyperbolic sub-reflector at the focus point, which reflects signals back into the feed.

**cassette tape** A slow-speed, low-capacity method of storing and retrieving data which uses a technology similar to audio cassettes.

**Casual Connection** An IBM Network Control Program (NCP) feature which allows resources of one SNA host using IBM's Virtual Telecommunications Access Method (VTAM) to call an SNA resource of another SNA host using LU6.2 without the VTAM's of either host knowing the location of the other's resource.

**cat whisker** A flexible wire that was positioned on a crystal by the use of an adjusting arm for tuning early radios.

**cataloged data set** A dataset that is represented in an index, or hierarchy of indexes, in the system catalog; the indexes provide the means for locating the data set.

**cataloged procedure** A set of job control statements that has been placed in a library and that can be retrieved by name.

**catanet** A collection of networks that are interconnected through gateways.

**Cathode Ray Tube (CRT)** The video imaging (picture) tube used in a television or video display data terminal.

**CAU** Channel Access Unit.

**CATV** Community Antenna Television.

**CB** Common Battery.

**CBEMA** Computer and Business Equipment Manufacturers Association.

**CBT** Computer-Based Terminal.

**CBTR** Carrier and Bit Timing Recovery.

**CBX** Computer Branch Exchange.

**CC** Control Check.

**CCA** Circuit Card Assembly.

**CCB** Central Control Box.

**CCC** Clear Channel Capability.

**CCEB** Combined Communications Electronics Board.

**CCIA** Computer and Communications Industry Association.

**CCIR** International Consultative Committee for Radio.

**CCIS** Common Channel Interoffice Signaling.

**CCITT** Consultative Committee for International Telegraph and Telephone. (Now renamed as the ITU)

**CCITT HIgh-Level Language (CHILL)** A telephony-oriented high-level language increasingly employed for switching system software. CCITT recommendation Z.200, Fascicle VI.12 are examples of CHILLs.

**CCITT #7 signaling** The newest standard for signaling within telecommunications networks being developed by CCITT. It will eventually replace the CCIS6 network in the United States.

**CCM** Channel Controller Module.

**CCP** In IBM's VTAM, Configuration Control Program.

**CCR** Customer Controlled Reconfiguration.

**CCS** Common Channel Signaling.

**CCS** Hundred-call seconds.

**CCSA** Common Control Switching Arrangement.

**CCSS** Common Channel Signaling System.

**CCT** Coupler Cut Through.

**CCTV** Closed Circuit Televsion.

**CCU** Cluster Control Unit.

**CCU** Communications Control Unit.

**CD** 1. Compact Disk. 2. Carrier Detect. 3. Collision Detection.

**CDAR** Customer-Dialed Account Recording.

**CDB** Circuit Descriptor Block.

**CDCCP** 1. Control Data Communications Control Procedure. 2. Control Data Communications Control Unit Procedure (Control Data Corp.).

**Cdev** Control Panel Device.

**CDFP** Centrex Facility Data Pooling.

**CDI** Control and Data Interface.

**CDMA** Call Division Multiple Access.

**CDNnet**   A Canadian academic network based upon the International Standards Organization Open Systems Interconnection Reference Model.

**CDR**   Call Detail Recording.

**CDRM**   In IBM's VTAM, Cross-Domain Resource Manager.

**CD-ROM**   Compact Panel Device.

**CDRSC**   In IBM's VTAM, Cross-Domain Re-SourCe.

**CE**   Customer Engineer.

**C-E**   Communications-Electronics.

**CEB**   In IBM's VTAM, Conditional End Bracket.

**CEI**   Communications-Electronics Installation.

**CEKS**   Centrex Electronic Key Set.

**Cell Relay**   The name used by the IEEE and Bellcore for Fast Packet. Cell Relay is a transmission method for voice, data and video in which addressed units of information known as cells are 53 bytes in length, using a 5-byte header. This technique requires variable length frames to be segmented into small, fixed length cells and enables such cells to be rapidly transported over permanent or dynamic logical links.

**CellBlazer**   A modem manufactured by Telebit Corporation of Mountain View, CA, for cellular mobile transmission applications.

**Cellnet**   The national cellular telephone network in the United Kingdom which is owned by a consortium headed by British Telecom.

**cells**   A subdivision of a mobile telephone service area; containing a low-powered radio communications system connected to the local telephone service; a block of fixed length.

**Cellular Access Digital Network (CADN)**   1. A network that provides customers a wide range of ISDN-type capabilities in a mobile environment. 2. A concept pioneered at AT&T Bell Laboratories to provide integrated digital voice and data transmission over a single cellular channel.

**cellular geographic service area**   In cellular communications a metropolitan area in which a cellular telephone company is authorized to operate.

**cellular mobile radio**   A system providing exchange telephone service to a station located in a mobile vehicle, using radio circuits to a base radio station which covers a specific geographical area. As the vehicle moves from one area to another different base radio stations handle the call.

**cellular radio communications**   A mobile radio system that consists of hexagonal geographic areas with groups of frequencies allocated to each cell. Seven cells make up a block, and no adjacent cell uses the same set of frequencies. The frequency allocation pattern, however, is the same in each successive block.

**cellular telephone technology**   A relatively new technology that vastly improves mobile telephone communications. Cellular towers receive and transmit telephone signals, and switch them to or from the nearest Central Office for further transmission. Cellular networks are being constructed in most major metropolitan areas; customers in most markets have or will have a choice of two cellular telephone companies.

**cellular telephone frequency**   In the United States cellular telephone is assigned the 825 to 890 MHz band, excluding 845 to 870 MHz. The mobile station transmits in the low band (825 to 845 MHz) while the fixed station transmits in the high band (870 to 890 MHz).

**CELP**   Code Excited Linear Predictive Speech.

**CEMS**   Communications-Electronics Management System.

**CEN**   European Committee for Standardization.

**CENELEC**   European Committee for Electrotechnical Standardization.

**CenPath**   A switched digital 56 and 64 Kbps service marketed by Pacific Bell of San Francisco, CA.

**Central Office**   The building where common carriers terminate customer circuits and where the switching equipment that interconnects those circuits is located. Sometimes also known as the central exchange—or just simply as exchange.

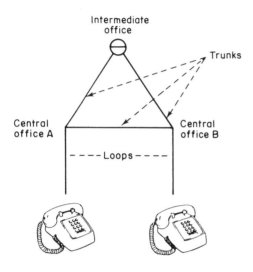

*Central Office*

**Central Office Exchange Service (CENTREX)** A service provided by telephone companies in which a portion of its central office is used to switch calls to and from individual stations at different user sites; functionally similar to an on-site PBX, typically providing direct inward dialing, direct distance dialing, and attendant switching.

**central office switching equipment** Electromechanical, or electronic equipment that routes a call to its destination.

**Central Office Terminal (COT)** The terminating equipment for a digital line closest to the central office.

**Central Office Local Area Network (COLAN)** A Centrex-like service marketed by several Bell operating companies (BOCs) in which LAN capabilities are provided to customers within a common local telephone serving area through the use of the BOC's central office switch.

**Central Processing Unit (CPU)** The unit that executes programmed instructions, performs the logical and arithmetic functions on data and controls input/output functions. Often used as a synonym for a computer.

**Central Processing Unit Master Switching Subsystem (CPMS)** The series of software codes, as developed at Telenet, that provides the operating environment of a particular Packet Switching Network (PSN). Software residing in the master Central Processing Unit (CPU) that performs all packet level processing system control and resource management functions. CPMS40 denotes the static-environment subsystem. Dynamic series include CPMS60 and above. CPMS is composed of three software components: PROTO, SWITCH, and the Telenet Processor Operating System (TPOS).

**central site customizing** The process of tailoring microcode for a device in a network at the central site.

**Centrale Strecke** A planned broadband, fiber optic cable to connect the cities of Moscow, Warsaw and Berlin with Frankfurt.

**centralized** Processing with one central processor, which may support remote terminals and/or remote job entry stations.

**centralized adaptive routing** A method of routing in which a network routing center dictates routing decisions, based on information supplied to it by each node.

**Centralized Attendant Service (CAS)** The capability of a centrally located attendant console to control a number of switches, some of which may be geographically remote.

**Centralized Automatic Message Accounting (CAMA)** A feature on an AT&T 5ESS switch which permits the switch office to collect and store toll information and message unit billing on calls originated by local offices served by the CAMA office.

**Centralized Automatic Reporting On Trunks (CAROT)** An AT&T computerized system that automatically accesses and tests trunks for a maximum of 14 offices simultaneously.

**centralized management** A form of network management where management is performed from a single point in the network.

**centralized processing** A data processing configuration where the processing for various locations is centralized in a single computer site (compare with distributed processing).

**Centralized Trunk Test Unit (CTTU)** An AT&T operational support system providing centralized trunk maintenance through a data link on a switch.

**CENTREX** Central Office Exchange Service.

**Centronics** Printer manufacturer that set the *de facto* interconnection standards for parallel printers, using a 36-pin, byte-wide connector.

**CEOI** Communications-Electronics Operation Instructions.

**CEPT** Conference of European Postal and Telecommunications Administrations.

**CERFNet** California Education and Research Federation Network.

**CERN** Corporation for Educational and Research Networking.

**CERT** Character Error Rate Testing.

**CESE** Communications Equipment Support Element.

**CEWI** Combat Electronic Warfare and Intelligence.

**CE-11** A U.S. military lightweight, portable, field wire reel designed to be carried by one man.

**CF** Call Forwarding.

**CFCA** Communications Fraud Control Association.

**CFDA** Call Forwarding Don't Answer.

**CFV** 1. Call for votes. 2. Call Forwarding Variable.

**CFR** Customer Formatted Reports.

**CG** Commanding General.

**CGI** Code Group Index.

**CGMA** Color Graphics Monitor Adapter.

**ch** Call hour.

**CH** Call Hour.

**CH SETTING** CHannel SETTING.

**chad** The material removed when forming a hole or notch in a storage medium such as punched tape or punched cards.

**chadless tape** Perforated tape with the chad partially attached, to facilitate interpretive printing on the tape.

**chain** A series of processing locations in which information must pass through each location on a store and forward basis to get to a destination.

**chaining I/O commands** The linking together (in a chain) of the commands which initiate input/output operations. When one command is finished the next one in the chain begins operation.

**Chameleon 32** A protocol analyzer with a T1 analysis option, produced by Tekelec, Inc., of Calabasas, CA.

**Change Of Frame Alignment (COFA)** A framing bit position shift resulting from an uncontrolled slip which causes an Out of Frame condition in a downstream T1 terminal.

**change-direction protocol** In SNA, a data flow control protocol in which the sending logical unit (LU) stops sending normal-flow requests, signals this fact to the receiving LU using the change-direction indicator (in the request header of the last request of the last chain), and prepares to receive requests.

**change-screen key** A key or sequence of keys on a display station keyboard used to change sessions, one at a time.

**channel** 1. (CCITT standard) A means of one-way transmission. Compare with circuit. 2. (Tariff and common usage) As used in tariffs, a path for electrical transmission between two or more points without common carrier-provided terminal equipment, such as a local connection to DTE. Also called circuit, line, data link, path, or facility. 3. In an IBM host system, a high-speed data link connecting the CPU and its peripheral devices. 4. (T1 usage) A North American T1 circuit carries 24 channels; a European T1 circuit carries 31 channels; an ISDN Basic Rate circuit carries three channels. The designation "channel" does not indicate a specific bandwidth or bit rate (e.g. the ISDN Basic Rate has two B-channels at 64 Kbps and one D-channel at 16 Kbps).

**channel, analog**  A channel on which the information transmitted can take any value between the limits defined by the channel. Typically, a channel that carries alternating current (AC) but not direct current (DC).

**channel, digital**  A channel that carries pulsed, direct-current signals.

**channel, four-wire**  A communications channel designed for full-duplex operation. Four wires are provided at each termination—two for sending data and two for receiving data.

**channel, grade**  A channel's relative bandwidth (i.e., narrowband, voice-grade, wideband).

**channel, primary**  The higher speed of two channels used for transmitting the data. Also called forward or main.

**channel, reverse**  The slower speed of two channels used for slow-speed data such as error-detection.

**channel, two-wire**  A communications circuit designed for simplex or half-duplex operation. Two wires are provided at each termination point, and information is transmitted in only one direction at a time.

**channel, voice-grade**  A channel suitable for transmission of speech, digital, or analog data, or facsimile, generally with a frequency range of about 300 to 400 hertz.

**Channel Access Unit (CAU)**  A device which monitors a T1 span for synchronization and proper signals and which provides access to individual DSO (64 Kbps) circuits for inserting test equipment and monitoring.

**channel adapter**  In IBM's VTAM, a communication controller hardware unit used to attach the controller to a System/360 or a System/370 channel.

**channel associated signaling**  Call control signaling transmitted within the bandwidth of the call it controls; also called in-band signaling. In T1 transmission, channel associated signaling is performed by bit-robbing.

**channel-attached**  Of devices attached directly to the input/output channels of a mainframe computer; devices attached by cables, not by communications (locally attached IBM).

**channel-attached communication controller**  In IBM's VTAM, An IBM communication controller that is attached to a host processor by means of a data channel.

**channel-attached cross-domain ncp**  In IBM's VTAM, An NCP that is channel-attached to a data host, but resides in the domain of another host. It has been contacted over the channel by the host, but it has not been activated. (That is, no SSCP-to-PU session exists for it.)

**channel attachment major node**  In VTAM: 1. For MVS and VSE operating systems, a major node whose minor node is an NCP that is channel-attached to a data host. 2. A major node that may include minor nodes that are the line groups and lines that represent a channel attachment to an adjacent (channel-attached) host. 3. For the VSE operating system, a major node that may include minor nodes that are resources (host processors, NCPs, line groups, lines, SNA physical units and logical units, cluster controllers, and terminals) attached through a communication adapter.

**channel bank**  A device for multiplexing and de-multiplexing T1 circuits at a telephone company's central office. The channel bank also transmits and detects signaling and framing information.

**channel bypass**  A T1 multiplexing technique similar to drop-and-insert, used when some channels in a DS1 stream must be demultiplexed at an intermediate node; also called "passthrough." With bypass, only those channels destined for the intermediate node are demultiplexed—ongoing

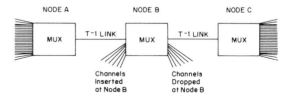

Channel bypass - Channels bound from Node A to Node C "bypass" the demultiplexing process at Node B

traffic remains in the T1 signal, and new traffic bound from the intermediate node to the final destination takes the place of the dropped traffic. With drop-and-insert, all channels are demultiplexed, those destined for the intermediate node are dropped, traffic from the intermediate node is added, and a new T1 stream is created to carry all channels to the final destination.

**channel capacity** The highest possible information-transmission rate through a channel at a specified error rate. Also the total individual channels in a system.

**channel coding with redundant multilevel signals** The original name used to describe Trellis Coded Modulation which produced a pattern of mapping that occurs within a finite state machine.

**channel demultiplexing** The extraction of one or more channels from a multiplexed bit stream.

**channel designation** In the U.S. military, one or more letters used to identify a station in conjunction with a channel number.

**channel group** A grouping of 12 standard telephone circuits in a single carrier system, with each circuit frequency-multiplexed on the channel group, occupying adjacent bands in the frequency spectrum.

**channel interface** See channel and interface.

**channel loopback** A diagnostic test that forms the loop at the multiplexer's channel interface (refer to loopback).

**Channel Service Unit (CSU)** A device on a digital circuit that manages such physical characteristics of the signal as one's density, clocking, and bipolar signal format. On a T1 circuit, the CSU is installed between the customer's multiplexer and the circuit provider's network; according to the current reading of the rules governing customer/network boundaries, the CSU is Customer Premises Equipment. The CSU may be integrated into a DSU (Digital Service Unit).

**channel terminal** That portion of the multiplexing equipment required to derive a subscriber channel from the carrier facility.

**channel-to-channel adapter** A hardware device that can be used to connect two channels on the same computing system or on different systems.

**CHAOSNET** A network protocol developed at the Massachusetts Institute of Technology which is used in LISP machines and other artificial intelligent devices.

**char** Character.

**character** A sequence of adjacent bits representing a unit of language.

1 Control character—A character used to convey status of functions, remote diagnostic information, or nonprint character information.

2 Data character—A character containing alphanumeric information mainly used in a communications link between terminals and/or computers.

3 Sync character—Used to establish and maintain synchronization between computers, multiplexers, and/or synchronous terminals.

**character code** One of several standard sets of binary representations for the alphabet, numbers, common symbols, and control functions. Common character codes include ASCII, BAUDOT, BCD, and EBCDIC.

**Character Error Rate Testing (CERT)** Testing a data line with test characters to determine error performance.

**character interleave** A TDM technique in which a channel slot contains a full character of data.

**character mode** A mode in which input is treated as alphanumeric data, rather than graphic data.

**character parity** The addition of a redundant overhead bit to a character to provide error detection capability.

**character position** 1. A location on the screen at which one character can be displayed. 2. An addressed location in the buffer at which one character can be stored.

**character set** A collection of characters, such as ASCII or EBCDIC used to represent data in a system. Usually includes special symbols and control functions. Often synonymous with code.

**character string** A sequence of consecutive characters.

73

**character synchronization**  A process used in a receiver to determine which bits, sent over a data link, should be grouped together into characters.

**character-coded**  In IBM's VTAM, pertaining to commands (such as LOGON or LOGOFF) entered by an end user and sent by a logical unit in character form. The character-coded command must be in the syntax defined in the user's unformatted system services definition table. Synonym for unformatted.

**characteristic distortion**  Distortion caused by transients which, as a result of modulation, are present in the transmission channel. The magnitude of distortion depends on the channel's transmission qualities.

**characteristic impedance**  The impedance termination of an approximately electrically uniform transmission line which minimizes reflections from the end of the line.

**character-oriented**  Term used to describe communications protocols, such as BSC, in which control information is coded in character-length fields. Contrast with bit-oriented.

**Characters Per Second (CPS)**  A measure of data rate.

**character-time**  The basic unit for coding time delays for flyback buffering delays. One character-time equals the reciprocal of the channel data rate in bits per second times the number of bits in the character.

**chassis**  The main structure of a computer cabinet that supports the electronic components and holds the circuit boards.

**Cheapernet**  Colloquial name for a thinner Ethernet coaxial cable based LAN.

**check bit**  An additional bit, such as a parity bit, that is added to a message. This check bit will be used in later error-detection procedures.

**check character**  A character transmitted for message validity purposes. If an error occurs, the check character will cause a check or compare procedure to fail in the receiver and an error will be reported.

**check notice**  A note displayed to a user at sign-on asking the user to check a monitored bulletin board or other type of subsystem.

**checkpoint**  1. A place in a routine where a check, or a recording of data for restart purposes, is performed. 2. A point at which information about the status of a job and the system can be recorded so that the job step can be restarted later. 3. The name that IBM uses for the continuous ARQ operation of their SDLC protocol.

**checkpointing**  Preserving processing information during a program's operation that allows such processing to be restarted and duplicated.

**checksum**  A block check character or sequence that is computed using binary addition to determine the validity of data that has been received. The receiving device computed the checksum based on the data stream and compares it to the received checksum to determine validity.

**child directory entry**  In Digital Equipment Corporation Network Architecture (DECnet), an entry in the Naming Service which points to a child directory of some directory in the namespace.

**CHILL**  CCITT HIgh-Level Language.

**chip**  A flat, fingernail-size piece of material, usually silicon, in which tiny amounts of other elements with desired electronic properties are deposited and etched to form integrated circuits. These tiny chips are the key to the microelectronic revolution in computers. Also called microchips.

**ChosenLAN**  A trademark of Moses Computers of Los Gatos, CA, as well as the name of a star topology local area network marketed by that vendor.

**CI**  Control Interface.

**CICNet**  Committee on Institutional Cooperation Network.

**CICS**  Customer Information Control System

**CICS/VS**  Customer Information Control System/Virtual Storage.

**CID**  In IBM's SNA, Communication IDentifier.

**CIM**  Computer Integrated Manufacturing.

**CINCEUR**   Commander IN Chief EURope.

**CINCPAC**   Commander IN Chief PACific.

**CINCSAC**   Commander IN Chief Strategic Air Command.

**Cindi**   A voice-mail system marketed by Genesis Electronics Corporation of Rancho Cordova, CA.

**CINIT**   In IBM's SNA, a network services request sent from an SSCP to an LU requesting that LU to establish a session with another LU and to act as the primary end of the session.

**cipher**   A form of cryptography depending only on the sequence of characters or bits. On the other hand, in a cryptographic code, meaningful words or phrases are the units that are encoded.

**ciphertext**   In IBM's SNA, synonym for enciphered data.

**circuit**   1. In data communications, a means of two-way communications between two points consisting of transmit and receive channels. 2. In electronic design, one or more components that act together to perform one or more functions.

**circuit, four-wire**   A communications path in which four wires (two for each direction of transmission) are connected to the station equipment.

**circuit, multipoint**   A circuit connecting three or more points.

**circuit, two-wire**   A transmission in which two wires are used to implement a half-duplex or simplex channel.

**circuit arrangements**   1. Two-wire circuits (2W) use a single cable pair for both transmit and receive. 2. Four-wire circuits (4W) use two individual cable pairs, one for transmit and one for receive.

Two- and four-wire circuits are used primarily for switched voice and data communications. In this configuration the circuit would start out as two-wire, change to a four-wire segment, and then change back to a two-wire section. Telephone message network circuits are configured in this way. A hybrid coil is at the two-wire/four-wire junction and is used to make the configuration and balance the circuit so that no energy is reflected back to the transmitter.

**circuit board**   Plastic base on which microchips and other components are mounted to make

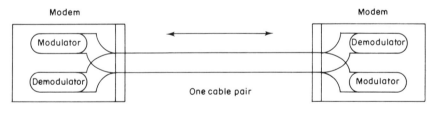

Two-wire circuit

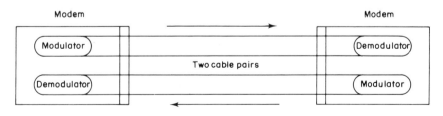

Four-wire circuit

*circuit arrangements*

75

up circuitry—for example, for computers and switching systems.

**Circuit Descriptor Block (CDB)**   A temporary list that summarizes an individual virtual call within the Telenet Processor (TP). This list includes the Logical Channel Numbers (LCNs), the Circuit Line Block (CLB) address, the X.25 Line Block (XLB) address, the number of the next frame expected, and timing information.

**Circuit Line Block (CLB)**   A table describing a non-X.25 port and line on the Telenet processor (TP). This table includes the X.25 address of the port, the Line Processing Unit (LPU) line number, and the status of the line. The CDB address is entered in the Circuit Line Block (CLB) for the duration of the call.

**circuit pack**   An AT&T term for a plug-in, printed-circuit board.

**Circuit Quality Monitoring System (CQMS)**   A method for monitoring critical parameters of data communications lines and circuits.

**circuit segment**   A single point-to-point circuit, usually one part of a multipoint circuit or network. Also called a Link.

**Circuit Switched Digital Capability (CSDC)**   A technique for making end-to-end digital connections. It lets customers place calls normally, then use the same connection to transmit high-speed data.

**circuit switched network**   A type of communications network that establishes a physical connection—a circuit—between sender and receiver for the duration of a call. Telephone systems are circuit switched networks.

**circuit switching**   A technique in which physical circuits, as opposed to virtual circuits, are transferred (switched) to complete connections. Sometimes called line switching.

**circular buffer**   A form of queue in which items are placed in successive locations in a store, and are taken from these locations in the same sequence. Two pointers keep track of the head and tail of the queue. When a pointer reaches the end of the available store it returns to the head. The items in the circular buffer may themselves be pointers to the items in the queue.

**circular polarization**   A mode of transmission in which signals are downlinked in a rotating corkscrew pattern. A satellite's transmission capacity can be doubled by using both right-hand and left-hand circular polarization.

**City Direct**   A switched voice communications service offered by British Telecom International for companies that want to communicate between London and New York.

**Cityruf**   A paging network in West Germany which enables page subscribers to be contacted in Germany, the United Kingdom (Europage), France (Alphapage), and Italy (Teldrin). Each network operates in the UHF 466 frequency band.

**City-wide Centrex**   A Centrex feature which uses software to provide a multilocation customer all the appearance and conveniences of a single-location customer.

**CIX**   Commercial Internet Exchange. A group formed to link commercial providers into an Internet subnet.

**CL**   Central Logic.

**cladding**   The internal layer within a fiber optic cable, outside the conducting core and inside the opaque sheath, which assists in guiding light waves down the length of the conductor.

**cladding mode**   Light contained within the cladding by the primary protective coating (buffer) placed on the fiber.

**clamping voltages**   The "sustained" voltage held by a clamp circuit at some desired level.

**ClariNews**   A fee-based Usenet newsfeed available from ClariNet Communications.

**Clark Belt**   A satellite orbit region approximately 22 300 miles above the earth's surface that enables satellites to be stationary relative to a position on earth. The region was named after the author, Arthur C. Clark, who first defined it in 1946.

**class**   Custom Local Area Signaling Services.

**class of office**   A classification scheme used to describe the relative position a telephone switching office occupies in the overall switching hierarchy. From the bottom of the hierarchy, they include Class 5 (end office); Class 4 (toll center); Class 3 (primary center); Class 2 (sectional center); and Class 1 (regional center).

**Class Of Service (COS)**   In SNA, a designation of the path control network characteristics, such as path security, transmission priority, and bandwidth, that apply to a particular session. The end user designates class of service at session initiation by using a symbolic name that is mapped into a list of virtual routes, any one of which can be selected for the session to provide the requested level of service.

**class X office**   Designation of a telephone switching facility in the overall telephone switching office hierarchy.

**CLB**   Circuit Line Block.

**CLEANUP**   In IBM's SNA, a network services request, sent by an SSCP to an LU, that causes a particular LU-LU session with that LU to be ended immediately without requiring the participation of either the other LU or its SSCP.

**clear channel**   A communications path with full bandwidth available to the user; control and other signals are typically transmitted on a separate channel.

**clear data**   Data that is not enciphered. Synonymous with plaintext.

**clear packet**   In the CCITT X.25 recommendation a packet which carries information required to terminate a previously established call.

**clear session**   A session in which only clear data is transmitted or received.

**clearinghouse**   1. In Digital Equipment Corporation Network Architecture (DECnet), a collection of directory replicas stored together in one location. 2. A collection of directories located on a node.

**Clearline**   The name used by Sprint Communications Company for a series of digital services to include fractional T1.

**Clear-To-Send (CTS)**   An RS-232 modem interface control signal (sent from the modem to the DTE on pin 5) which indicates that the attached DTE may begin transmitting; issued in response to the DTE's RTS. Called Ready-For-Sending in CCITT V.24.

**Clear-To-Send (CTS) Delay**   The time required by a data set to inform a terminal device that it is ready to send or reply to information just received. Also called modem turnaround delay.

**CLESN**   Continuing Legal Education Satellite Network.

**CLI**   1. Change Level Indicator. 2. Change Line Identification.

**client**   1. The user of a network service. 2. In Digital Equipment Corporation Network Architecture (DECnet), the user of the service provided by a module or layer in the architecture. 3. Workstation software which uses the services of any type of server on a local area network.

**Client/Server Computing**   Computing which requires a network linking at least two computers and the interaction between those computers which provides synergy.

**CLIST**   In IBM's SNA, Command List.

**CLK**   CLocK.

**CLNS**   ConnectionLess-mode Network Service.

**clock**   1. The timing signal used in synchronous transmission. 2. The source of such timing signals.

**clock, external**   A clock or timing signal from another device; a modem can provide an external clock (clocking).

**clock interrupt**   A type of interrupt which occurs at regular intervals and is used to initiate processes such as polling, which must happen regularly.

**clock recovery**   The extraction of the clock signal which may accompany data. This recovery occurs only when the signal is received on a synchronous channel.

**clock slip**   In T1 networking, the phase variation equal to a single T1 clock cycle (1/1.544 MHz) or 0.648 $\mu$s.

77

**clocking** Time synchronizing of communications information.

**closed architecture** An architecture that is compatible only with hardware and software from a single vendor.

**Closed User Group (CUG)** In public data networks, a selected collection of terminal users that do not accept calls from sources not in their group; also, often restricted from sending messages outside the group.

**Closed-Circuit Television (CCTV)** In LAN technology, one of the many services often found on broadband networks.

**closedown** In IBM's SNA, the deactivation of a device, program, or system.

**close-up** A remote connectivity program marketed by Norton Lambert of Santa Barbara, CA. The program is remarketed by several vendors to provide off-site technicians with the ability to take control of a remote personal computer that is normally connected to a local area network or mainframe computer.

**CLTA** Liaison committee for transatlantic telecommunications of CEPT.

**cluster** A collection of two or more terminals or other devices in a single location.

**Cluster Control Unit (CCU)** In IBM 3270 systems, a device that controls the input/output operations of a group (cluster) of display stations. Also called terminal control unit.

**cluster controller** In IBM's SNA, a device that can control the input/output operations of more than one device connected to it. A cluster controller may be controlled by a program stored and executed in the unit; for example, the IBM 3601 Finance Communication Controller. Or it may be controlled entirely by hardware; for example, the IBM 3272 Control Unit.

**cm** Call minutes.

**CM** Communications Module.

**CMA** Cellular Mobile Carrier.

**CMA** Communications Managers Association.

**CMC** In IBM's SNA, Communication Management Configuration.

**CMCS** COMSEC Material Control System.

**CMDSA** Communications Security Material Direct Support Activity.

**C-message noise** Filtered measurement of noise similar to the response of the human ear to various frequencies over a telephone circuit.

**Cmin** Call minute.

**CMIP** Common Management Information Protocol.

**CMOS** Complementary Metal Oxide Semiconductor.

**CMS** Conversational Monitor System.

**CMS router** A program running under IBM's VM/SP that uses the Server–Requester Programming Interface (SRPI) to route requests from the PC to the corresponding server on the host. The CMS router is part of the CMSSERV command processor in VM/SP Release 4.

**CMSS** Customer Message Service System.

**CMSSERV** 1. An IBM program that provides the Server–Requester Programming Interface (SRPI) and a server–requester manager on an IBM System/370 using VM/CMS. 2. The implementation of the enhanced connectivity facilities on a VM/SP system CMS installed.

**CMTT** Joint CCITT/CCIR study group on transmission of sound broadcasting and television signals over long distances.

**CMV** Joint CCITT/CCIR study group on vocabulary.

**C/N** Carrier to Noise Ratio.

**CNCE** Communications Nodal Planning Element.

**CNCE(M)** Communications Nodal Control Element—Management.

**CNCE(T)** Communications Nodal Control Element—Technical Control.

**CNCP** Canadian National/Canadian Pacific Telecommunications.

**CNG** CalliNG tone.

**CNM** Communications Network Management.

**CNMA** Communications Network for Manufacturing Applications.

**CNM/CS** Communications Network Management/Controller Support.

**C-notched noise** Measurement of metallic noise when a 1000 Hz tone is present through active components on a circuit. The tone is filtered or notched out during measurement.

**C-notched noise test** Used to determine the unwanted power present when a channel is carrying a normal signal. Thus, it actually measures the signal/noise ratio, which is most important with respect to data transmission. This test is a true measure of noise in circuits which have compandors.

**CO** Central Office.

**co- and contra-directional interface** The CCITT G.703 recommendation that specifies the relationship between the direction of the information and its associated timing signals. Co-directional specifies that the information and its timing are transmitted in the same direction while contra-directional specifies that the information and its timing are transmitted in the opposite directions from one another.

**COAX** Coaxial cable.

**coaxial cable** A transmission medium noted for its wide bandwidth and for its low susceptibility to interference; signals are transmitted inside a fully enclosed environment—an outer conductor or screen which surrounds the inner conductor. The conductors are commonly separated by a solid insulating material.

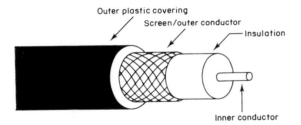

Outer plostic covering
Screen/outer conductor
Insulation
Inner conductor

**coaxial converter** In IBM 3270 systems, a protocol converter designed to be used between 3270 control units and attached to asynchronous devices; uses coaxial cable to connect to the control unit.

**COBOL** COmmon Business-Oriented Language.

**COCOM** CoOrdinating COMmittee.

**COCOT** Customer-Owned Coin-Operated Telephone.

**code** 1. A set of rules specifying the way in which data can be represented (e.g., the set of correspondences in the American Standard Code for Information Interchange) 2. In data communications, a system of rules and conventions specifying the way in which the signals representing data can be formed, transmitted, received, and processed. 3. The statements that make up a computer program.

**code conversion** The process of changing from one character coded representation into the corresponding character in a second code.

**Code Excited Linear Predictive Speech Processor (CELP)** An AT&T speech coder which digitizes voice at a rate of 8000 bits per second while maintaining the quality of current systems.

**code level** The number of bits used to represent characters.

**CODEC** An abbreviation of coder/decoder. The coder is used to encode Pulse Amplitude Modulation or PAM samples into PCM pulses while the decoder modifies the received PCM information and puts it in a form which can be understood by the receiver.

**Code-Division Multiple-Access (CDMA)** A spread-sprectrum communications technique which allows multiple users to transmit simultaneously in time within the same frequency band. Also referred to as spread-spectrum multiple-access (SSMA).

**coding scheme** Plan for changing human-usable language (letters and/or numbers) into computer-usable form.

**codress** A type of message in which the complete address is contained only in the encrypted text.

**COFA**   Change Of Frame Alignment.

**cohesion**   The cohesion of a connected network is the least number of lines which must be removed to separate the network into disconnected parts.

**coin first**   A type of coin station which requires the presence of a coin in order to provide an off-hook indication to the local office.

**COLAN**   Central Office Local Area Network.

**cold boot**   Reloading a computer's operating system, by turning the electricity to the computer off and back on.

**cold start**   A system restart that ignores previous data areas and accounting information in main storage and purges the queues upon starting.

**collect call**   A call in which the originating party requests that the receiving party accept responsibility for call payment.

**collect connection**   A procedure whereby the charge for a virtual connection is billed to the destination Data Terminal Equipment (DTE).

**collision**   1. In LAN technology, the result of two stations attempting to use a shared transmission medium simultaneously. 2. In a half-duplex system, the result of both ends trying to transmit at the same time.

**Collision Detection (CD)**   The ability of a transmitting node to detect simultaneous attempts on a shared medium.

**colocation**   1. The practice of installing another organization's equipment at the central office of a Bell operating company (BOC) or an interexchange carrier (IX). Equipment is normally colocated to permit the BOC or IX to provide maintenance for the organization. 2. Placement of serveral satellites near each other in orbit. This allows a single fixed antenna to receive signals from all of the satellites without tracking.

**COM**   Computer Output Microfilm.

**combined station**   In High-Level Data Link Control (HDLC) protocol, a station capable of assuming either the role of a primary or a secondary station (balanced station).

**COMBO**   A trademark of National Semiconductor as well as a chip from that vendor combines a CODEC and filtering functions.

**COMCODE**   AT&T ordering code.

**COMINT**   COMmand INTerpreter.

**COMINT**   COMmunications INTelligence.

**Comité Consultatif Internationale de Télégraphique et Téléphonique**   An international consultative committee that sets international communications recommendations, which are frequently adopted as standards; develops interface, modem, and data network recommendations. Membership includes PTTs, scientific and trade associations, and private companies. CCITT is part of the International Telecommunications Union (a United Nations treaty organization in Geneva) and is now known as the ITU. The following table is a selected list of CCITT recommendations. The reader is referred to the entries V-Series and X-Series for additional information concerning CCITT recommendations.

| NUMBER | GENERAL SUBJECT |
|---|---|
| | *V Series* |
| V.10 | Interchange circuits |
| V.11 | Interchange circuits |
| V.21 | 300 bps dial modem |
| V.22 | 1200 bps dial or 2-wire leased line modem |
| V.22 bis | 2400 bps dial 2-wire leased line modem |
| V.23 | 600/1200 bps dial modem |
| V.24 | 25-pin interface circuits |
| V.25 | Dial parallel interface |
| V.25 bis | Dial serial interface |
| V.26 | 2400/1200 bps leased line modem |
| V.26 bis | 2400/1200 bps dial modem |
| V.26 ter | 2400 bps dial or 2-wire leased line modem |
| V.27 | 4800 bps leased line modem/manual equalizer |
| V.27 bis | 4800 bps leased line modem/auto equalizer |
| V.27 ter | 4800 bps dial modem |
| V.28 | 25-pin interface circuits |
| V.29 | 9600 bps leased line modem |
| V.32 | 9600 bps dial or 2-wire leased line modem |
| V.35 | Wideband interface |

NUMBER  GENERAL SUBJECT

*X Series*

| | |
|---|---|
| X.3 | PAD/terminal interface |
| X.21 | DTE/DCE 15-pin interface |
| X.21 bis | DTE/DCE 25-pin interface |
| X.25 | Packet switching |
| X.28 | PAD/terminal interface |
| X.29 | PAD/PDN interchange |
| X.32 | Dial PDN access |
| X.121 | PDN numbering plan |

**comm**  Communication.

**command**  A code that will cause a device to perform an electronic or mechanical action.

**command file**  A file of commands that can be carried out by a computing device after initial setup.

**COMand INTerpreter (COMINT)**  Tandem's Guardian Command Interpreter program.

**Command LIST (CLIST)**  In IBM's SNA (NCCF), a sequential list of commands and control statements that is assigned a name. When the name is invoked (as a command) the commands in the list are executed.

**command port**  The console used to control and monitor a network or system; also, the interface to which the console is connected.

**command processor**  A software component that interprets control commands issued by an operator or user and invokes the requested function.

**Command Routing**  A communications carrier 800 service feature which allows customers to define preprogrammed rerouting for 800 number calls for use in emergency situations.

**command set**  Commands that control an intelligent modem. Most modems support the Hayes AT command set, but many high-speed modems have their own enhanced command sets.

**command state**  Operating mode wherein the modem will accept commands entered from a computer or terminal device. Most modems allow users to enter the command state while on-line without breaking the connection. In the AT command set, typing three "+" characters transfers the modem to command state.

**Command Terminal Protocol (CTERM)**  The sub-protocol component of the Digital Network Architecture (DNA) Network Virtual Terminal Protocol (NVT) which does the read and write mapping between a user terminal and the virtual terminal.

**COMMCEN**  COMMunications CENter.

**Commission of the European Community (CEC)**  The European Economic Community (EEC) administrative ministries which coordinate telecommunications developments.

**Common Applications Service Elements (CASE)**  The Manufacturing Automation Protocol (MAP) 2.X equivalent of Association Control Service Elements (ACSE). CASE was MAP 2.X specific and was not an International Standards Organization (ISO) standard.

**Common Battery Telephone System**  A telephone system in which power is supplied to each substation by a switchboard.

**Common Business-Oriented Language (COBOL)**  COBOL is a common procedural language designed for commercial processing as developed and defined by a national committee of computer manufacturers and users. It is a specific language by which business-oriented data processing procedures may be precisely described in a standard form. COBOL is intended to communicate data processing procedures to users as well as providing a means to write business-oriented programs for computers.

**common carrier**  A private data communications utility company or government organization that furnishes communications services to the general public and that is usually regulated by local, state, or federal agencies. Often, PTTs provide these services outside the USA; telcos provide them inside.

81

**common carrier principle** The regulatory concept limiting the number of companies that provide needed services in a specific geographic area.

**Common Channel Signaling (CCS)** Any form of signaling system where the signals related to a number of traffic circuits are sent via a separate common channel.

**common communications carrier (also common carrier)** A firm engaged in the business of supplying communications lines for general public utility.

**Common Control Switching Arrangement (CCSA)** Switching facilities located in a common carrier's central offices. These facilities are used for the interconnection of corporate foreign exchange networks.

**common line** A general term for describing the switched access service charges associated with the transport of calls between a customer location and the central office switch.

**common logic redundancy** See redundancy.

**Common Management Information Protocol (CMIP)** 1. An International Standards Organization network management protocol. 2. In Digital Equipment Corporation Network Architecture (DECnet), a management protocol which encompasses the Management Information Control and Exchange (MICE) and Management Event Notification (MEN) protocols.

**common user circuit** A circuit designated to furnish a communications service to a number of users.

**common-battery** A DC power source in the central office that supplies power to switching equipment and to subscribers.

**Common-Channel Interoffice Signaling (CCIS)** A technique by which the signaling information for a group of trunks is sent between switching offices over a separate voice channel. This technique uses time-division multiplexing.

**Common-Channel Signaling System (CCSS)** A system where all signaling for a group of circuits is carried over a common channel.

**Common-Channel Signaling System No.7** The current international standard for out-of-band call control signaling on digital networks.

**communication adapter** A hardware component inserted into a personal computer or a mini-computer to permit it to interface with a communication circuit.

**Communication Control Unit (CCU)** A non-programmed communications interface to a computer. This control unit will perform such functions as parity checking, block checking, and automatic answering of incoming calls.

**communication controller** IBM term for a front end processor.

**Communication IDentifier (CID)** In IBM's VTAM, a key for locating the control blocks that represent a session. The key is created during the session-establishment procedure and deleted when the session ends.

**communication line** Deprecated term for telecommunication line and transmission line.

**communication link** The software and hardware, to include cables, connectors, converters, etc., required for two devices such as a computer and terminal to communicate.

**communication macro instructions** In IBM's VTAM, the set of RPL-based macro instructions used to communicate during a session.

**Communication Management Configuration (CMC)** In IBM's SNA: 1. In VTAM, a technique for configuring a network that allows for the consolidation of many network management functions for the entire network in a single host processor. 2. A multiple-domain network configuration in which one of the hosts, called the communication management host, performs most of the controlling functions for the network, thus allowing the other hosts, called data hosts, to process applications. This is accomplished by configuring the network so that the communication management host owns most of the resources in the network that are not application programs. The resources that are not owned by the communication management host are the resouces that are channel-attached stations of data hosts.

**communication management host** In IBM's SNA, the host processor in a communication management configuration that does all network-control functions in the network except for the control of devices channel-attached to data hosts.

**Communication Network Management (CNM)** In IBM's SNA: 1. The process of designing, installing, operating, and managing the distribution of information and controls among end users of communication systems. 2. NCCF is an IBM program base that provides functions collectively called CNM. CNM programs include NLDM, NPDA, MVS/OCCF and VSE/OCCF.

**Communication Network Management (cnm) Application Program** In IBM's VTAM, an application program that is authorized to issue formatted management services request units containing physical-unit-related requests and to receive formatted management services request units containing information from physical units.

**Communication Network Management (CNM) Interface** In IBM's SNA, the interface that allows an application program to send Forward request/response units to an access method and to receive Deliver request/response units from an access method. These request/response units contain network services request/response units (data and commands).

**Communication Network Management (CNM) Processor** In IBM's SNA, a program that manages one of the functions of a communications system. A CNM processor is executed under control of NCCF and requires NCCF as a prerequisite program.

**Communication Scanner Processor (CSP)** In IBM's SNA, a processor in the 3725 communication controller that contains a microprocessor with control code. The code controls transmission of data over links attached to the CSP.

**communication server** An intelligent device normally located on the node of a local area network which performs communications functions.

**communications** The conveyance of information between two points, without alteration of the structure or content of the message.

**Communications Act of 1934** This Act, passed by Congress in 1934, established a national telecommunications goal of high quality, universally available telephone service at reasonable cost. The act also established the FCC and transferred federal regulation of all interstate and foreign wire and radio communications to this commission. It requires that prices and regulations for service be just, reasonable, and not unduly discriminatory.

**communications adapter** In IBM's Network Problem Determination Application (NPDA), refers to the hardware feature on the IBM 43X1 that performs some of the functions of other IBM communications controllers.

**communications common carrier** A government-regulated private company which provides the general public with telecommunications services and facilities.

**communications control character** A character used to control transmission over data networks. ASCII specifies 10 control characters that are used in byte-oriented protocols.

**Communications Control Unit (CCU)** In IBM 3270 systems, a communications computer, often a minicomputer, associated with a host mainframe computer. It may perform communications protocol control, message handling, code conversion, error control, and application functions.

**Communications Fraud Control Association (CFCA)** A nonprofit organization established as a national clearinghouse for telecommunications crime information.

**communications line control** Computer device which, with associated software, accepts data and performs necessary control functions.

**communications link** The physical medium connecting two systems.

**Communications Manager** Software which is part of IBM's OS/2 Extended Edition and which provides support for 3270 emulation using one of IBM's 3270 adapter cards, as well as IBM 3101 and DEC VT-100 emulation using asynchronous communications.

**Communications Module (CM)** A component of an AT&T 5ESS switch which acts as the focal point for voice, data, and message switching. The CM provides the communication paths between the administrative module (AM) and the switching module (SM).

**communications monitor** Software in mainframe that provides shared interface between applications programs and communications devices; includes communication access methods.

**communications network** The collection of transmission facilities, switches, and terminal devices that, when combined, provide the capability to communicate between two points.

**Communications Network for Manufacturing Applications (CNMA)** A consortium backed by European carmakers and aircraft and computer vendors to upgrade MAP functional standards and develop testing tools.

**Communications Network Management (CNM)** The process of designing, installing, operating, and managing the distribution of information and controls among end users of communications systems.

**Communications Network Management/Controller Support (CNM/CS)** A feature of the IBM 3600 and IBM 4700 systems.

**Communications Network Management (CNM) Interface** The interfaces provided to application programs by the access method for handling data and commands associated with communications systems management. CNM data and commands are handled across this interface.

**Communications Network Management (CNM) Processor** An IBM program that manages one of the functions of a communications system. A CNM processor is executed under control of IBM's Net-work Communications Control Facility (NCCF) and requires NCCF as a prerequisite program.

**communications processing** The function of operating upon the information or upon control characters that precede, accompany, or follow the information to ensure its successful entry, transmission, and delivery.

**communications protocol** The means used to control the orderly exchange of information between stations on a data link or on a data communications network or system. Also called line discipline, or protocol, for short.

**communications satellite** An orbiting microwave repeater that relays signals between communications stations. Typically geosynchronous.

**Communications Satellite Corp. (COMSAT)** Private U.S. satellite carrier, established by Congress in 1962, for the coordination and construction of satellite facilities used for international voice and data communications (U.S. counterpart of Intelsat).

**Communications Security (COMSEC)** The protection resulting from all measures designed to deny to unauthorized persons information of value that might be derived from a study of communications. Communications security includes transmission, cryptosecurity, and physical security.

**communications terminal** Any device which generates electrical or tone signals that can be transmitted over a communications channel.

**Communications-Electronics Installation (CEI)** The name given to a battalion of U.S. Army personnel with technicians to do on-site installation of communications-electronics equipment. The CEI battalion is under the control of the U.S. Army Information Systems Engineering Command.

**Communicator** A voice-mail system marketed by Voice Systems & Services, Inc., of Mannford, OK.

**Community Antenna Television (CATV)** A cable-based broadcast television distribution system typically covering an entire city or community based upon radio frequency (RF) transmission;

generally using 75-ohm coaxial cable as the transmission medium.

**Commute**   A trademark of Central Point Software of Beavertown, OR, as well as the name of a program which enables a remote personal computer to access a local area network.

**COMMZ**   COMMunications Zone.

**COMNET II.5**   A telecommunications analysis tool marketed by CACI of La Jolla, CA, which accepts voice or data network descriptions and provides such measures of network performance as circuit group utilization, packet delay statistics, and end-to-end blocking probabilities.

**Compact**   A voice-mail system marketed by Boston Technology, Inc., of Cambridge, MA.

**Compact XL**   A BT Tymnet packet switch which has a single data communications processor and can support a maximum of 144 asynchronous or 56 synchronous ports.

**compaction**   See compression.

**companding**   Compressing/expanding. The process in which an analog signal is reduced in bandwidth to permit it to be transmitted over a smaller bandwidth channel and then reconstructed (expanded) to its original form.

**compandor**   A compandor is a combination of a compressor at one point in a communications path for reducing the volume range of signals, followed by an expandor at another point for restoring the original volume range. Usually its purpose is to improve the ratio of the signal to the interference entering in the path between the compressor and expandor.

**Compania Anonima Nacional de Telefonos de Venezuela (CANTV)**   A telephone comany in Venezuela.

**compatibility**   The ability to receive and process programs and data from one computer system on another computer system without modification.

**compiler**   A computer program used to convert symbols meaningful to a human operator into codes meaningful to a computer.

**Complementary   Metal-Oxide   Semiconductor (CMOS)**   A circuit technology that uses n-channel and p-channel MOS transistors. N-channel and p-channel refer to the negatively charged electrons and the positively charged "holes" that make the circuit work. Such devices consume very little power.

**complementing music-on-hold**   A PABX feature in which callers waiting after an announcement hear either a quiet tone or music-on-hold until an agent is assigned to the call.

**Compmail+**   An electronic mail and bulletin board service operated by the Institute for Electrical and Electronic Engineers for IEEE members.

**component**   1. Hardware or software that is part of a functional unit. 2. A functional part of an operating system, for example, the scheduler or supervisor. 3. Any part of a local area network other than an attaching device, such as an IBM 8228 Multistation Access Unit.

**composite**   The line side signal of a concentrator or multiplexer that includes all the multiplexed data.

**composite link**   The line or circuit connecting a pair of multiplexers or concentrators; the circuit carrying multiplexed data.

**composite loopback**   A diagnostic test that forms the loop at the line side (output) of a multiplexer.

**Composite   Signaling   (CS)**   A direct current signaling system which requires a single line conductor for each signaling channel. The signaling system provides full-duplex operation.

**Comprehensive Surface Analysis (COMPSURF)**   A utility program included in some versions of Novell NetWare which formats and tests file server disk drives.

**compression**   Two types are available: data compression, which reduces the number of bits required to represent data (accomplished in many ways, including using special coding to represent strings of repeated characters or using fewer bits to represent the more frequently used characters); and analog compression, which reduces the bandwidth needed to transmit an analog signal. Also called compaction and companding.

**compressor** A device that performs analog compression.

**compromise equalizer** An equalizer set for best overall operation for a given range of line conditions; often fixed, but may be manually adjustable.

**COMPSURF** COMPrehensive SURFace analysis.

**CompuServe** A subsidiary of H&R Block which operates a packet switching network and an information utility which provides subscribers access to different data bases and electronic mail.

**CompuServe Mail Hub** A facility of CompuServe which enables users on a local area network operating Novell Netware MHS software to exchange electronic messages with other MHS users, CompuServe Mail subscribers and users of other E-mail services that can be reached via a CompuServe gateway.

**computer** An electronic system which, in accordance with its programming, will store and process information as well as perform high-speed mathematical or logical operations.

**Computer Aided Design (CAD)** Automated design operations. CAD allows users—via keyboard, light pen, and video display terminal—to manipulate design data in a computer's memory and generate varying video images or hard-copy printouts of an object.

**Computer Aided Manufacturing (CAM)** Automated manufacturing operations. CAM uses computers to control manufacturing processes.

**Computer Aided Software Engineering (CASE** A term used to denote the automation of various aspects of the software development process.

**computer branch exchange** A computer-controlled PABX system.

**computer conferencing** A visual form of conference telephone call.

**Computer Inquiry II** More formally known as the Second Computer Inquiry, FCC Docket Number 20808, the final decision in 1980 resulted in a policy that resulted in competition for and deregulation of the telecommunications industry in the United States.

**Computer Inquiry III** Computer Inquiry III removed the structural separation requirement between basic and enhanced services provided by Bell operating companies (BOCs) and AT&T. Adopted by the US Federal Communications Commission in May 1986.

**computer port** The physical location where a communications channel interfaces to a computer.

**Computer System for Mainframe Operations (COSMOS)** AT&T software which assists telephone company network administration and frame control centers in managing, controlling, and assigning terminations on the main distribution frame (MDF), and central office equipment, facilities and circuits.

**computer-controlled keys** In the PC 3270 emulation environment, the IBM PC keyboard keys whose function is determined by the computer program.

**Computerized Branch Exchange (CBX)** A PBX that uses a computer with an electronic switching network.

**Computer-to-PBX Interface (CPI)** A specification originated by Northern Telecom, for a T1 based data communications interface between a mainframe computer and a Private Branch Exchange. Incompatible with both ISDN and AT&T's DMI specifications.

**COMSAT** COMmunications SATellite Corp.

**COMSEC** Communications Security.

**Comsphere** A trademark of AT&T Paradyne as well as the name of products from that vendor which unify the previous separate modem products of Paradyne and AT&T.

**Comstar** A joint venture between GEC Plessey Telecommunications and the Moscow Telephone Network which is to provide international and local telephone service all over the former Soviet Union.

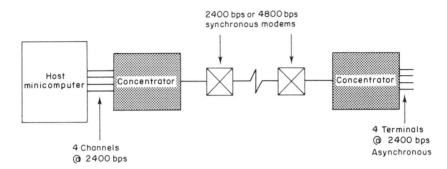

2400 bps or 4800 bps
synchronous modems

Host
minicomputer

Concentrator

Concentrator

4 Channels
@ 2400 bps

4 Terminals
@ 2400 bps
Asynchronous

Typical concentrator configuration

*concatenation*

**ComWatch** An off-site network monitoring service marketed by Timeplex, Inc., of Woodcliff Lake, NJ.

**concatenation** 1. The linking of transmission channels or subnetworks end to end. 2. The linking of blocks of user data or protocol transmissions. 3. In fiber optic technology, the interconnection of two or more fibers into one continuous length.

**concentration** Collection of data at an intermediate point from several low- and medium-speed lines for transmission across one high-speed line.

**concentrator** A device used to divide a data channel into two or more channels of average lower speed, dynamically allocating space according to the demand in order to maximize data throughput at all times. Also called an intelligent TDM, STDM, or statistical multiplexer.

**concurrent** Pertaining to the occurrence of two or more events or activities within the same specified interval of time.

**condensed print** A print width approximately 60 percent of the width of standard characters.

**Conditional End Bracket (CEB)** In IBM's SNA, the value (binary 1) of the conditional end bracket indicator in the request header (RH) of the last request of the last chain of a bracket; the value denotes the end of the bracket.

**conditioned circuit** A circuit that has been electrically altered to obtain the desired characteristics for voice and data transmission. The reader is referred to the entries C-1 through C-8 and D-1 and D-2 for specific information on C-level and D-level conditioning.

**conditioned loop** A loop that has conditioning equipment, usually equalizers, attached to obtain a desired line characteristic, to facilitate voice or data transmission.

**conditioning** The "tuning" or addition of equipment to improve the transmission characteristics or quality of a leased voice-grade line so that it meets specifications for data transmission. (See figure below.)

COMMON CARRIER CLASSIFICATION OF
LINE PARAMETERS

C conditioning applies only to frequency response and delay distortion characteristics.

Actual measurements are at 4 Hz higher, such as 1004, etc.

The + gain or − loss is with respect to a 1004 Hz reference

D1 Point to point channels

D2 Two or three point channels

All D conditioned channels must meet the following:

    1. Signal to C-notched noise = 28 dB

    2. Signal to second harmonic distortion = 35 dB

    3. Signal to third harmonic distortion = 40 dB

87

Bandwidth parameter limits

| Channel condition | Frequency response relative to 1004 Hz | | Envelope delay distortion | |
|---|---|---|---|---|
| | Frequency range | Variation in dB | Frequency range | Variation microseconds |
| Basic | 500-2500<br>300-3000 | -2 to +8<br>-3 to +12 | 800-2600 | 1750 |
| C1 | 1000-2400<br>300-2700<br>300-3000 | -1 to +3<br>-2 to +6<br>-3 to +12 | 1000-2400<br>800-2600 | 1000<br>1750 |
| C2 | 500-2800<br>300-3000 | -1 to +3<br>-2 to +6 | 1000-2600<br>600-2600<br>500-2800 | 500<br>1500<br>3000 |
| C4 | 500-3000<br>300-3200 | -2 to +3<br>-2 to +6 | 1000-2600<br>800-2800<br>600-3000<br>500-3000 | 300<br>500<br>1500<br>3000 |
| C5 | 500-2800<br>300-3000 | -0.5 to +1.5<br>-1 to +3 | 1000-2600<br>600-2600<br>500-2800 | 100<br>300<br>600 |

*conditioning*

**conductivity** The relative ability of a material to allow the flow or passage of an electrical current.

**conduit metal** Tubing through which telephone cables or other wiring are passed.

**Conference** An interactive discussion forum which enables users to post statements, comments, questions and opinions. Although a conference has a general topic user comments are not restricted to that topic.

**conference call** A telephone call among three or more locations.

**Conference Européene des Postes et Télécommunications (CEPT)** A regulatory body formed in 1959 which holds plenary sessions every two years. The resolutions concerning technical standards from these sessions apply only to its 26 nation members. Because the rollcall of CEPT comprises many telecommunication administrators from the CCITT, both organizations influence each other.

**configuration** An arrangement of parts to achieve some purpose; for example, network configuration or node configuration; hardware configuration—the equipment to be used and the way it is to be connected; software configuration—a procedure performed to prepare a software program for operation or define a workstation's resources to the file server.

**Configuration Control Program (CCP)** In IBM's SNA, an SSP interactive application program by which configuration definitions for the IBM 3710 Network Controller can be created, modified, and maintained.

**configuration file** A file that specifies the characteristics of a system or subsystem.

**Configuration Report Program (CRP)** In IBM's SNA, an SSP utility program that creates a configuration report listing network resources and resource attributes for networks with NCP, EP, PEP, or VTAM.

**configuration restart** In IBM's ACF/VTAM, the recovery facility that can be used after a failure or deactivation of a major node, ACF/VTAM, or the host processor to restore the domain to its status at the time of the failure or deactivation.

**configuration services** In IBM's SNA, one of the types of network services in the system services control point (SSCP) and in the physical unit (PU); configuration services activate, deactivate, and maintain the status of physical units, links, and link stations. Configuration services also shut down and restart network elements and modify path-control routing tables and address-transformation tables.

**configuration table** A table of program statements used to describe the topology, equipment characteristics, and operating parameters of a device in a network. The table enables the device to interface with other devices in the network.

**conformal array** An antenna that aligns to the surface of the structure carrying it. Conformal arrays are lighter and flatter than conventional arrays.

**conformance testing** The process of verifying a protocol implementation to check its conformance to standards. The objective of conformance testing is that the protocol implementation operates correctly in all valid, invalid and inopportune modes as specified by the standard.

**congestion** Congestion occurs when a network, or part of a network, is overloaded and has insufficient communications resources for the volume of traffic.

**congestion avoidance** A mechanism used to adjust the load on the network to prevent congestion.

**connect time** 1. A measure of system usage; the interval during which the user was on-line for a session. 2. The interval during which a request for a connection is being completed.

**connected** In IBM's ACF/VTAM, pertaining to a PU or LU that has an active physical path to the host processor containing the SSCP that controls the PU or LU.

**connecting arrangement** Interface equipment that is required by the common carrier when connecting customer-provided equipment to the public network.

**connecting block** Flame retardant plastic blocks containing quick clips for terminating cable and wire without the removal of insulation from the conductor. In addition, they provide an electrically tight connection between the cable and the cross-connect wire.

**connection** 1. An established data communications path. 2. The process of establishing that path. 3. A point of attachment for that path.

**Connection Control** In Digital Equipment Corporation Network Architecture (DECnet), the Session Control function concerned with the system-dependent functions related to creating, maintaining, and destroying Transport Connections.

**connection menu** In an IBM 3270 network environment, a menu on the screen of a display station attached to the 3174 Subsystem Control Unit, from which a user can select an available host.

**connection point manager** In IBM's SNA, a component of the transmission control layer that: 1. performs session-level pacing of normal-flow requests, 2. checks sequence numbers of received request units, 3. verifies that request units do not exceed the maximum permissible size, 4. routes incoming request units to their destinations within the half-session, and 5. enciphers and deciphers FMD request units when cryptography is selected. The connection point manager coordinates the normal and expedited flows for one half-session. *Note*. The sending connection point manager within a half-session builds the request/response header (RH) for the outgoing request/response units, and the receiving connection point manager interprets the request/response headers that precede incoming request/response units.

**connectionless delivery service** The routing of messages from one computer to another based upon address information carried in the message.

**connectionless service** Transport of a single quantum of information not set up by a signaling or administrative procedure (i.e., a packetgram).

A characteristic of the packet delivery service offered by token-ring local area network hardware and by the Internet TCP/IP protocol. A connectionless service treats each packet or datagram as a separate entity that contains the source and destination address.

**Connectionless-Mode Network Service** In Digital Equipment Corporation Network Architecture (DECnet), a network service which operates according to a datagram model. Each message is routed and delivered to its destination independently of any other. The Network Layer of DNA provides this type of service.

**Connection-Mode Network Service (CONS)** In Digital Equipment Corporation Network Architecture (DECnet), a network service which operates according to a connection-oriented model. Before data can be exchanged, a connection must first be established.

**connectivity** The connections of communications lines in a network.

**connector** A physical interface, such as an RJ plug/jack or DP25P/S plug/socket.

**ConnNet** The name of Southern New England Telephone's packet network.

**CONS** COnnection-mode Network Service.

**console** The device used by the operator, system manager, or maintenance technician to monitor or control computer, system, or network performance.

**constant** A value that does not change.

**constant carrier** A type of modem that supports only full-duplex handshaking. CD is always on.

**Consultative Committee on International Telephony and Telegraphy (CCITT)** An international consultative committee that sets international communications recommendations, which are frequently adopted as standards; develops interface, modem, and data network recommendations such as V.22, V.27, and V.25. Membership includes PTTs, scientific and trade associations, and private companies. CCITT is part of the International Telecommunications Union (a United Nations treaty organization in Geneva) and was renamed the ITU although still commonly referred to as the CCITT. The CCITT meets every four years (last in 1992) to settle, revise, and publish current standards. Each meeting's version of the complete set of standards is published as a multi-volume document identified by color (e.g. the 1980 CCITT Yellow Book; the 1984 CCITT Red Book).

**contact bounce** An imperfection in relay switching resulting from the physical properties of relay contacts which may initially move apart when closed prior to forming a complete closure. A coating of mercury can be used as a conductive coating that will bridge the contacts together during the bouncing period.

**contended access** In LAN technology, a shared access method that allows stations to use the medium on a first-come, first-served basis.

**contending port** A programmable port type which can initiate a connection only to a preprogrammed port or group of ports.

**contention** The facility provided by the dial network or a data PABX which allows multiple terminals to compete on a first-come, first-served basis for a smaller number of computer ports.

**contention delay** Time spent waiting to use a facility due to sharing with other users.

**contention group** A number of ports functioning as a group. Port connection request are addressed to the group, and the call is routed to any available port within the contention group.

**contiguous ports** Ports occurring in unbroken numeric sequence.

**Continue—Any Mode** In IBM's ACF/VTAM, a state into which a session is placed that allows its input to satisfy a Receive request issued in any-mode. While this state exists, input on the session can also satisfy Receive requests issued in specific-mode.

**Continue—Specific Mode** In IBM's ACF/VTAM, a state into which a session is placed that allows its input to satisfy only Receive requests issued in specific-mode.

90

**Continuing Legal Education Satellite Network (CLESN)** A network used to provide on-site legal training via satellite-based business television services.

**continuity of traffic** In the U.S. military, the means, by use of station serial numbers and channel numbers, that insure that the receiving communications center receives all messages transmitted by the connected communications center.

**Continuous ARQ** A protocol in which error checking and correction procedures consist of the receiving station notifying the sending station that blocks or frames of data were received in error. Continuous ARQ allows the sending station to send more than one block or frame before stopping to wait for acknowledgment.

**continuous form** Paper that is supplied to the printer as a continuous sheet, generally with perforations for separating.

**continuous paper** Paper that has pin-fed holes on each side and which is perforated between pages. Also called fan-fold paper.

**Continuous Variable-Slope Delta Modulation (CVSD)** A method of digitizing analog voice signals that uses 16 000 to 64 000 bps bandwidth depending on the sampling rate. CVSD reduces the bandwidth needed for a voice channel from 64 Kbps to 32 Kbps or less.

**Continuous Waves (CW)** Radio waves having a constant amplitude and a constant frequency.

**continuously variable** Capable of having one of an infinite number of values, differing from each other by an arbitrary small amount; usually used to describe analog signals or analog transmission.

**Continuously Variable Slope Delta (CVSD)** A method of converting an analog (voice) signal into a digital stream by sampling the signal and encoding the delta or change that has occurred since the previous sample.

**control block** In the IBM Token-Ring Network, a specifically formatted block of information provided from the application program to the Adapter Support Interface to request an operation.

**control center** A site from which a network administrator can exercise distributed systems administration and control functions.

**control character** A nonprinting character used to start, stop, or modify a function; the carriage return (CR) is an example of a control character.

**control codes** Special nonprinting codes which cause the terminal or computer to perform specific electronic or mechanical actions (such as setting tabs, etc.).

**control console** A computer device used for control of system operation.

**Control Data Communications Control Procedure (CDCCP)** Control Data Corporation's bit-oriented protocol.

**control disk** In an IBM 3270 network environment, a customized diskette or fixed disk containing the microcode that describes a particular control unit's attached terminals, and its method of attachment to the host.

**control line** A line in an interface which is used for signaling between devices but is not used for exchanging data or clock signals.

**control line timing** Clock signals between a modem and communications control unit.

**control mode** The state that all terminals on a line must be in to allow line control actions, or terminal selection to occur. When all terminals on a line are in the control mode, characters on the line are viewed as control characters performing line discipline, e.g. polling or addressing.

**Control Panel Device (Cdev)** A file on a Macintosh computer whose contents control such parameters as speaker volume, cursor blink rate, system date and time and desktop color.

**Control Program (CP)** In an IBM environment, the Control Program in a Virtual Memory system manages the real resources of the computer to include memory and direct access storage devices.

**Control Program for Microcomputers (CP/M)** A once popular, 8 bit, microcomputer operating system developed by Digital Research in the mid 1970s.

**control programs** Control programs contain many routines that would otherwise have to be put into each individual program. Such routines include those for handling error conditions, interruptions from the console, or interruptions from a communications terminal.

**control signal** An interface signal used to announce, start, stop, or modify a function; for example, CD is an RS-232 control signal that announces the presence of a carrier.

EXAMPLE OF RS-232 CONTROL SIGNALS

| Pin | Control signal | From | To |
|-----|----------------|------|-----|
| 4 | Request-To-Send (RTS) | DTE | DCE |
| 5 | Clear-To-Send (CTS) | DCE | DTE |
| 6 | Data Set Ready (DSR) | DCE | DTE |
| 8 | Carrier Detect (CD) | DCE | DTE |
| 20 | Data Terminal Ready (DTR) | DTE | DCE |
| 22 | Ring Indicator (RI) | DCE | DTE |

**control statement** In IBM's SNA (NCCF), a statement in a command list that controls the processing sequence of the command list or allows the command list to send messages to the operator and receive input from the operator.

**control station** A station on a network which superses network control procedures such as polling, selecting, and recovery. The control station is also responsible for establishing order on the line in the event of contention, or any other abnormal situation that may arise between stations on the network. A control station manages the flow of data on a multipoint line.

**Control Switching Points (CSP)** The regional, sectional, and primary switching centers for telephone dialing.

**control unit** In an IBM host system, equipment coordinating the operation of an input/output device and the CPU.

**control unit terminal (CUT)** In an IBM 3270 network environment, a terminal that relies on a control unit to interpret the data stream. Examples are the IBM 3178, 3179, 3278, and 3279.

**control unit terminal (CUT) mode** In an IBM 3270 network environment, a host-interactive mode that enables an IBM 3270 Personal Computer customized in this mode to run only one session emulating a 3178, 3179, 3278 Model 2, or 3279 Model S2A terminal.

**controlled** A terminal and communications system design in which a central host or switching system exerts control over terminals to prevent them from transmitting when the host or switch is unable to service the traffic.

**controlled carrier** A feature of a modem which allows the modem carrier signal to be turned on or off under command of the DTE. Controlled carrier is necessary on multipoint lines.

**controlled SLIP** The overflow or underflow of a T1 Frame buffer typically caused by frequency departures, channel banks or digital switches that are not properly loop timed.

**controller** A unit that controls input/output operations for one or more devices, such as terminals and printers. Also known as a control unit.

**controlling application program** In IBM's ACF/VTAM, an application program with which a logical unit (other than a secondary application program) is automatically put in session whenever the logical unit is available.

**controlling link** In a local area network, the reporting link between a bridge and a network manager program that is authorized to change bridge configuration parameters and to disable and enable certain bridge functions.

**controlling logical unit** In IBM's VTAM, a logical unit with which a secondary logical unit (other than an application program) is automatically put in session whenever the secondary logical unit is available. A controlling logical unit can be either an application program or a device-type logical unit.

**CONUS** CONtinental United States.

**conventional memory** A term used to refer to random access memory up to 1024 Kbytes (1 Mbyte) that are accessable by an IBM PC or compatible computer using DOS.

**Conversant** A registered trademark of AT&T used for a series of voice reponse systems. Callers interact with the system by pressing keys on a touch-tone phone or by speaking words or numbers.

**conversational mode** A mode of communications between terminals. In this mode, an entry by one terminal elicits a reply by the other terminal.

**Conversational Monitor System (CMS)** CMS is an IBM operating system that is used to manage a virtual machine. It is a component of VM/SP and enables application programs to be run interactively from a terminal.

**conversion** 1. In programming languages, the transformation between values that represent the same data item but belong to different data types. Information may be lost as a result of conversion because accuracy of data representation varies among different data types. 2. The process of changing from one method of data processing to another or from one data processing system to another. 3. The process of changing from one form of representation to another, for example, to change from decimal representation to binary representation.

**converted command** In IBM's SNA, an intermediate form of a character-coded command produced by ACF/VTAM through use of an unformatted system services definition table. The format of a converted commmand is fixed; the unformatted system services definition table must be constructed in such a manner that the character-coded command (as entered by a logical unit) is converted into the predefined, converted command format.

**COP** Character-Oriented Protocol.

**copper wire** Literally, the wire formed from copper which carries the electrical signals which comprise telephone transmission. Frequently used as slang for coaxial cable.

**core** 1. The memory elements in older computer systems where bits were stored as magnetic "charges" in tiny ferrous beads. Today, some people still refer to semiconductor memories as "core". 2. The region of a fiber optic waveguide through which light is transmitted—typically 8 to 12 micrometres in diameter for (single-mode fiber), and from 50 to 200 micrometres for (multimode fiber).

**coresident** Pertaining to the condition in which two or more modules are located in main storage at the same time.

**Cornet** An ISD networking upgrade for the Siemens SATURN family of digital voice/data business communications systems.

**Corporation for Open Systems (COS)** An American consortium of companies, founded in 1986, dedicated to the development of worldwide standards through interoperability and conformance testing of OSI products.

**COS** 1. Call Originate Status. 2. Class Of Service. 3. Corporation for Open Systems.

**COSAC** COmmunications SAns Connections.

**COSIR** Cite Our Service In Reply.

**COSMOS** Computer System for Mainframe Operations.

**COSPAS-1** A Russian satellite which is part of a satellite program called COSPAS-SARSAT Spaced-Based Global Search and Rescue System. The COSPAS-1 was launched on June 30, 1982.

**COST** In Digital Equipment Corporation Network Architecture (DECnet), a metric used by the routing algorithm. Each link is assigned a cost, and the routing algorithm selects paths with minimum cost.

**COSTAB** Class Of Service TABle (VTAM).

**COT** Central Office Terminal.

**cotton bond** A type of paper where a certain percentage of cotton fibers are included in the manufacturing process.

**counterpoise**  A conductor or system of conductors used as a substitute for ground in an antenna system.

**country code**  Second set of digits to place an international call following the international access code. Country codes are listed in the following table. (See table below.)

**coupler**  1. A device which interconnects a circuit to a terminal. 2. In fiber optic technology, a passive optical device used for splitting and/or combining the light at two or more ports.

**coupling**  The association of two circuits in such a way that energy may be transferred from one to the other.

| COUNTRY | COUNTRY CODE | COUNTRY | COUNTRY CODE | COUNTRY | COUNTRY CODE |
|---|---|---|---|---|---|
| Afghanistan | 93 | Costa Rica | 506 | Iran | 98 |
| Albania | 355 | Croatia | 385 | Iraq | 964 |
| Algeria | 213 | Cyprus | 357 | Ireland | 353 |
| American Samoa | 684 | Czech Republic | 42 | Israel | 972 |
| Andorra | 33 | Denmark | 45 | Italy | 39 |
| Angola | 244 | (including Faroe Islands) | | Ivory Coast | 225 |
| Anguilla | (809) | Dominica | (809) | Jamaica | (809) |
| Antigua (including Barbuda) | (809) | Dominican Republic | (809) | Japan | 81 |
| Argentina | 54 | Ecuador | 593 | Jordan | 962 |
| Aruba | 297 | Egypt | 20 | Kenya | 254 |
| Ascension Island | 247 | El Salvador | 503 | Korea | 82 |
| Australia | 61 | Ethiopia | 251 | Kuwait | 965 |
| (including Tasmania) | | Fiji | 679 | Latvia | 371 |
| Austria | 43 | Finland | 358 | Lebanon | 961 |
| Bahamas | (809) | France | 33 | Lesotho | 266 |
| Bahrain | 973 | French Antilles | 596 | Liberia | 231 |
| Bangladesh | 880 | (including Martinique, | | Libya | 218 |
| Barbados | (809) | St. Barthelemy, and | | Liechtenstein | 41 |
| Belgium | 32 | St. Martin) | | Lithuania | 370 |
| Belize | 501 | French Guiana | 594 | Luxembourg | 352 |
| Benin | 229 | French Polynesia | 689 | Macao | 853 |
| Bermuda | (809) | Gabon | 241 | Malawi | 265 |
| Bolivia | 591 | Gambia | 220 | Malaysia | 60 |
| Bosnia–Hercogovinia | 387 | Germany | 49 | Mali | 223 |
| Botswana | 276 | Ghana | 233 | Malta | 356 |
| Brazil | 55 | Gibraltar | 350 | Marshall Islands | 692 |
| British Virgin Islands | (809) | Greece | 30 | Mexico | 52 |
| Brunei | 673 | Greenland | 299 | Micronesia | 691 |
| Bulgaria | 359 | Grenada (including Carriacou) | (809) | Monaco | 33 |
| Burma | 95 | Guadeloupe | 590 | Montserrat | (809) |
| Burundi | 257 | Guam | 671 | Morocco | 212 |
| Cambodia | 855 | Guantanamo Bay | 53 | Mozambique | 258 |
| Cameroon | 237 | Guatemala | 502 | Namibia | 264 |
| Canada | 1 | Guyana | 592 | Nepal | 977 |
| Cayman Islands | (809) | Haiti | 509 | Netherlands | 31 |
| Central African Republic | 236 | Honduras | 504 | Netherlands Antilles | 599 |
| Chad | 235 | Hong Kong | 852 | Nevis | (809) |
| Chile | 56 | Hungary | 36 | New Caledonia | 687 |
| China | 86 | Iceland | 354 | New Zealand | 64 |
| Colombia | 57 | India | 91 | Nicaragua | 505 |
| Congo | 242 | Indonesia | 62 | | |

*continued*

| COUNTRY | COUNTRY CODE | COUNTRY | COUNTRY CODE | COUNTRY | COUNTRY CODE |
|---|---|---|---|---|---|
| Niger | 227 | Serbia | 381 | Tanzania | 255 |
| Nigeria | 234 | Sierra Leone | 232 | Thailand | 66 |
| Norway | 47 | Singapore | 65 | Togo | 228 |
| Oman | 968 | Slovakia | 42 | Trinidad and Tobago | (809) |
| Pakistan | 92 | Slovenia | 386 | Tunisia | 216 |
| Panama | 507 | Somalia | 252 | Turkey | 90 |
| Papua New Guinea | 675 | South Africa | 27 | Turks Islands and | |
| Paraguay | 595 | Spain (including | 34 | Caicos Islands | (809) |
| Peru | 51 | Balearic Islands, | | Uganda | 256 |
| Philippines | 63 | Canary Islands, | | United Arab Emirates | 971 |
| Poland | 48 | Ceuta, and Melilla) | | United Kingdom | 44 |
| Portugal (including | 351 | Sri Lanka | 94 | (England, N.Ireland, | |
| Madeira Islands) | | St. Kitts | (809) | Scotland, Wales) | |
| Qatar | 974 | St. Lucia | (809) | Uruguay | 598 |
| Romania | 40 | St. Pierre/Miquelon | 508 | USA | 1 |
| Russian Federation | 7 | St. Vincent and | | Vatican City | 39 |
| Rwanda | 250 | the Grenadines | (809) | Venezuela | 58 |
| Saipan (including Rota | 670 | Surinam | 597 | Vietnam | 84 |
| and Tinian) | | Swaziland | 268 | Yemen Arab Republic | 967 |
| San Marino | 39 | Sweden | 46 | Zaire | 243 |
| Saudi Arabia | 966 | Switzerland | 41 | Zambia | 260 |
| Senegal | 221 | Taiwan | 886 | Zimbabwe | 263 |
| | | | | ( )indicates area code. | |

**coupling loss** The price in power a light pulse pays when it jumps between two optical devices such as a laser and a fiber lightguide.

**Courier** A name used by U.S. Robotics of Skokie, IL, for a series of modems marketed by that vendor.

**Courier HST** A trademark of U.S. Robotics of Skokie, IL, as well as an asymmetrical modem that can operate at data rates up to 28,400 bps on the switched telephone network.

**COVIA** The name of the airline reservation system owned by UAL Corp, USAir Group Inc. and a consortium of four foreign airlines.

**CP** 1. Command Port. 2. Control Program. 3. Copy protection.

**CPE** Customer Premises Equipment.

**CPH** Characters Per Hour.

**CPI** Computer to PBX Interface.

**CPM** 1.Characters Per Minute. 2. Control Program for Microcomputers.

**CP/M** Control Program for Microcomputers.

**CPMS** Central Processing Unit Master Switching Subsystem.

**CPNI** Customer Proprietary Network Information.

**CPS** 1. Characters Per Second. 2. Customer Premises System.

**CPU** Central Processing Unit.

**CPUO** Call PickUp Originating.

**CPUT** Call PickUp Terminating.

**CQMS** Circuit Quality Monitoring System.

**CR** 1. Carriage Return. 2. Command Reject.

**Cracker** A person who breaks or attempts to break the security of a computer system.

**crash or crashed**   A term which means hardware or software has stopped functioning properly.

**CRC**   Cyclic Redundancy Check.

**CRC-6**   The North American standard for Cyclic Redundancy Checking on 1.544 Mbps circuits that use the Extended Superframe Format. The 6-bit CRC value, calculated on the previous Superframe, is transmitted in the Framing Bits of the 4th, 8th, 12th, 16th, 20th, and 24th frames of an Extended Superframe. The CRC-6 is used as a measure of signal quality over time, not as a check on the validity of individual Superframes.

**CRC-8**   The European standard for Cyclic Redundancy Checking on 2.048 Mbps circuits. The CRC-8 value, calculated on the previous Multiframe, is transmitted in the Framing Bits of the 1st, 3rd, 5th, 7th, 9th, 11th, 13th, and 15th frames of the 16-frame Multiframe. The CRC-8 is used as a measure of signal quality over time, not as a check on the validity of individual Multiframes.

**CRE**   Call Reference Equivalent.

**credit**   In Digital Equipment Corporation Network Architecture (DECnet), a flow control mechanism whereby the receiver of data tells the transmitter how many messages it is prepared to receive at a given point.

**credit window**   In Digital Equipment Corporation Network Architecture (DECnet), the credit window identifies the range of message numbers which the receiver is prepared to receive at a given point.

**criss-cross directory**   A telephone directory, usually published by Coles or Polk, which contains a section which lists each street address in a city and the name and telephone number of persons living there. The directory also contains a section that lists all telephone numbers in sequence and the person who is listed for each number.

**critical angle**   The minimum angle at which a light wave striking the cladding in an optical fiber will be reflected back into the core.

**CRL**   Circuit Routing List.

**Cross Box Inventory System**   A telephone cable verification system marketed by Lee Data Corporation.

**cross connection**   The wire connections running between terminals on the two sides of a distribution frame, or between binding posts in a terminal.

**cross keys**   Synonym for cross-domain keys.

**cross modulation**   A type of crosstalk in which the carrier frequency being received is interfered with by an adjacent carrier, so that the modulated signals of both are heard at the same time.

**crossbar**   A circuit switch having several horizontal and vertical paths that are electromagnetically or electronically interconnected.

**cross bar switch**   An electomechanical switch used in older telephone Central Offices for switching and connecting circuits.

**cross bar system**   A type of line switching which uses crossbar switches.

**cross-connect**   1. A piece of hardware used to interconnect multiplexers with line terminating equipment and other multiplexers. 2. Distribution system equipment where communication circuits are administered (that is, added or rearranged using jumper wires or patch cords). In a wire cross-connect, jumper wires or patch cords are used to make circuit connections. In an optical cross-connect, fiber patch cords are used. The cross-connect is located in an equipment room, backbone closet, or satellite closet.

**cross-connect field**   Wire terminations grouped to provide cross-connect capability. The groups are identified by color-coded sections of satellite closets, or by designation strips placed on the wiring block or unit. The color coding identifies the type of circuit that terminates at the field.

**cross-domain**   A networking term and in IBM's SNA, pertaining to control of resources involving more than one domain.

**cross-domain keys**   In IBM's SNA, a pair of cryptographic keys used by a system services control point (SSCP) to encipher the session cryptography key that is sent to another SSCP and to decipher the session cryptography key that is received from the other SSCP during initiation of cross-domain LU–LU sessions that use session-level cryptography. Synonymous with cross keys.

**cross-domain link** In IBM's SNA: 1. A subarea link connecting two subareas that are in different domains. 2. A link physically connecting two domains.

**cross-domain LU–LU session** In IBM's SNA, a session between logical units (LUs) in different domains.

**Cross-Domain Resource (CDRSC)** In IBM's SNA, a resource owned by a cross-domain resource manager (CDRM) in another domain but known by the CDRM in this domain by network name and associated cross-domain resource manager.

**Cross-Domain Resource Manager (CDRM)** In IBM's VTAM, the function in the system services control point (SSCP) that controls initiation and termination of cross-domain session.

**crossed pinning** Configuration that allows two DTE devices or two DCE devices to communicate. Also called cross-over cable and null-modem cable.

**cross-modulation** Interference caused by two or more carriers in a transmission system that interact through non-linearities in the system.

**cross-network** In IBM's SNA, pertaining to control or resources involving more than one SNA network.

**cross-network LU–LU session** In IBM's SNA, a session between logical units (LUs) in different networks.

**cross-network session** In IBM's SNA, a LU–LU or SSCP–SSCP session whose path traverses more than one SNA network.

**cross-over cable** A cable which permits a terminal to be directly connected to a computer port (DTE to DTE) or a tail circuit modem to a modem port (DCE to DCE). Also called null-modem cable.

**cross-subarea** In IBM's SNA, pertaining to control or resources involving more than one subarea node.

**cross-subarea link** In IBM's SNA, a link between two adjacent subarea nodes.

**crosstalk** The unwanted transfer of a signal from one circuit, called the disturbing circuit, to another, called the disturbed circuit. 2. A communications program for use on IBM PCs and compatible computers from the Crosstalk Communications Division of Digital Communications Associates, Inc. of Roswell, GA. The program includes numerous file transfer protocols and allows up to 15 concurrent communications sessions.

**crosstalk, far-end** Crosstalk which travels along the disturbed circuit in the same direction as the signals in that circuit. To determine the far-end crosstalk between two pairs, 1 and 2, signals are transmitted on pair 1 at station A, and the level of crosstalk is measured on pair 2 at station B.

**crosstalk, near-end** Crosstalk which is propagated in a distributed channel in the direction opposite to the direction of propagation of the current in the distributing channel. Ordinarily, the terminal of the disturbed channel at which the near-end crosstalk is present is near or coincides with the energized terminal of the disturbing channel.

**CRP** In IBM's SNA, Configuration Report Program.

**CRQ** Call ReQuest.

**CRT** Cathode Ray Tube.

**CRV** In IBM's SNA, CRyptography Verification.

**CRYPTO** CRYPTOgraphy or CRYPTOgraphic.

**cryptocenter** An establishment maintained for the encrypting and decrypting of messages.

**cryptographic** Pertaining to the transformation of data to conceal its meaning.

**cryptographic algorithm** A set of rules that specify the mathematical steps required to encipher and decipher data.

**cryptographic key** In systems using the Data Encryption Standard (DES), a 64-bit value (containing 56 independent bits and 8 parity bits) provided as input to the algorithm in determining the output of the algorithm.

**cryptographic session** In IBM's SNA products, a LU–LU session in which a function management data (FMD) request may be enciphered before it is transmitted and deciphered after it is received.

**cryptographic session key** In IBM's SNA, deprecated term for session cryptography key.

**cryptography** The process of hiding information from unauthorized disclosure by conversion through a secret algorithm or key.

**Cryptography Verification Request (CRV)** In IBM's SNA, a request unit sent by the primary logical unit (PLU) to the secondary logical unit (SLU) as part of cryptographic session establishment, to allow the SLU to verify that the PLU is using the correct cryptographic session key.

**cryptosecurity** Communications security which deals with the proper use of authorized codes, cipher devices, and machines used for encrypting and decrypting messages.

**crystal** A natural substance, such as quartz or tournaline, that is used to control the frequency of radio transmitters.

**CS** Composite Signaling.

**CSA** 1. Carrier Serving Area. 2. Communications System Agency.

**CSCE** Communications System Control Element.

**CSDC** Circuit Switched Digital Capability.

**CSIRONET** Commonwealth Scientific & Industrial Research Organization Network.

**CSMA** Carrier Sense Multiple Access.

**CSMA/CA** Carrier Sense Multiple Access with Collision Avoidance.

**CSMA/CD** Carrier Sense Multiple Access with Collision Detection.

**CSM–ASU** Centralized Systems Management–Application Start-Up.

**CSNET** Computer Science NETwork.

**CSP** 1. Communication Scanner Processor (IBM's SNA). 2. Control Switching Points.

**CSPE** Communications System Planning Element.

**CSR** Customer Station Rearrangement.

**CSSR** Communications System Segment Replacement.

**CSU** 1. Channel Service Unit. 2 Circuit Switching Unit.

**CTAK** Cipher Text Auto Key.

**CTDB** Terminal Descriptor Block (Common).

**CTIA** Cellular Telecommunications Industry Association.

**CTIX/386** A trademark of Convergent, Inc. of San Jose, CA, as well as that firm's implementation of AT&T's UNIX System V Release 3.0 operating system.

**ctl** Control.

**CT1** The term used to describe residential cordless telephones which use radio waves to transmit voice signals to base stations which are plugged into household telephone sockets.

**CTS** Clear To Send.

**CTS flow control** A procedure for a communicating device to signal its readiness to receive data by raising the CTS lead on an EIA RS-232D interface.

**CTTU** Central Trunk Test Unit.

**C-TYPE CONDITIONING** Conditioning used to control attenuation distortion and envelope delay distortion.

**CT2** A mobile telephone service based upon the use of cordless phones. CT2 users placing calls must be within range of a local transceiver; however, they cannot receive calls. The CT2 system in the United Kingdon is called Telepoint.

**CT3** A proposed digital European cordless telephone standard.

**CU** Control Unit.

**CUD** Call User Data.

**CUG** Closed User Group.

**current**   The amount of electrical charge flowing past a specified circuit point per unit time, measured in amperes.

**current loop**   1. (Single-current signaling, used in USA) Method of interconnecting Teletype terminals and transmitting signals that represent a mark by current on the line and a space by the absence of current. 2. (Double-current signaling, used everywhere else) A mark is represented by current in one direction and a space by current in the other direction.

**cursor**   A movable underline, rectangular-shaped block of light, or an alternating block of reversed video on the screen of a display device, usually indicating where the next character will be entered.

**cursor under cron**   In IBM's PROFS cursor under cron is the capability to place the cursor under the document number and view a document that is referenced in a note or another document.

**custom chip**   A microchip designed for a specific job or function; e.g. the echo canceler chip.

**Custom Local Area Signaling Services (CLASS)**   Number translation services provided by a communications carrier based upon the availability of common channel interoffice signaling. Examples of CLASS include call-forwarding and caller identification.

**customer**   The person, firm, corporation, or other entity that orders services or products and is responsible for payment of charges and for compliance with carrier tariff regulations. Except for duly authorized and regulated common carriers, no one may be a customer who does not have a communications requirement of his own for service. Also referred to as a subscriber.

**Customer Controlled Reconfiguration/Rerouting (CCR)**   A feature offered on AT&T T1 circuits that allows customers to change the channel assignments (e.g. destinations of individual channels) on a T1 circuit by placing a data call to the central office on a separate line.

**Customer Formatted Reports (CFR)**   A term which references the software capability of a PBX or other communications equipment that enables the customer to tailor reports to meet their specific needs.

**Customer Information Control System (CICS)**   IBM communications monitor program product that provides an interface between the operating system access method and applications programs to allow remote or local display terminal interaction with a data base in the central processor.

**Customer Information Control System/Virtual Storage (CICS/VS)**   IBM Host application program/Operating System.

**Customer Premises Equipment (CPE)**   Devices at the customer location for interfacing between public transmission facilities and other equipment such as telephones, terminals, multiplexers, and computers. Devices classified at CPE can be supplied by third party vendors as well as the telephone companies.

**Customer Proprietary Network Information (CPNI)**   Information to include a company's calling patterns, billing, network design and use of network services telephone companies keep in their existing customer data bases.

**customer retrials**   The number of times a telephone user will retry a call after finding a busy signal.

**customer station rearrangement**   A feature that allows business customers to change, display, and verify data affecting their telephone service from an on-premises terminal.

**Customer-Provided Equipment (CPE)**   Any device connected to a common carrier facility which is not provided by the common carrier.

**customization**   In an IBM 3270 network environment, procedures that tailor the control unit microcode to fit the various types of diaplay stations and printers and the method of host attachment that a particular control unit will handle.

**customize** To describe (to the system) the devices, programs, users, and user options for a particular data processing system.

**CUT** Control Unit Terminal.

**cut down** A method of securing a wire to a wiring terminal. The insulated wire is placed in the terminal groove and pushed down with a special tool. As the wire is seated the terminal cuts through the insulation to make an electrical connection, and the tool's spring-loaded blade trims the wire flush with the terminal. Also called "punch down."

**cut sheet** Separate sheets used for printing.

**cut sheet feeder** A device which attaches to certain printers and which automatically feeds single sheets of paper into the printer.

**cutoff wavelength** The wavelength on a single-mode fibre above which only one mode can propagate.

**cutset** A minimal set of elements of a connected graph (nodes, lines or both) which, when removed from the graph, disconnects it. It must be a minimal set, which means that any proper subset of the cutset does not disconnect the graph. The graph is "disconnected" when it is separated into at least two parts without connecting links between them. Cutsets are used to measure cohesion and connectivity.

**CVSD** Continuous Variable Slope Delta Modulation.

**CV-425/U** A U.S. military telegraph–telephone signal converter. The CV-425/U converts 20 hertz ringing signals to a higher audio frequency for transmission over circuits that will not pass 20 hertz.

**CV-1548/G** A U.S. military telephone signal converter which provides 20 Hz to 1600 Hz and 1600 Hz to 20 Hz conversion for two-way transmission over 12 voice-frequency channels.

**CW** 1. Call Waiting. 2. Continuous Wave.

**CWC** City-Wide Centrex.

**CXR carrier** A communications signal used to indicate the intention to transmit data on a line. Also called Carrier Detect.

**Cybernet** Network of Control Data Corp.

**Cyberspace** A term used to refer to electronic communications.

**cycle** 1. A complete sequence of a wave pattern that recurs at regular intervals. 2. One iteration or loop through a set of logical points.

**Cycles Per Second (CPS)** Measure of frequency.

**cyclic access to store** Access to a store (for reading or writing) in which successive addresses are accessed in turn returning eventually to the first address.

**Cyclic Redundancy Check (CRC)** An error detection scheme in which the block check character is the remainder after dividing all the serialized bits in a transmission block by a predetermined binary number—or a polynomial based on the transmitted data.

**C&D** Cause and Diagnostic (codes).

**C-1 conditioning** AT&T conditioning for two-point or multipoint channels. The attenuation distortion and envelope delay distortion parameters for C-1 conditioning are:

*Attenuation distortion*

| Frequency range (Hz) | Variation (2B) |
|---|---|
| 1004–2404 | −1 to +3 |
| 304–2704 | −2 to +6 |
| 2704–3004 | −1 to +12 |

*Envelope delay distortion*

| Frequency range (Hz) | Variation (ms) |
|---|---|
| 1004–2404 | 1000 |
| 804–2604 | 1750 |

**C-2 conditioning** AT&T conditioning for two-point or mutipoint channels. The attenuation distortion and envelope delay distortion parameters for C-2 conditioning are:

*Attenuation distortion*

| Frequency range (Hz) | Variation (2B) |
|---|---|
| 504–2804 | −1 to +3 |
| 304–3004 | −2 to +6 |

*Envelope*

| Frequency range (Hz) | Variation (ms) |
|---|---|
| 1004–2604 | 500 |
| 604–2604 | 1500 |
| 504–2804 | 3000 |

**C3I** Command, Control, Communications and Intelligence.

**C-3 conditioning** AT&T conditioning for switched network access lines and switched network trunks. The attenuation distortion and envelope delay distortion parameters for switched network interoffice access lines are:

*Attenuation distortion*

| Frequency range (Hz) | Variation (2B) |
|---|---|
| 504–2804 | −5 to +1.5 |
| 304–3004 | −0.8 to +3 |

*Envelope delay distortion*

| Frequency range (Hz) | Variation (ms) |
|---|---|
| 1004–2604 | 110 |
| 604–2604 | 300 |
| 504–2804 | 650 |

The attenuation distortion and envelope delay distortion parameters for switched network trunks are:

*Attenuation distortion*

| Frequency range (Hz) | Variation (2B) |
|---|---|
| 504–2804 | −0.5 to +1 |
| 304–3004 | −0.8 to +2 |

*Envelope delay distortion*

| Frequency range (Hz) | Variation (ms) |
|---|---|
| 1004–2604 | 80 |
| 604–2604 | 260 |
| 504–2804 | 500 |

**C-4 conditioning** AT&T conditioning for two-point or multipoint channels having four or fewer

customer's premises. The attenuation distortion and envelope delay distortion parameters for C-4 conditioning are:

*Attenuation distortion*

| Frequency range (Hz) | Variation (2B) |
|---|---|
| 504–3004 | −2 to +3 |
| 304–3204 | −2 to +6 |

*Envelope delay distortion*

| Frequency range (Hz) | Variation (ms) |
|---|---|
| 1004–2604 | 300 |
| 804–2804 | 500 |
| 604–3004 | 1500 |
| 504–3004 | 3000 |

**C-5 conditioning** AT&T conditioning available for two-point channels where both customer premises are located in the mainland. The attenuation distortion and envelope delay distortion parameters for C-5 conditioning are:

*Attenuation distortion*

| Frequency range (Hz) | Variation (2B) |
|---|---|
| 504–2804 | −0.5 to +1.5 |
| 304–3004 | −1 to +3 |

*Envelope delay distortion*

| Frequency range (Hz) | Variation (ms) |
|---|---|
| 1004–2604 | 100 |
| 604–2604 | 300 |
| 504–2804 | 600 |

**C-7 conditioning** AT&T conditioning available for two-point channels where both customer's premises are located in the mainland. The attenuation distortion and envelope delay distortion parameters for C-7 conditioning are:

*Attenuation distortion*

| Frequency range (Hz) | Variation (2B) |
|---|---|
| 404–2804 | −1 to +4.5 |

*Envelope delay distortion*

| Frequency range (Hz) | Variation (ms) |
|---|---|
| 1004–2604 | 550 |

**C-8 conditioning** AT&T conditioning available

for two-point channels where both customer's premises are located in the mainland. The attenuation distortion and envelope delay distortion parameters for C-7 conditioning are:

*Attenuation distortion*

| Frequency range (Hz) | Variation (2B) |
| --- | --- |
| 404–2804 | −1 to +3.0 |

*Envelope delay distortion*

| Frequency range (Hz) | Variation (ms) |
| --- | --- |
| 1004–2604 | 125 |

# D

**D** Display.

**DA** Data Available.

**DAA** Data Access Arrangement.

**DAB** Demand Assigned Bus.

**D/A conversion** Digital-to-Analog (D/A) conversion.

**DACCS** Department of the Army Command and Control.

**DACOM MHS** An electronic messaging system operated by DACOM in Korea.

**DACS** 1. Data Acquisition and Control System. 2. Digital Access Cross-connect System.

**DACS/CCR** Digital Access Cross-connect System/ Customer Controlled Rerouting.

**DAF** 1. Dedicated Access Facility. 2. Direct Access Facility.

**Daini-Denden** A Tokyo based alternative long distance communications carrier.

**daisy chaining** The physical connection of cables that allows the cascading of signals between devices.

**DAL** 1. Data Access Language. 2. Data Access Line.

**DAMA** Demand Assignment Multiple Access.

**DAMES** Demonstration of Avionics Module Exchangeability via Simulation.

**DAP** Data Access Protocol.

**dark fiber** Fiber for which the customer and not the carrier has control of both ends with respect to transmission format and optics.

**DARPA** Defense Advanced Research Projects Agency.

**DART** A file transfer protocol available to users of Crosstalk MK.4 and Crosstalk for Windows communications programs.

**DAS** Digit AnalysiS.

**DAS/C** Directory Assistance System/Computer.

**DASD** Data Access Storage Device.

**DASS** 1. Demand Assignment Signaling and Switching. 2. Digital Access Signaling System.

**data** Any material which is represented in a formalized manner so that it can be stored, manipulated, and transmitted by machine.

**Data Access Arrangement (DAA)** DCE furnished or approved by a common carrier that permits privately owned DCE or DTE to be attached to the common carrier's network; all modems now built for the public telephone network have integral DAAs.

**Data Access Language (DAL)** Software marketed by Apple Computer, Inc., which lets users of networked Macintosh desktop systems obtain a

103

standard method to access host data without requiring a terminal emulator.

**Data Access Protocol (DAP)**  A Digital Network Architecture (DNA) protocol which permits remote file access and file transfers.

**data acquisition**  A system for measurement and recording of data from physical entities and devices.

**Data Acquisition and Control System (DACS)**  System designed to handle a variety of real-time applications, process control, and high-speed data acquisition. Each system is individually tailored with modularity allowing satisfaction of specific system requirements.

**data bank**  A centralized collection of computer information usually accessible by dial-up.

**data base**  A collection of data stored electronically in a predefined format and according to an established set of rules, often called a schema.

**data blocks**  Logical units of information that are being sent over computer I/O channels.

**Data Call Plus**  A service mark of Telenet, now part of Sprint, which enables users to transmit data anywhere in the United States for a low hourly rate.

**data call queueing standby**  A PABX feature that places a data caller into a "standby" queue for shared facilities, such as computer ports and modem pools.

**Data Carrier Detect (DCD)**  A signal sent from a dataset which informs the terminal that a carrier waveform is being received. The signal may also be called Carrier Detected, Carrier Found, Carrier On, etc.

**data channel**  The communication path along which data can be transmitted.

**data character**  A character containing alphanumeric information mainly used in a communications link between terminals and/or computers.

**data circuit**  Communication facility permitting transmission of information in digital or analog form.

**Data Circuit-terminating Equipment (DCE)**  Equipment that provides the functions required to establish, maintain, and terminate a connection, the signal conversion, and coding for communications between data terminal equipment (DTE). Also called data communications equipment.

**data class of service**  A PABX feature that allows the communications manager to customize service by controlling access to data facilities, devices, and features.

**data collection**  Procedure in which data from various sources is accumulated at one location (in a file or queue) before being processed.

**data communication service**  A specified user information transfer capability provided by a data communication system to two or more end users.

**data communication session**  A coordinated sequence of user and system activities whose purpose is to cause digital user information present at one or more source users to be transported and delivered to one or more destination users. A normal data communications session between a user pair comprises: (1) an access function, (2) a user information transfer function, (3) a disengagement function for each user. A data communication session is formally defined by a data communication session profile.

**data communication session profile**  The exact sequence of user/system interface signals by which data communication service is provided in a typical (successful) instance. A complete data communication session profile should also include any possible blocking (service refusal) sequences for a particular set of users.

**data communication subsystem**  A group of data communication system elements terminated within the end user interfaces.

**data communication system**  A collection of transmission facilities and associated switches, data terminals, and protocols that provide data communication service between two or more end users. The data communication system includes all functional and physical elements that participate in transferring information between end users. The system element that interfaces with the end user is a data terminal or a computer operating system. A computer operating system normally serves as the first point of contact for application programs requiring data communication service.

**data communication user**   Either 1. an end user of a data communication system, or 2. an aggregate user of a data communication subsystem.

**data communications**   The processes, equipment, and/or facilities used to transport signals from one data processing device at one location to another data processing device at another location. 2. In IBM's SNA, the transmission and reception of data.

**Data Communications Equipment, or Data Circuit-Terminating Equipment (DCE)**   In common usage, synonymous with modem; the equipment that provides the functions required to establish, maintain, and terminate a connection as well as the signal conversion required for communications between the DTE and the telephone line or data circuit.

**data compression**   A method to reduce the amount of transmitted data by applying an algorithm to the basic data at the point of transmission. A decompression algorithm expands the data back at the receiving end into its original format.

**data dictionary**   A list of all the data names and data elements in a system.

**data dump mode**   A feature incorporated into some printers to facilitate the isolation of communication problems between a computer and the printer. When placed in the data dump mode the printer prints the exact codes received from the computer, ignoring any printer codes that may be embedded in the data.

**data encrypting key**   A cryptographic key used to encipher and decipher data transmitted in a cryptographic session.

**Data Encryption Algorithm—1 (DEA—1)**   The first international standard for encryption, more commonly known as the Data Encryption Standard (DES).

**Data Encryption Standard (DES)**   Cryptographic algorithm endorsed by the National Bureau of Standards (NBS), now the National Institute of Science and Technology (NIST), to encrypt data using a 56-bit key. Specified in the Federal Information Processing Standard Publication (FIPS PUB) 46 dated January 15, 1977.

**data entry**   Introducing data into a data processing or information processing system for input.

**Data Flow Control (DFC)**   In IBM's SNA, a request/response unit (RU) category used for requests and responses exchanged between the data flow control layer in one half-session and the data flow control layer in the session partner.

**Data Flow Control (DFC) Layer**   In IBM's SNA, the layer within a half-session that (1) controls whether the half-session can send, receive, or concurrently send and receive request units (RUs); (2) groups related RUs into RU chains; (3) delimits transactions via the bracket protocol; (4) controls the interlocking of requests and responses in accordance with control modes specified at session activation; (5) generates sequence numbers; and 6. correlates requests and responses.

**data flow control protocol**   In IBM's SNA, the sequencing rules for requests and responses by which network addressable units in a session coordinate and control data transfer and other operations.

**data host**   In IBM's SNA communication management configuration, a host that is dedicated to processing applications and does not control network resources, except for its channel-attached or communication adapter-attached devices.

**data hot line**   A feature on a PABX that allows the user to automatically place a data call to a particular terminal.

**data integrity**   A measure of data communications

performance, indicating a sparsity (or, ideally, the absence) of undetected errors.

**data integrity point** The generic name given to the point in an IBM 3800 model 3 printing process at which the data is known to be secure. Also called the stacker.

**data least cost routing** A PABX feature that automatically selects the most economical route (WATS, FX, tie-line, switched network) for each outgoing call.

**data line group** A feature on a PABX that allows a single number to find a free port in a group of numbers.

**data line interface** The point where a data line is connected to a telephone line or communications circuit.

**data link** A serial communications path between nodes or devices without any intermediate switching nodes.

**Data Link Connection Identifier (DLCI)** A 10-bit address in the frame relay header which identifies a virtual circuit on the frame relay link.

**data link connector** A type of single fiber optical connector used to connect fiber cable to a host computer or other equipment.

**data link control** The combination of software and hardware that manages the transmission of data over the communications line.

**Data Link Control (DLC) Layer** In IBM's SNA, the layer that consists of the link stations that schedule data transfer over a link between two nodes and perform error control for the link. Examples of data link control are SDLC for serial-by-bit link connection and data link control for the IBM System/370 channel.

**data link control protocol** A set of rules used by two nodes on a data link to accomplish an orderly exchange of information. Synonymous with line control.

**data link layer** Second layer in OSI model; takes data from the network layer and passes it on to the physical layer; responsible for transmission

and reception of packets, datagram service, local addressing, and error detection (but not error correction).

**data medium** Either 1. the material in which or on which a specific physical variable may represent data or 2. the physical quantity that may be varied to represent data.

**data message detail recording** A PABX feature which provides a record of all external and internal data calls. Information recorded can include the calling party's extension number, the destination number, time of call, and call duration.

**data mode** A condition of a modem or a DSU with respect to the transmitter in which its Data Set Ready and Request to Send circuits are on and it is ready to send data.

**data network** A system consisting of a number of terminal points that are able to access one another through a series of communication lines and switching arrangements.

**Data Network Identification Code (DNIC)** A four-digit number assigned to public data networks as well as specific services on some of those networks. The first three digits denote the country while the fourth digit denotes the network number within the country. The first digit is determined from the following table:

0 reserved
1 reserved
2 used for network identification codes
3 used for network identification codes
4 used for network identification codes
5 used for network identification codes
6 used for network identification codes
7 used for network identification codes
8 used for interworking with telex networks
9 used for interworking with telephone networks

The DNIC is a subset of a 14-digit, global address defined in X.121 as illustrated below:

| F/S | B1 channel | D | B2 channel |
|-----|-----------|---|-----------|

ISDN basic access channel format

The table below lists current worldwide DNICs.

## DATA NETWORK IDENTIFICATION CODE (DNIC)

| Country | Network | DNIC |
|---|---|---|
| Australia | Austpac | 5052 |
| Australia | Midas | 5053 |
| Austria | Radio Austria | 2329 |
| Belgium | DCS | 2062 |
| Brazil | Interdata | 7240 |
| Canada | Datapac | 3020 |
| Canada | Globedat | 3025 |
| Canada | Infoswitch | 3029 |
| Denmark | Datapak | 2382 |
| Eire | Eirpac | 2724 |
| Finland | Finnpak | 2442 |
| France | NTI | 2081 |
| France | Transpac | 2080 |
| French Polynesia | Tompac | 5470 |
| Gabon | Gabopac | 6282 |
| Germany (FDR) | Datex P | 2624 |
| Great Britain | IPSS | 2341 |
| Great Britain | PS | S2342 |
| Greece | Helpac | 2022 |
| Guadeloupe | Dompac | 3400 |
| Hong Kong | Intelpac | 4542 |
| Hong Kong | Datapac | 4545 |
| Israel | Isranet | 4251 |
| Italy | Itapac | 2222 |
| Ivory Coast | Sytranpac | 6122 |
| Japan | DDX P | 4401 |
| Japan | Venus P | 4408 |
| Luxembourg | Luxpac | 2704 |
| Martinique | Dompac | 3400 |
| Netherlands | DABAS | 2044 |
| Netherlands | Datanet 1 | 2041 |
| New Zealand | Pacnet | 5301 |
| Norway | Datapak | 2422 |
| Singapore | Telepac | 5252 |
| South Africa | Saponet | 6550 |
| Spain | TIDA | 2141 |
| Spain | Iberpac | 2145 |
| Sweden | Datapak | 2402 |
| Switzerland | Datalink | 2289 |
| Switzerland | Telepac | 2284 |
| USA | Autonet | 3126 |
| USA | Compuserve | 3132 |
| USA | DBS (WUI) | 3104 |
| USA | Datapak | 3119 |
| USA | LSDS (RCA) | 3113 |
| USA | Marknet | 3136 |
| USA | Telenet | 3110 |
| USA | Tymnet | 3106 |
| USA | UDTS (ITT) | 3103 |
| USA | Uninet | 3125 |
| USA | Wutco | 3101 |

**data networking**  A capability that will allow users to combine separate data bases, telecommunication systems, and specialized computer operations into a single integrated system, so that data communication can be handled as easily as voice messages.

**Data Over Voice (DOV)**  Technology used to transmit data and voice simultaneously over twisted-pair copper wire. Primarily used with local Centrex services or special customer premises PBXs. An FDM technique which combines data and voice on the same line by assigning a portion of the unused bandwidth to the data; usually implemented on the twisted pair cables used for in-house telephone system wiring. (See illustration on the following page.)

**Data Packets Per Second (DPPS)**  The rate at which data packets are processed through a communications processor.

**Data PBX**  A switch that allows a user's circuit to select other circuits for the purpose of establishing a connection. Only digital transmission, and not analog voice is switched.

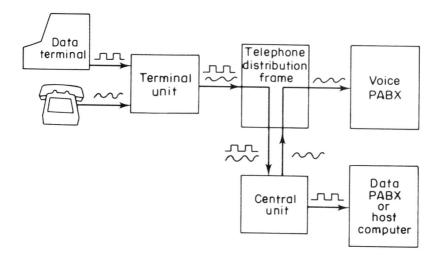

Typical data-over-voice configuration

*Data Over Voice (DOV)*

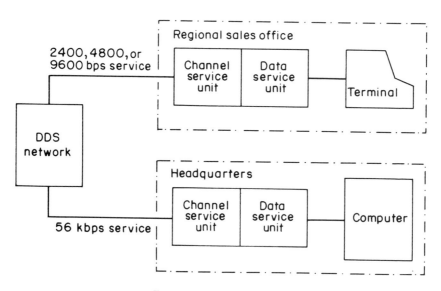

Typical DDS connection

*Data Service Unit (DSU)*

**data processing (DP)**   The systematic performance of operations upon data; such as sorting and computing.

**Data Qualifier (DQ) Packet**   Defined in X.25 for signaling overhead information across an X.25 virtual circuit. In Telenet, the DQ bit is set to 1 to signal that an X.29 Packet Assembler/Disassembler (PAD) message is contained in the data field of the X.25 data packet. High-Level Data Link Control (HDLC) and Synchronous Data Link Control (SDLC) also use the DQ packet.

**Data Quality Monitor (DQM)**   A device used to measure data bias distortion above or below a threshold.

**Data Race**   A company located in San Antonio, TX, which specializes in the manufacture of high-speed modems designed for use on the public switched telephone network.

**data rate, data signaling rate**   A measure of how quickly data is transmitted, expressed in bps. Also commonly, but often incorrectly, expressed in baud. Synonymous with speed.

**data restraint**   A synonym for flow control.

**Data Service Unit (DSU)**   A device used to provide a digital data services interface. Located on the user's premises, the DSU interfaces directly with the data terminal equipment. The DSU provides loop equalization, remote and local testing capabilities, and the logic and timing necessary to provide a standard EIA/CCITT interface. May have an integrated Channel Service Unit (CSU).

**data services command processor (DSCP)**   An IBM Network Communications Control Facility (NCCF) component that structures the request for recording and retrieving data in the application program's data base, and also structures the request to solicit data from a network component. IBM's Network Problem Determination Application (NPDA) supplies a DSCP for managing problem determination data.

**Data Services Manager (DSM)**   A function in IBM's Network Communications Control Facility (NCCF) that provides Virtual Storage Access Method (VSAM) services for data storage and retrieval and provides the interface between Data Services Command Processors (DSCPs) and the Communications Network Management (CNM) interface.

**data set**   A synonym for modem (coined by AT&T); a software term for a certain type of data file.

**data set separator pages**   Those pages of printed output that delimit data sets.

**Data Set Ready (DSR)**   An RS-232 modem interface control signal (sent from the modem to the DTE on pin 6) which indicates that the modem is connected to the telephone circuit. Usually a prerequisite to the DTE issuing RTS.

**data signalling rate**   Used in communications to define the rate at which signal elements are transmitted or received over a transmission path by data terminal equipment. The data signalling rate is expressed in bits per second (bps) and baud.

**data sink**   A device that can accept data signals from a transmission device.

**data stream**   The collection of characters and data bits transmitted through a channel.

**data stream format**   In SNA, the format of the data elements (end user data) in the request unit (RU).

**data structure**   A system of relationships between items of data. To express these relationships when a data structure is stored, lists or other systems using pointers etc. may be used.

**data switch**   A system that connects network lines to a specific input/output port of a computer or front end processor (FEP).

**Data Switching Exchange (DSE)**   A local switching station for shared trunk lines in a switched telecommunications network.

**data terminal**   A device associated with a computer system for data input and output that may be at a location remote from the computer system, thus requiring data transmission.

**Data Terminal Equipment (DTE)** In data communications notation, a user device such as a terminal or computer, connected to a circuit.

**Data Terminal Ready (DTR)** An RS-232 modem interface control signal (sent from the DTE to the modem on pin 20) which indicates that the DTE is ready for data transmission and which requests that the modem be connected to the telephone circuit.

**data traffic reset state** In IBM's VTAM, the state usually entered after Bind Session, if Cryptography Verification is used, and after Clear, but prior to Start Data Traffic. While a session is in this state, requests and responses for data and data flow control cannot be sent. Only certain session control requests can be sent.

**data transfer mode** In packet switching, data packets carrying customer data are exchanged in this virtual circuit mode. A virtual circuit is in data transfer mode after a call is successfully established. It remains in data transfer mode until it is cleared or until an escape to command mode is accomplished.

**data transfer rate** The average number of bits, characters, or blocks per unit of time transferred from a data source to a data link.

**data transmission** The technology of transmitting and receiving information over communication channels.

**data type** A category that identifies the internal representation of data.

**data types** In IBM's Network Problem Determination Application (NPDA), a concept to describe the organization of data displays. Data types are defined as alerts, events, and statistics. Data types are combined with resource types and display types to describe NPDA display organization.

**Data Under Voice (DUV)** A technique used for combining data and voice transmission which exploits the unused part of the transmission channel's bandwidth for data transmission. On microwave links, the low frequency bands are used for data while the higher frequency bands are used for voice. Hence, data is "under" voice.

**Data User Part (DUP)** In CCITT Signaling System 7, the level 4 protocols for use in data switching and transmission.

**database server** A combination of hardware and software which functions as a hub on a local area network. The database server is used to store the shared network data base files as well as manage the access and use of those files for network users.

**datacommonality** A term used by General DataComm, Inc., to describe a unique packaging technique which provides the following benefits: (1) high density modular packaging, (2) a broad array of versatile data sets and accessories, (3) system flexibility and ease of expansion, (4) low power consumption and heat dissipation, (5) quick and simple installation, (6) at-a-glance monitoring of system operation, (7) convenient, low-cost maintenance and (8) high reliability.

**datagram** In a local area network a particular type of information encapsulation at the Network layer of the adapter protocol for NETBIOS. No explicit acknowledgment for the information is sent by the receiver. Instead, transmission relies on the "best effort" of the link layer. Rarely, if ever, implemented on current PDNs.

**Datakit** An AT&T virtual circuit switch used by telephone companies to support such data services as central office local networks.

**Datakit II** A more flexible version of AT&T's Datakit virtual circuit switch. The Datakit II supports Ethernet and Starlan local network bridging as well as a variety of mainframe computer interfaces and synchronous data transmission.

**DataLAN** A local area network operating system developed by Datapoint Corporation. DataLAN incorporates the Xerox Networking Standard and

Net-BIOS, enabling any DOS-compatible personal computer to be connected to that network.

**Dataline** The name used by British Telecom to reference a fixed link to its packet siwtching exchange from a customer's premises.

**DATALYNX/3174** A protocol converter manufactured by Local Data Corporation (now Andrew Corporation) that provides as many as 32 serial ports allowing ASCII terminals, PCs, minicomputers, microcomputers, and printers, to emulate IBM full screen 327x, 318x, 319x terminal, and 328x printer devices.

**DATALYNX/3274** A protocol converter manufactured by Local Data Corporation (now Andrew Corporation) that provides up to nine serial ports, allowing asynchronous terminals, PCs, minicomputers, microcomputers, printers, etc. to emulate IBM 3278 models 1–5 terminals and 327x printer devices.

**Datamizer** A family of data compression products manufactured by Symplex Communications Corporation of Ann Arbor, MI.

**DATAPAC** A public data network using packet switching techniques operated by Telecom Canada.

**Dataphone** Both a service mark and a trademark of AT&T. As a service mark, it indicates the transmission of data over the telephone network. As a trademark, it identifies the communications equipment furnished by AT&T for data communication service. The generic name given to the modems marketed by AT&T.

**Dataphone Digital Service (DDS)** AT&T's private line service, filed in 1974, for transmitting data over a digital system. The digital transmission system transmits electrical signals directly, instead of translating the signals into tones of varied frequencies as with the traditional analog transmission system. The digital technique provides more efficient use of transmission facilities, resulting in lower error rates and costs than analog systems. In 1987, AT&T introduced switched 56 Kbps Dataphone Digital Service.

**Dataphone Digital Service with Secondary Channel** A private line service offered by AT&T and several Bell Operating Companies (BOCs) that allows 64Kbps clear channel data with a secondary channel that provides end-to-end supervisory, diagnostic and control functions. Also referred to as DDS II.

**Dataphone II** The generic name given to the diagnostic modems marketed by AT&T.

**DATAROUTE** A digital data transmission service offered by Telecom Canada.

**DATAS II** The computer reservation system operated by Delta Airlines.

**Dataspeed** An AT&T marketing term for a family of medium-speed paper tape transmitting and receiving units.

**DataVoice** An adapter card designed for use in an IBM PC or compatible personal computer from the French company SCII (Services et Conseils en Informatique et Ingenierie) that can be configured either for simultaneous B-channel transmission of voice and data at 64 Kbps each, or for multiplexed D-channel transmission of up to eight terminals at 16 Kbps, using the ISDN S protocol.

**Datel** A British Telecom service which provides a means for transmitting data over the public telephone network in the United Kingdom and from the United Kingdom to 70 countries worldwide.

**Date-Time Group (DTG)** In U.S. military communications, the date, time, month, and year a message is prepared for transmission. The date-time group is normally assigned by either the drafter or the releasing authority, as determined by local command. *Example*: 250800Z FEB 89.

**Datex-P** The packet switching network operated by Deutsche Bundespost in Germany.

**DA-15** A 15-pin D-type physical connector used in T1 multiplexing equipment as an alternative to WECO 310 or Bantam Connector.

**dB, db** Decibel.

**Dba (dBrn adjusted)**   A unit of noise measurement for a circuit having F1A-line or HA-1-receiver weighting. 0 dBa is noise power of 3.16 picowatts (-85 dBm), equivalent to 0 dBrn ($-90$ dBm or 1 picowatt with C-message weighting) adjusted to F1A weighting.

**DBa (F1A)**   The noise level measured on a line by a noise measuring set having F1A-line weighting.

**DBa (HA-1)**   The noise level measured across the receiver of a 302-type or similar telephone handset measured by a test set having HA-1-receiver weighting.

**DBAAU**   Dial Backup Auto Answer Unit.

**DBaO**   The power level of noise expected at a particular location in a circuit, called "dBrn adjusted reference level."

**dBI**   The ratio of the gain of an actual antenna to an isotropic radiator using decibels.

**D-BIT**   The delivery confirmation bit in a X.25 packet that is used to indicate whether or not the DTE wishes to receive an end-to-end acknowledgment of delivery.

**dBm**   A decibel referenced to 1 milliwatt. Absolute measure of signal power where 0 dBm is equal to one milliwatt. In the telephone industry, dBm is based on 600 ohms impedance and 1000 hertz frequency. 0 dBm is 1 milliwatt at 1000 Hz terminated by 600 ohms impedance.

**dBm0**   The abbreviation commonly used to indicate the signal magnitude in dBm at the 0TLP. The Transmission Level Point (TLP) is the ratio (in dB) of the power of a signal at any particular point to the power of the same signal at the reference point:

   TLP (dB) + Power at 0TLP
   = Measured power at TLP (dBm)

For example: the receive TLP of a data circuit is -3 dB. Data is normally transmitted at $-13$ dB below 0TLP or $-13$ dBm0. Using this information, what

is the expected measured receive level on our circuit? If we used the formula:

$$-3 + (-13 \text{ dBm0}) = -16 \text{ dBm}$$

**DBMS**   Data base Management System.

**dBmV**   Decibel millivolt. This measurement of signal in dB relative to 1 mV across 75 ohms is used in cable systems.

**dBrn**   The abbreviation for decibel reference noise. In their desire to make things easier for the technician, the early telco engineers created a separate scale for noise that generated larger positive numbers for larger numbers of noise. The zero difference used in describing noise is $-90$ dBm or 0 dBrn. The relationship between these two scales is shown in the following table. It is important to remember that dBrn is a measurement of noise. It is also important to remember the relationship, for example 60 dBrn is equal to $-30$ dBm.

| dBrn | dBm | dBrn | dBm |
|------|-----|------|-----|
| 90   |     | 40   | $-5$ |
| 80   | $-1$ | 30   | $-6$ |
| 70   | $-2$ | 20   | $-7$ |
| 60   | $-3$ | 10   | $-8$ |
| 50   | $-4$ | 0    | $-9$ |

**dBrnC**   The abbreviation for decibel reference noise C-message weighted. Noise contains numerous irregular waveforms which have a wide range of frequencies and powers. Although any noise superimposed upon a conversation has a degrading effect, experiments have shown that this effect is greatest at the mid-range of the voice frequency band. To obtain a useful measurement of the interfering effect of noise, the various frequencies contributing to the overall noise are weighted in accordance to their relative interference effect. This weighting is accomplished through a weighting network or filter. The filter which closely emulates these characteristics is called a C-message weighted filter.

**dBrnC0**   The abbreviation for decibel reference noise C-message weighted zero transmission level.

Noise may also be referenced to the 0TLP. For example, the noise measured at a −3 TLP is 33 dBrnc. To determine the noise at the 0TLP:

(TLP) − (Actual measured noise) = (Noise at 0 TLP) −3 − 33 dBrnc = 36 dBrnC0

**DBS**   Direct Broadcasting via Satellite.

**DBSS**   Direct Broadcast Satellite Service.

**DBU**   Dial Back Up.

**dBW**   Decibel watts, the ratio of the power received to the transmission of one watt, expressed in decibels.

**DB25P**   EIA designation for the male plug used to connect communications cables which fit into a DB25S socket plug.

**DB25S**   EIA designation for the female socket used to connect communications cables.

**DC**   1. Device Control. 2. Direct Current.

**DC continuity**   A circuit that appears to be a continuous circuit. A metallic circuit composed of wire not interrupted by amplifiers, transformers, etc.

**DC current loop**   See current loop.

**DCA**   1. Defense Communications Agency. 2. Distributed Communications Architecture.

**DCACOC**   DCA Operations Center.

**DCC**   Double Coax Conversion.

**DCD**   Data Carrier Detect.

**DCE**   Data Circuit-Terminating Equipment.

**DCEC**   Defense Communications Engineering Center.

**DCEO**Division Communications-Electronics Officer.

**DCF**   Document Composition Facility.

**D-channel (delta channel)**   The Common Channel Signaling channel of an ISDN circuit. In the Basic Rate service, the D-channel occupies 16 Kbps of 144 Kbps bandwidth, carries signaling for 2 B-channels, and can be used for low-speed, end user packet data as well as for signaling. In the Primary Rate service, the D-channel occupies 64 Kbps of 1.544 Mbps (or 2.048 Mbps) of bandwidth, and carries signaling for 24 (or 30) B-channels. In both ISDN services, the D-channel uses the LAP-D link protocol. The following illustration shows the ISDN basic access channel format, where F/S (frame and synchronization) accounts for 48 Kbps. This results in a total basic access channel rate of 192 Kbps.

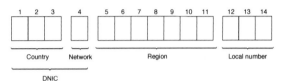

**DCL**   Digilog Command Language.

**DCLU**   Digital Carrier Line Unit.

**DCM**   1. Digital Circuit Multiplexing. 2. Digital Circuit Multiplication. 3. Dynamic Connection Management.

**D-conditioning**   Service offering from telephone companies to control the signal-to-noise ratio and harmonic or non-linear distortion. It is currently used for transmission above 9600 bps with complex modems on voice-grade private lines.

**DCP**   Digital Communications Protocol.

**DCP telephone**   A digital voice telephone of the AT&T Model 7400 series, operated with the Digital Communications Protocol (DCP) used in System 75/85 digital PBXs.

**DCPSK**   Differentially Coherent Phase Shift Keying.

**DCS**   1. Defense Communications System. 2. Digital Cellular Switch. 3. Digital Crossconnect System.

**DCT**   Discrete Cosine Transforms.

**DCTU**   Directly Connected Test Unit.

**DDA**   1. Directory Database Administration. 2. Domain Defined Attribute.

**DDCMP**  Digital Data Communications Message Protocol.

**DDD**  Direct Distance Dialing.

**DDD network**  The long-distance Direct Distance Dialing (DDD) telephone network.

**DDI**  Direct Digital Interface.

**DDN**  Defense Data Network.

**DDP**  Distributive Data Processing.

**DDR**  Dual Dial Restore.

**DDS**  Dataphone Digital Service.

**DDS S/C**  DDS with Secondary Channel.

**DDS II**  Dataphone Digital Service with Secondary Channel.

**DDS-SC**  Dataphone Digital Service with Secondary Channel.

**DDT/DDS**  Digital Data Throughput/Dataphone Digital Service.

**DDX-C**  A circuit switched network operated by NTT in Japan.

**dead front**  An insulated surface that eliminates any exposure of metal parts. Eliminates inadvertent disruptions and noise.

**dead letter message**  A notification sent to the originator of an electronic mail message indicating that the message could not be delivered.

**dead spot**  In cellular communications an area in a system where service is not available.

**deadly embrace**  A state of a system of cooperating processes in which it is logically impossible for the activity of some or all of these processes to continue. A deadly embrace may result, for example, when the existence of a critical section is not recognized.

**deallocate**  To release a resource that is assigned to a specific task.

**DEA—1**  Data Encryption Algorithm—1.

**DEC**  Digital Equipment Corporation.

**decoder**  A device used to unscramble a signal previously encoded.

**decoding**  The restoration of a coded signal to its original format. For example, changing a digital voice signal to an analog voice signal so the listener can understand it.

**decentralized**  Processing in which intelligence is located at several remote sites of the same processing system.

**decentralized network**  An information processing system in which processing and data base storage are distributed among two or more locations.

**deceptive jamming**  Covert jamming which exploits specific characteristics of a targeted system's receiver and radio frequency. One example is the use of squelch tone against tactical radios in which the jammer transmits a tone that simulates the one used by the radio being jammed.

**decibel (dB)**  A tenth of a bel. A unit of measuring relative strength of a signal parameter such as power, voltage, etc. The number of decibels is ten times the logarithm (base 10) of the ratio of the measured quantity of the reference level. The reference level must always be indicated, such as 1 milliwatt for power ratio.

**decimal**  A digital system that has ten states, 0 through 9.

**decipher**  In IBM's VTAM, to convert enciphered data into clear data. Synonymous with decrypt.

**DECnet**  Trademark for DEC's communications network architecture that permits interconnection of DEC computers using DDCMP.

**DECNET E-net**  DEC Engineering Network.

**decoding**  Changing a digital signal into analog form or into another type of digital signal.

**DECrouter 250**  A dedicated router marketed by Digital Equipment Corp. which routes traffic between Ethernets or host computers with synchronous or X.25 packet switching connections.

**decrypt** In IBM's VTAM, to convert encrypted data into clear data. Synonym for decipher.

**decryption** The reversal of encryption. Encryption and decryption are carried out by mathematically applying a KEY (a 64-bit number) to an algorithm to scramble and descramble the data.

**DECserver 200** A specialized terminal server manufactured by Digital Equipment Corp. which converts asynchronous traffic for eight terminals and printers to Ethernet traffic to and from VAX computers.

**DECT** Digital European Cordless Telephone.

**DECUS** Digital Equipment Computer Users Society.

**DED** Dynamically Established Data Link.

**dedicated** Committed to one specific use, such as a dedicated port on a computer to a specified terminal or microcomputer.

**Dedicated Access Facility (DAF)** Connects customer-premises devices to a public network via leased access channels.

**dedicated circuit** A circuit offered on a full-period basis, as opposed to on an on-demand switched basis. A leased circuit.

**dedicated communications link** A link that is continuously available to the user. Also called private or leased.

**dedicated connection** A nonswitched circuit that is always available to the user.

**dedicated line** A communications line which is not dialed. Also called a leased line or private line.

**dedicated network** A network confined to the use of one customer. Also called private network.

**dedicated port** Port which is assigned to communicate with one other port.

**default** Condition which exists from POWER ON or RESET if no instructions to the contrary are given to a device; a value or option assumed by a computer or a network program, when no other value is specified.

**default destination** A destination for display stations and printers that is defined during customization.

**default drive** The disk or diskette drive currently in use by a workstation.

**default server** The server in a local area network to which your default drive is mapped.

**default SSCP list** In IBM's VTAM, a list of SSCPs, either in VTAM's network or another network, that can be used when no predefined CDRSC or name translation function is provided specifying an LU's owning CDRM. This list is filed as a part of an adjacent SSCP table in the VTAM definition library.

**default SSCP selection** In IBM's VTAM, a function that selects a set of one or more SSCPs to which a session request can be routed when there is no predefined CDRSC or name translation function provided that specifies an LU's owning CDRM.

**Defense Advanced Research Projects Agency (DARPA)** A funding agency for computer networking experiments performed on the ARPANET.

**deferred print** The ability to use a disk or diskette to be the print destination in the 3270 emulation environment.

**definite response** In SNA, a value in the form-of-response-requested field of the request header. The value directs the receiver of the request to return a response unconditionally, whether positive or negative, to that request.

**definition statement** IBM's SNA: 1. In VTAM, the means of describing an element of the network. 2. In NCP, types of instructions that define a resource to the NCP.

**Definity** A trademark of AT&T as well as the name for a modular series of Private Branch Exchanges (PBXs) from that vendor.

**Definity Manager I**  An AT&T system management tool for the vendor's Generic 1 switch that is identical to the existing System 75 System Access Terminal. It enables users to access switch-based administrative programs to perform station moves, adds and changes, and it can be used to produce management reports that detail alarms, traffic by trunk group and the use of attendant consoles.

**Definity Manager II**  A basic switch administration application for AT&T and other MS-DOS-based personal computers that provides a user-friendly interface, English language field descriptions and on-line help for such AT&T switches as Definity, Generic 2, Dimension 600, Dimension 2000 and all versions of the System 85.

**Definity Manager III**  An AT&T system management tool which builds on Definity Manager II by adding an Informix data base for reports, scheduling and modeling and which lets users extract performance data from the Definity Generic 2's data base. The software runs on an AT&T 6386E Work Group System (WGS) or 3B2 600, and it has a feature that lets up to three people manage the switch simultaneously.

**Definity Manager IV**  An AT&T system management tool which is a modular version of that vendor's Centralized System Management offering. Definity Manager IV consists of two modules: one supporting facilities management and a second for terminal change management. Definity Manager IV runs on an AT&T 3B2 600 computer system.

**Definity Monitor I**  An AT&T traffic application that can be used alone or with that vendor's Definity Manager III and Definity Manager IV. It enables customers to monitor the trunk groups of Generic 1 and Generic 2 switches.

**degradation**  Deterioration in the quality or speed of data transmission, caused as more users access a computer or computer network.

**degree**  In the application of graph theory to networks, the degree of a node is the number of lines which connect to it.

**DEL**  The ASCII DELETE code used in some instances to delete transmitted characters or to exit modes of operation.

**delay**  The waiting time between two events in a communications system.

**delay, absolute**  The length of time taken for a signal to travel from one point to another in a communications system. Absolute delay is dependent on the length, frequency, and the medium of transmission.

**delay distortion**  A distortion occurring on a communication line due to the different propagation speeds of signals at different frequencies. Some frequencies travel more slowly than others in a given transmission medium and therefore arrive at the receiver at different times. Delay distortion is measured in microseconds of delay relative to the delay at 1700 Hz. This type of distortion can have a serious effect on data transmissions because the data bits going down a transmission line cover a wide bandwidth. Also known as envelope delay.

**delay equalizer**  A corrective device which is designed to make the phase delay or envelope delay of a circuit or system substantially constant over a desired frequency range.

**delay modulation**  A modulation scheme that uses different forms of delay in a signal element. Frequently used in radio, microwave, and fiber optic systems.

**delay vector**  Associated with one node of a packet switched network, the delay vector has as its elements the estimated transit times of packets starting there and destined for other nodes in the network. Nodes may send copies of their delay vector to their neighbors as part of an adaptive routing scheme.

**delayed-request mode**  In SNA, an operational mode in which the sender may continue sending request units on the normal flow after sending a definite-response request chain on that flow, without waiting to receive the response to that chain.

**delayed-response mode** In SNA, an operational mode in which the receiver of normal-flow request units can return responses to the sender in a sequence different from that in which the corresponding request units (RUs) were sent. *Note.* An exception is the response to the DFC request CHASE. All resonses to normal-flow request units received before CHASE must be sent before the response to CHASE is sent.

**delimit** The process of structuring data with specific boundaries, such as control characters which define the beginning and end of a block of data.

**delimiter** A special character or group of characters or word that enables a computer to recognize the beginning or end of a portion of a program or segment of data.

**Delivery Reports** In the CCITT Message Handling System (MHS) delivery reports are system-generated messages that verify the deliverability of a message.

**delivery time** Time from start of transmission at the transmitting terminal to reception at the receiving terminal.

**DELNI** An acronym for a Local Network Interconnect manufactured by Digital Equipment Corporation which functions as a concentrator for up to eight Ethernet-compatible non-terminal devices.

**DELQA** An acronym for an Ethernet-to-Q-Bus communications controller manufactured by Digital Equipment Corporation that connects Q-bus MicroPDP-11 and MicroVAX systems to an Ethernet V2.0 or IEEE 802.3 LAN.

**delta** A multi-node network topology that consists of three multiplexer nodes each interconnected with one another with at least one aggregate link.

**delta channel (D-channel)** In Integrated Services Digital Networks, a channel, multiplexed with data

channels which is used for signaling and other administrative purposes.

**delta channel, D-channel** (i.e. the channel carrying information about charges (delta) in service). The Common Channel Signaling channel of an ISDN circuit. In the Basic Rate service, the D-channel occupies 16 Kbps of 144 Kbps bandwidth, carries signaling for 2 B-channels, and can be used for low-speed, end user packet data as well as for signaling. In the Primary Rate service, the D-channel occupies 64 Kbps of 1.544 Mbps (or 2.048 Mbps) of bandwidth, and carries signaling for 24 (or 30) B-channels. In both ISDN services, the D-channel uses LAP-D link protocol. See D-channel.

**Delta Modulation (DM)** A signal-difference sampling technique which uses a one-bit if the amplitude difference between two samples is increasing and a zero-bit if it is decreasing.

**delta routing** A method of routing in which a central routing controller receives information from nodes and issues routing instructions, but leaves a degree of discretion to individual nodes.

**DELUA** An acronym for an Ethernet-to-Unibus synchronous communications controller which connects Unibus systems (VAX and PDP) to an Ethernet V2.0 or IEEE 802.3 LAN.

**demand assignment** A technique whereby various users share a communications channel on a real-time demand basis. Upon completion of the call, the circuit is deactivated and the capacity becomes available for use by other calls.

**demand multiplexing** A form of time division multiplexing in which the allocation of time to subchannels is made according to their need to carry data. A subchannel with no data to carry is given no time slot. Other names are "dynamic multiplexing" and "statistical multiplexing."

**DEMARC** Rate DEMARCation point.

**demarcation** The physical point where a regulated service is provided to a user, e.g., the point where the wires associated with a dedicated circuit terminate in a connection strip or connector block.

**demarcation strip**  A terminal board used to provide a physical point of interconnection between communications equipment on the user's premises and the transmission facilities of the communications line vendor.

**DEMARK**  Demarcation.

**demodulation**  The process of retrieving data from a carrier; the reverse of modulation.

**Demonstration of Avionics Module Exchangeability via Simulation (DAMES)**  A program established by the U.S. Department of Defense in 1987 to provide common avionics for the Air Force and Navy advanced tactical aircraft and the Army's light helicopter.

**DEMPR**  An acronym for a ThinWire Ethernet multiport repeater manufactured by Digital Equipment Corporation which connects eight ThinWire Ethernet segments to a standard Ethernet cable.

**demultiplexer**  In voice transmission, a device used to unbundle a group of circuits that were grouped together, so they can be switched or locally distributed.

**demultiplexing**  The process of breaking a composite signal into its component channels; the reverse of multiplexing.

**D-encapsulant**  A pourable, fast-setting, all-weather, two-part, reenterable splice encapsulant. It is used to encapsulate splices on aircore and waterproof cables as well as the interface between aircore and waterproof cables.

**Department Of Defense (DOD)**  Part of USA government executive branch that handles military matters, including data communications; responsible for some LAN-associated protocols and standards, such as TCP/IP, as well as selected FIPS.

**Department of Trade and Industry (DTI)**  The UK government department responsible for telecommunications.

**DEPCA**  An acronym for an Ethernet communications controller manufactured by Digital Equipment Corporation that connects the IBM PC, XT and AT to Ethernet and IEEE 802.3 LANs.

**DEPIC**  Dual Expanded Plastic Insulated Conductors consisting of a dual plastic coating made up of a foamed core within a thin, solid skin.

**Deputy Internet Architect**  A member of the Internet Activities Board who is also the Request for Comment (RFC) editor.

**deregulation**  The governmental process which results in competitive and technological pressures in the telecommunications industry as a result of making market entry easier and promoting tariff reductions. Called privatization in the UK.

**DEREP**  An acronym for an Ethernet repeater manufactured by Digital Equipment Corporation which connects up to three segments of baseband Ethernet cable.

**DES**  Data Encryption Standard.

**Designated gateway SSCP**  In IBM's VTAM, a gateway SSCP designated to perform all the gateway control functions during LU–LU session setup.

**Designet**  A network design tool of BBN Communication Corp. of Cambridge, MA.

**Deskpro**  A trademark and the name used by Compaq Computer Corporation for a series of desk top, personal computers.

**despooling**  The process of reading records off the spool into main storage. During the despooling process, the physical track addresses of the spool records are determined.

**DESPR**  An acronym for a ThinWire Ethernet single-port repeater manufactured by Digital Equipment Corporation which connects on ThinWire Ethernet segment to a standard Ethernet network.

**DESTA** An acronym for a ThinWire station adapter marketed by Digital Equipment Corporation which converts a standard 15-pin AUI Ethernet station connector to a ThinWire cabling.

**destination** A logical location to which data is moved (a data receiver). A destination can be a terminal, a person, or a program.

**destination field** A field contained in a message header which contains the address of the station to which a message is being directed.

**destination group** Same as a rotary.

**Destination Logical Unit (DLU)** In IBM's VTAM, the logical unit to which data is to be sent.

**Destination Service Access Point (DSAP)** In Digital Equipment Corporation Network Architecture (DECnet), the one-byte field in an LLC frame on a LAN that identifies the receiving Data Link client protocol.

**destination user** The user to whom a data communication system is to deliver a particular user information block (or unit).

**destination user information bits** The binary representation of user information transferred from a data communications system to a destination user. When the user information is output as nonbinary symbols (e.g., alphanumeric characters), the destination user information bits are the bits on which a final decoding is performed to generate the delivered symbols.

**DESVA** An acronym for a synchronous Ethernet communications controller marketed by Digital Equipment Corporation that connects MicroVAX 2000s and VAXstation 2000s to ThinWire Ethernet networks.

**detection** The process of recovering the audio component (audible signal) from a modulated radio frequency (RF) carrier wave.

**detector** 1. A transducer that provides an electrical output signal in response to an incident optical signal. The current is dependent on the amount of light received and the type of device. 2. In fiber optic technology, a device, usually a semiconductor, which converts incoming light into electric current.

**deterministic** An access method that requires each station on a local area network to wait its turn to transmit.

**deterministic network** A network having the characteristic of predictable access delay.

**Deutsche Bundespost** The PTT of the Federal Republic of Germany (FRG).

**Deutsche Bundespost Telekom** The telecommunications operating portion of Deutsche Bundespost.

**deviation** A term used in frequency modulation to indicate the amount by which the carrier frequency increases or decreases when modulated. It is usually expressed in kilohertz.

**device** A piece of computer equipment, such as a display or a printer, that performs a specific task.

**Device Control (DC)** A category of control characters primarily intended for turning on or off a subordinate device. (DC1, DC2, DC3, and DC4) also (X-ON and X-OFF).

**device control character** In IBM's VTAM, (ISO) a control character used for the control of ancillary devices associated with a data processing system or data communication system, for example, for switching such devices on or off.

**device driver** A software component that controls all data transfers to or from a peripheral or communications device.

**device partitioning** A pool of devices (called a fence) to be used exclusively by a set of jobs in a specific job class allowing an installation to tailor its device usage to its anticipated workload.

**device-type logical unit** In IBM's VTAM, a logical unit that has a session limit of one and usually acts as the secondary end of a session. It is typically an SNA terminal (such as a logical unit for a 3270 terminal or a logical unit for a 3790 application program). It could be the primary end of a session, for example, the logical unit representing the Network Routing Facility (NRF) logical unit.

**DF**   Direction Finding.

**DFC**   Data Flow Control (IBM's VTAM).

**DFI**   Digital Facilities Interface.

**DFN**   Deutsche Forschungsnetz.

**DFSYN RESPONSE**   In IBM's VTAM, a data flow synchronous (DFSYN) response which is a normal-flow response that is treated as a normal-flow request so that it may be received in sequence with normal-flow requests.

**DFT**   Distributed Function Terminal.

**DFTAC**   Distributing Frame Test Access Controller.

**DGM**   Digital Group Multiplexer.

**DHNR**   Dynamically Hierarchical Network Routing.

**DIA**   Document Interface Architecture.

**diagnostic modems**   Modems that communicate with each other over a low-speed channel for the purposes of communicating control and diagnostic information.

**diagnostic port**   A port on a communications device that can be used for testing and/or receiving statistics generated by the device.

**diagnostic programs**   Computer programs used to check equipment malfunctions and pinpoint faulty components; can be used manually or can be automatically called in by supervisory programs.

**diagnostic unit**   A device used on a conventional computer to detect faults in the system. Separate unit diagnostics will check such items as arithmetic circuitry, transfer instructions, each input–output unit, and so on.

**diagnostics**   Programs or procedures used to test a piece of equipment, a communications link or network, or any similar system.

**dial backup**   The process of using the public switched telephone network to reestablish communications if a leased line facility becomes inoperative.

**Dial Backup Auto Answer Unit (DBAAU)**   An ancillary device for automatic phone answering.

**dial line (dial-up line)**   A line that requires signaling, such as tone or pulse, to alarm a potential receiving station that a call is pending.

**dial network**   A network that is shared among many users, any one of whom can establish communication between desired points by use of a dial or pushbutton telephone. Synonymous with public switched telephone network.

**dial pulse**   A current interruption in the DC loop of a calling telephone. It is produced by the breaking and making of the dial pulse contacts of a calling telephone when a digit is dialed. The loop current is interrupted once for each unit of value of the digit.

**dial speed**   The number of pulses that a rotating dial can transfer in a given amount of time.

**dial tone**   A tone indicating that automatic switching equipment is ready to receive dial signals.

**dial train**   The series of pulses or tones that contain the information necessary to route a call over the public switched network.

**dial transparency**   The dial transparency option (EIA passthrough) allows a multiplexer to pass RS-232C interface signals from a remote, dial-up modem to the central-site computer as if the remote modem were connected directly to the computer.

**Dial X.25**   Packet mode communications carried out by dial-in X.25 Data Terminal Equipment (DTEs) that are temporarily connected to the network (as distinct from a hard-wired connection).

**dial-back security**   A security technique whereby the remote-access system will limit access to users who have the proper password and logon and are dialing from a prearranged location. After providing the proper logon sequence, the system will hang up and dial a number that was previously stored in the system from which the user was supposed to be originating the call.

**Dialcom Incorporated** A large electronic-mail service provider. Formerly owned by ITT, Dialcom is now a subsidiary of British Telecom.

**Dialed Number Identification Service (DNIS)** A service offered by MCI which identifies specific 800 number categories to 800 number subscribers.

**dial-in** The ability of a modem to answer an incoming call from a remote terminal.

**dialing** Establishing a connection through common communications lines over the switched telephone network.

**Dial-it 900 service** An AT&T mass announcement service for larger volumes of calls over short time-frames, used to provide a pre-recorded or live message or to register an audience's opinion. Can be designed to have a certain frequency of calls forwarded to a live operator or another message to award a prize, speak to a celebrity, collect a donation, etc. Callers ordinarily pay for the calls.

**Dialog** A comprehensive on-line information service that has over 300 data bases covering such subjects as business, science, medicine, law, energy, agriculture, and the humanities. For additional information telephone 1-800-334-2564.

**dial-out** The ability of a modem to call a remote device.

**dial-tone first service** A coin telephone service that permits callers to reach the operator and to dial directory assistance or 911 without charge.

**dial-up** The use of a dial or pushbutton telephone to initiate a station-to-station telephone call.

**dial-up line, dial-in line, dial line** A circuit or connection on the public telephone network.

**dial-up network** The use of a dial or pushbutton instrument, or an automatic machine, to initiate a call on the public switched telephone network.

**diatomic encoding** A data compression process whereby a pair of characters is replaced by a special character. The special character is normally a character not defined in a code set and its use to represent a pair of characters permits a 50 percent data reduction.

**dibit** A group of two bits. In 4-phase, phase modulation such as DPSK, each possible value of a dibit is encoded as a unique carrier phase shift; the four possible values of a dibit are 00, 01, 10, and 11. The following table illustrates one common assignment of dibit values to phase shifts.

DPSK PHASE SHIFTS

| Dibit value | Phase shift |
| --- | --- |
| 00 | 45 degrees |
| 01 | 135 degrees |
| 10 | 315 degrees |
| 11 | 225 degrees |

**DID** Direct Inward Dialing.

**dielectric** An electrically nonconductive material through which signals can be induced by magnetic flux.

**differential encoding** An encoding technique in which binary data is represented by the presence or absence of a signal transition at the beginning of the bit time. NRZI (non-return-to-zero-inverted) is an example of differential encoding.

**differential modulation** A type of modulation in which the absolute state of the carrier for the current signal element is dependent on the state after the previous signal element.

**Differential Phase Shift Keying (DPSK)** Modulation technique used in Bell 201 modem.

**differential signaling** A method of signaling that requires a cable pair to convey information with respect to the voltage levels in two wires. As an example, to transmit a logical "1," one wire is driven more positive while the second wire is driven more negative. At the receiver a comparator examines the relative voltage levels of the two wires. Differential signaling is specified in the RS-442 interface standard.

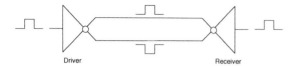

*differential signaling*

**Diffraction**  The deviation of a wavefront from the path predicted by geometric optics when a wavefront is restricted by an opening or an edge of an object. *Note*. Diffraction is usually most noticeable for openings of the order of a wavelength. However, diffraction may still be important for apertures many orders of magnitude larger than the wavelength.

**Digilog**  A company located in Montgomeryville, PA, which specializes in the manufacture of protocol analyzers.

**Digilog Command Language (DCL)**  A programming language used to prepare setup and testing programs for a Digilog protocol analyzer.

**digit**  One of the numerals in a number system.

**digital**  Discretely variable as opposed to continuously variable. Data characters are coded in discrete, separate pulses or signal levels.

**Digital Access and Cross-connect System/ Customer Controlled Rerouting (DACS/CCR)**  Allows DACS routing functionality to be under the control of the T1 user.

**Digital Access Cross-connect Switch (DACS)**  An electronic switch which is capable of switching data at either the DS0 or DS1 levels. DACS allows T1 carrier facilities or any of its subchannels to be switched or cross-connected to another T1 carrier.

**Digital Access Cross-connect System (DACS)**  A software-based, multi-processor-controlled system for taking North American T1 (1.544 Mbps) or European T1 (2.048 Mbps) facility terminations and cross-connecting them at the DSO (64 Kbps) level. DACS is the acronym first coined by AT&T.

**digital adaptive equalization**  A repeating pattern is transmitted by a modem on one side of a circuit which enables a modem receiver at the opposite end of the circuit to measure and build up compensation to correct line distortion.

**digital–analog converter**  A device that converts a representative number sequence into a signal that is a function of a continuous waveform.

**Digital Business System (DBS)**  A term used by Panasonic Co. for a midrange digital hybrid product line.

**Digital Carrier Line Unit (DCLU)**  Provides integrated (i.e., no central office terminal) access to an AT&T 5ESS switch for a SLC 96 carrier in a pre-ISDN environment. The DCLU-5 extends that integrated access to ISDN service.

**Digital Cellular Switch (DCS)**  A major element of the AT&T Autoplex third generation cellular telecommunications system which is used to supervise and switch trunks.

**Digital Circuit Multiplexing (DCM)**  A proprietary speech compression technique developed by ECI Telecom, Inc., which increases the voice capacity over the TAT-8 transatlantic cable from 10 000 to 50 000 voice channels.

**Digital Circuit Multiplication (DCM)**  A method for increasing the effective capacity of primary-rate and higher-level PCM hierarchies based upon speech coding at 64 Kbps.

**digital city**  A city in which a Principal Telephone Company Central Office is located and serves a specific geographical area for Dataphone Digital Service (DDS).

**Digital Color Code (DCC)**  In cellular telephone communications a code used to prevent a control channel from a nonadjacent cell from being accepted as the correct control channel. The mobile station transmits its DCC to the land station which returns a coded version of the DCC. The coded version serves to verify that the mobile unit is locked onto the correct signal.

**Digital Communications Protocol (DCP)** The protocol used in AT&T's System 75/85 digital PBXs.

**digital computer** A programmable electronic device that handles information in the form of discrete binary numbers. The user's program and data are stored in the memory of a computer.

**Digital Cross Connect (DSX)** A Bell standard transmission interface which specifies signal levels, signal format, and connectivity.

**Digital Cross Connect, Level 1 (DSX-1)** A Bell standard transmission interface which specifies signal levels, format, and connectivity for DS-1.

**Digital Crossconnect System (DCS)** A software-based, multi-processor-controlled system for taking North American T1 (1.544 Mbps) or European (2.048 Mbps) facility terminations and cross-connecting them at the DSO (64 Kbps) level. DCS is the acronym most used by manufacturers other than AT&T Network Systems.

**digital data** Information represented by a code consisting of a sequence of discrete elements.

**Digital Data Communications Message Protocol (DDCMP)** A communications protocol used in DEC computer-to-computer communications.

**digital data network** A network specifically designed for the transmission of data, wherever possible in digital form, as distinct from analog networks such as telephone systems, on which data transmission is an exception.

**Digital Data Throughput (DDT/DDS)** A service offered by carriers for terminating dataphone digital service circuits at a central office via a T1 connection. DDS circuits are 56-kb/s data circuits offered by telcos. Data rates of 2.4 kb/s, 4.8 kb/s, 9.6 kb/s, 19.2 kb/s, and 56 kb/s can be used on DDS circuits.

**digital display module** Snap-on or integrated module for telephones and terminals connected to AT&T System 75 and 85 digital PBX systems. Each module has two lines of alphanumeric display, with a total capacity of 80 characters, which can be scrolled for longer messages.

**Digital Equipment Corporation (DEC)** A leading manufacturer of minicomputers and related hardware and software.

**digital error** Digital transmission where a zero is interpreted as one or vice versa.

**digital hierarchy** A series of multiplexing schemes over digital circuits in which signals within the bandwidth of any circuit at a given bit rate can be multiplexed into the bandwidth of a circuit at the next higher rate. The current North American digital hierarchy incorporates the following multiplexing levels: DS-0, 64 Kbps; DS-1, 1.544 Mbps, octet interleaved, carries 24 DS-0 circuits; DS-1C, 3.152 Mbps, bit interleaved, carries two DS-1 circuits or 48 DS-0 circuits; DS-2, 6.312 Mbps, bit interleaved, carries two DS-1C circuits, four DS-1 circuits, or 96 DS-0 circuits; DS-3, 44.746 Mbps, bit interleaved, carries seven DS-2 circuits, 14 DS-1C circuits, 28 DS-1 circuits, or 672 DS-0 circuits; and DS-4, 274.176 Mbps, bit interleaved, carries six DS-3 circuits, 42 DS-2 circuits, 84 DS-1C circuits, 168 DS-1 circuits, or 2016 DS-0 circuits.

DIGITAL HIERARCHY:

The six Levels of Transmission in the current North American Digital Hierarchy

| Circuit Type | Bit rate | Equivalent number of circuits | | | | |
|---|---|---|---|---|---|---|
| | | DS-0 | DS-1 | DS-1C | DS-2 | DS-3 |
| DS-0 | 64 K | 1 | — | — | — | — |
| DS-1 | 1.544 M | 24 | 1 | — | — | — |
| DS-1C | 3.152 M | 48 | 2 | 1 | — | — |
| DS-2 | 6.312 M | 96 | 4 | 2 | 1 | – |
| DS-3 | 44.736 M | 672 | 28 | 14 | 7 | 1 |
| DS-4 | 274.176 M | 2016 | 168 | 84 | 42 | 6 |

**digital loopback** Technique for testing the digital circuitry of a communications device; may be initiated locally, or remotely via communications circuits. Device tested will transmit back a received

123

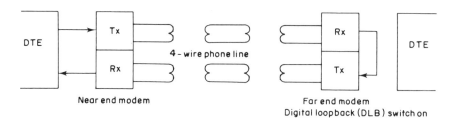

DTE — Tx / Rx — 4-wire phone line — Rx / Tx — DTE

Near end modem

Far end modem
Digital loopback (DLB) switch on

*digital loopback testing*

test message, after decoding. Results are compared with original transmission.

**digital loopback testing**   In data communications, a technique whereby a test device transmits a known pattern through a local modem and a communications channel to a remote modem. The test pattern is demodulated, remodulated, then internally looped back in the remote modem and transmitted back to the test device. Any bit errors induced in the loopback path will be detected and counted.

(1) Tests both modems and phone lines from the near end. (2) Loopback is initiated from the near end if the modem is capable, otherwise far-end operator intervention is required.

**digital matrix switch**   A switch which reconfigures its connections based upon instructions from a supervisory terminal normally located within a network control center.

**digital milliwatt**   A pulse code modulated signal equivalent to an analog signal of 1 kHz at 1 mW; used as a reference pattern for testing digital circuits. A digital milliwatt signal fulfills the one's density requirements for digital transmission through repeaters.

**Digital Multiplexed Interface (DMI)**   A specification, originated by AT&T, for a T1 based data communications interface between a mainframe computer and a private branch exchange. Unlike Northern Telecom's CPI, the DMI interface conforms to current ISDN specifications.

**Digital Multiplexed Interface Bit-Oriented Signaling (DMI-BOS)**   A simple common channel signaling protocol for AT&T's Digital Multiplexed Interface. Designed as an interim step between channel-associated signaling and the more complex DMI Message-Oriented Signaling (DMI-MOS, based on Common Channel Signaling System No. 7), DMI-BOS uses bits in the signaling channel to emulate the E- and M-wires of an analog telephone.

**Digital Multiplexed Interface Message-Oriented Signaling (DMI-MOS)**   A common channel signaling protocol for mature installations of AT&T's Digital Multiplexed Interface; based on Common Channel Signaling System No. 7.

**digital multiplexers**   Digital multiplexers form the interface between digital transmission facilities of different rates. They combine digital signals from several lines in the same level of the hierarchy into a single pulse stream suitable for connection to a facility of the next higher level in the hierarchy.

**digital PBX (private branch exchange)**   A private, on premises, digital telephone switching exchange, owned and managed by the end user.

**digital preference software**   AT&T software which lets its network automatically select digital trunks for traffic in a mixed analog and digital environment.

**digital repeaters**   Devices inserted in a transmission medium to regenerate a digital signal sent over the medium.

**digital routing**   A software-defined network (SDN)

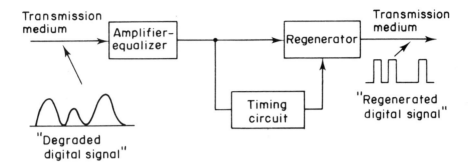

*digital repeaters*

service feature offered by AT&T which enables an SDN user to transmit digital data at 56 Kbps between locations.

**Digital Service Area (DSA)** A metropolitan area that is served by digital service.

**Digital Service, Europe** A digital circuit of 2.048 Mbps, the European standard bit rate for DS1 (T1) transmissions.

**Digital Service, Level n** A general descriptor for a level of the North American Digital Hierarchy.

**Digital Service Unit, Data Service Unit (DSU)** 1. A digital format converter that converts the communications line format into the digital format of the associated terminal or computer. 2. Connects DTE to a digital transmission line such as AT&T's DDS.

**Digital Service Unit/Channel Service Unit (DSU/CSU)** A combined digital service unit and channel service unit. The DSU portion converts the digital signals from the data terminal equipment to the form required for operation with the digital network, while the CSU portion is the circuitry designed as the interface to the digital lines.

**digital signal** A discrete or discontinuous signal; one whose various signals are discrete intervals apart.

**Digital Signal Processing (DSP)** Refers to the technology that enables a modem to convert digital signals to analog signals and analog signals to digital signals.

**digital signaling** The transmission of digital data over a communications medium by means of electrical signaling directly driven by the digital data stream, as opposed to analog signaling in which an indirect representation is used.

**Digital Speech Interpolation (DSI)** A voice compression technique which takes advantage of the pauses inherent in human speech to multiplex additional voice conversations onto the same transmission link.

**digital station terminal** A path for digital transmission furnished within the serving area of a digital city between the principal telephone company central office and a station.

**Digital Subscriber Line (DSL)** The two-wire line which supports full duplex transmission for ISDN's basic rate access (2B+D channels plus overhead bits)

**digital switch** A solid-state device that routes information in a digital form. The term is also used to describe a digital switching system.

**digital switching** The process of establishing and maintaining a connection, under program control, where digital information is routed between input and output. Generally, a "virtual circuit."

**digital system cross-connect frame (DSX)** A bay or panel to which T1 lines and DFI circuit packs are wired that permits cross-connections by patch cords and plugs. A DSX panel is used in small office applications where only a few digital trunks are to be installed.

**Digital Termination System (DTS)** An FCC-proposed system that integrates cellular radio for local access and microwave or satellite for long distance.

**Digital to Analog (D/A)** The conversion of a digital (discrete) signal to an analog (continuous) signal.

**digital transmission** A transmission in which signal elements are in the form of discrete pulses of constant rather than variable (analog) amplitude.

**digital transmission group** A grouping of voice channels that are combined into a digital bit stream for transmission over various communications media.

**Digital Trunk and Line Unit (DTLU)** Provides system access for T1 carrier lines used for inter-office trunks or remote switching module umbilicals.

**Digital Video Interactive (DVI)** A technology which reduces television-quality bandwidths by a factor of 150:1, permitting an hour's worth of motion video to be recorded on a CD-ROM.

**digitize** To convert information from its nominal state into a digital code.

**digitizer** A tablet of switches or a position sensing device that converts movement into digital data elements that a computer can use.

**digroup** Represents the information content of 24 pulse code modulated, or PCM, channels.

**diode** A one-way device that controls the flow of electrical charge; the diode lets current flow in one direction.

**DIP** Dual In-line Package/Pins.

**dip switches** Switches for opening and closing leads between two devices. When a lead is opened, a jumper wire can be used to cross or tie it to another lead.

**dipole antenna** An antenna having an electrical length equal to a half wavelength at the frequency for which it is designed. It may be a single conductor, or it may consist of two elements, whose total length is one half wavelength, separated by an insulator or an air space at the point of connection to the transmitter.

**Direccion General de Telecomunicaciones** Spain's telecommunications administration.

**direct access** 1. The process of direct interaction between an on-line user program and a terminal, with the user program controlling requests for data and the immediate output of data. 2. A method for reading or writing a record in a file by supplying its key or index value.

**direct access memory** A method of computer storage that allows a particular address to be accessed independently of the location of that address; therefore, the items stored in the memory can be accessed in the same amount of time, regardless of location.

**direct activation** In IBM's VTAM, the activation of a resource as a result of an activation command specifically naming the resource.

**Direct Broadcast Satellite (DBS)** Refers to service that uses satellites to broadcast multiple channels of television programming directly to small-dish antennas.

**Direct Broadcast Satellite Service (DBSS)** A TV transmission system in which video is broadcast via satellite directly to subscribers.

**direct conduit method** A ceiling distribution method in which cables are run in conduit from a satellite location directly to the desired information outlets.

**Direct Current (DC)** An electrical current that is used for digital signaling.

**direct current loop** A data transmission interface technique usually used with teleprinters, which transmit a digital signal usually at 20 milliamperes DC. (Also current loop.)

**direct current signaling** A method whereby the signaling circuit E&M (ear and mouth) leads use the same cable pair(s) as the voice circuit and no filter is needed to separate the signaling frequency from the voice transmission.

**direct deactivation** In IBM's VTAM, the deactivation of a resource as a result of a deactivation command specifically naming the resource.

**direct delivery** A type of delivery in which an electronic message is sent directly to a device.

**direct dialing service** A service feature that permits a user to place credit card calls, collect calls, and special billing calls information into the public telephone network without operator assistance.

**Direct Distance Dialing (DDD)** A telephone service in North America which enables users to call their subscribers outside their local area without operator assistance. In the United Kingdom and some other countries, this service is known as STD, subscriber trunk dialing.

**Direct Inward Dialing (DID)** A feature of a PBX and some Centrex systems that allows outside callers to call an extension directly without local operator assistance.

**Direct Inward System Access (DISA)** A PBX feature which allows a caller to dial a PBX and after entering an identification code gain access to the system's features to include the ability to dial long distance calls.

**Direct Memory Access (DMA)** A method of transferring information to or from a computer memory device independently of the central processing unit.

**Direct Outward Dialing (DOD)** A feature of a PBX and Centrex systems that allows an inside caller to call an external number without local operator assistance.

**direct print** The ability to use a printer as the print destination in the 3270 emulation environment.

**Direct Termination Overflow** A service mark of MCI Communications as well as a routing feature from that vendor which automatically switches 800 calls to a customer's regular business lines when MCI 800 lines to the customer are occupied.

**Direction Finding (DF)** The process of locating the position of a transmitter by observing its maximum signal power from two separated receivers. By tracing the angle to the maximum signal power from the two receiving sites the point of intersection provides the location of the transmitter.

**directional antenna** An antenna which limits the beamwidth of a received radio signal to the direction from which the radio signals arrive.

**directional couplers** Passive components that divide a signal unequally into two or more parts.

**directive** In Digital Equipment Corporation Network Architecture (DECnet), a management request sent by a director to an entity.

**Director** In Digital Equipment Corporation Network Architecture (DECnet), the management software used by a network manager.

**directory** 1. A structure containing one or more files and a description of those files. 2. In DECnet, a container in the Naming Service which stores a set of names.

**directory assistance** A service provided by telephone companies to aid subscribers in obtaining unknown phone numbers.

**directory routing** A message-routing system which uses a directory at each switching node. Each directory contains the preferred outgoing link for each destination.

**dirty loop** A two-wire telephone loop that has impairments that results in a normal voice connection becoming noisy.

**DIS** Draft International Standard.

**DISA**   Direct Inward System Access.

**disable**   To turn off a device or prevent certain interrupts from occurring in a processing unit (such as a network communications card). Interrupts may be disabled by several means to include setting a switch or a jumper, or issuing a command.

**disabled**   In IBM's VTAM, pertaining to an LU that has indicated to its SSCP that it is temporarily not ready to establish LU-LU sessions. An Initiate request for a session with a disabled LU can specify that the session be queued by the SSCP until the LU becomes enabled. The LU can separately indicate whether this applies to its ability to act as a primary logical unit (PLU) or a secondary logical unit (SLU).

**disc**   See disk.

**disconnect**   The process of disengaging or releasing a circuit between two stations.

**disconnect sequence**   In a switching network, a sequence of characters used to terminate a call connection and exit the network.

**disconnect signal**   A signal transmitted from one end of a switched circuit that the established connection should be broken. The signal is sent to the other end of the circuit.

**disconnection**   The termination of a physical connection.

**discrete access**   In LAN technology, an access method used in star LANs. Each station has a separate (discrete) connection through which it makes use of the LAN's switching capability.

**Discrete Cosine Transform (DCT)**   A mathematical process used in video compression. DCT reorganizes pixel information into a more compact form and generates a series of numbers that depict pixel values.

**discretely variable**   Capable of one of a limited number of values; usually used to describe digital signals or digital data transmission.

**discrimination**   In frequency modulation systems, the process at a receiver of detecting frequency changes of a carrier signal and deriving the information originally used to cause the frequency shifts; equivalent to frequency demodulation.

**discriminator**   A circuit that has an output voltage which varies in amplitude and polarity in accordance with the frequency of the applied signal. It is used primarily as a detector in a frequency modulation (FM) receiver.

**disengagement confirmation**   The vent that verifies that a particular user's participation in an established data communications session has been terminated. For that user, it completes the disengagement function and stops the counting of disengagement time. Disengagement confirmation occurs, for a particular user, when (1) disengagement of that user has been requested; and (2) the user is able to initiate a new access attempt. Most data communication systems notify the user that a new access attempt can be initiated by issuing an explicit disengagement confirmation signal. In cases where no disengagement confirmation signal is issued, the user may initiate a new access attempt to confirm disengagement.

**disengagement request**   The event that notifies the system of a user's desire to terminate an established data communication system. It is complementary to the access request in most systems. Each disengagement request begins the disengagement function and starts the counting of disengagement time for one or both users. Disengagement is normally requested simultaneously for both users in connection-oriented sessions, and independently for each end user in connectionless sessions.

**dish**   Synonym for an antenna or earth station.

**disk**   An electromagnetic storage medium for digital data.

**disk controller**   Electronics mounted on an adapter card or on a computer's motherboard which controls the operation of a disk drive.

**disk mirroring**   The process of writing data onto two disk drives instead of one. Disk mirroring is a safety feature which protects data from disk drive or disk controller failure.

**Disk Operating System (DOS)** A program or set of programs that instruct a disk-based computing system to schedule/supervise work, manage computer resources, and operate/control peripheral devices.

**disk server** A node on a local area network that allows other nodes to create and store files on its disks.

**DISKED** A Novell NetWare utility program which can be used to modify the disk drive subsystem on a byte-by-byte basis.

**diskette** A thin, flexible magnetic plate that is permanently sealed in a protective cover. It can be used to store information.

**diskless workstation** A device connected to a local area network that does not have a local disk. The diskless workstation uses the disk of a server connected to the network.

**DISOSS** DIStributed Office Support System.

**dispatcher** A solid-state microprocessor controlled alarm reporting system manufactured by ComDev of Sarasota, FL. The dispatcher is capable of dialing user-programmable telephone numbers and delivering user-specified text messages to a printer or PC-based alarm reporting center.

**dispersed network** A communications network consisting of more than one processor, where processing may take place in many locations. This network may have both host and satellite processors.

**dispersion** In an optical system, dispersion is the spreading of photons in a light pulse as they speed down a lightguide. This "smears" the pulses and reduces the rate at which they can be transmitted. Most common are: (1) chromatic dispersion (different wavelengths of light within a pulse travel at different speeds); (2) modal dispersion (some parts of the pulse follow longer paths through the lightguide than other parts).

**display** In IBM's Network Problem Determination Application (NPDA), refers to the title page and all following pages of a display. It is also referred to as a screen or panel.

**display converter** In IBM 3270 systems, a coaxial converter that allows asynchronous display terminals to emulate IBM 3278 display stations.

**display field** 1. An area in a display buffer that contains a set of characters that can be manipulated or operated upon as a unit. 2. A group of consecutive characters in a display buffer that starts with an attribute character which defines the characteristics of the field and contains one or more alphanumeric characters. The field continues to, but does not include, the next attribute character.

**display frame** 1. In computer graphics, an area in storage in which a display image can be recorded. 2. In computer micrographics, an area on a microform in which a display image can be recorded.

**display station, display terminal** A device consisting of a keyboard and video or CRT display. In the IBM 3270 Information Display System, a 3278 is an example of a display station; an ASCII CRT terminal is an example of a display terminal.

**Display System Protocol (DSP)** Packet Assembler/Disassembler (PAD) designed to support 3270 devices. 3270 Binary Synchronous Communications (BSC) DSP supports 3270 BSC devices. Systems Network Architecture (SNA) DSP supports 3270 Synchronous Data Link Control (SDLC) devices.

**display types** In IBM's Network Problem Determination Application (NPDA), a concept to describe the orgnaization of data displays. Display types are defined as total, most recent, user action, and detail. Display types are also referred to as display levels. Display types are combined with resource types and data types to describe NPDA display organization.

**display unit, video** A device that provides a visual representation of data.

**DisplayWrite/370**  The DisplayWrite Editor of the IBM System/370 which provides users with the ability to create revisable-form text (RFT) documents and format them as they are created.

**distance sensitivity**  A method of changing for circuit services based on the distance traveled, as opposed to distance insensitivity, in which other bases than distance, such as duration or data volume are used.

**Distance Vector Protocol**  A routing protocol which periodically transmits broadcasts which propagate routing tables across a network.

**distinctive ringing**  A telephone feature offered by many communications carriers and supported by most PBX's which lets a subscriber specify numbers from which incoming calls will have a distinctive ring. Also known as call selector and priority call.

**distortion**  The unwanted changes in signal or signal shape that occurs during transmission between two points.

**distributed adaptive routing**  A method of routing in which the decisions are made on the basis of exchange of information between the nodes of a network.

**distributed application**  An application in which some of the processing, storage, and control functions are situated in different places and connected by transmission facilities.

**distributed architecture**  In LAN technology, a LAN that uses a shared communications medium; used on bus or ring LANs; uses shared access methods.

**distributed computing**  The name of the trend to move computing resources such as minicomputers, microcomputers, or personal computers to individual workstations.

**distributed function terminal (DFT)**  In an IBM 3270 network environment, a programmable terminal that can perform operations previously performed by a control unit. DFT terminals can interpret the 3270 data stream. Examples of DFT terminals include the IBM 3270 Personal Computer and the 3290 Information Panel.

**distributed management**  In Digital Equipment Corporation Network Architecture (DECnet), a form of management where network managers and directors are dispersed across many systems.

**distributed network**  A communications network made up of two or more processing centers. The processing may be distributed between each center and each processor must be in communication with all others.

**Distributed Office Systems Support (DISOSS)**  One of IBM's strategic Systems Application Architecture (SAA) products for document library storage and corporate electronic messaging.

**distributed processing**  An arrangement that allows separate computers to share work on the same application program. Often erroneously used to mean distributed computing.

**Distributed Queue Dual Bus (DQDB)**  The cell relay IEEE 802.6 standard for metropolitan area networks.

**distributed system management model**  The framework within which DNA network management is designed.

**distributing frame**  A structure for terminating permanent wires of a telephone central office, private branch exchange, and for permitting the easy change of connections between them by means of cross-connecting wires.

**distributing rings**  Rings used to provide support for wires in building apparatus and distribution closets.

**distribution**
**block**  Centralized connection equipment where telephone or data terminal wiring is terminated and cross-connections are made.

**distribution list**  Contains references to subsets of electronic mail users that can be used for broadcast addressing. A distribution list is used to send the same message simultaneously to a group of users.

**distribution panel** A wiring board that provides a patch panel function and mounts in a rack.

**distribution switch** An X.25 processor, used to route and switch data packets through a Packet Switching Network (PSN), which can be equipped with Packet Assembler/Disassembler (PAD) software to interface terminal and host devices to the network.

**Dithering** The process of adding random noise to a signal to make it difficult or impossible to use.

**Diversi-T** A T1 multiplexer manufactured by Larse Corporation that supports access to switched ISDN as well as other switched voice and data services.

**diversity** A technique employing two or more channels over which a single transmission is carried, thus improving reliability, security, etc.

**diversity routing** A routing scheme marketed by some communication carriers in which pairs of lines are routed through different carrier offices and use different carrier facilities. Diversity routing is designed to ensure one route remains operational if a problem occurs on the other route.

**diversity system** A system of radio communications in which a single received signal is derived from a combination of, or selected from, a plurality of transmission channels or paths. The system employed may include space diversity, polarization diversity, or frequency diversity. The diversity principle takes advantage of the fact that fading characteristics of a given signal generally vary widely, at any given instant, at different receiving antenna locations, and with different frequencies.

**divestiture** The breakup of AT&T by the Federal Court based on an antitrust agreement reached between AT&T and the U.S. Department of Justice, effective January 1, 1984.

**DI3270** 1. A gateway emulation product from Data Interface Systems Corporation of Austin, TX, which enables a Novell NetWare LAN to emulate one or more remote IBM 3X74 cluster controllers and their associated terminals and/or printers. 2. A dial-in 3270 communications program from Doradus Corporation of Minneapolis, MN, for use on Data General/One (DG/One) laptop personal computers. DI3270 permits the DG/One to be configured as a bisynchronous dial-in, bisynchronous dedicated or SNA/SDLC dial-in 3270 terminal.

**DKTS** Digital Key Telephone System.

**DL** Digital Loop.

**DLB** Digital LoopBack.

**DLC** Data Link Control.

**DLCF** Data Link Control Field.

**DLCI** Data Link Connection Identifier.

**DLE** Data Link Escape.

**DLI** Dual Link Interface.

**DLO** Data Line Occupied.

**DLTU** Digital Line Trunk Unit.

**DLU** Destination Logical Unit (IBM's SNA).

**DM** 1. Delta Modulation. 2. Digital Multiplex.

**DMA** Direct Memory Access.

**DMI** Digital Multiplexed Interface.

**DMI-BOS** Digital Multiplexed Interface Bit-Oriented Signaling.

**DMI-MOS** Digital Multiplexed Interface Message-Oriented Signaling.

**DN** Directory Number.

**DNA** Digital Network Architecture.

**DNC** Director of Naval Communications.

**DNHR** Dynamic Non-Hierarchical Routing.

**DNIC** Data Network Identification Code.

**DNIS** Dialed Number Identification Service.

**DNVT** Digital Non-secure Voice Terminal.

**DOCC** DCA Operations Control Complex.

**Document Composition Facility (DCF)** An IBM program product that formats documents with its SCRIPT/VS formatter component.

**document content architecture** An IBM architecture, part of SNA which defines a standard for the transmission and storage of documents over networks to include voice, data, and video.

**document interface architecture** An IBM architecture which is included in SNA for controlling the exchange of document information between dissimilar editing systems.

**document style** In IBM PROFS and OFFICE VISION a file in which certain elements are present, such as margin, tabs, and controls. The file is used as the basis for creating a new document or a new document style. A document style is also called a document format file.

**documentation** A written description of a program that includes its name, purpose, how it works, and, frequently, operating instructions. Good documentation reduces both labor and ulcers for those who later use and debug a program.

**DOCVIEW** A command in IBM's PROFS and OFFICE VISION which allows a user to view a document referenced in another document.

**DOD** Department Of Defense (United States).

**DOD** Direct Outward Dialing.

**domain** 1. The owner or controller of computer resources. 2. In IBM's SNA, a host-based System Services Control Point (SSCP) and the physical units (PUs), logical units (LUs), links, link stations, and associated resources that the host SSCP has the capability to control. 3. The name of Apollo Computer's network protocol.

**domain name** In the Internet a domain name consists of a sequence of names or words separated by dots.

**Domain Defined Attribute (DDA)** Under X.400 addressing the DDA is a special field that may be required to assist a receiving E-mail system in delivering a message to the intended recipient. Up to four DDAs are allowed per address, with each DDA address entry made up of two parts, a Type and a Value.

**domain naming** A LAN feature which allows a network administrator to create groups of end users that have the same access rights to resources.

**domain operator** In IBM's SNA multiple domain network, the person or program that controls the operation of the resources controlled by one system services control point (SSCP).

**domain specific part** In Digital Equipment Corporation Network Architecture (DECnet), the part of an NSAP address assigned by the addressing authority identified by the IDP.

**DON** Digital Overlay Network.

**Door** Any application which bulletin board software can "shell out" to execute. Most doors provide access to games and data bases.

**dormant state** A state of inactivity in which there is no current request for a task.

**DOS** Disk Operating System.

**DOS command** An instruction to the Disk Operating System (DOS) to carry out a task.

**DOS command line** The area of a display screen where DOS commands are entered; the line on which the DOS prompt displays.

**DOS prompt** A letter, followed by the > character, displayed by DOS to indicate that it is ready to process commands from the standard input device (usually the keyboard).

**dot** One point in a printer or display block matrix.

**dot graphics** A graphic design formed by patterns of dots.

**dot-matrix printer** A printing method using wires or pins to strike a ribbon to form charcters and graphics using a dot pattern.

**Dots-Per-Inch (DPI)** A measure of the printed quality for dot-matrix, ink-jet, thermal and laser printers.

**Double Coax Conversion (DCC)** A device that provides IBM 3270 coax to ASCII to IBM 3270 coax conversion.

**double-current transmission, polar direct-current system** A form of binary telegraph transmission in which positive and negative direct currents denote the significant conditions.

**double recording** In IBM's Network Problem Determination Application (NPDA), refers to the recording of certain individual events under two resource levels.

**double sideband transmission** A method of communications in which the frequencies produced by the process of modulation are symmetrically spaced, both above and below the carrier frequency and are all transmitted.

**double width** A print width in which each character is twice as wide as normal characters.

**DOV** Data Over Voice.

**DOVPath** A service provided by NYNEX operating companies that permits simultaneous voice and data rates up to 19.2 Kbps on a single telephone line.

**Dow Jones News/Retrieval Service** An on-line information service that specializes in providing subscribers with business news to include stock prices, annual report information, mergers, and other news of interest to investors. For additional information, telephone 1-800-257-5114.

**down converter** A device used to lower the frequency of a signal or group of signals.

**downline loading** The process of sending configuration parameters, operating software, or related data from a central source to individual stations.

**downlink** The transmission path of radio frequency signals from a satellite to a ground station.

**down-loop** A fixed pattern (repeating 100) which forces a receiving CSU to loopback the signal to its transmit side.

**downstream** 1. IBM's SNA, in the direction of data flow from the host to the end user. 2. On a ring network, the direction of data flow.

**downstream device** For the IBM 3710 Network Controller, a device located in a network, such that the 3710 is positioned between the device and a host. A display terminal downstream from the 3710 is an example of a downstream device.

**downstream line** In IBM's SNA, a telecommunication line attaching a downstream device to an IBM 3710 Network Controller.

**Downstream Load Utility (DSLU)** In IBM's SNA, a program product that uses the communication network management (CNM) interface to support the load requirements of certain type 2 physical units, such as the IBM 3644 Automatic Data Unit and the IBM 8775 Display Terminal.

**down-time** Period when all or part of a system or network is not available to end users due to failure or maintenance.

**DP** 1. Data Processing. 2. Dial Port. 3. Distribution Panel. 4. Draft Proposal.

**DPBX** Digital Private Branch Exchange.

**DPC** Destination Point Code.

**DPCM** Differential Pulse Code Modulation.

**DPDU** Data Link Protocol Data Unit.

**DPI** Dots Per Inch.

**DPMA** Data Processing Managers Associaton.

**DPNSS** Digital Private Network Signaling System.

**DPPS** Data Packets Per Second.

**DPSK** Differential Phase Shift Keying.

**DPST** Double Pole, Single Throw Switch.

**DQ PACKET** Data Qualifier Packet.

**DQDB**   Distributed Queue Dual Bus.

**DR**   1. Distinctive Ringing. 2. Dynamic Reconfiguration (IBM's SNA).

**Draft International Standard (DIS)**   The last stage in the OSI standardization process before a standard receives final approval.

**draft print mode**   A print mode of a near letter quality (NLQ) dot-matrix printer in which a minimum number of dots per character is used to permit high-speed printing.

**Draft Proposal (DP)**   An ISO standard document that has been registered and numbered but not yet given final approval. The stage in the OSI standardization process before the Draft International Standard (DIS).

**drafter**   In U.S. military communications, the person who actually composes a message for release by the originator or releasing officer.

**DRAM module**   Dynamic Random Access Memory Module.

**Dr.BuB**   The name of an electronic bulletin board operated by Motorola. Dr.Bub is dedicated to providing information on Motorola's Digital Signal Processing (DSP) products.

**DRDS**   Dynamic Reconfiguration Data Set (IBM's SNA).

**drive specification**   In DOS, the letter–colon pair that specifies a drive.

**driver**   1. A software module that controls an input/output port or external device. 2. Short for line driver.

**drop**   1. The place on a multipoint line where a tap is installed so that a station can be connected. 2. In IBM Cabling System, a cable that runs from a faceplate to the distribution panel in a wiring closet.

**drop, subscriber's**   The line from a telephone cable to a subscriber's building.

**drop and insert**   A process in which a part of the data carried in a transmission system is dropped at an intermediate point on a transmission line, channel, or circuit, and different data is inserted for subsequent or continued transmission in the same space, time, frequency, or phase position previously occupied by the terminated data.

**drop cable**   In local area networks, a short cable connecting a network device with the network cable.

**drop sync**   In synchronous formats, the action taken by the receiving terminal when it stops framing a message and begins searching for a new sync sequence.

**drop-and-insert**   A T1 multiplexing technique similar to Channel Bypass, used when some channels in a DS1 stream must be demultiplexed at an intermediate node. With drop-and-insert, all channels are demultiplexed, those destined for the intermediate node are dropped, traffic from the intermediate node is added, and a new T1 stream is created to carry all channels to the final destination. With bypass, only those channels destined for the intermediate node are demultiplexed—ongoing traffic remains in the T1 signal, and new traffic bound from the intermediate node to the final destination takes the place of the dropped traffic.

**drop-and-insert emulation**   A technique for T1 protocol analysis, in which the protocol analyzer terminates one channel of the DS-1 bit stream in one or both directions, and inserts test data into that channel while maintaining the DS-1 line protocol (useful for testing terminal equipment through a T1 multiplexer or PBX).

**dropouts**   Cause of errors and loss of synchronization with telephone data transmission; where signal level unexpectedly drops at least 12 dB for more than 4 milliseconds; AT&T standards suggest only two dropouts per 15-minute period.

**dropped channels**   In a multinode multiplexer network dropped channels are channels which are terminated at a node.

**DRS**   Data Rate Selector.

**dry fiber service** Fiber optic cable installed by a communications carrier without an equipment hookup to customers.

**DRY T1** T1 with an unpowered interface.

**DR-5** A U.S. military spool-type container used to store, transport, lay or recover field wire. The DR-5 holds $2\frac{1}{2}$ miles (4 km) of field wire.

**DSA** 1. Digital Service Area. 2. Directory Service Agent.

**DSC** Direct Satellite Communications.

**DSCP** Data Services Command Processor.

**DSCS** Defense Satellite Communication System.

**DSE** 1. Data Switching Exchange. 2. Digital Service Europe. 3. Distributive System Environment.

**DSI** Digital Speech Interpolation.

**DSIRnet** Government network in New Zealand.

**DSL** 1. Data Set Label. 2. Digital Subscriber Line. 3. Downstream Load.

**DSM** Data Service Manager.

**DSN** 1. Defense Switched Network. 2. Digital Service, level N.

**DSP** 1. Digital Signal Processing. 2. Domain Specific Part.

**DSR** Data Set Ready.

**DSSCS** Defense Special Security Communications System.

**DSU** Data Service Unit.

**DSU/CSU** Digital Service Unit/Channel Service Unit.

**DSVT** Digital Secure Voice Terminal.

**DSX** Digital Cross Connect.

**DSX-1** Digital Cross Connect, Level 1.

**DSX-1** Digital signal cross-connect level 1. The set of parameters used where DS-1 digital signals are cross-connected (T1 carrier).

**DS-0** Digital Signal zero. A 64 Kbps channel containing a PCM encoded digitized voice conversation.

**DS-0** Digital System, Level 0. An individual 64 Kbps data or digitized voice channel. Voice digitization is 8-bit PCM.

**DS-1** Digital Signal one, 1.544 Mpbs.

**DS-1 (Digital Signal Level 1)** The type of signal carried by a T1 facility, normally at 1.544 Mbps.

**DS-1C** Digital Signal Level lC, 3.152 Mkbps digital signal.

**DS-1C (Digital Signal Level 1C)** The type of signal carried by a 3.152Mbps digital transmission facility. A DS-1C circuit carries the equivalent of two DS1 circuits, or 448 DS-0 circuits.

**DS-2 (Digital Signal Level 2)** The type of signal carried by a 6.312 Mbps T2 digital transmission facility. A DS-2 circuit carries the equivalent of two DS-1C circuits, 4 DS-1 circuits, or 96 DS-0 circuits.

**DS-3 (Digital Service, Level 3)** A digital circuit of 44.736 Mbps. A DS-3 circuit carries the equivalent of seven DS-2 circuits, 14 DS-1C circuits, 28 DS-1 circuits, or 672 DS-0 circuits. DS-3 circuits are now the largest units of digital transmission available for private use.

**DS-4 (Digital Service, Level 4)** A digital circuit of 274.176 Mbps, currently the largest unit of transmission in the North American Digital Hierarchy. A DS-4 circuit carries the equivalent of six DS-3 circuits, 42 DS-2 circuits, 84 DS-1C circuits, 168 DS-1 circuits, or 4032 DS-0 circuits. *Note.* Do not confuse the DS-4 transmission level with D4 framing, a technique used only on DS-1 circuits.

**DT** Distant Terminal.

**DTACCS** Director Telecommunications and Command and Control Systems.

**DTE** Data Terminal Equipment.

**DTE-C** A character mode DTE. A DTE-C only

communicates with other DTEs on a character-by-character basis.

**DTE-P**   A packet mode DTE. A DTE-P has a processing capability that enables it to transmit and receive packets.

**DTG**   Date-Time Group.

**DTI**   Department of Trade and Industry.

**DTLU**   Digital Trunk and Line Unit.

**DTMF**   Dual Tone Multi-Frequency.

**DTMF/MF COMPELLED OPERATION**   The method of using DTMF/MF tones to communicate commands and responses to and from a master controlling unit.

**DTO**   Direct Termination Overflow.

**DTR**   Data Terminal Ready.

**DTR flow control**   A procedure for a communicating device to signal its readiness to receive data by raising the DTR lead on an EIA RS-232D interface.

**DTS**   1. Digital Tandem Switch. 2. Digital Terminal System. 3. Digital Termination System.

**DTSC**   Defense Telecommunication System Center.

**D-type conditioning**   Conditioning used to control the signal to C-notched noise ratio and intermodulation distortion.

**dual cable**   A type of broadband cable system in which separate cables are used for transmission and reception.

**dual chassis**   Two interconnected chassis in a cabinet with one network address.

**dual compact XL**   A Tymnet packet switch which houses two communication processors. The Dual Compact XL can support up to 80 asynchronous or a maximum of 48 synchronous ports.

**Dual Dial Restoral (DDR)**   An option that provides public switched telephone network backup to the

modem's normal leased line channel in the event of failure of that channel.

**dual fiber cable**   A type of optical fiber cable that has two single fiber cables enclosed in an extruded overjacket of polyvinyl chloride, with a rip cord for peeling back the overjacket to access the fibers.

**Dual In-Line Pins (DIP)**   Term used to describe the pin arrangement on an integrated circuit (IC) or a multiple (electric) switch.

**dual time diversity**   The transmission of a message twice using a time lapse of $N$ minutes between messages. In the event the first message is not received the second message may be successful.

**Dual-Tone Multiple-Frequency (DTMF)**   Term used to describe the audio signaling frequencies on touch-tone, pushbutton telephones. The telephone handset generates a composite audio signal made of the superposition of two tones selected by line-and-column addressing of a keyboard. This scheme is shown in the illustration. These frequencies were chosen in such a way that neither their harmonics nor their intermodulation products fall in one of the tone bands. Separation between tones is typically 10 percent.

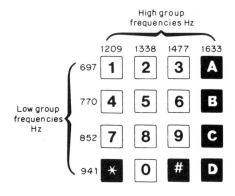

DTMF keypad encoding

The fourth column of buttons (1633 Hz tone) is not usually present in today's telephone sets. However, it is available for future use. A 4 × 4 matrix of buttons is presently used in some military phones and is suitable for implementation in telephone sets intended for multiple functions such as financial transaction terminals, credit checking machines, and some PBX or CBX systems.

The dual-tone signal is heard in the earpiece and is sent over the telephone lines to the private branch exchange (PBX) or directly to the central office, where it must be decoded into the digit that it represents. The decoder circuitry—or receiver—will convert the DTMF signal to a binary format or to a 2-of-8 format. For the traditional dial-pulse equipment, a further conversion is needed to achieve a complete touch-tone to dial-pulse interface, allowing telephone companies to offer subscribers touch-tone service in areas where the equipment is not compatible. Such an application is termed a tone-to-pulse converter.

**dualview** An option for Codex 2600 series of leased line modems which enables them to be managed directly from IBM's Netview or from a Codex network management system.

**duct system** The tubular enclosures that hold wires and cables, enabling them to pass through the floors or ceilings of a building.

**dumb terminal** A term used to describe a Teletype or Teletype-compatible terminal. The dumb terminal is an asynchronous terminal that may operate at speeds as high as 9600 bps or higher. The dumb terminal is an ASCII terminal that, although it may be "intelligent" in many of the functions it provides, uses no communications protocol.

**dump** 1. Computer printout of storage. 2. To write the contents of all or part of storage to an external medium as a safeguard against errors or in connection with debugging. 3. (ISO) Data that have been dumped.

**duobinary signalling** A method of transmitting asynchronous binary waveform in which neighboring signal segments influence one another in a controlled manner.

**DUP** Data User Part.

**duplex** A facility which permits transmission in both

directions simultaneously (sometimes referred to as full-duplex, contrasted with half-duplex).

**duplex chassis** Two independent chassis in a cabinet with two network addresses.

**duplex circuit** A circuit used for transmission in both directions at the same time. It may be called "full duplex" to distinguish it from half-duplex.

**duplex signaling (DX)** A signaling system which occupies the same cable pair as a voice path, yet does not require filters.

**duplex transmission** Simultaneous two-way independent transmission in both directions. Also called full-duplex.

**duplexing** An outdated term referring to the use of duplicate computers, files or circuitry, so that in the event of one component failing, an alternative one can enable the system to carry on its work.

**duty cycle** The relationship between the time a device is on and the time it is off. For example, a terminal that is powered on for one hour a day is said to have a low duty cycle.

**DUV** Data Under Voice.

**DVI** Digital Video Interactive.

**DX** 1. Duplex Signaling. 2. Duplex Transmission.

**DXMCOMOD.SYS** The file name which contains IBM's token-ring adapter handler software which operates on IBM personal computers.

**DYAD** A Siemens trademark for the vendor's series of digital telephones that are used with the firm's SATURN digital voice/data business communications systems.

**dynamic allocation** A technique in which the resources assigned are determined by criteria applied at the moment of need.

**dynamic assignment** In Digital Equipment Corporation Network Architecture (DECnet), the use of a Dynamically Established Data Link by the Network layer, in such a way that a connection is only made when data is required to be transferred, and the subnetwork address to which the connection is made is determined by the destination NSAP address.

**Dynamic Connection Management (DCM)** In Digital Equipment Corporation Network Architecture (DECnet), the use of a Dynamically Established Link by the Network layer, in such a way that a connection is only made when data is required to be transferred.

**dynamic LPDA** In IBM's SNA, a function enabling the NCCF application to set/query the LPDA status for a link/station.

**dynamic multiplexing** Time-division multiplexing in which the allocation of time to constituent channels is based on the demands of these channels.

**Dynamically Non-Hierarchical Routing (DNHR)** An AT&T routing procedure for 800-type calls which provides 21 alternate routes to complete every call.

**Dynamic Random Access Memory (DRAM)** A type of computer memory in which information can be stored and retrieved in miscellaneous order, but which must be maintained or "refreshed" by a periodic electrical charge if the memory is not "read out" or used immediately.

**Dynamic Reconfiguration (DR)** In IBM's VTAM, the process of changing the network configuration (peripheral PUs and LUs) associated with a boundary node, without regenerating the boundary node's complete configuration tables.

**Dynamic Reconfiguration Data Set (DRDS)** In IBM's VTAM, a data set used for storing definition data that can be applied to a generated communication controller configuration at the operator's request.

**dynamic tables** A configuration in which tables can be downline-loaded into a network device while the device is in operation.

**dynamic threshold alteration** In IBM's SNA, an NCP function to allow an NCCF operator to dynamically change the traffic count and temporary

error threshold values associated with SDLC and BSC devices.

**dynamic threshold query** In IBM's SNA, an NCP function to allow an NCCF operator to query the current settings of a traffic count or temporary error threshold value associated with an SDLC or BSC device.

**dynamically established data link** In Digital Equipment Corporation Network Architecture (DECnet), a connection-oriented subnetwork used by the Network layer.

**Dynamically Hierarchical Network Routing (DHNR)** Software which allows the U.S. Sprint network to be partitioned to support the FTS2000 network as a defined partition of the entire U.S. Sprint network.

**Dynamically Redefinable Character Set** The ability to download dot patterns of text or graphic characters to a terminal's memory instead of transmitting them every time they are required.

**D&I** Drop and Insert.

**D1** A digital cellular mobile telephone network in Germany operated by Deutsch Bundespost Telekom.

**D-1 conditioning** AT&T conditioning for two-point services not arranged for switching and for two-point services arranged for through switching, which controls the signal to C-notched noise ratio and intermodulation distortion.

**D1D** A channel placement and counting sequence in which 64 Kbps data streams from channels 1 to 12 of a T1 signal are placed into odd-numbered time slots 1 to 23 while data from channels 13 to 24 are placed into even-numbered time slots 0 to 24.

**D-2 conditioning** AT&T conditioning for two-point switched services, which controls the signal to C-notched noise ratio and intermodulation distortion.

**D-3 conditioning** AT&T conditioning for SCAN System B access lines terminated in a four-wire telephone arrangement to control the signal to C-notched noise ratio and intermodulation distortion.

**D4**   An AT&T specified frame format that designates every 193rd bit position in an AT&T supplied T1 facility as reserved for D4 allows continuous monitoring and non-destructive diagnostic framing to be implemented by the carrier.

**D4 channel bank**   The current generation of central office T1 multiplexing equipment. Capable of handling D4 and Extended Superframe formats. The standard PCM channel bank was originally made by Western Electric and commonly found in Bell Operating Company central offices.

**D4 framing**   A framing technique for T1 circuits using D4 channel banks. D4 framing uses a 12-frame Superframe with bit-robbing for AB signaling in each channel octet of the 6th and 12th frames of each superframe. D4 framing is now being superseded by the Extended Superframe Format.

**D-5 conditioning**   AT&T conditioning for polled multipoint data services not arranged for switching, which controls the signal to C-notched noise ratio and intermodulation distortion.

**D-6 conditioning**   AT&T high performance data conditioning available for two-point applications within the contiguous United States. D-6 conditioning provides more stringent control over intermodulation distortion, signal-to-C-notched noise ratio, phase jitter, attenuation distortion and envelope delay distortion than other types of conditioning. The following table lists D-6 conditioning parameters.

| *Parameter* | *Requirement* |
|---|---|
| Intermodulation distortion (4-tone) | |
|    Signal to second order (R2) | $\geq$ 45 dB |
|    Signal to third order (R3) | $\geq$ 46 dB |
| Phase jitter | |
|    20–300 Hz | $\leq$ degree peak to peak |
| Signal to C-notched noise ratio | $\geq$ 32 dB |
| Envelope delay distortion | |
|    604–2804 Hz | 1400 $\mu$s |

139

# E

**EAASY SABRE** An on-line travel service marketed by American Airlines which augments the firm's SABRE computerized reservation system.

**EADAS** Engineering and Administrative Data Acquisition System.

**EAEO** Equal Access End Office.

**EAM** Emergency Action Message.

**Early Token Release** An enhancement to IBM's original token-ring architecture which allows multiple "frames," or units, of data to be on the ring at the same time. This capability enables a workstation on the ring to transmit a token immediately after sending a frame of data instead of waiting for the frame to return.

**EARN** European Academic Research Network.

**earth stations** Ground terminals that use antennas and other related electronic equipment designed to transmit, receive, and process satellite communications.

**EAS** 1. Equal Access Signaling. 2. Extended Area Service.

**Eastern Mediterranean Optical System (EMOS)** A fiber optic undersea cable which is designed to link the European Continent with Sicily, Greece, Turkey, and Israel.

**Easylink** An electronic mail service formally marketed by Western Union Corporation of Upper Saddle River, NJ and now part of AT&T.

**Easynet** The internal computer network of Digital Equipment Corporation.

**Easyplex** An electronic mail system offered by CompuServe Inc. of Columbus, OH.

**Easytalk** A trademark of Nady Systems, Inc. of Oakland, CA as well as a series of telephone headsets.

**Easy-View** A trademark of AR Telenex of Springfield, VA, for a menu system incorporated into a series of protocol analyzers manufactured by that company.

**EBCDIC** Extended Binary Coded Decimal Interchange Code. The EBCDIC code implementcd for the IBM 3270 information display system is illustrated on page 142.

**EC** European Commission.

**ECAC** Electronmagnetic Compatibility Analysis Center.

**ECB** Event Control Block (IBM's SNA).

**ECC** Error Correcting Code.

**ECCM** Electronic Counter-CounterMeasure.

**ECD** Equipment Configuration Database.

**echo** 1. Distortion caused when a signal being reflected back to the originating station is perceptibly delayed in time. 2.The transmission of messages between bulletin board systems. A group of bulletin board systems which exchange messages using a predefined format is known as an echo network.

| Bits 4567 | Hex 1 | 00 | | | | 01 | | | | 10 | | | | 11 | | | | Bits ← 0, 1 |
|---|---|---|---|---|---|---|---|---|---|---|---|---|---|---|---|---|---|---|
| | | 00 | 01 | 10 | 11 | 00 | 01 | 10 | 11 | 00 | 01 | 10 | 11 | 00 | 01 | 10 | 11 | ← 2, 3 |
| | | 0 | 1 | 2 | 3 | 4 | 5 | 6 | 7 | 8 | 9 | A | B | C | D | E | F | ← Hex 0 |
| 0000 | 0 | NUL | DLE | | | SP | & | · | | | | | | | | | 0 | |
| 0001 | 1 | SOH | SBA | | | | / | | | a | j | | | A | J | | 1 | |
| 0010 | 2 | STX | EUA | | SYN | | | | | b | k | s | | B | K | S | 2 | |
| 0011 | 3 | ETX | IC | | | | | | | c | l | t | | C | L | T | 3 | |
| 0100 | 4 | | | | | | | | | d | m | u | | D | M | U | 4 | |
| 0101 | 5 | PT | NL | | | | | | | e | n | v | | E | N | V | 5 | |
| 0110 | 6 | | | ETB | | | | | | f | o | w | | F | O | W | 6 | |
| 0111 | 7 | | | ESC | EOT | | | | | g | p | x | | G | P | X | 7 | |
| 1000 | 8 | | | | | | | | | h | q | y | | H | Q | Y | 8 | |
| 1001 | 9 | | EM | | | | | | | i | r | z | | I | R | Z | 9 | |
| 1010 | A | | | | | ¢ | ! | ¦ | : | | | | | | | | | |
| 1011 | B | | | | | . | $ | , | # | | | | | | | | | |
| 1100 | C | | DUP | | RA | < | * | % | @ | | | | | | | | | |
| 1101 | D | | SF | ENQ | NAK | ( | ) | _ | ' | | | | | | | | | |
| 1110 | E | | FM | | | + | ; | > | = | | | | | | | | | |
| 1111 | F | | ITB | | SUB | | | ¬ | ? | " | | | | | | | | |

*EBCDIC*

**echo area** In IBM's PROFS and Office Vision, the echo area is the area on the main menu screen below the calendar and list of tasks where messages and reminders are displayed.

**echo cancellation** The technique used in high-speed modems which isolates and removes unwanted signal energy resulting from echoes of the transmitted signal.

**echo canceller** A chip that identifies and erases echoes occurring in satellite and terrestrial long distance phone calls of about 2000 miles and beyond. Essentially, it creates a mirror image of an echo which then blots up the real echo when it comes along.

**echo canceller testing** Testing of an echo canceller's ability to attenuate unwanted echo signals, signal distortion, and verify disable operation.

**echo check** A method of checking data transmission accuracy whereby the received data are returned to the sending end for comparison with the original data.

**echo distortion** A telephone line impairment caused by electrical reflections at distant points where line impedances are dissimilar.

**echo modulation** A method of producing a shaped pulse in which a main pulse with "echos" is generated and put through a simple band-limiting filter. The "echos" are pulses of controlled amplitude occurring both before and after the main pulse.

**echo plug** A standard DB25 connector wired to echo transmitted data and modem signals back to received-data and modem-signal inputs for diagnostic purposes. Also called a loopback plug.

**echo suppressor** A device used by telcos or PTTs that blocks the receive side of the line during the time that the transmit side is in use, preventing energy from being reflected back (echoed) to the transmitter. Because of impedance irregularities in the terminating networks, a small amount of the energy transmitted in one direction over the switched network is often reflected at the receiving end back toward the originating end. This reflection is commonly referred to as a talker echo. If the talker echo encounters another impedance irregularity at the originating end, a second echo called a listener echo is produced.

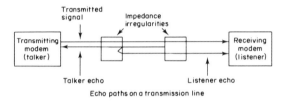

Echo paths on a transmission line

Most modems are designed to ignore talker echoes. On the other hand, listener echoes can interfere directly with receiving data if the time delay is sufficient and the echo is of sufficient amplitude. As a consequence, echo suppressors are selectively employed in the switched network. The echo suppressor blocks the reflections.

Characteristics of the echo suppressors are:

(1) There is only one echo suppressor in each transmission path and it is between regional centers.

(2) It is normally used where the echo signal path delay is greater than 45 milliseconds.

(3) The echo suppressor has a turnaround time of 100 milliseconds.

(4) Echo suppressors are disabled with a signal operating in the range of 2010 to 2240 Hz and the signal must be present for at least 400 milliseconds.

(5) Echo suppressors are enabled when no power is present, in the 300 to 3000 Hz range for a period greater than 100 milliseconds.

**echo testing** A method of checking data transmission for accuracy by returning received data to the sending end for comparison with the original data. (1) Immediate Echo Data is retransmitted character by character as soon as it is received by the echoing device. (2) Buffered Echo Data is captured and stored by the echoing device until a predefined message end character is received. The accumulated data is then retransmitted to the original transmitter.

**EchoMail** A public message area or conference on a bulletin board system that is "echoed" to other systems in a BBS network. EchoMail is organized into different groups, each with a different topic and the term normally refers to communications on a FidoNet network.

**echoplex** A method of checking data integrity by returning characters to the sending station for verification of data integrity. Also called echo check.

**ECL** Emitter Coupled Logic.

**eclipse** In satellite communications the blockage of the sun or another satellite when a satellite passes through the line between the earth and the sun or another satellite.

**ECM** Electronic CounterMeasures.

**ECMA** European Computer Manufacturers Association.

**ECN** Engineering Change Notice.

**ECOM** Electronic Computer Originated Mail.

**ECOM 25** A family of multiprotocol switches with integrated Packet Assembly/Disassembly (PAD) functionality marketed by OST, Inc., of Chantilly, VA. An ECOM 25 node supports up to 72 channels that can be configured to support X.25, X.32, X.75, PAD, Videotex, VIP7700, or SDLC protocols.

**ECP** 1. Emergency Command Precedence. 2. Executive Cellular Processor.

**ECS** European Communications Satellite.

**ECSA** Exchange Carriers Standards Association.

**ECSC** Exchange Carriers Standards Commission.

**EDI**   Electronic Data Interchange.

**EDIA**   Electronic Data Interchange Association, formerly called the Transportation Data Co-ordinating Committee (TDCC).

**EDIFACT**   Electronic Data Interchange For Administration, Commerce and Transport.

**EDP**   Electronic Data Processing.

**EDS**   Electronic Directory Service.

**EDTV**   Enhanced Definition TV.

**EEC**   European Economic Community.

**EECT**   End-to-End Call Trace.

**EEPROM**   Electrically Erasable Programmable Read-Only Memory.

**Effective Isotropic Radiated Power (EIRP)**   The measurement of the signal strength of a satellite at a location on earth.

**EFS**   Error Free Seconds.

**EFT**   Electronic Funds Transfer.

**EFTO**   Encrypted For Transmission Only.

**EFTPOS**   Electronic Funds Transfer at Point Of Sale.

**EGA**   Enhanced Graphics Adapter.

**EGP**   Exterior Gateway Protocol.

**EHF**   Extremely High Frequency.

**EIA**   Electronic Industries Association.

**EIA communication adapter**   A communication adapter conforming to EIA standards that can combine and send information at speeds up to 19.2 Kbps.

**EIA interface**   A set of signal characteristics (time duration, voltage, and current) specified by the Electronic Industries Association and accepted by the American National Standards Institute (ANSI).

**EIA standard equipment rack unit**   The distance between mounting holes (center to center) in front vertical rails of an EIA standard equipment rack/cabinet. One unit equals 1.75 inches.

## EIA STANDARDS

| | |
|---|---|
| RS-232-C | An EIA-specified physical interface, with associated electrical signaling, between data circuit-terminating equipment (DCE) and data terminal equipment (DTE); the most commonly employed interface between computers and modems. |
| RS-269-B | Synchronous signaling rates of data transmission. |
| RS-334-B | Signal quality at DTE/synchronous DCE interface. |
| RS-363 | Signal quality at DTE/non-synchronous DCE interface. |
| RS-366-A | DTE/Automatic Call Equipment interface. |
| RS-404 | Start–stop signal quality between DTE and non-synchronous DCE. |
| RS-422-A | Electrical characteristics of balanced-voltage digital interface circuits. |
| RS-423-A | Electrical characteristics of unbalanced-voltage digital interface circuits. |
| RS-485 | The recommended standard of the Electronic Industry Association that specifies the electrical characteristics of generators and receivers for use in balanced digital multi-point systems. |
| RS-496 | DCE/PSTN interface. |
| RS-499 | General-purpose 37-position and 9-position interface for data terminal equipment and data circuit-terminating equipment employing serial binary data interchange. |
| RS-499-1 | Addendum 1 to RS-499 |

**EIN**   European Infomatics Network.

**EIRP**   Effective Isotropic Radiated Power.

**EISA**   Extended Industry Standard Architecture.

**EIT**   Encoded Information Type.

**EKRAN satellite**   The second Russian geostationary communications satellite that was launched in 1976. This satellite was the world's first direct television broadcast satellite.

**EKS**   Electronic Key System.

**ELA**  Extended Line Adapter.

**elastic buffer**  A storage device which compensates for speed or throughput differences. The buffer is used to store data temporarily until the receiving device is ready to receive it.

**elastic store**  A storage buffer designed to accept data under one clock and deliver it under another, to accommodate short-term instabilities (jitter) in either clock. Elastic stores are extensively used in multiplexing and clock recovery applications.

**Electra**  A series of key telephone systems marketed by NEC America, Inc. The Electra 8/24 is an analog key system that supports up to eight lines and 24 stations.

**Electrically Erasable Programmable Read Only Memory (EEPROM)**  A nonvolatile memory device in which the contents can be changed by special circuitry incorporated in the parent system or device.

**Electrically Programmable Read Only Memory (EPROM)**  Memory which is initially programmed by the manufacturer. Changes can be made to conform to customer requirements at their premises.

**ElectroMagnetic Compatibility (EMC)**  The ability of suitable designed electronic equipment to operate in the same location without mutual interference.

**ElectroMagnetic Interference (EMI)**  The interference in signal transmission or reception caused by the radiation of electrical and magnetic fields.

**ElectroMagnetic Pulse (EMP)**  An intense, short-duration burst of electromagnetic waves generated by a nuclear explosion. EMP is many times stronger than the static pulse generated by lightning and can cause such circuit components as transistors and integrated circuits to fail.

**electromagnetic wave**  A wave of energy consisting of interrelated electric and magnetic waves. Both light and radio are examples of electromagnetic waves.

**electron**  An elementary particle made of a tiny charge of negative electricity.

**electronic banking**  Any banking transaction which takes place electronically, including automatic teller machines, bank-to-bank wire transfers, etc.

**Electronic Data Interchange for Administration, Commerce and Transport (EDIFACT)**  An international EDI standard effort of the United Nations.

**electronic funds transfer**  The use of communications to simultaneously debit an account in one location and credit an account in another, thus "transferring" funds.

**Electronic Industries Association (EIA)**  An American group whose members include manufacturers that formulates technical standards and marketing data. The EIA often represents the telecommunications industry in governmental concerns. United States contributions to the CCITT (now known as the ITU) are made through the EIA.

**Electronic Key System (EKS)**  A microprocessor based key system which provides users with more station features than 1A2 key systems.

**electronic mail**  A store and forward service for text and graphic based messages from one computer terminal or system to another. The text is stored for the recipient until that person logs into the system to retrieve messages.

**Electronic Order Exchange (EOE)**  A data communications approach to inter company transactions between buyers and sellers. EOE can be employed to transmit purchase orders, price and product listings, and order-related information.

**electronic patch panel**  Programming mechanism that enables the user to change dedicated port routing using program commands.

**Electronic Switched Network (ESN)**  A comprehensive private networking package developed by Northern Telecom which connects geographically dispersed PBXs into a single, unified, private communications network.

**Electronic Switching System (ESS)** A type of telephone switching system which uses a special purpose stored program digital computer to direct and control the switching operation. ESS permits the provision of custom calling services such as speed dialing, call transfer, three-way calling, etc.

**Electronic Switching System (ESS)** AT&T manufactured, stored-program-control, central office switches. Common ESSS include Nos 1, 1A, 4 and 5 switches.

**Electronic Tandem Network (ETN)** An AT&T PBX-to-PBX network facility for large users that was popular in the 1970s.

**electronic wire transfer** In banking, a means by which funds from an account in one bank are transferred to another via telephone.

**ElectroStatic Discharge (ESD)** A surge of electric current from an accidental source of stored electric charge. This is usually an accidental event resulting from a user moving around in a dry environment.

**Elektroimpek** The purchasing agent for the Hungarian telecommunications administration.

**element** In IBM's SNA: 1. A field in the network address. 2. The particular resource within a subarea identified by the element field.

**element address** In IBM's SNA, a value in the element address field of the network address identifying a specific resource within a subarea.

**Element Management System (EMS)** A system which manages network elements, such as modems, multiplexers and digital service units.

**elevation** In satellite communications, the angle between an antenna beam and the horizontal plane.

**ELF** Extremely Low Frequency.

**ELINT** ELectronic INTelligence.

**ELISA** An electronic messaging system operated by Helsinki Telephone Company in Finland.

**ELSEC** ELectronic SECurity.

**ELT** Emergency Locator Transmitter.

**EM** End of Medium.

**EMA** Enterprise Management Architecture.

**E-mail** Electronic mail.

**embedded base equipment** All customer-premises equipment provided by Bell Operating Companies (BOCs) prior to January 1, 1984 that was ordered transferred from the BOCs to AT&T by court order.

**embedded blanks** Blank characters that are surrounded by any other characters.

**Embedded Operations Channel (EOC)** 1. The ISDN channel which carries commands and responses between network elements and the NT1. 2. A portion of the bandwidth of a circuit used for the transmission of maintenance and/or administrative information.

**embedded signalling** A technique for transmitting signalling information on the same channel as the customer information.

**Embratel** Brazil's federal telecommunications operator.

**EMC** Electromagnetic Compatibility.

**Emergency Power System (EPS)** Consists of an engine and a generator set. This system, designed to start within 30 seconds of a commercial power failure, carries the electrical load for the duration of the commercial power failure.

**EMETF** ElectroMagnetic Environmental Test Facility.

**EMI** ElectroMagnetic Interference.

**EMI/RFI** ElectroMagnetic Interference/Radio Frequency interference.

**emission code** A three position code used by the U.S. military to define the type of electromagnetic emission produced by an electronic device. The emission code is of the form: NMI where: N=bandwidth in kHz, M=modulation type (A—amplitude, F—frequency or phase, P—pulse) and I=intelligence;

where: 0—none

1—telegraphy (CW, FSK, NSK)

2—modulated CW (MCW)

3—telephone (voice)

3A—single sideband, reduced carrier

3J—single sideband, suppressed carrier

3B—two independent sidebands, reduced carrier

3H—single sideband, full carrier

4—facsimile

5—television

9—composite, or not otherwise covered

**emission security** Communications security measures taken to deny unauthorized persons information of value which is derived from the intercept and analysis of compromising emanations from cryptographic equipment and telecommunications systems.

**Emitter Coupled Logic (ECL)** A type of transistor circuit characterized by fast operation and high heat dissipation.

**EMOS** Eastern Mediterranean Optical System.

**EMOS-1** The Eastern Mediterranean fiber optic cable system put into service during 1990.

**Emoticons** Figures created with the symbols on the keyboard and sent with electronic mail to express the emotions of normal voice communications. They are used to convey the spirit in which a line of text is typed. (See table below)

**EMP** ElectroMagnetic Pulse.

**EMPI** Electrical Pulse Interference.

**empty slot ring** In LAN technology, a ring LAN in which a free packet circulates past (or, more precisely, through) every station; a bit in the packet's header indicates whether it contains any messages (if it contains messages, it also contains source and destination addresses).

**EMS** 1. ElectroMagnetic Sprectrum. 2. Element Management System.

**EMSEC** EMission SECurity.

**emulation** A device or computer program, or combination that acts like another device or computer program. For example, a protocol analyzer can emulate DTE or DCE.

**emulation mode** In IBM's SNA, the function of a network control program that enables it to perform activities equivalent to those performed by a transmission control unit.

**Emulation Program (EP)** In IBM's SNA, a control program that allows a local 3705 Communications Controller to emulate the fuction of an IBM 2701 Data Adapter Unit, an IBM 2702 Transmission Control, or an IBM 2703 Transmission Control.

**EM4010** A Tektronix 4010/4014 graphics terminal emulator marketed by Diversified Computer Systems of Boulder, CO, for use on IBM PC and compatible personal computers.

**EM220** A DEC VT220 terminal emulator marketed by Diversified Computer Systems of Boulder, CO, for use on IBM PC and compatible personal computers.

**enable** To turn on a device or place it in a state which will allow certain interrupts to occur in a processing unit. Interrupts are usually enabled by issuing a command or setting a switch or jumper.

**enabled** 1. In IBM's ACF/VTAM, pertaining to an LU that has indicated to its SSCP that it is now ready to establish LU–U sessions. The LU can separately indicate whether this applies to its ability to act as a PLU or an SLU. 2. Active, operational, and can receive frames from the network. (Servers and functional addresses may be enabled by programs running on the Token-Ring Network.)

**Encapsulation** 1. The process of adding control information to data segments. Such control information can include addressing, destination receipt acknowledgment and error correction. 2. The term used to describe the carrying

## EMOTICON SYMBOLS

| Emoticon | Meaning | Emoticon | Meaning | Emoticon | Meaning |
|---|---|---|---|---|---|
| .-] | User has one eye | :-' | User spitting out chewing tobacco | {(:-) | User is wearing toupee |
| .-) | User has one eye | :-* | User after eating something bitter | }(:-( | User, wearing toupee in wind |
| :8) | User is a pig | :-> | Hey hey | +-(:-) | User is the pope |
| :-i | Semi-Smiley | :-E | User has a dental problem | +:-) | Smiley priest |
| :-] | Smiley blockhead | :-X | User is wearing a bow tie | *:o) | User is a Bozo |
| :-% | User has beard | :-0 | User is an orator | *<I:-) | User is Santa Claus |
| :-o | user singing national anthem | :-7 | User after a wry statement | <:I | Dunce |
| :-t | User is cross | :-#| | User with bushy mustache | I-)= | User is Chinese |
| :-: | User is mutant | :-@ | User face screaming | =:-) | User is a hosehead |
| :-( | Drama | :-% | User is a banker | =:-#} | Smiley punk with a moustache |
| :-) | Comedy | :-} | User wears lipstick | =:-) | Smiley punk-rocker |
| :- | User is male | :-c | Bummed out Smiley | >- | Female |
| :-? | User is smoking pipe | :-J | I'm being tongue-in-cheek | >:-< | Mad |
| :-=) | Older user with mustache | :-x | "My lips are sealed" Smiley | %-) | User is cross-eyed |
| :-\ | Undecided user | :-l | "Have an ordinary day" Smiley | #-) | User partied all night |
| :-p | User is sticking tongue out (at you!) | :-< | Real sad Smiley | @:I | Turban |
| :-)' | User tends to drool | :-( | User is crying | :-) | Hee hee! |
| :-'l | User has a cold | :-I | Hmm! | :-D | Ho ho! |
| :-)8 | User is well dressed | :-8( | Condescending stare | I-) | User is asleep (boredom) |
| :-D | User talks too much | :-O | Uh oh! | I-P | Yuk! |
| :-\ | Popeye smiling face, for people who look like popeye | :-Q | User is smoking/smoker | 3:-o | User is a cow |
| :-# | User's lips are sealed | :=) | User has two noses | 5:-) | I'm Elvis |
| :-o | User is bored (yawn) | :> | Midget Smiley | 8-) | User wears glasses |
| :-& | User is tongue-tied | :>) | User has a big nose | 8:-) | Glasses on forehead |
| :-* | User just ate a sour pickle | :%)% | User has acne | 8] | Normal smiling face except that user is a gorilla |
| :-)-{8 | User is a big girl | :-))) | User is overweight | B-) | Horn-rims |
| :-s | User after a bizarre comment | ;-) | Winking Smiley | B-l | User is wearing cheap sunglasses |
| :-o | User is surprised | ;-\ | Popeye gets his lights punched out | o-) | User is a cyclops |
| :-l | No expression face, "that comment doesn't faze me" | '-) | User only has a left eye, which is closed | L:-) | User just graduated |
| -:-) | User sports a mohawk and admires Mr. T | (-: | User is left-handed | P-) | User is getting fresh |
| :-& | User is tongue-tied | (-) | User needing a haircut | [:-) | User is listening to Walkman radio |
| :-9 | User licking his or her lips | (:-) | Smiley big-face | [:l] | User is a robot |
| :-( | Sad | (:I | Egghead | CI:-= | I'm Charlie Chaplin |
| | | {:-) | Smiley with its hair parted in the middle | | |

of information in one protocol by another protocol. For example, an Ethernet frame transporting TCP/IP datagrams would be said to have encapsulated the datagrams.

**encipher**  In IBM's SNA, 1. To scramble data or convert it, prior to transmission, to a secret code that masks the meaning of the data to any unauthorized recipient. 2. In ACF/VTAM, to convert clear data into enciphered data.

**enciphered data**  In IBM's SNA, data that is intended to be illegible to all except those who legitimately possess the means to reproduce the clear data. Synonymous with ciphertext.

**encode**  To code.

**Encoded Information Type (EIT)**  In the CCITT Message Handling System the EIT in the envelope defines the type of encoded information, such as text, teletex or facsimile.

**encoder**  A portion of a multi-bit encoded modem, the encoder groups bits for modulation.

**encoding/decoding**  The process of reforming information into a format suitable for transmission.

**encrypt**  In IBM's VTAM, to convert clear data into enciphered data. Synonym for encipher.

**Encrypted For Transmission Only**  A U.S. military communciations procedure designed to protect unclassified portions of communications by encryption.

**encryption**  The process of altering information in such a way that it is unintelligible to unauthorized parties. The process involves transforming a clear message into cipher text to conceal its meaning.

**end bracket**  In IBM's SNA, the value (binary 1) of the end bracket indicator in the request header (RH) of the first request of the last chain of a bracket; the value denotes the end of the bracket.

**end distortion**  Communications signal distortion usually caused by speed error. It is the percentage difference in time between two consecutive mark to space transitions as compared to the ideal time.

End distortion = ((ideal pulse width − actual pulse width)/ideal pulse width) * 100%

If the received speed is too fast the result is positive. If the received speed is too slow the result is negative.

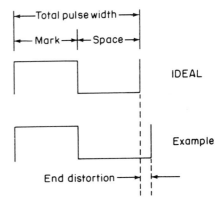

**end of address**  The character used to separate the address or routing part of a message from the subsequent parts of a message. Sometimes called end-of-routing symbol.

**End of Medium (EM)**  A control character that may be used to identify the physical end of data recorded on a medium. Also end of file.

**End of Text (ETX)**  Indicates end of message. If multiple transmission blocks are contained in the message in Binary Synchronous Communications, ETX terminates the last block.

**End Of Transmission (EOT)**  An ASCII code that means "end of transmission" (EOT); used in the EOT/ACK handshaking protocol. The computer sends an EOT at the end of each transmission to the terminal. When the terminal is ready to receive more data, it transmits an acknowledge (ACK) back to the computer.

**End of Transmission Block (ETB)**  A transmission control character used to indicate the end of a transmission block of data.

149

**end office**  A Class 5 telephone central office; local subscribers are connected to an end office.

**end system**  In Digital Equipment Corporation Network Architecture (DECnet), a system which transmits and receives NPDUs from other systems, but which does not relay NPDUs between other systems.

**end tag**  In IBM's PROFS and Office Vision, the end tag is a tag that marks the end of index information in Revisable Form Test (RFT) document.

**end user**  In IBM's SNA, the ultimate source or destination of application data flowing through an SNA network. An end user may be an application program or a terminal operator.

**End-of-Message (EOM)**  In synchronous formats, a character or bit sequence to tell the receiving terminal to drop sync.

**end-of-routing signal**  In U.S. military communications, a switching control signal consisting of a period (.), which is inserted in the position immediately following the last addressee's routing indicator to separate the message address from the text.

**End-Of-Transmission (EOT)**  In U.S. military communications, the final card of a card message containing a repeat of most of the header information and the end-of-transmission signal (EOTS).

**End-Of-Transmission Signal (EOTS)**  In U.S. military communications, a switching control signal used to terminate a data transmission in an automatic electronic switching center.

**endpoint**  A named source and/or destination activity, which can send and/or receive information through the distributed system. Generally synonymous with Mailbox.

**end-to-end digital connectivity**  The linking of one terminal point (telephone, computer, facsimile unit, etc.) to another by a switched digital circuit. There are no analog sections along the circuit's path.

**end-to-end responsibility**  The principle that assigns to communications common carriers complete responsibility for all equipment and facilities involved in providing a telecommunications service from one end of the connection to the other. Also End-to-End Accountability.

**end-user**  An individual who utilizes an information system during the performance of his or her normal duties. End-users can include but are not limited to bank tellers, clerks, factory workers, engineers, and managers.

**Engineering Change Notice (ECN)**  A document that identifies and authorizes a hardware change and explains why the change is necessary. ECNs are used to maintain effective review and control of engineering changes.

**Engineering Change Order (ECO)**  Any design change that will require revision to specifications, drawings, documents, or configurations.

**Enhanced Call Router (ECR)**  An MCI Communications Corporation 800 service feature which allows users of that service to save money by customizing call routing and messaging capabilities supported in MCI's network. ECR capabilities include busy/no answer rerouting, message announcements and menu routing.

**enhanced connectivity facilities**  A set of programs for interconnecting IBM Personal Computers and IBM System/370 host computers operating with the MVS/XA or VM/SP environment.

**Enhanced Definition TV (EDTV)**  An improved quality of television signal that is compatible with existing transmission and receiving equipment but has a lower quality than HDTV (high definition TV).

**Enhanced Interactive Network Optimization System (E-INOS)**  A network design tool product of AT&T Communications of Bedminster, NJ.

**Enhanced Position Location Reporting System (EPLRS)**  An upgraded version of the Position

Location Reporting System (PLRS) used by the U.S. Army.

**enhanced service**  A term used to refer to the combination of "basic" service with computer processing applications that results in a modification of the transmitted information.

**ENQ**  ENQuiry.

**ENQ/ACK protocol**  A Hewlett-Packard communications protocol. The HP3000 computer follows each transmission block with ENQ to determine if the destination terminal is ready to receive more data; the destination terminal indicates its readiness by responding with ACK.

**enquiry (ENQ)**  A control character (control E in ASCII) used as a request to obtain identification or status.

**Entel Chile**  The Chilean intercity telephone company.

**enter**  To send information to the computer by pressing the ENTER key.

**Enterprise Management Architecture (EMA)**  Digital Equipment Corporation's OSI-based network management strategy. EMA provides multivendor network management using an interface based on OSI standards.

**enterprise number**  A synonym for a toll-free number.

**Enterprise Systems Connection Architecture (ESCON)**  The IBM name for the use of fiber optic technology in that vendor's Enterprise 19000 family of computers which speeds up and extends the channels between the computer's main memory, its control unit and peripherals.

**entity**  A network component that can be managed using network management.

**entity attribute**  A piece of management information maintained by an entity. DNA defines Identification, Characteristic, Status, and Counter attributes.

**entity hierarchy**  The logical structure of manageable components in a system.

**entity instance**  An occurrence of entity.

**entity name**  The name of an entity consisting of a global name part and a local name part.

**entry**  A single input operation on a workstation.

**entry point**  The location in an IBM SNA network from which network management and session data is transported to a host over the same link as the one to which the device is attached.

**envelope, analog**  In an amplitude modulated signal the waveform has maxima and minima. The location of these maxima and minima can be joined by two smooth curves which form the envelope of the waveform.

**envelope, digital**  A group of bits in a specific format which usually has a data field as well as qualifier addresses.

**envelope delay**  Characteristics of a circuit which result in some frequencies arriving ahead of other frequencies, even though the frequencies were transmitted together.

**envelope encoding**  The process of providing signaling and control information as a prefix and suffix to a group of information. British Telecom's Kilostream service places customer data into a 6+2 envelope encoded data group as illustrated in the following figure.

Kilostream envelope encoding

8 bit envelope

| A | I | I | I | I | I | I | S |
|---|---|---|---|---|---|---|---|

A = Alignment bit which alternates between "1" and "0" in successive envelopes to indicate the start and stop of each 8-bit envelope.

S = Status bit which is set or reset by the control circuit and checked by the indicator circuit.

I = Information bits

**Envoy 100**  An electronic messaging system operated by Telecom Canada.

**EOA**   End Of Address.

**EOB**   End Of Block; a control character.

**EOC**   Embedded Operations Channel.

**EOF**   End Of File.

**EOL**   End Of Line.

**EOM**   End Of Message.

**EOMS**   End Of Message Signal.

**EON**   End Of Number.

**EORS**   End Of Routing Signal.

**EOT**   End Of Transmission.

**EOTS**   End Of Transmission Signal.

**EP**   1. Emulation Program (IBM's SNA). 2. Emulator Program.

**EPIRB**   Emergency Position-Indicating Radio Beacon.

**EPLRS**   Enhanced Position Location Reporting System

**EPROM**   Erasable Programmable Read-Only Memory.

**EPS**   Emergency Power System.

**EPSCS**   Enhanced Private Switched Communication Service.

**EQ**   Equalizer.

**equal access**   The requirement, due to the divestiture of AT&T, for the Bell Operating Companies (BOCs) to provide interexchange carriers other than AT&T with the same level of access to their central office switches as that provided to AT&T. This permits, as an example, telephone company subscribers to dial long distance using their preferred interexchange carrier without having to dial extra digits.

**Equal Access End Office (EAEO)**   An EES switch used to provide carrier access to collocated stations.

**equalization**   A method of compensating for the distortion introduced during data transmission over telephone channels. Usually a combination of adjustable coils, capacitors, and resistors are used to compensate for the difference in attenuation and delay at different frequencies.

**equalizer**   A device used by modems to compensate for distortions caused by telephone line conditions.

**equalizer, adaptive**   An equalizer that can constantly change its equalization to compensate for changes on the line. Also called automatic equalizer.

**equalizer, automatic**   An equalizer circuit that can adjust itself to compensate for distortion on the line. Also called adaptive equalizer.

**equalizer, fixed**   An equalizer circuit that is permanently set.

**equalizer, manual**   An equalizer that must be adjusted by the user.

**equipment rack**   A metal stand for mounting components.

**equipment room**   The room in which voice and data common equipment is housed, protected, and maintained, and where circuit administration is done using the trunk and distribution cross-connects.

**Equipment Serial Number (ESN)**   In cellular telephone communications a 32-bit binary code which, when transmitted by a mobile station, identifies the manufacturer and specific unit of the telephone.

**equipment wiring subsystem**   The cable and distribution components in an equipment room that interconnect system common equipment, other associated equipment, and cross-connects.

**equipotential bonding**   A means of limiting ground potential differences within a building.

**equivalent four-wire system**   A transmission system using frequency division to obtain full-duplex operation over only one pair of wires.

**ER**   Explicit Route (IBM's SNA).

**ERADCOM**   Electronics Research And Development COMmand.

**Erasable Programmable Read-Only Memory (EPROM)** A nonvolatile semiconductor PROM that can have its current contents cleared (usually through exposure to ultraviolet light) and then accept new contents for storage.

**ergonomics** The science of determining a proper relation between mechanical and computerized devices and personal comfort and convenience. Determining the height and shape of a display is an ergonomic issue.

**ERIC** Educational Resources Information Center.

**erlang** A standard unit of measurement for communications traffic, equal to 36 CCS. Used for throughput and capacity planning, the erlang represents the full-time use of a circuit.

**Erlang B** A formula used to compute the probability that a call will be blocked based on the assumption that blocked traffic is lost.

**Erland C** A formula used to compute the probability that a call will be blocked based upon the assumption that blocked calls are held until a channel becomes free.

**EROS** Earth Resource Observation Satellite.

**ERP** Effective Radiated Power.

**error** Any inconsistency between the true state of what is transmitted and the information that is received.

**error burst** A sudden increase in errors over a short period of time, as compared to the period immediately preceding and following the occurrence.

**error control** An arrangement that detects and may correct errors. In some systems, refinements are added that correct the detected errors, either by operation on the received data or by retransmission from the source.

**error controller** A device that provides error control, usually installed in pairs between the modem and DTE at each end of a data link.

**error correction** An arrangement that restores data integrity in received data, either by manipulating the received data or by requesting retransmission from the source.

**error detection** An arrangement that senses flaws in received data by examining parity bits, verifying block check characters, or using other techniques.

**error detection with retransmission** Error handling technique where the receiver can detect errors, then request retransmission of the message from the sender. Also ARQ.

**error injection** In BERT testing, the process of inserting known violations of the test pattern into the transmitted bit stream, either on command (on low-speed circuits) or at a predetermined rate (on high-speed circuits, such as T1).

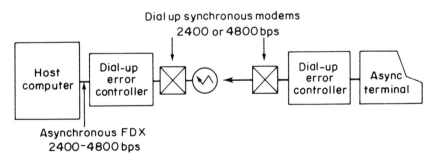

Dial-up error controller configuration

*error controller*

**error injection rate** In BERT testing, the rate at which the test set can be programmed to inject errors into the transmitted test pattern.

**error rate** A measure of data integrity, given as the fraction of bits which are flawed. Often expressed as a negative power of 10—as in 1E-6 (a rate of one error in every one million bits).

**Error-Correcting Code (ECC)** An error-detecting code that incorporates additional signaling elements and enables errors to be detected and corrected at the receiving end. Also called Forward Error Correction (FEC).

**error-detecting telegraph code** A telegraph code in which each telegraph signal conforms to specific rules of construction, so that departures from this construction in the received signals can be automatically detected.

**Error-Free Second (EFS)** A second of activity on a communications circuit during which no bit errors occur.

**ERS-1** The first microwave remote sensing satellite to be launched by the European Space Agency.

**ERTS** Earth Resources Technology Satellite.

**ES** Errored-Second.

**ESA** European Space Agency.

**ESC** ESCape Character.

**ESCape character (ESC)** A code meaning "escape" which is used to control various electronic and mechanical functions of the terminal or the software program currently in use. It is used to denote the termination of some action or to identify the prefix of a command, such as performing a printer operation. An ASCII 0011011 (decimal 27) is an Escape (ESC) Character.

**escape sequence** In a switching network, a sequence of characters entered during a call to access commands without disconnecting the call.

**ESCON** Enterprise Systems Connection Architecture.

**ESD** 1. Electronic Systems Division. 2. ElectroStatic Discharge.

**ESDI** Enhanced Small Device Interface.

**ESF** Extended Superframe Format.

**ESM** 1. Electronic Support Measures. 2. Electronic Warfare Support Measures.

**ESN** Equipment Serial Number.

**ESP** Essential Service Protection.

**ESPRIT** European Strategic Program for Research and development in Information Technology.

**ESS** Electronic Switching System.

**essential facilities** In packet switched networks, essential facilities are standard network facilities which are on all networks.

**ESTAE** Extended Specify Task Abnormal Exit (IBM's SNA).

**ET** Exchange Termination.

**ETB** End of Transmitted Block.

**Ethernet** In LAN technology, a *de facto* standard, developed first by Xerox and then sponsored by Xerox, Intel, and DEC. An Ethernet LAN uses coaxial cables and CSMA/CD. Ethernet is similar to an IEEE 802.3 LAN (they can share the same cable and can communicate with each other).

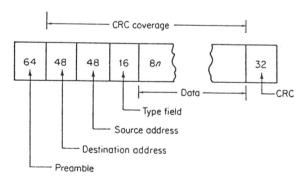

*Ethernet Frame*

On an Ethernet network data is packetized, with packets ranging in size from 64 to 1518 bytes and contains six fields. The preamble allows the receiving station to synchronize with the transmitted message. The destination and source address fields contain the network ID of the nodes receiving and initiating the message. The type field indicates the type of data in the data field. The data field contains the actual data. The CRC field helps the receiving node perform a cyclical redundancy check against the received packet contents.

**Ethernet cable**   A generic term used to refer to a type of 50 ohm, coaxial cable.

**ETN**   Electronic Tandem Network.

**ETO**   Equalized Transmission Only.

**ETR**   Early Token Release.

**ETS**   1. Electronic Tandem Switching. 2. European Telephone System.

**ETSI**   European Telecommunications Standardization Institute.

**ETX**   End of TeXt character.

**EUnet**   European Unix network. A subscription-funded research oriented network.

**Europage**   A paging network in the United Kingdom which enables page subscribers to be contacted in the United Kingdom, France (Alphapage), West Germany (Cityruf), and Italy (Teldrin). Each network operates in the UHF 466 frequency band.

**European Academic Research Network (EARN)** A general purpose computer network connecting universities and research sites throughout Europe, Africa and the Middle East.

**European Computer Manufacturers Association (ECMA)**   A Western European trade organization that issues its own standards and belongs to ISO. Membership include Western European computer suppliers and manufacturers.

**European framing**   The international standard framing format for 2.048 Mbps circuits is also known as the CEPT-30 frame. In European framing,

each frame contains 256 bits, and each multiframe contains 16 frames, divided into two sub-multiframes of eight frames each. The first eight bits of each frame are reserved; the remaining 248 bits contain 31 octet channels. Within the multiframe, frames are numbered 0 through 15, the second through eighth bits of frames 0, 2, 4, 6, 8, 10, 12, and 14 of each multiframe carry the Frame Synchronization Pattern (0011011). The first bit of each frame is reserved for national use, and on networks crossing international borders, its value is fixed at 1. In most national networks, the first bit of each frame alternates among three patterns; the first bits of frames 0, 2, 4, 6, 8, 10, 12, and 14 carry a CRC-8 computed over the previous multiframe; the first bits of frames 1, 3, 5, 7, 9, and 11 carry the pattern 001011; and the first bits of frames 13 and 15 are spare bits reserved further for national use. In frames not carrying the framing synchronization pattern (i.e. frames 1, 3, 5, 7, 9, 11, 13, and 15), the value of the second bit is fixed at 1 to avoid confusion with the framing synchronization pattern, and bit 3 carries a remote alarm indication (0 if no alarm, 1 if alarm condition is indicated); bits 4 through 8 of these frames are spare bits reserved for national use. The CEPT-30 frame and multiframe composition is illustrated in the diagram.

**European Informatics Network (EIN)**   A European project to facilitate research into networks and promote agreement on standards. Its first task was to construct a packet-switched subnet and this became operational in 1976. The five centers on the network are in London, Paris, Zurich, Milan and Ispra and there are also "secondary centers" which join in the research on network protocols. The transport protocol of the EIN is the basis of the one adopted by IFIP Working Group 6.1.

**European Telecommunications Standardization Institute (ETSI)**   An Association under French law whose members can be PTT administrations, public network companies, telecommunication industries, user organizations and research institutes. Headquartered in Antipolis, a suburb of Nice, France. ETSI produces technical standards

CEPT PCM-30 frame and multiframe composition

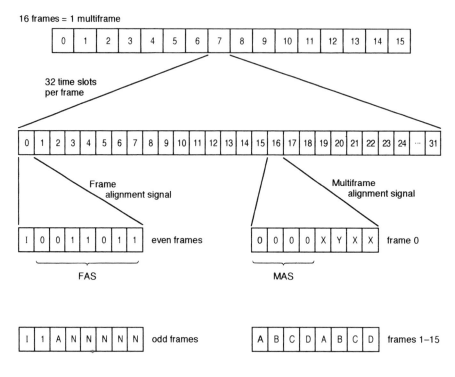

*European framing*

in the area of telecommunications and such related fields as radio and television.

**Euroset Plus**   A trademark of Siemens for that firm's quality telephones designed for home and business use.

**EUT**   Equipment Under Test.

**EUTELSAT**   European Satellite Communications System.

**EUUG**   European Unix Systems Group.

**EVD**   EVent Dispatcher.

**EVE**   Extreme Value Engineering.

156

**even parity check (odd parity check)** This is a check which tests whether the number of digits in a group of binary digits is even (even parity check) or odd (odd parity check).

**event** 1. An occurrence of a normal or abnormal condition detected by an entity and of interest to network management. 2. In IBM's Network Problem Determination Application (NPDA), a resource-generated data base record containing a description of some unusual occurrence that has been detected at the resource.

**Event Control Block (ECB)** In IBM's SNA, a control block used to represent the status of an event.

**EVent Dispatcher (EVD)** The intermediary between event sources and event sinks.

**event driven reconfiguration** The ability of an unattended multiplexer to automatically install a predetermined alternate configuration in response to a predetermined event, such as a link or node failure.

**event sink** An entity which is a consumer of event reports.

**event source** An entity that detects events and generates event reports.

**EVS** Enhanced Videotex Services.

**EW** Electronic Warfare.

**EWC** Electronic Warfare/Cryptologic.

**EX** EXception.

**exception message** A message displayed when an abnormal network condition exists or an illegal command is given.

**EXception Request (EXR)** In IBM's SNA, a request that replaces another message unit in which an error has been detected. *Note.* The exception request contains a 4-byte sense field that identifies the error in the original message unit and, except for some path errors, is sent to the destination of the original message unit; if possible, the sense data is returned in a negative response to the originator of the replaced message unit.

**exception response** In IBM's SNA, a value in the form-of-response-requested field of a request header. The receiver is requested to return a response only if the request is unacceptable as received or cannot be processed; that is, a negative response, but not a positive one, may be returned.

**excess zeros** More consecutive zeros received than are permitted for the selected transmission coding technique. For AMI encoded DS1 signals, 15 or more consecutive zeros are defined as excessive. For B8ZS encoded signals, 8 or more zeros are defined as excessive.

**excessive bipolar violations** An error condition on an AT&T T1 circuit defined as the occurrence of 1544 bipolar violations in a period of 1000 seconds that includes no period of 85 or more seconds without a bipolar violation.

**exchange** A unit established by a common carrier for the administration of communications services in a specified geographical area such as a city. It consists of one or more central offices together with the equipment used in providing the communications services. Frequently used as a synonym for central office.

**exchange, classes of** Class 1—Regional Center; Class 2—Sectional Center; Class 3—Primary Center; Class 4—Toll Center; Class 5—End Office. (ATT definition.)

**exchange, private automatic (PAX)** A dial telephone exchange that provides private telephone service to an organization and that does not allow calls to be transmitted to or from the public telephone network.

**exchange, private automatic branch (PABX)** A private automatic telephone exchange that provides for the transmission of calls to and from the public telephone network.

**exchange, private branch (PBX)** A manual exchange connected to the public telephone network on the user's premises and operated by an attendant supplied by the user. PBX is today commonly used to refer to an automatic exchange.

**exchange, trunk**   An exchange devoted primarily to interconnecting trunks.

**exchange area**   A geographical area that has a single uniform set of telephone charges.

**Exchange Carriers Standards Association (ESCA)**   An ANSI standards group that existed prior to divestiture of AT&T and which now consists of Regional Bell Operating Companies (RBOCs) and independent telephone companies.

**Exchange Carriers Standards Commission (ECSC)**   Committee organized to carry out national integration formerly assumed by AT&T.

**exchange service**   A service permitting interconnection of any two customers' stations through the use of the exchange system.

**exclusive hold**   A restriction in which only the station that has placed a line circuit on hold is capable of breaking the hold condition and reestablishing conversation.

**EXEC**   In an IBM environment, an EXEC file contains a sequence of CMS commands that can be invoked by typing the name of the file and pressing the ENTER key.

**EXECU-BILL**   A service mark of AT&T as well as a service from that vendor which assists business to manage their calling card expenses. EXECU-BILL provide a variety of billing and reporting options, summary billing capabilities, fraud control and corporate card program administration.

**exit list (EXLST)**   In IBM's VSAM and VTAM, a control block that contains the addresses of routines that receive control when specified events occur during execution; for example, routines that handle session-establishment request processing or I/O errors.

**exit routine**   In IBM's SNA, any of several types of special-purpose user-written routines.

**EXLST exit routine**   In IBM's VTAM, a routine whose address has been placed in an exit list (EXLST) control block. The addresses are placed there with the EXLST macro instruction, and the routines are named according to their corresponding operand; hence DFASY exit routine, TPEND exit routine, RELREQ exit routine, and so forth. All exit list routines are coded by the VTAM application programmer.

**Expanded Memory**   A term to refer to memory over 1 Mbyte in an IBM PC or compatible personal computer that conforms to a particular specification, such as the Lotus/Intel/Microsoft (LIM) Expanded Memory Specification.

**expandor**   A device that reverses the effect of analog compression.

**expedited flow**   In IBM's SNA, a data flow designated in the transmission header (TH) that is used to carry network control, session control, and various data flow control request/response units (RUs); the expedited flow is separate from the normal flow (which carries primarily end-user data) and can be used for commands that affect the normal flow. *Note.* The normal and expedited flows move in both the primary-to-secondary and secondary-to-primary directions. Requests and responses on a given flow (normal or expedited) usually are processed sequentially within the path, but the expedited flow traffic may be moved ahead of the normal-flow traffic within the path at queueing points in the half-sessions and for half-session support in boundary functions.

**explicit access**   In LAN technology, a shared access method that allows stations to use the transmission medium individually for a specific time period; every station is guaranteed a turn, but every station must also wait for its turn.

**explicit command**   An IBM Network Problem Determination Application (NPDA) user request designed to circumvent the step-down error tracking sequences that use selection number entry. Commands consist of a user-defined operator, or verb, followed by operands that establish the data type, display level, resource level, and other allowed variations of the command.

**Explicit Route (ER)**   In IBM's SNA, the path control

network components, including a specific set of one or more transmission groups, that connect two subarea nodes. An explicit route is identified by an origin subarea address, a destination subarea address, an explicit route number, and a reverse explicit route number.

**explicit route length** In IBM's SNA, the number of transmission groups in an explicit route.

**Explorer Packet** In local area networking source routing an explorer packet contains a unique envelope which is recognized by a source routing bridge. The source routing bridge enters the connection on which the explorer packet was received and its own name in the routing information field of the explorer packet. That packet is then transmitted on all links except the connection from which it was received, a process known as flooding. The recipient then selects either the earliest received or most direct routed explorer packet and sends a response to the originator. The response lists the specific route between source and destination which is used for subsequent transmission.

**exponential backoff** The practice of waiting a random period of time after a collision is detected on a CSMA/CD local area network prior to attempting to transmit again on the network.

**exposed** Cable, wire, strand, etc., which are subject to disturbance by lightning, possible contact, or induction from electrical currents in excess of 300 volts to ground, or ground potential rises from nearby power generating stations, substations, or higher voltage industrial transformers. A building is considered exposed or unexposed according to the classification of the outside plant which serves it.

**Express Transfer Protocol (XTP)** A protocol which corresponds to the network and transports layers of the International Standards Organization Open Systems Interconnection (OSI) network model and which is designed to be implemented in silicon to maximize the throughput of high-speed local networks.

**ExpressView** A trademark of DAVID Systems, Inc., as well as the name of that firm's network management software.

**EXR** EXception Request (IBM's SNA).

**EXT** EXTernal.

**EXT TXC** EXTernal Transmit Clock (control signal).

**extended addressing** In bit-oriented protocol, a facility allowing a larger address to be used. IBM SNA adds two high-order bits to the basic address.

**eXtended Architecture (XA)** An extension to the IBM MVS operating systems for large mainframe computers.

**Extended Area Service (EAS)** A communication carrier offering which allows the subscriber to have a wider geographical coverage without extra per-call charges.

**Extended Binary Coded Decimal (EBCD)** A code using six bits plus one parity bit (IBM).

**Extended Binary Coded Decimal Interchange Code (EBCDIC)** An 8-bit character code used primarily in IBM equipment. The code provides for 256 different bit patterns. See EBCDIC.

**Extended Digital Subscriber Line (EDSL)** An ISDN trunk line connnecting two Primary Rate Interfaces (PRIs) with a combination of 23 B-channels and one D-channel, each channel operating at 64 kilobits per second.

**extended distance cables** Data communication cables which operate at longer distances than standard cables.

**extended highlighting** 1. A function that provides blink, reverse video, and underscore for emphasizing fields or characters on devices supporting extended field attributes and character attributes. 2. An attribute type in the extended field attribute and character attribute. 3. An attribute passed between session partners in the Start Field Extended, Modify Field, and Set Attribute orders.

**Extended Industry Standard Architecture (EISA)**
A proposed Intel 80386 microprocessor-based personal computer bus standard supported by an industry consortium led by Compaq Computer Corp., AST, Epson, Hewlett-Packard, NEC, Olivetti, Tandy, Wyse and Zenith.

**Extended Memory**   A term used to refer to memory above 1 Mbyte that is directly accessable on an IBM PC or compatible personal computer that uses an Intel 80286, 80386 or 80486 microprocessor.

**extended network addressing**   In IBM's SNA, the network addressing system that splits the address into a subarea and an element portion. The subarea portion of the address uses a fixed number of bits to address host processors or communication controllers. The element portion uses the remaining bits to permit processors or controllers to address resources.

**extended numbering**   A feature of CCITT Recommendation X.25 that provides an extension of frame sequence numbering. With extended numbering, for example, frame sequence numbers range from 0 through 127, allowing for larger window sizes to compensate for satellite circuit delays. Also known as Modulo 128.

**extended sequence numbering**   An option in HDLC which permits sequence numbers up to 127 to be used rather than up to seven.

**Extended Specify Task Abnormal Exit (ESTAE)**
In IBM'S VTAM, an MVS macro instruction that provides recovery capability and gives control to the user-specified exit routine for processing, diagnosing an ABEND, or specifying a retry address.

**Extended Superframe Format (ESF)**   A framing format, specified by AT&T, and used on North American T1 circuits, in which each superframe contains 24 frames, and in which the framing bits of sequential frames carry four signaling patterns (Frame Synchronization, Signaling Synchronization, CRC-6, and Link Data Channel (LDC). The Extended Superframe Format is gradually

replacing the 12-frame, D4 format, in North American telephony. The Interview 7000 Series can test both ESF- and D4-framed T1 circuits.

**extent**   A group of contiguous allocated sectors on a disk.

**Exterior Gateway Protocol (EGP)**   A protocol used by a gateway in one autonomous system to advertise the Internet addresses of networks in that system to a gateway in another autonomous system.

**External Data Representation (XDR)**   A protocol within Sun Microsystems' Network File System which enables computers to represent internal data in different ways to facilitate its exchange. XDR includes the ability to represent 32-bit signed and unsigned integers, 64-bit signed and unsigned integers, IEEE single and double precision floating-point numbers, text strings and opaque data types, the later being blocks of bytes whose contents are not required to be interpreted by any computer other than the one that supplied them.

**external domain**   In IBM's SNA, the part of the network that is controlled by an SSCP other than the SSCP that controls this part.

**external drive**   A disk drive that contains its own housing and power supply and which is designed to be cabled to a computer.

**external gateways**   Gateways into electronic mail systems on computers not included in the LAN.

**external modem**   A stand-alone modem, as opposed to a modem board that is inserted into the system unit of a personal computer.

**external procedure**   A routine that is assembled or compiled separately from the program that called it.

**external transmit clock**   An interface timing signal provided by a DTE device, which synchronizes the transfer of transmit data.

**external writer**   In OS/VS2, a program that supports the ability to write SYSOUT data in ways and to devices not supported by the job entry subsystem.

**Extremely High Frequency (EHF)** That portion of the electromagnetic spectrum consisting of frequencies in the microwave range of approximately 30 to 300 GHz.

**eye diagram** The diagram produced by superimposing many received waveforms, taken with a wide range of bit patterns (e.g. a pseudo-random bit sequence). A useful eye diagram shows the waveforms at the point at which regeneration takes place. The "eye" is the hole in the pattern which is necessary for successful regeneration.

**EZTym** An asynchronous communications and workstation applications development toolkit marketed by BT Tymnet.

**E&M Signaling** A transmission system that uses separate paths for signaling and voice signals. The "M", derived from "mouth" lead, transmits ground or battery to the distant end of the circuit. The "E", derived from "ear" lead, accepts the signal. This technique is commonly used between PBXs to signal on-hook and off-hook conditions. Commonly referred to as ear and mouth.

**E1** A circuit that operates at 2.048 Mbps which is the European equivalent of a North American T1 circuit.

**E.163** The CCITT public telephone network address standard.

**E.164** The CCITT numbering plan standard that defines the addressing scheme for ISDN. This numbering plan standard uses a total of 15 digits.

**E3TV** A TV system from Ear Three Systems Manufacturing Company which has a 2625-line resolution with 3500 crosspoints and scan frequencies of 30 Hz, 59.94 Hz, 60 Hz and higher.

**E78** A software program marketed by Digital Communications Associates which enables a personal computer to emulate an IBM 3278 or 3279 terminal when an appropriate adapter card is installed in the PC.

# F

**F** Fahrenheit.

**faceplate (FP)** A plate for connecting data and voice connectors to a cabling system. It may be wall-mounted or surface-mounted.

**facility** 1. In general, a feature or capability offered by a system, item of hardware, or software. 2. In telco environments, line and equipment used to furnish a completed circuit. 3. In packet switched networks, nonstandard facilities selected for a given national network which may or may not be found on other networks.

**Facility Restriction Level (FRL)** These levels define the calling privileges associated with a line; for example, intragroup calling only in the warehouse, but unrestricted calling in the board room.

**FACS** Facilities Assignment and Control System.

**facsimile (FAX)** The transmission of image via communications channels by means of a device which scans the original document and transforms the image into coded signals. Facsimile device operation typically follow one of four CCITT standards for information representation and transmission:

Group 1— analog with page transmission in four or six minutes

Group 2— analog with page transmission in two or three minutes

Group 3— digital with page transmission in less than one minute

Group 4— digital with page transmission in less than 10 seconds

**factory default** Default settings for a device which were set at the factory. These settings are usually established by positioning dip switch elements or via programming entries into memory.

**fading** A loss in transmission intensity caused by changes in the transmission medium. Typically encountered in microwave and radio transmission.

**fail safe** Describing a circuit or device which fails in such a way as to maintain circuit continuity or prevent damage.

**fail soft** The ability of a system to detect component failures and modify its processing quickly and temporarily to prevent irretrievable loss of data or equipment.

**fail softly** When a piece of equipment fails, a fail softly program lets the system fall back to a degraded mode of operation rather than let it fail catastrophically and give no response to its users.

**failed second** A second during which a digital circuit is in a Failed Signal State.

**failed signal state** An error condition on a digital circuit, defined as the occurrence of 10 consecutive Severely Errored Seconds, and considered to be cleared after 10 consecutive seconds without a Severely Errored Second. Each second of a Failed Signal State is called a Failed Second.

**fallback, double** Fallback in which two separate equipment failures have to be contended with.

**fallback data rate** A feature in a modem where, if a communications line deteriorates to the point where reception is difficult, the modem can be switched to transmit at a lower speed.

**fallback procedures** Predefined operations (manual or automatic) invoked when a fault or failure is detected in a system.

**fan-fold paper** Paper that has pin-feed holes on each side and which is perforated between pages. Also called continuous paper.

**fanning strip** A narrow strip having smooth holes or slots through which cross-connecting can be brought for orderly termination on a connecting block, terminal block, or terminal strip.

**fanout** A modem feature that permits more than one data source to be attached to the device.

**fan-out cable** A cable used to provide a connection between a 50-pin PBX connector and multiple RS-232 connectors.

**FAQ** Frequently Asked Question; refers to a document containing common questions and their answers.

**Far-End Block Error (FEBE)** The binary indicator sent by the receiver to the data originator which allows the originator to determine whether or not the receiver detected a CRC error in the frame transmitted by the originator.

**far-end crosstalk** Crosstalk that travels along a circuit in the same direction as the signals in the circuit.

**far-end echo** The reflection of a signal at the four-wire to two-wire interface. This is where the signal leaves the carriers' backbone network and enters the local loop. Contrast with near-end echo.

**FARNET** Federation of Academic Research Networks.

**FAS** 1. Flexible Access System. 2. Frame Acquisition and Sychronization. 3. Frequency Assignment Subcommittee, IRAC.

**Fascinator** A trademark of Motorola, Inc. of Schaumburg, IL as well as the name for a series of digital encrpytion radios manufactured by that vendor.

**fast packet** A high-speed, packet-mode technology for transmitting voice, data and video as fixed size packets, with each packet containing source and destination information.

**fast packet switching** A packet switching architecture which provides a mechanism for transporting voice, data, and image via a packetized format.

**Fastalk** The name of a series of modems manufactured by Universal Data Systems of Huntsville, Al.

**Fast Select** In packet switching a method used to transmit up to 128 data bytes in call-setup and call-

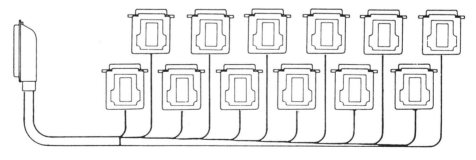

*fan-out cable*

clearing packets. Fast Select allows economical transmission of credit card purchase information from dial-up retail point-of-sale terminals to authorised host computers.

**Fastext** A teletex transmission method in which broadcasters transmit extra signals that provide links between pages on related topics. When a page is being read a fastext compatible TV set stores the linked pages in memory, which allows them to be instantly accessed.

**FASTLANE** A trademark of Telco Systems Network Access Corporation of Fremont, CA, as well as a compression performing bridge marketed by that vendor which performs adaptive data compression.

**FastPath** An Ethernet to LocalTalk gateway marketed by Kinetics, Inc., a division of Excelan, Inc. The FastPath gateway lets users connect 230 Kbps LocalTalk networks to 10 Mbps Ethernet networks.

**fastrun** In IBM's SNA, one of several options available with NCP/EP Definition Facility (NDF) that indicates only the syntax is to be checked in generation definition statements.

**FAT** File Allocation Table.

**fault domain** In IBM Token-Ring Network problem determination, the portion of a ring that is involved with an indicated error.

**FAX** Facsimile.

**FaxBox** Facsimile software for Alpha Micro Systems marketed by ICS Software of Hatfield, Herts, England.

**FaxDat** An adapter card manufactured by Xecom Inc. of Milipitas, CA, for use in laptop computers. This card performs both facsimile and data communications.

**FaxPress** The name of a LAN facsimile server marketed by Castelle Corporation of Santa Clara, CA.

**Faxway** A trademark of Compfax Software Inter-

national, Inc., of New York, NY, as well as a program that allows local area network users to transmit electronic mail, documents, or spreadsheets to a fax gateway.

**F-bits (framing bits)** In multiplexed digital transmission, the leading bit of each frame, not included in any octet channel of user data. On any T1 circuit, framing bits occupy 8 Kbps bandwidth. In D4 framing, the framing bits of successive frames carry the Frame Synchronization Pattern and the Signaling Synchronization Pattern. In ESF, the framing bits of sequential frames carry four signaling patterns—Frame Synchronization, Signaling Synchronization, CRC-6, and Link Data Channel (LDC).

**FCB** File Control Block.

**FCC** Federal Communications Commission.

**FCC Registration Number** The last three digits of the FCC Registration Number identify the dialing capability (letter) and general category of a telephone or telephone accessory (two letters) as follows:

| Last letter | Capability |
|---|---|
| N | Device has no dialing capability |
| R | Device has only rotary dialing capability |
| T | Device has only touch-tone dialing capability |
| E | Device has both rotary and touch-tone capability |

Two-letter code preceding last letter:

| Code | Capability |
|---|---|
| AL | alarm dialer |
| AN | answering machine |
| BR | bridge to set up conference between two or more lines |
| CI | call forwarding device |
| DI | automatic dialer |
| DM | modem |
| DT | modem or data terminal |
| FA | telecopier |
| MD | computer modem |
| MU | music on hold accessory |

| MT | multifunction telephone |
|----|----|
| RC | conversation recorder |
| RG | extension ringer |
| RT | pen register or other device that records numbers dialed |
| SP | speakerphone |
| TE | telephone |
| TR | toll restriction |
| WT | cordless telephone |

**FCCSET** Federal Coordinating Council on Science, Engineering and Technology.

**FCS** Frame Check Sequence.

**FCT** Frame Creation Terminal.

**FD** Full-Duplex.

**F/D** The ratio of antenna focal length to antenna diameter. A higher ratio means a shallower dish.

**FDCS** Flight Deck Communications System.

**FDDI** Fiber Distributed Data Interface.

**FDDI token** An FDDI token is a special sequence of control signals that do not occur in data portions of the FDDI frame. The FDDI frame includes preamble, starting delimiter, frame control, destination address, source address, information field, error check, and end-of-frame delimiter.

**FDM** Frequency Division Multiplexing.

**FDP** Field Developed Program.

**FDX** Full-Duplex.

**Fe** A common abbreviation for Extended Superframe Format, a framing format, specified by AT&T, and used on North American T1 circuits, in which each Superframe contains 24 frames, and in which the framing bits of sequential frames carry four signaling patterns—Frame Synchronization, Signaling Synchronization, CRC-6, and Link Data Channel (LDC). The Extended Superframe Format is gradually replacing the 12-frame, D4 format, in North American telephony. The Interview 7000 Series can test both Fe- and D4-framed T1 circuits.

**Fe** Format Effector.

**FE bits (framing bits extended)** Framing bits in Extended Superframe Format.

**FEAL-8** A data communications security code developed by Nippon Telegraph and Telephone Corporation of Tokyo, Japan. FEAL-8 is widely used in Japan to protect data communications systems.

**Feature Interactive Verification Environment (FIVE)** An AT&T test facility which can be used by telephone companies and their customers to test, verify and evaluate the AT&T 5ESS digital central office switch.

**FEBE** Far End Block Error.

**FEC** Forward Error Correction.

**FECN** Forward Explicit Congestion Notification.

FDDI token format

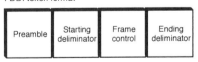

FDDI frame format

*FDDI token*

**FEDAC** A trademark of Frederick Engineering, Inc. of Columbia, MD, as well as a dataline monitor/protocol analyzer on an IBM PC adapter card which permits on-site and unattended remote operations.

**Federal Communications Commission (FCC)** A USA government board of seven presidential appointees established by the Communications Act of 1934 that has the power to regulate all USA interstate communications systems as well as all international communications systems that originate or terminate in the USA.

**Federal Coordinating Council on Science, Engineering and Technology (FCCSET)** A committee under the U.S. President's Office of Science and Technology Policy (OSTP) that studies communications networks.

**Federal Information Processing Standards (FIPS)** The standards formulated by Telecommunications Standards program of the Federal Government—usually in close coordination with industry—and published by the National Bureau of Standards.

**Federal Networking Council (FNC)** Representatives from several U.S. Government Agencies which coordinate activities of those agencies that fund research or development of TCP/IP and the Internet.

**FED-STD** Federal Standard.

**FED-STD-1001** Synchronous high-speed data signaling between data terminal equipment (DTE) and data communications equipment (DCE).

**FED-STD-1002** Time and frequency reference information in telecommunications systems.

**FED-STD-1003** A synchronous bit-oriented datalink control procedure; Advanced Data Communications Procedures (ADCCP).

**FED-STD-1005** Coding and modulation requirements for 2400 bps modems.

**FED-STD-1006** Coding and modulation requirements for 4800 bps modems.

**FED-STD-1007** Coding and modulation requirements for duplex 9600 bps modems.

**FED-STD-1008** Coding and modulation requirements for duplex 600 bps and 1200 bps modems.

**FED-STD-1010** Bit sequencing of the American National Standard Code for Information Interchange in serial-by-bit data transmission. FED-STD 1010 is a joint standard with FIPS 16-1.

**FED-STD-1011** Character structure and character parity sense for serial-by-bit data communication in the American National Standard Code for Information Interchange. FED-STD 1011 is a joint standard with FIPS 17-1.

**FED-STD-1012** Character structure and character parity sense for parallel-by-bit data communication in the Code for Information Interchange. FED-STD 1012 is a joint standard with FIPS PUB 18-1.

**FED-STD-1013** Synchronous signaling rates between data terminal equipment and data circuit-terminating equipment utilizing 4 kHz circuits. FED-STD 1013 is a joint standard with FIPS PUB 22-1.

**FED-STD-1015** Analog to digital conversion of voice by 2400 bit/second linear predictive coding.

**FED-STD-1020A** Electrical characteristics of balanced voltage digital interface circuits.

**FED-STD-1026** Interoperability and security requirements for use of the data encryption standard in the physical layer of data communication.

**FED-STD-1027** General security requirements for equipment using the data encryption standard (DES). The DES algorithm for digital encryption is described in FIPS PUB 46.

**FED-STD-1028** Interoperability and security requirements for use of the data encryption standard with CCITT Group 3 facsimile equipment.

**FED-STD-1030A** Electrical characteristics of unbalanced voltage digital interface circuits.

**FED-STD-1031** General purpose 37-position and

9-position interface between data terminal equipment and data circuit-terminating equipment.

**FED-STD-1033** Data communications systems and service-user oriented performance parameters.

**FED-STD-1037A** Glossary of telecommunications terms.

**FED-STD-1041** Interface between data terminal equipment (DTE) and data circuit-terminating equipment (DCE) for operation with packet switched data communication networks. FED-STD 1041 is a joint standard with FIPS PUB 100.

**FED-STD-1061** Group 2 facsimile apparatus for document transmission.

**FED-STD-1062** Group 3 facsimile apparatus for document transmission.

**FED-STD-1063** Procedures for document facsimile transmission.

**feedback information** In IBM's VTAM, information that is placed in certain RPL fields when an RPL-based macro instruction is completed.

**feedline** A coaxial cable or waveguide connecting a receiver or transmitter to an earth station antenna.

**feeds** In satellite communications, a device mounted at the focus point of an antenna which gathers the signals reflected from the dish.

**FELINE** A general purpose protocol analyzer constructed on an adapter card that is installed in an IBM PC or compatible computer. FELINE is manufactured by Frederick Engineering, Inc., of Columbia, MD.

**FELINE-LT** A trademark of Frederick Engineering of Columbia, MD, as well as a dataline monitor/protocol analyzer constructed on an adapter card that can be inserted into laptop computers.

**femtosecond** A quadrillionth, or million billionth of a second ($10^{-15}$ 5).

**FENICS** Fijitsu's domestic packet network.

**FEP** Front-End Processor.

**FET** Field Effect Transistor.

**Fetex-150** A modular central office telephone switch manufactured by Fujitsu. The Fetex-150 serves as a digital central office, a tandem switch, a combined central office/tandem switch, or an international gateway. In a small configuration the Fetex-150 can serve as a remote, portable or compact switch.

**FF** Form Feed.

**FFTSO** File Transfer Time Sharing Option.

**FGND** Frame GrouND.

**FIB** Forward Indicator Bit.

**fiber bandwidth** The lowest frequency at which the magnitude of the transfer function of an optical waveguide has fallen to half the zero frequency value; in other words, the frequency at which the signal loss has increased by 3 dB (decibels). Since the bandwidth of an optical waveguide is approximately reciprocal to its length (mode mixing), the bandwidth–length product (MHz–km) is often quoted as a quality characteristic.

**fiber buffer** Material used to protect an optical fiber or cable from physical damage, providing mechanical isolation or protection. Fabrication techniques include both tight jacket and loose tube buffering, as well as multiple buffer layers.

**Fiber Distributed Data Interface (FDDI)** A specialized local area network concept developed under American National Standards Institute (ANSI) auspices for use in computer-to-computer exchanges via fiber optic links. The set of ANSI standards defines a LAN topology consisting of a dual-fiber, token-ring network with counter-rotating rings that supports a data transfer rate of 100 Mbps. The standard specifies multimode fiber 50/125, 62.5/125, or 85/125 core-cladding; a LED or laser light source; and 2 kilometers for unrepeated data transmission at 40 Mbps. Each station on the ring terminates both fiber rings, which can accommodate up to 500 stations. Stations can be repeaters, bridges to other FDDI rings or gateways to other networks, such as Ethernet or Token Ring, enabling the fiber ring to be used as a backbone network to interconnect slower LANs. FDDI is partitioned into four ANSI standards: PMD (Physical Media Dependent)—ANSI X3T9.5; PHY (Physical)—ANSI X3T9.5; MAC (Media Access Control)—ANSI X3.139; SMT (Station Management)—ANSI X3T9.5. The PMD standard defines operating specifications for duplex connectors, optical transceivers and optical bypass switches. The Physical standard defines clock synchronization and the encoding scheme. The MAC standard controls media access and provides peer-to-peer communications services to the higher layers, Logical Link Control and SMT. The SMT standard defines station management procedures, ring map generation and synchronous bandwidth allocation.

**fiber loss** Attenuation of light signal in optical fiber transmission.

**fiber miles** The number of cable miles times the number of fiber strands in a cable.

**fiber optic cable** A transmission medium composed of small strands of glass, each of which provides a path or light rays which act as a carrier.

**fiber optic facility** A transmission facility in which information is transmitted as light pulses over fiberglass threads. Fiber optic facilities are immune to electrical interference, require very little power and can transmit at very high rates of speed.

**fiber optic waveguides** Thin filaments of glass or other transparent materials through which light beams can be transmitted for long distances through multiple internal reflections.

**fiber (optical)** Any filament or fiber, made of dielectric materials, that guides light.

**fiber optics** A signal-conducting medium that conveys light waves through transparent fibers. These fibers act as waveguides for light waves.

**fiber SLC carrier system** A system developed by AT&T Bell Laboratories that now carries up to 96 voice circuits on a pair of fiber lightguides between a central office and a remote terminal. The first installation was in Chester Heights, Pennsylvania, in 1982.

**Fiber-To-The-Curb (FTTC)** The process by which a local exchange carrier installs fiber optic cable to a nearby curbside or pedestal location from which it is connected by copper and in most cases, coaxial cable to the home.

**Fiberworld** A term used by Northern Telecom for a family of fiber optic access, transport and switching products.

**FIC** First-In-Chain (IBM's SNA).

**FID** Format IDentification (IBM's SNA).

**FIDB** Facility Interface Data Bus.

**FidoNet** An amateur network of bulletin board systems that exchange messages electronically.

**field** A group of bits that describes a specified characteristic; displayed on a reserved area of a CRT or located in a specific part of a record.

**field formatted** In IBM's SNA, pertaining to a request or response that is encoded into fields, each having a specified format such as binary codes, bit-significant flags, and symbolic names.

**field formatted request** In IBM's SNA, a request that is encoded into fields, each having a specified format such as binary codes, binary counts, bit-significant flags, and symbolic names; a format indicator in the request/response header (RH) for a request is set to zero.

**Field Programmable Logic Array (FPLA)** A logic element that can replace a conventional Read-Only Memory (ROM), provided that only some of the possible words in the storage matrix and address decoder are required. Both the address decoder and the storage matrix are programmed to obtain the maximum reduction in chip size. Devices in which the programming can be carried out by the user are termed FPLA.

**Field Replaceable Unit (FRU)**   Same as service parts.

**field-wire line**   Simple pairs of insulated wire twisted together. Field-wire lines are typically used in military applications for emergency and temporary short distance connections due to their high transmission loss.

**FIFO (First In, First Out)**   A method used to process an item in a queue according to which item has been in the queue longest.

**FIGS**   FIGures Shift.

**figures shift (FIGS)**   1. A physical shift in a terminal using Baudot Code that enables the printing of numbers and symbols. 2. The character that causes the shift.

**file**   A collection of records. A file might include all the names and addresses of a company's employees.

**File Allocation Table (FAT)**   An area on a diskette or fixed disk that acts as an index, or directory and which is used by the operating system to determine where data is stored and where data can be stored.

**file attribute**   Any of the set of file characteristics which determines its accessibility and degree of protection from other than its current owner. Common file attributes are read, write, modify, and share.

**File Control Block (FCB)**   A data structure that controls a user's access to a file.

**file extension**   A suffix to a file name which further identifies the contents of the file. Common file extensions include:

| | |
|---|---|
| .bas | Basic language file |
| .com | executable file |
| .exe | executable file |
| .gif | graphic interchange format |
| .ps | PostScript file |
| .tar | Unix tape archive format |
| .z | Unix compressed file |
| .zip | DOS compressed file, compressed via use of PKZIP program |

**file name**   1. A name assigned to a collection of related data records or to a peripheral or communications device. 2. The name under which a file is stored on disk. MS-DOS file names have an 8.3 format consisting of an eight-character name followed by a three character extension where the name and extension are separated by a period. Apple ProDOS file names consist of up to 15 alphanumeric characters, the first character of which must be alphabetic. Macintosh file names can be up to 32 characters in length.

**file name extension**   An optional short name that starts with a period and has 1, 2 or 3 (PC DOS) or more characters, and follows immediately after the file name. The file name extension can be used to further define the type of file. Common file name extensions used by popular software are listed in the following table.

| *Graphics formats* | *File extension* |
|---|---|
| AutoCAD | .DXF |
| CompuServe Graphic | .GIF |
| Computer Graphics Metafile | .CGM |
| Encapsulated PostScript | .EPS |
| GEM Metafile | .GDI |
| HP Graphics Language | .HPG |
| Halo DPE | .IMG |
| Lotus Picture | .PIC |
| Macintosh | .PICT |
| MacPaint | .PICT |
| Micrografx | .DRW |
| PC Paintbrush | .PCX |
| Publisher's Paintbrush | .PCX |
| Tagged Image File | .TIF |
| Windows | .PIC |
| Windows Metafile | .WMF |

| *Text formats* | |
|---|---|
| Data Interchange Format | .DIF |
| dBASE | .DBF |
| Lotus 1-2-3 | .WK1 |
| Microsoft Word | .DOC |
| MultiMate | .DOC |
| Windows Write | .WRI |
| WordPerfect | Any |
| WordStar | Any |
| XyWrite | Any |

**file naming** Assigning a specific name to a data file.

**file organization** A method that establishes the relationship between a record and its location in a file.

**File Separator (FS)** A control character used to separate and qualify data logically; normally delimits data item called a file.

**file server** A device or station, usually in a local network, that provides file and storage services to other stations on the network.

**file server protocol** In LAN technology, a communications protocol that allows application programs to share files.

**file specification (filespec)** The name and location of a file. A file specification consists of a driver specifier, a path name, and a file name.

**File Transfer, Access, and Management (FTAM)** An ISO Application layer protocol for transferring, accessing and managing files among dissimilar systems.

**File Transfer Customer Monitor System (FTCMS)** A term used to describe third party software that permits file transfers between a personal computer and an IBM mainframe's CMS subsystem.

**File Transfer Protocol (FTP)** The Internet standard, high-level protocol used for transferring files from one computer to another.

**File Transfer Time Sharing Option (FTTSO)** A term used to describe third party software that permits file transfers between a personal computer and an IBM mainframe time sharing subsystem.

**FileTalk** A trademark of Mountain Computer as well as software from that vendor which permits LAN users to backup files to a common tape backup unit.

**filing time** In U.S. military communications, the date and time, expressed in Julian day and Greenwich Mean Time, a message is received from an originator by a communications center for transmission. *Example*: 050758Z.

**Fill-In Signal Unit (FISU)** In CCITT Signaling System 7, a FISU is sent when no other information is available for sending.

**filter** 1. An arrangement of electronic components designed to pass signals in one or several frequency bands and to attenuate signals in other frequency bands. 2. An IBM Network Problem Determination Application (NPDA) facility that allows the user to prevent certain data from being recorded or from being viewed on a display. A filter can also be used to generate alert data from an event.

**filter item** An IBM Network Problem Determination Application (NPDA) filter element that defines a single, specific decision statement that passes or blocks data being presented to the filter.

**final-form document** An electronic document that can only be printed or displayed.

**final route** The last-choice route in an automatic switching system.

**Find America** An on-line directory assistance service of AT&T.

**finger** A TCP/IP application which is used to obtain a list of persons logged onto other hosts or the staus of a specific user on another host.

**FIPS** Federal Information Processing Standards.

**FIPS PUB 1-1** Code for information interchange.

**FIPS PUB 1-2** Code for information interchange, its representations, subsets and extensions.

**FIPS PUB 2-1** Perforated tape code for information interchange.

**FIPS PUB 3-1** Recorded magnetic tape information interchange (800 CPI, NRZI).

**FIPS PUB 4-1** Representation for calendar date and ordinal date for information interchange.

**FIPS PUB 5-2** Codes for the identification of the states, District of Columbia, outlying areas of the United States and associated areas.

**FIPS PUB 6-3** Counties and county equivalents of the states of the United States.

**FIPS PUB 7** Implementation of the code for information interchange and related standards.

**FIPS PUB 8-5** Metropolitan statistical areas including CMSAs, PMSAs, and NECMAs.

**FIPS PUB 9** Congressional districts of the United States.

**FIPS PUB 10-3** Countries, dependencies and areas of special sovereignty, and their principal administrative divisions.

**FIPS PUB 13** Rectangular holes in 12-row punched cards.

**FIPS PUBS 14-1** Hollerith punched card code.

**FIPS PUB 15** Subset of the standard code for information interchange.

**FIPS PUB 16-1/FED-STD 1010** Bit sequencing of the code for information interchange in serial-by-bit data transmission.

**FIPS PUB 17-1/FED-STD 1011** Character structure and character parity sense for serial-by-bit data communications in the code for information interchange.

**FIPS PUB 18-1/FED-STD 1012** Character structure and character parity sense for parallel-by-bit data communication in the Code for information interchange.

**FIPS PUB 21-2** COBOL.

**FIPS 22-1/FED-STD 1013** Synchronous signaling rates between data terminal equipment (DTE) and data circuit-terminating equipment (DCE).

**FIPS PUB 24** Flowchart symbols and their usage in information processing.

**FIPS PUB 25** Recorded magnetic tape for information interchange (1600 CPI, phase encoded).

**FIPS PUB 26** One-inch perforated paper tape for information interchange.

**FIPS PUB 27** Takeup reels for one-inch perforated tape for information interchange.

**FIPS PUB 30** Software summary for describing computer programs and automated data systems.

**FIPS PUB 32-1** Character sets for optical character recognition character sets.

**FIPS PUB 33-1** Character sets for handprinting.

**FIPS PUB 37/FED-STD 1001** Synchronous high-speed data signaling rates between data terminal equipment (DTE) and data circuit-terminating equipment (DCE).

**FIPS PUB 46** Specifies the 64-bit key, Data Encryption Standard (DES).

**FIPS PUB 50** Recorded magnetic tape for information interchange, 6250 CPI (246 CPMM), group coded recording.

**FIPS PUB 51** Magnetic tape cassettes for information interchange (3.810 mm [0.150 in] tape at 32 bpmm [800 bpi], PE).

**FIPS PUB 52** Recorded magnetic tape cartridge for information interchange, 4-track 6.30 mm (0.250 in), 63 bpmm (1600 bpi), phase encoded.

**FIPS PUB 53** Transmittal form for describing computer magnetic tape file properties.

**FIPS PUB 54** Computer Output Microform (COM) formats and reduction ratios, 16 mm and 105 mm.

**FIPS PUB 58-1** Representations of local time of the day for information interchange.

**FIPS PUB 59** Representations of universal time, local time differentials and United States time zone references for information interchange.

**FIPS PUB 60-2** Input/output (I/O) channel interface.

**FIPS PUB 61-1** Channel level power control interface.

**FIPS PUB 62** Operational specifications for magnetic tape subsystems.

**FIPS PUB 63-1** Operational specifications for variable block, rotating mass storage subsystems.

**FIPS PUB 66** Standard Industrial Classification (SIC) codes.

**FIPS PUB 68-1** Minimal BASIC.

**FIPS PUB 69-1** FORTRAN.

**FIPS PUB 70-1** Representation of geographic point locations for information interchange.

**FIPS PUB 71/FED-STD 1003A** Advanced Data Communications Control Procedures (ADCCP).

**FIPS PUB 78** Guidelines for implementing Advanced Data Communications Control Procedures (ADCCP).

**FIPS PUB 79** Magnetic tape labels and file structure for information interchange.

**FIPS PUB 81** Data Encryption Standard (DES) modes of operation.

**FIPS PUB 84** Microfilm readers.

**FIPS PUB 85** Optical Character Recognition (OCR) inks.

**FIPS PUB 86** Additional controls for use with American National Standard Code for Information Interchange.

**FIPS PUB 89** Optical Character Recognition (OCR) character positioning.

**FIPS PUB 91** Magnetic tape cassettes for information interchange, dual track Complementary Return-to-Bias (CRB), four-states recording on 3.81 mm (0.150 in) tape.

**FIPS PUB 93** Parallel recorded magnetic tape cartridge for information interchange, 4-track 6.30 mm (0.250 in), 63 bpmm (1600 bpi), phase encoded.

**FIPS PUB 95** Code for the identification of federal and federally assisted organizations.

**FIPS PUB 97** Operational specifications for fixed block, rotating mass storage subsystems.

**FIPS PUB 100/FED-STD 1041** Interface between data terminal equipment (DTE) and data circuit-terminating equipment (DCE) for operation with packet switched data communication networks.

**FIPS PUB 103** Codes for the identification of hydrologic units in the United States and Caribbean outlying areas.

**FIPS PUB 104-1** American National Standard codes for the representation of names of countries, dependencies, and areas of special sovereignty for information interchange.

**FIPS PUB 107** Local area networks: baseband carrier sense multiple access with collision detection access method and physical layer specifications and link layer protocol.

**FIPS PUB 108** Alphanumeric computer output microform quality test slide.

**FIPS PUB 109** PASCAL.

**FIPS PUB 111** Storage module interfaces (with extensions for enhanced storage module interfaces).

**FIPS PUB 112** Password usage.

**FIPS PUB 113** Computer data authentication.

**FIPS PUB 114** 200 mm (8 in) flexible disk cartridge track format using two-frequency recording at 6631 bprad on one side 2.9 tpmm (48 tpi) for information interchange.

**FIPS PUB 115** 200 mm (8 in) flexible disk cartridge track format using modified frequency modulation recording at 13 262 bprad on two sides—1.9 tpmm (48 tpi) for information interchange.

**FIPS PUB 116** 130 mm (5.25 in) flexible disk cartridge track format using two-frequency recording at 3979 bprad on one side—1.9 tpmm (48 tpi) for information interchange.

**FIPS PUB 117** 130 mm (5.25 in) flexible disk cartridge track format using modified frequency modulation recording at 7958 bprad on two sides—1.9 tpmm (48 tpi) for information interchange.

**FIPS PUB 118** Flexible disk cartridge labelling and file structure for information interchange.

**FIPS PUB 119** Ada.

**FIPS PUB 120**   Graphical Kernel System (GKS).

**FIPS PUB 121**   Videotex/teletext presentation level protocol syntax (North America PLPS).

**FIPS PUB 123**   Specification for a Data Descriptive File for Information Interchange (DDF).

**FIPS PUB 125**   MUMPS programming language.

**FIPS PUB 126**   Database Language (NDL).

**FIPS PUB 127**   Database Language (SQL).

**FIPS PUB 128**   Computer Graphics Metafile (CGM).

**FIPS PUB 129**   Optical Character Recognition (OCR) dot-matrix character sets for OCR-MA.

**FIPS PUB 130**   Intelligent Peripheral Interface (IPI).

**FIPS PUB 131**   Small Computer System Interface (SCSI).

**Fireberd**   A trademark of Telecommunications Techniques Corporation. A widely used BERT tester, capable of testing high-speed digital circuits such as T1 lines.

**firmware**   A computer program or software stored permanently in PROM or ROM or semi-permanently in EPROM or EEPROM.

**First Advanced Satellite Technology (FAST)**   The satellite reservation system marketed by Days Inns to franchises. FAST uses AT&T's SkyNet Clearline network for transmission via VSAT terminals to the firm's headquarters in Atlanta, GA.

**First-In-Chain (FIC)**   In IBM's SNA, a request unit (RU) whose request header (RH) begin chain indicator is on and whose RH end chain indicator is off.

**first-in/first-out operation**   A PABX feature which answers the longest-waiting call first.

**first-party maintenance**   End-users of equipment doing self-maintenance.

**first speaker**   In IBM's SNA, the LU–LU half-session defined at session activation as: (1) able to begin a bracket without requesting permission from the other LU–LU half-session to do so, and (2) winning contention if both half-sessions attempt to begin a bracket simultaneously.

**FISINT**   Foreign Instrumentation and Signals INTelligence.

**FISU**   Fill-In Signal Unit.

**FIU**   Facilities Interface Unit.

**FIVE**   Feature Interactive Verification Environment.

**five-pair rubber cable**   A cable which contains ten conductors arranged in five pairs.

**Fix Area**   A triangular location formed by the line of bearing of three direction finders taking simultaneous bearings which provides the probable location of a transmitting antenna.

**fixed disk**   A flat, circular, non-removable plate with a magnetized surface layer on which data can be stored by magnetic recording.

**fixed-disk drive**   The mechanism used to read and write information on fixed disk.

**fixed equalization**   A simple equalization technique for modems where the amount of compensation is set by the user via strap settings.

**fixed-length record**   A record stored in a file in which all of the records are the same length.

**fixed night service**   Routing of a PBX feature which results in the incoming exchange calls to preselected stations within the PBX system when the attendant is not on duty.

**Fixed Loss Loop (FLL)**   A classification of FCC registered modems that limits output to 4 dB. In the fixed loss loop jack, it is required that the modem does not exceed a transmit level of −4 dBm. Once the loop loss of the subscriber line is measured, a resistor is selected to "build out" the loss of the loop so that the data signal hits the telephone company's central office at a predescribed level. This process is designed so that data hits the message telephone network 13 dB below the normal voice level. The

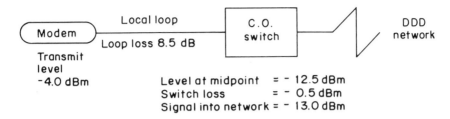

*Fixed Loss Loop (FLL)*

voice TLP at the input to the Message Telephone System network is 0 dBm. If data is 13 dB below voice, then we would expect the data level to be -13 dBm at this point. First, the installer must measure the insertion loss of the circuit from the customer's location to the telephone company central office. Let's assume that the installer measures a loss of 4 dB.

We already know that the modem transmits at −4 dBm. With a little addition:

−4 dBm   (modem's transmit level)
−4 dB    (loop loss)
−8 dBm   (total measured signal at the central
         office)

Notice that the objective was to be at a -13 dB at the central office. In order to accomplish this, the installer selects a resistive pad (attenuator) to reduce the signal in the data jack to meet the requirements. Now we have:

−4 dBM   (modem's transmit level)
−5 dB    (loss of resistive pad in data jack)
−4 dB    (loss of loop)
−13 dBm  (total measured signal at the central
         office)

**fixed routing**   A method of routing in which the behavior is predetermined, taking no account of changes in traffic or of network component outages.

**Fixed Satellite Service (FSS)**   International designation for satellite transmissions intended for point-to-point services in the 11.7 GHz to 12.2 GHz band.

**flag**   A bit pattern of six consecutive "1" bits with a prefix and suffix "0" bit (character representation

01111110) used in many bit-oriented protocols to mark the beginning and end of a frame.

**flame**   An electronic mail message or Usenet posting which is violently argumentative.

**flash**   A signal generated by the momentary depression of the telephone switchhook or other device. The flash is used to request additional services.

**FlashBox**   A product of Sun Microsystems, Inc., designed to upgrade a LocalTalk network to 80 percent of Ethernet performance for approximately 20 percent of the cost. FlashBox permits Macintosh computers to communicate at 770 Kbps over existing twisted-pair wire.

**flat rate service**   Service where the user is entitled to an unlimited number of telephone calls within a specified local area for a fixed rate.

**flat weighting**   A type of telephone noise weighting in which measurements are conducted using an input filter to obtain a flat amplitude-frequency response over a specified frequency range.

**Fleet Satellite Communications System (FLTSAT-COM)**   The US Navy's network of UHF communications satellites that links aircraft, ships, submarines, ground stations, the US Air Force Strategic Air Command, and national command authorities.

**Flexible Access System (FAS)**   A fiber optic transmission system that spans the United Kingdom. Designed for completion in the mid-1990s, FAS is an overlay network that is part of the UK's public network. FAS allows large numbers of leased lines

175

to be controlled by software, permitting business customers to mix connections of various circuits to the network as requirements change.

**flexible routing**   A Software-Defined Network (SDN) service feature offered by AT&T which lets the user reroute calls automatically to another location or be answered with a recorded announcement that instructs the caller to dial another site.

**Flexgel**   Trademark name for AT&T developed filling compound used in waterproof cable. Waterproof material consisting of Extended Thermoplastic Rubber (ETPR).

**FlexOS**
A trademark of Digital Research Corporation of Monterey, CA as well as a real-time, multitasking, protected mode operating system from that vendor. FlexOS was first released in 1985 to IBM for use in their 4800 point-of-sale system.

**FLL**   Fixed Loss Loop.

**floating command line**   In IBM's PROFS and Office Vision the floating command line is a command line a user calls up by pressing a PF key whenever they want to type a command without moving the cursor from the text.

**flooding**   A packet routing method which replicates packets and sends them to all nodes, thus ensuring that the actual destination is reached.

**flooding, network layer**   A means of propagating a message throughout a network. The message is transmitted to each neighboring system except the one from which it was received.

**floppy disk**   A flexible plastic $5\frac{1}{4}$-inch disk coated with magnetic material and used to store data. Newer $3\frac{1}{2}$-inch diskettes are encased in a hard plastic case.

**flow control**   The procedure for controlling the transfer of messages or characters between two points in a data network—such as between a protocol converter and a printer—to prevent loss of data when the receiving device's buffer begins to reach its capacity. Also called pacing.

**flowchart**   A chart to represent the flow of data or instructions through a process, including decision points and loopbacks where appropriate. Often used for designing and documentating computer programs.

**FLTSATCOM**   FLeeT SATellite COMmunications System.

**Flushot**   A program marketed by Executive Network of Mt. Vernon, NY, that can be used to discover a computer virus or prevent one from entering a system.

**flyback buffering delay**   The time delay provided by the statistical multiplexer to prevent loss of data during mechanical functions of terminal equipment such as printer carriage return and line feed. The unit of time delay is in character-times.

**FM**   1. Frequency Modulation. 2. Function Management (IBM's SNA).

**FM subcarrier**   One-way data transmission using signals modulated in the unused portions of the FM radio frequency spectrum.

**FMD**   Function Management Data (IBM's SNA).

**FMH**   Function Management Header (IBM's SNA).

**FMProf**   In IBM's ACF/VTAM, a macro which defines the function management profile to be used for an SNA session.

**FMS**   Facilities Management System.

**FNC**   Federal Networking Council.

**FOC**   Fiber Optic Communications.

**focal length**   In satellite communications, the distance from the feed to the center of the dish.

**Focal point**   The location in an IBM SNA network that provides central network management for the domain.

**FOGM**   Fiber Optic Guided Missile.

**folder dipole antenna**   An antenna which uses a single rod bent into a flattened circle.

**Follow Me Roaming**   A service marketed by GTE Mobilnet which allows cellular telephone subscribers to receive calls made to their telephone when they are located in another city.

**Fonline**   An 800 service offered by U.S. Sprint Communications Company.

**font**   A collection of characters that have a consistent size and style.

**font cartridge**   A cartridge containing read-only memory (ROM) that defines one or more fonts. A font cartridge is installed in a printer to expand its printing capability.

**Fonview**   Sprint Communication Company billing management software that allows users to analyze monthly bills sent to customers on a diskette.

**footprint**   1. The area of the earth's surface within which the signals of a specific satellite can be received. 2. The amount of desktop space that a device occupies.

**footprints**   Compatible configuration of two mating pieces of apparatus. Specifically, an exclusion feature which is molded into the base of a protector unit which matches a raised detail in the protector.

**foreground**   In a 3270 PC environment, foreground is the color of characters inside a window, one of the ways to set colors on the 3270 PC.

**foreign**   In IBM's Network Problem Determination Application (NPDA), used to describe a network domain controlled by an System Services Control Point (SSCP) other than that which controls the user's terminal. Also called remote domain.

**foreign attachment**   Equipment attached to a common carrier facility which is not provided by the common carrier.

**foreign exchange (FX) line**   A communications common carrier line in which a termination in one central office is assigned a number belonging to a remote central office.

**foreign exchange service**   A service which connects

a customer's telephone to a telephone company central office normally not serving the customer's location.

**Form Feed (FF)**   A printer control character used to skip to the top of the next page (or form).

**format**   1. Message structure that may include the code set and protocol characters. 2. A defined arrangement of such things as characters, fields, and lines, usually used for displays, printouts, or files. 3. The pattern which determines how data is recorded.

**Format Effectors (FE)**   Category of control characters mainly intended for the control of the layout and positioning of information on printing and/or display devices. Samples of FE characters are: back space (BS), carriage return (CR), form feed (FF), horizontal tab (HT), line feed (LF), and vertical tab (VT).

**Format Identification (FID) field**   In IBM's SNA, a field in each transmission header (TH) that indicates the format of the TH; that is, the presence or absence of certain fields. TH formats differ in accordance with the types of nodes between which they pass. *Note.* There are six FID types:

FID0  used for traffic involving non-SNA devices between adjacent subarea nodes when either or both nodes do not support explicit route and virtual route protocols.

FID1  used for traffic between adjacent subarea nodes when either or both nodes do not support explicit route and virtual route protocols.

FID2  used for traffic between a subarea node and an adjacent PU type-2 peripheral node.

FID3  used for traffic between a subarea node and an adjacent PU type 1 peripheral node.

FID4  used for traffic between adjacent subarea nodes when both nodes support explicit route and virtual route protocols.

FIDF  used for certain commands (for example, for transmission group control) sent between adjacent subarea nodes when both nodes support explicit route and virtual route protocols.

**formatted diskette**   A diskette on which track and

sector control information has been written and which may or may not contain data. A diskette must be formatted before it can receive data.

**formatted document**  In IBM's PROFS and Office Vision a formatted document is one that has been laid out by one of two programs: either the Document Composition Facility (DCF) Text Processing program or the Revisable-Form Text (RFT) Document program. It means the text has been arranged with preset margins and page lengths. Any DCF or RFT commands that were included in the text are translated and used in the formatted document.

**formatted system services**  In IBM's SNA, a portion of VTAM that provides certain system services as a result of receiving a field-formatted command, such as an Initiate or Terminate command.

**formatting control**  An instruction embedded into the text that tells IBM's DisplayWrite/370 program and a printer how to present the text of a Revisable-Form Text (RFT) document.

**FormFeed**  A control code or a button on a printer that advances the paper to the top of the next page.

**Forms Control Buffer (FCB)**  A buffer that is used to store vertical formatting information for printing, each position corresponding to a line on the form.

**FORmula TRANslator (FORTRAN)**  A compiler language developed by the IBM Corporation. FORTRAN was originally conceived for use in scientific problems; however, it has now been adopted for commercial application as well. FORTRAN programs consist of sets of equation statements.

**FORTRAN**  FORmula TRANslator.

**fortuitous distortion**  Distortion resulting from causes generally subject to random laws (accidental irregularities in the operation of the apparatus and of the moving parts, disturbances affecting the transmission channel, etc.).

**forward busying**  A PBX or central office feature where supervisory signals are forwarded in advance of address signals to seize assets of the system before attempting to establish a call.

**forward channel**  The communications path carrying data from the call initiator to the called party; opposite of reverse channel. Also main channel.

**Forward Error Correction (FEC)**  Technique allowing the receiver to correct errors occurring in a transmission channel without requiring retransmission of the data.

**Forward Explicit Congestion Notification (FECN)**  A bit in the frame relay header which when set to a 1 indicates that the frame has encountered a congested path on its way across the network.

**Forward Indicator Bit (FIB)**  In CCITT Signaling System 7, the FIB is inverted by the transmitter when a retransmission occurs.

**Forward Sequence Number (FSN)**  In CCITT Signaling System 7, the forward sequence number is incremented by 1 each time a new message is transmitted.

**Forwarding table**  A data base constructed by a bridge which lists each data link source address connected to the bridge.

**FOSSIL**  Fideo Opus SeaDog Serial Interface Layer. A program which takes over the serial port of a computer in a standard manner, acting as a buffer between some bulletin board system programs and the serial port.

**FOTE**  Follow-On Test and Evaluation.

**four-wire circuit**  A circuit using two pairs of conductors, one pair for the "transmit" channel and the other pair for the "receive" channel.

**four-wire equivalent circuit**  A circuit using the same pair of conductors to give "transmit" and "receive" channels by means of different carrier frequencies for the two channels.

**Fourth-Generation Language (4GL)**  A software productivity tool which assists programmers in the design and development of data base management systems.

**fox message**  A diagnostic test message that uses all the letters (and that sometimes includes numerals): "THE QUICK BROWN FOX JUMPED OVER A LAZY DOG'S BACK 1234567890." (In French, "VOYEZ LE BRICK GEANT QUE J'EXAMINE PRES DU WHARF." Often run continuously during system testing and fault isolation.

**FP**  FacePlate.

**FPLA**  Field Programmable Logic Array.

**FRACTIONAL T1**  A communications service which subscribers can obtain a portion of a T1 channel's bandwidth, such as 2, 4, 6, 8 or 10 64 Kbps channels.

**Fragmentation**  The process of dividing a basic unit of information transfer into smaller units for transmission from one network to another.

**frame**  1. Same as transmission block. 2. The sequence of bits and bytes in a transmission block. 3. The overhead bits and bytes which surround the information bits in a transmission block. 4. In T1/DS1 signals, the frame is the collection of 8 bits from each of the 24 channels plus one frame bit for a total of 193 bits. 5. The unit of transmission in some local area networks, including the IBM Token-Ring Network. It includes delimiters, control characters, information, and checking characters. A frame is created from a token when the token has data appended to it. 6. Equipment in a common carrier office where physical cross connections are made between circuits.

**frame alignment pattern**  In a T1 signal, a logical sequence of six bit values carried in the framing bits of the 1st, 3rd, 5th, 7th, 9th, and 11th frames of a D4 framed circuit, or the 1st, 5th, 9th, 13th, 17th, and 21st frames of an ESF circuit. Also called the "Frame Synchronization Sequence." The Frame Alignment Pattern is the pattern 101010. The European Frame Alignment Pattern contains 7 bits—0011011.

**frame bits**  Digital bits used in the frame which are added to the basic data rate to identify the start of a collection of individual channel data.

**Frame Check Sequence (FCS)**  A 16-bit field, usually appended at the end of a frame, used for error detection in bit-oriented communications protocols.

**frame errors**  In a T1 environment, errors in the 12-bit, D4 Frame word. An error is counted when the 12-bit Frame word received does not conform to the standard 12-bit frame word pattern.

**frame level**  In packet switching, Level 2 of the CCITT X.25 Recommendation which defined the link access procedure for reliable data exchange over a link between a DTE and a DCE.

**frame mnemonics**  A set of acronyms used to identify the frame types as defined by the appropriate bits in the control field.

**frame relay**  A high-speed packet switching technology which achieves up to 10 times the speed of conventional X.25 packet switching networks using the same hardware. This still-evolving CCITT ISDN interface standard (I.122), is considered as a precursor of broadband or B-ISDN. Frame relay can be described as a packet-mode interface layered on top of narrowband ISDN. Frame relay can run over a B, D, or H channel, and allows packet-mode devices to exchange information over an ISDN. Frame relay eliminates all error handling and flow control procedures associated with such protocols as X.25 and SDLC, requiring a relatively error-free transmission path for its effective utilization.

**frame synchronization sequence**  A logical sequence of six bit values carried in the framing bits of the 1st, 3rd, 5th, 7th, 9th, and 11th frames of a D4 framed circuit, or the 1st, 5th, 9th, 13th, 17th, and 21st frames of an ESF circuit. The Frame Synchronization Pattern is the pattern 101010. The European Frame Synchronization Pattern contains 7 bits—0011011. Also called the "Frame Alignment Pattern." See illustration under European framing entry.

**framed BERT**  A Bit Error Rate Test in which the D4 Framing or Extended Superframe Format line discipline is imposed on the test pattern.

179

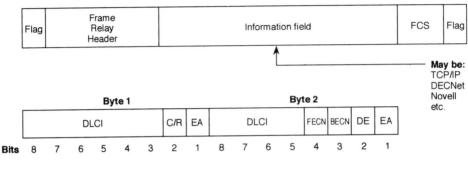

DLCI = Data Link Connection Identifier
C/R = Command/Response field bit (not modified by network)
FECN = Forward Explicit Congestion Notification
BECN = Backward Explicit Congestion Notification
DE = Discard Eligibility Indicator
EA = Address Extension bit

*frame relay*

**FrameNet** The name used by CompuServe for its frame relay service.

**framing** 1. The process of establishing a reference so that time slots or elements within the frame can be identified. 2. Process of inserting control bits to identify channels; used in TDM signals such as the formatted version of T1.

**framing bits** In multiplexed digital transmissions, the leading bit of each frame, not included in any octet channel of user data; also known as "F-Bits." On any T1 circuit, framing bits occupy 8 Kbps bandwidth. In D4 framing, the framing bits of successive frames carry the Frame Synchronization Pattern and the Signaling Synchronization Pattern. In ESF, the framing bits of sequential frames carry four signaling patterns: Frame Synchronization; Signaling Synchronization; CRC-6; and Link Data Channel (LDC).

**free list** A list structure holding all the units of store which are not currently in use. It may be operated as a kind of queue so that store units which become free are added to the tail, while new requests for store are met from the head of the queue, but other regimes are possible.

**free running mode** This mode of using a virtual terminal allows two associated systems or users to have access to the data structures simultaneously. The possibility that conflict between them may cause difficulties must be taken account of by the users. This problem is avoided in alternate mode.

**free-wheeling** An old approach to terminal and communications system design in which terminals were able to transmit an operator initiative, without regard to whether the host or switching system was able to receive and handle all the data sent.

**Freefone** A British Telecom communications service which bills the cost of a public switched telephone call to the destination party. Similar to AT&T's WATS.

**Freenet** An open access, free to use, community computer system.

**freespace** Transmission of information using lightwave signals directly through the air, rather than through a conductor.

**freeze frame** A television transmission technique in which pictures are captured, transmitted, and displayed every few fractions of a second at a rate considerably slower than full-motion television, thus lowering the required transmission bandwidth and storage. Commonly used in video-based teleconferencing systems.

**Freqing (Freq)** A term or abbreviation for a "file request" for a file from another node in a network. In FidoNet a node user usually freqs a file through mailer software which sends an appropriate request to a distant node that has the desired file.

**frequency** The rate at which a signal alternates; typically, the number of complete cycles per second, normally expressed in hertz (Hz).

**frequency, carrier** The frequency of the unmodulated carrier.

**frequency band** An arbitrarily defined bandwidth of the electromagnetic spectrum for which the boundary frequencies are stated.

**frequency bands** Frequency bands are defined arbitrarily as follows:

| Range (MHz) | Name |
|---|---|
| 0.03–0.3 | Low frequency (LF) |
| 0.3–3.0 | Medium frequency (MF) |
| 3–30 | High frequency (HF) |
| 30–300 | Very high frequency (VHF) |
| 300–3000 | Ultra high frequency (UHF) |
| 3000–30000 | Super high frequency (SHF) (microwave) |
| 30000–300000 | Extremely high frequency (EHF) (millimeterwave) |

**frequency-derived channel** Any of the channels obtained from multiplexing a channel by frequency division.

**frequency distortion** Distortion that occurs as a result of failure to amplify or attenuate equally all frequencies present in a complex wave.

**Frequency Division Multiple Access (FDMA)** A technique for sharing a multipoint or broadcast channel. This technique allocates different frequencies to different users.

**Frequency-Division Multiplexer (FDM)** A device that divides the available transmission frequency range into narrower bands, each of which is used for a separate channel.

**Frequency Division Multiplexing (FDM)** A multiplexing technique that partitions the composite bandwidth into channels, assigning a specific range of frequencies to each channel.

**frequency hopping** A spread spectrum technique by which the information is hopped between several communications channels. The carrier frequency is periodically changed according to a schedule based upon a code sequence.

**Frequency Modulation (FM)** One of three basic ways to add information to a sine wave signal; the frequency of the sine wave, or carrier, is modified in accordance with the information to be transmitted.

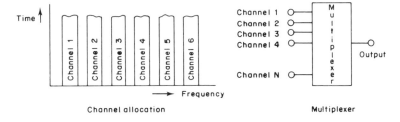

*Frequency Division Multiplexing (FDM)*

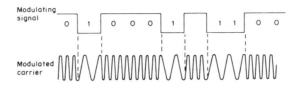

*Frequency Modulation (FM)*

**frequency offset** Analog line frequency changes which is one of the impairments encountered on a communications line.

**frequency range** The lowest to the highest frequency that can be transmitted over a band.

**frequency response** The variation in relative strength (measured in decibels) between frequencies.

**Frequency Shift Keyed Modulation (FSK)** A modulation technique where frequency shifts occur due to binary digital level changes. The carrier shifts between two predetermined frequencies (commonly 1200 Hz and 2200 Hz). Typically when the binary modulating signal is positive, the lower of the two frequencies is generated. The higher of the two frequencies is generated when the modulating signal is negative.

**Frequency Shift Keying (FSK)** Frequency shift key (FSK) modulation varies the number of waves per unit of time using different tones to represent a binary "1" and a binary "0." It is used with low-speed, asynchronous transmission. It cannot transmit more than one bit per baud.

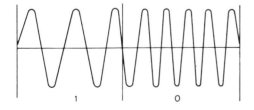

**frequency translation** The amount of movement of a single frequency signal at the input end of the transmission line to a different frequency when received at the output end.

**frequency translator** See head end.

**Fresnel reflection** The optical reflections off the endface(s) of a fiber at a connection, splice, or break point, due to the glass to air boundary.

**FRL** Facility Restriction Level.

**Frobnicate** To manipulate, adjust or tweak something. Often abbreviated as frob.

**frogged** A term used to indicate that the same channel is not used twice for a patch through a trunk, regardless of the system's length. Frogging is used in FDM systems to prevent carrier noise buildup.

**front end mailer** A program that operates on a bulletin board system and which determines if a caller is another computer that wants to exchange mail or a human that wants to access the BBS resources. Usually the mailer transmits the prompt "Press Esc.." and upon receiving an ESC character or the passing of a timeout period considers the caller to be human and gives it the resources of the BBS. Also known as a mailer.

**Front End Processor (FEP)** A specialized computer processor generally used in conjunction with a larger mainframe computer that interfaces the computer to communications facilities and remote users. The FEP performs data communications functions which serve to preprocess and offload the attached computer(s) of network processing functions. In an IBM network, an FEP is called a communications controller.

**FRS** 1. Facilities Restoral System. 2. Flexible Route Selection.

**FRU** Field Replaceable Unit.

**FS** File Separator.

**FS bits (framing bit signaling)** In D4 or ESF framing, framing bits that carry the Signaling Synchronization Sequence.

**FSA startup** That part of system initialization when the FSA is loaded into the functional subsystem address space and begins initializing itself.

**FSD** Full Scale Development.

**FSED** Full Scale Engineering Development.

**F-series** A set of CCITT (now ITU) recommedations relating to the operation of telegraph and telematic service to include facsimile, teletex and videotex. F-series recommendations include:

F72 International telex store and forward—general principles and operational aspects

F122 Operational procedures for the maritime satellite data transmission service

F150 Provisions applying to the operation of an international public automatic message-switching service for equipments utilizing the international telegraph alphabet 2

F161 International group 4 facsimile service (FAX 4)

F162 Operational requirements of an international store-and-forward facsimile switching service

F190 Operational provisions for the international facsimile service between public bureaux and sub-scriber stations and vice versa

F201 Interworking between teletex service and the telex service

F350 Application of Series T Recommendations

**FSI connect** The FSI communication service which establishes communication between IBM's JES2 and the FSA or functional subsystem.

**FSI disconnect** The FSI communication service which severs the communication between IBM's JES2 and the FSA or functional subsystem.

**FSI services** A collection of services available to users (JES2) of the FSI. These services compromise communication services, data set services, and control services.

**FSK** Frequency-Shift Keying.

**FSVS** Future Secure Voice System.

**FTAM** File Transfer, Access, and Management.

**FT-bits (framing bits tracking)** In D4 or ESF framing, framing bits that carry the Frame Synchronization Sequence.

**FTCMS** File Transfer Customer Monitor System.

**FTP** File Transfer Protocol. The Internet protocol used for transferring files from one computer to another.

**FTS** Federal Telecommunications System.

**FTTC** Fiber-To-The-Curb.

**FTTERM** File Transfer and Terminal Emulator Program.

**FT1** Fractional T1.

**FS-2000** An AT&T lightwave system which provides a synchronous digital transmission which meets both ANSI and SONET requirements.

**full-duplex (FD, FDX)** Synonym for duplex.

**full-duplex transmission (FDX)** Simultaneous two-way independent transmission in both directions. Compare with half-duplex transmission. Also used to describe terminals in the echoplex mode.

**full-duplex channel** A channel capable of trans-mitting data in both directions at the same time.

**full motion video** Television transmission by which images are captured, transmitted and displayed at a repetition rate fast enough that a human observer perceives full motion. Compare with freeze frame.

**full screen**  Use of the entire display screen rather than part of it, such as a panel or window.

**fully connected network**  A network topology in which each node is directly connected by other nodes, forming a mesh structure. Normally impracticable as the number of nodes in the network increases.

**function**  A standard, prepackaged set of coded instructions for carrying out a computer operation.

**function key**  1. A term associated with specific keys on a teletypewriter (such as CR, LF, LTRS, FIGS, etc.). These keys, when operated, cause the teletypewriter to perform mechanical functions in order that a message may be received in proper form. 2. A term associated with keys on intelligent terminals and personal computers that are programmable to perform predefined functions when pressed.

**Function Management Data (FMD)**  In IBM's SNA, an RU category used for end-user data exchanged between logical units (LUs) and for requests and responses exchanged between network services components of LUs, PUs, and SSCPs.

**Function Management Header (FMH)**  In IBM's SNA, one or more headers, optionally present in the leading request units (RUs) of an RU chain, that allow one half-session in an LU–LU session to: (1) select a destination at the session partner and control the way in which the end-user data it sends is handled at the destination, (2) change the destination or the characteristics of the data during the session, and (3) transmit between session partners status or user information about the destination (for example, a program or device). *Note*. FM headers can be used on LU–LU session types 0, 1, 4, and 6.

**Function Management (FM) PROFILE**  In IBM's SNA, a specification of various data flow control protocols (such as RU chains and data flow control requests) and FMD options (such as use of

FM headers compression, and alternate codes) supported for a particular session. Each function management profile is identified by a number.

**Functional Subsystem (FSS)**  An address space uniquely identified as performing a specific function related to the JES. For IBM's JES2, an example of an FSS is the program Print Services Facility that operates the 3800 Model 3 and 3820 printers.

**Functional Subsystem Application (FSA)**  The functional application program managed by the functional subsystem.

**Functional Subsystem Interface (FSI)**  The interface through which IBM's JES2 or JES3 communicate with the functional subsystem.

**functional subsystem startup**  That process part of system initialization when the functional subsystem address space is created.

**functional test**  A test carried out under normal working conditions to verify that a circuit or a particular device functions correctly.

**fundamental mode**  In fiber optic technology, the lowest-order mode. In a single-mode fiber the fundamental mode is the only mode guided by the fiber above the cutoff wavelength.

**fuse**  A device used for the protection against excessive currents. Consists of a short length of fusible metal wire or strip which melts when the current through it exceeds the rated amount for a definite time. Placed in series with the circuit it is to protect.

**fusible links**  Short lengths (about 25 feet) of fine gauge wire pairs inside metallic sheath cable that melt to interrupt an electrical circuit preventing overheating in building wiring and equipment.

**fusion splice**  A permanent joinint of two glass or plastic fibers by applying heat sufficient to fuse or melt the ends of separate fibers to form a continuous fiber.

**Fuzzball** Software which runs on Digital Equipment Corporation LSI-11 microcomputers in the NSF-NET which enables data transmitted on leased lines to access devices connected to Ethernet LANs.

**FW** Field Wire.

**FX** Foreign Exchange.

**FXO** Foreign Exchange Office.

**FYI** For Your Information. A series of documents put out by the Internet NIC which addresses common user questions.

**F1A-line weighting** A telephone noise measuring class of weighting used to measure noise on a line terminated by a 302-type or similar telephone handset.

**F1F2** A modem that operates over a half-duplex line to produce two subchannels at two different frequencies for low-speed, full-duplex operation.

# G

**G**  Giga.

**G**  Shunt conductance.

**gain**  The degree the amplitude is increased. The amplification realized when a signal passes through an amplifier, repeater, or antenna. Normally measured in decibels. Opposite of loss, or attenuation (negative gain).

**gain hits**  A cause of errors in telephone line data transmission, usually when the signal surges more than 3 dB and lasts for more than 4 milliseconds. Bell standards call for eight or fewer gain hits per 15-minute period.

**gain/slope**  The measurement of frequency levels over a specific frequency bandwidth on an analog circuit. Typically measured at frequencies of 404, 1004, and 2804 Hz.

**Galaxy**  A trademark of Rockwell International Corporation as well as the name of an all-digital automatic call distribution system from that vendor.

**galloping**  High-amplitude, low-frequency vibration.

**garbage**  An informal term used to refer to corrupted data.

**garble**  An error in transmission, reception, or encryption that renders the message or a portion thereof incorrect or undecipherable.

**gas plasma display**  A type of display screen mainly used in laptop computers. Gas plasma provides a better contrast ratio than most LCD displays, but uses up to six times more power, which results in most laptops with this type of display requiring AC power.

**gas tube surge protector**  Surge-limiting device similar in operation to the Carbon Block Surge Protector except that it has specially configured electrode with a more precise narrow gap and a sealed gas composition. The gas tube results in a more accurate and precise operating voltage range and extended service life under conditions of repeated operation.

**gate**  A basic logic circuit.

**gateway**  1. A combination of hardware and software used to interconnect otherwise incompatible networks, network nodes, subnets, or other network devices. Two examples are PADs and protocol converters. Gateways operate at the fourth through seventh layers of the OSI model. 2. A computer that functions as a node in two or more networks, forwarding mail and messages from one network to addresses in the other network.

**gateway access protocol**  The protocol used between a host system and a system that is a DTE on a PSDN, to provide the X.25 gateway access facility to a user on the host.

**gateway control functions**  In IBM's SNA, functions performed by a gateway SSCP in conjunction with the gateway NCP to assign alias network address pairs for LU–LU sessions, assign virtual routes for the LU–LU sessions in adjacent networks, and translate network names within BIND RUs.

**gateway facilities**   A device that connects two systems, especially if the systems use different protocols. For example, a gateway is needed to connect two independent local networks or to connect a local network to a long-haul network.

**gateway HOST**   In IBM's SNA, a host node that contains a gateway SSCP.

**gateway NCP**   In IBM's SNA, an NCP that performs address translation to allow cross-network session traffic. The gateway NCP connects two or more independent SNA networks.

**gateway node**   In IBM's SNA, synonym for gateway NCP.

**gateway SSCP**   In IBM's SNA, an SSCP that is capable of cross-network session initiation, termination, takedown, and session outage notification. A gateway SSCP is in session with the gateway NCP; it provides network name translation and assists the gateway NCP in setting up alias network addresses for cross-network sessions.

**gateway with network station**   A configuration option which lets an IBM PC perform both a gateway's communications server functions for IBM PCs attached to it and a network station's workstation functions for an operator.

**Gateway-to-Gateway Protocol (GGP)**   The TCP/IP protocol that core gateways use to exchange routing information. GGP implements a distributed shortest path routing computation.

**gaussian noise**   Undesirable, random, low-level background electrical energy introduced into a transmission, consisting of frequency components such that, over a period of time, a continuous frequency spectrum within the band limits of the system is observed. Also called white noise, ambient noise, and hiss.

**GBH**   Group Busy Hour.

**Gbytes (gigabytes)**   1 024 000 000 bytes (1000 Mbytes).

**GCE**   Ground Communications Equipment.

**GCE**   Ground Control Equipment.

**GCOS**   General Comprehensive Operating Supervisor.

**GCOS 6**   Operating system that executes in a Honeywell Level 6 processor and supports multiple interrupt-driven activities.

**GCT**   Greenwich Civil Time.

**GDSU**   Global Digital Service Unit.

**GE American Communications, Inc.**   The successor to RCA's American, GE American Communications is headquartered in Princeton, NJ, and operates five C-band and two K-band Satcom communications satellites as well as East- and West-Coast tracking, telemetry and command earth stations. The following table identifies the traffic carried on GE American Communications satellites.

| *Satellites* | *Traffic carried* |
| --- | --- |
| Ku-band (12/14 GHz) Satcom K-1 | Entertainment TV, business and government networks |
| Satcom K-2 | Network TV and TV program distribution; news and sports, business and government video/data networks |
| C-band (4/6 GHz) Satcom I-R | Entertainment and network TV; digital audio transmission and SCPC radio programming services; international interconnect |
| Satcom II-R | Government video and data traffic entertainment and occasional TV, international interconnect |
| Satcom III-R | Cable net 1 |
| Satcom IV | Cable net 2 |
| Satcom V | Alascom, Inc., service (Aurora) |

**GE network for information exchange (GEnie)** A popular on-line computer information network for personal computer enthusiasts.

**GE Remote Terminal Supervisor (GRTS)** Originally developed by General Electric (GE) when that firm was manufacturing computers. It now references a software system resident in a Honeywell (now Bull) DATANET 6600 or 335 Front End Network Processor that controls communications. Now more commonly known as Remote Terminal Supervisor; however, still abbreviated as GRTS.

**GEIS** General Electric Information Services.

**Gemini** A family of earth stations marketed by Hughes Network Systems of Germantown, MD.

**GEN** GENerate.

**GEN II** A trademark of Varian Microwave of Santa Clara, CA, as well as the name of a Ku-band klystron high-power amplifier manufactured by that vendor.

**gender mender** A pair of connectors of the same gender (male/plug or female/socket) connected back to back which permits a cable end and device having the same connector genders to connect.

**General Comprehensive Operating Supervisor (GCOS)** The Honeywell operating system for that vendor's Series 60 and 6000 mainframe computers.

**general format identifier** The four high-order bits of the first octet in an X.25 packet header, containing the qualifier bit, the delivery confirmation bit, and modulus value.

**general message** A message assigned an identifying title and usually a sequential serial number and which usually requires a wide distribution.

**general poll** Used in BSC for abbreviated addressing on lines that include cluster controllers. The cluster controller responds to a general poll and indicates if any of its connected devices has data to send.

**general purpose cable** Cable specifically used to connect computers or telephones inside a building. This type of cable is not for use in risers (vertical shafts) and plenums unless enclosed in non-combustible tubing.

**General Switched Telephone Network (GSTN)** Same as the public switched telephone network.

**general topology subnetwork** A non-broadcast subnetwork.

**Generalized Path Information Unit Trace (GPT)** In IBM's SNA, a record of the flow of path information units (PIU's) exchanged between the network control program and its attached resources. PIU trace records consist of up to 44 bytes of transmission header (TH), request/response header (RH), and request/response unit (RU) data.

**generic** A term with various definitions. It may refer to the fundamental version of a widely distributed program package, or to such a package and accompanying hardware. Generic also may refer to the latest version of a program package.

**generic bind** In IBM's SNA, a synonym for a session activation request.

**generic unbind** In IBM's SNA, a synonym for a session deactivation request.

**GEnie** GE network for information exchange.

**geostationary** Refers to a geosynchronous satellite angle with zero inclination, so the satellite appears to hover over one spot on the earth's equator.

**geosynchronous orbit** The orbit where communications satellites will remain stationary over the same earth location. The geosynchronous orbit altitude is approximately 23 300 miles above the equator.

**GFE** Government Furnished Equipment.

**GFI** Group Format Identifier.

**GGP** Gateway-to-Gateway.

**GHz (gigahertz)** A frequency unit equal to 1E+9 hertz.

**GIF**   Graphic Interchange Format file format.

**giga (G)**   A prefix for one billion (1E+9) times a specific unit.

**gigabyte**   One billion bytes.

**gigahertz**   A term used to state 1 000 000 000 (giga) cycles per second (hertz).

**GIGO**   An acronym for "Garbage In, Garbage Out."

**glare**   A condition that results in both ends of a trunk being seized at the same time by different users, thereby blocking a call.

**Global Multidrop Service**   A service marketed by Telenet Communications Corporation of Reston, VA, in conjunction with British Telecom International which enables users to send SDLC traffic beeen sites in the U.S. and the UK via packet networks operated by both companies. British Telecom International markets this service under the name International MultiStream Synchronous Service.

**global name**   That part of the entity name which identifies the node in the network.

**Global Naming**   A LAN feature which allows users to use one logon to access any resource on the network that they have permission to use.

**global network addressing domain**   In Digital Equipment Corporation Network Architecture (DECnet), the addressing domain consisting of all NSAP addresses in the OSI environment.

**Global Network Navigator (GNN)**   An application developed at CERN in Switzerland which provides information about new services available on the Internet, articles about existing services and an on-line version of Internet related books. The GNN is a World Wide Web (WWW) based information service.

**Global Positioning Satellite (GPS)**   A group of 40 U.S. satellites, each of which will produce an encrypted P code used by the military to guide missiles and a Clear/Acquisition signal available for commercial uses such as surveying and timing for communications networks.

**Global Virtual Private Network (GVPN)**   A service offering by Sprint Communications which extends that carrier's domestic VPN service to foreign locations, such as the UK and Hong Kong.

**GlobalNet**   A free, electronic amateur bulletin board system network which operates based upon Fido-Net technical standards. GlobalNet nodes are located in North America and Europe.

**GlobalView**   A network management system from Universal Data Systems, Inc., of Huntsville, AL, that supports dial-up modems.

**Globenet**   A value-added network that uses a strategy of interconnecting Bell operating company packet networks to form a nationwide network.

**G-Net**   The packet switching network operated by Generale Bank of Belgium.

**go-back-n**   A form of continuous ARQ where all blocks or frames following a block received in error are discarded and must be retransmitted.

**GOCP**   Government Outbound Calling Plan.

**GOES**   Geostationary Operational Environmental Satellite.

**Gopher**   A distributed menuing system developed at the University of Minnesota for information access on the Internet. Gopher servers store a wide range of information, from news and phone books to weather reports and recipes. To access the Gopher system your host must have a Gopher client package.

**GOS**   Grade Of Service.

**GOSIP**   Government Open Systems Interconnection (OSI) Profile.

**Government Data Network (GDC)**   A packet switching network being installed by Racal Data Group Ltd. in the United Kingdom. The company is to sell packet switching network services to the UK Government.

**Government Open Systems Interconnection Profile (GOSIP)**   A selected suite of Open System Interconnection (OSI) standards that major computer and communications purchases of the U.S. Federal Government must support beginning in August 1990.

**Government Outbound Calling Plan (GOCP)**  A Tariff 16 offering which allows U.S. government customers to purchase AT&T Megacom services at discounted rates.

**GPS**  Global Positioning System.

**G-P-T**  The name of a fiber optic cable route connecting Guam and the Philippines to Taiwan.

**GPT**  Generalized Path information unit Trace (IBM's SNA).

**Grade Of Service (GOS)**  1. In telephone transmission, a measure of users' options of the quality of speech heard during a telephone conversation. GOS is measured on a scale of 1 (unacceptable) to 5 (excellent). 2. When used for access measurement, GOS refers to the probability of a call being blocked or delayed, expressed as a percentage.

**graded-index fiber**  A fiber lightguide with a graduated refractive index higher at the center of the core than at the cladding. Because light travels faster where the refractive index is lower, it is speeded along the longer, outer paths and is slowed over the shorter inner paths, thus equalizing the time needed to travel the length of the fiber lightguide. This is a way of reducing dispersion.

**graded index profile**  Any refractive index profile that varies with radius in the core. Distinguished from a step index profile.

**grandfathered installation**  Any existing system, serving existing customers where existing intra-building network cable and network terminating wire are currently being used to provide both network service and intra-system or inside wire-like service is considered to be "grandfathered." Such use will continue to be allowed for the in-service life of the associated equipment.

**Graphical User Interface (GUI)**  A personal computer display which runs in the computer's graphics mode and enables users to point at icons on the screen and click with a mouse to initiate an activity. This type of interface eliminates the necessity of typing commands.

**graphics station**  A collection of hardware and software which operates a high-resolution video display to present information to the user in the form of both text and graphic images.

**groom-and-fill**  A process by which channels (DS-0s) from several DS-1s can be segregated for transmission to different remote locations and channels from separate DS-1s are combined to share a DS-1 for transmission to a remote location.

**Grosch's Law**  The amount of computational power obtained is proportional to the square of the price paid. First stated by Herbert Grosch.

**ground**  An electrical common conductor that at some point connects to the earth.

**ground sheet**  A method of signaling from a station to an exchange by connecting one side of the line to ground.

**ground start**  Signaling method which detects a circuit is grounded at the other end.

**ground station**  An assemblage of communications equipment to include a signal generator, transmitter, receiver, and antenna which receives signals to/from a communications satellite (earth station).

**ground wave**  A radio wave that is not refracted from the ionosphere.

**grounding**  The act of connecting a circuit or metal parts to earth or other pieces of equipment that are connected to earth.

**group access**  A method of granting rights to several users so they may access computer ports or files on a LAN.

**group address**  Address of two or more stations in a group.

**group addressing**  In transmission, the use of an address which is common to two or more stations.

**Group Busy Hour (GBH)** The busy hour offered to a given trunk group.

**group channel** A unit on telephone carrier (multiplex) systems. A full group is a channel equivalent to 12 voice-grade channels (48 kHz). A half-group has the equivalent bandwidth of six voice-grade channels (24 kHz). Group channels can be used for high-speed data communication (wideband).

**group delay** If a complex signal wih a narrow bandwidth is sent down a transmission path, at the receiving end the envelope of the signal will appear to have suffered a delay, called the group or envelope delay.

**Group Format Identifier (GFI)** In X.25 packet switched networks, the first four bits in a packet header; contains the Q bit, D bit, and modulus value.

**Group Separation (GS)** A control character used to separate and qualify data logically; normally delimits data item called a group.

**Group Switching Center (GSC)** The name given in the UK to the telephone exchange that forms the entry point to the national subscriber dialing network. The GSC holds the register translator equipment to provide the originating routing for national and international calls.

**Group 1** The CCITT standard for four-to-six minute analog facsimile.

**Group 2** The CCITT standard for two-to-three minute analog facsimile.

**Group 3** The CCITT standard for digital facsimile.

**Group 4** The CCITT standard for digital facsimile oriented to public data networks.

**GRP** GRouP.

**GRTS** GE Remote Terminal Supervisor.

**GS** Group Separator.

**GSA** General Services Administration.

**GSC** Group Switching Centre.

**GSTN** General Switched Telephone Network.

**G/T** Ratio of gain to noise temperature, expressed in decibels per kelvin. Zero kelvins implies no molecular motion, hence no noise.

**guard band** The unused bandwidth separating channels to prevent crosstalk in a frequency division multiplexing (FDM) system.

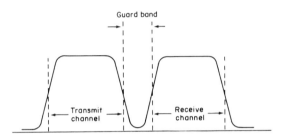

**guard frequency** 1. A single frequency carrier tone used to indicate a line is prepared to send data. 2. The frequencies between subchannels in FDM systems used to guard against adjacent channel interferences. Also guard band.

**Guardian** Tandem Computer Company Operating System.

**guardrail** A U.S. military airborne and ground communications intelligence system.

**GVPN** Global Virtual Private Network.

**GWEN** GroundWave Emergency Net.

**G.702** The CCITT (now ITU) standard for digital hierarchy bit rates.

**G.703** The CCITT (now ITU) standard (current version, 1984) for the physical and logical characteristics of transmissions over digital circuits. G.703 includes specifications for both the North American (1.544 Mbps) and European (2.048 Mbps) T1 circuits, as well as for circuits of larger bandwidth in the North American and European digital hierarchies. *Note.* In common usage, the term "G.703" usually refers only to the standard for 2.048 Mbps circuits.

**G.703 64K** Those transmission facilities at the 64 Kbps rate that use the CCITT (now

Bull) recommended physical/electrical interface specified under CCITT (now Bull) publication G.703.

**G.703 2.048 mbps** Those transmission facilities at the 2.048 Mbps rate that use the CCITT (now Bull) recommended physical/electrical interface specified under CCITT (now Bull) publication G.703.

**G.704** The CCITT (now ITU) standard for physical and electrical characteristics of hierarchical digital interfaces.

**G.706** The CCITT (now ITU) standard for frame alignment and cyclic redundancy check procedures related to basic frame structures defined in G.704.

**G.707** The CCITT (now ITU) standard for synchronous digital hierarchy bit rates.

**G.708** The CCITT (now ITU) standard for network mode interface for the synchronous digital hierarchy.

**G.709** The CCITT (now ITU) standard for synchronous multiplexing.

**G.721** The CCITT (now ITU) standard for 32 Kbps adaptive differential pulse code modulation (ADPCM).

**G.722** The CCITT (now ITU) standard for 7 kHz audio-coding within 64 Kbps.

**G.723** The CCITT (now ITU) standard which extends G.721 ADPCM to 24 and 40 Kbps for DCME applications.

**G.725** The CCITT (now ITU) standard for the use of 7 kHz audio codec within 64 Kbps.

**G.747** The CCITT (now ITU) standard for second-order digital multiplex equipment operating at 6312 Kbps and multiplexing three 2.048 Mbps data sources.

**G.755** The CCITT (now ITU) standard for digital multiplex equipment operating at 139.264 Mbps and multiplexing three 44.736 Mbps data sources.

**G.802** The CCITT (now ITU) standard for internetworking between networks based on different digital hierarchies and speech encoding laws.

**G.811** The CCITT (now ITU) standard for timing requirements at the outputs of primary reference clocks suitable for plesiochronous operation of international data links.

**G.812** The CCITT (now ITU) standard for timing requirements at the output of slave clocks suitable for plesiochronous operation of international data links.

**G.813** The CCITT (now ITU) standard for error performance of an international digital connection forming part of an integrated services digital network.

**G.821** The CCITT (now ITU) recommendation that specifies performance criteria for digital circuits in the ISDN.

**G.824** The CCITT (now ITU) standard for the control of jitter and wander within digital networks that are based on the 1.544 Mbps hierarchy.

**G.901** The CCITT (now ITU) standard covering general considerations on digital sections and digital line systems.

**G.921** The CCITT (now ITU) standard for digital sections based upon the 2.048 Mbps hierarchy.

**G.960** The CCITT (now ITU) standard for digital sections for ISDN basic rate access.

**G.961** The CCITT (now ITU) standard for digital transmission systems on metallic local lines for ISDN basic rate access.

# H

**A channel** A hierarchy of ISDN broadband channels for the transmission of voice, data and video. The following table summarizes the CCITT ISDN H-channel hierarchy.

| Level | Gross bit rate | Effective bit rate |
|-------|----------------|--------------------|
| H0    | 384 Kbps       | 384 Kbps           |
| H11   | 1.544 Mbps     | 1.536 Mbps         |
| H12   | 2.048 Mbps     | 1.920 Mbps         |
| H21   | 34.368 Mbps    | 32.768 Mbps        |
| H22   | 139.264 Mbps   | 135.168 Mbps       |

**hacker** An illegal intruder into a network.

**half-card** Of an adapter board that can be installed in a system expansion slot whose access is one-half the length of a full-size expansion slot.

**half-duplex** In data communication, pertaining to an alternate, one way at a time, independent transmission.

**half-duplex (HD or HDX) circuit** 1. CCITT definition: A circuit designed for duplex operation, but which, on account of the nature of the terminal equipments, can be operated alternately only. 2. Definition in common usage (the normal meaning in computer literature): A circuit designed for transmission in either direction but not both directions simultaneously.

**half-duplex channel** A channel capable of transmitting data in either direction but only in one direction at any time.

**half-duplex operation** The use of a circuit only in one direction at a time. The alternative word "simplex" is best avoided because it is used in two different senses.

**half-duplex terminal operation** A terminal whose transmitted data is printed locally.

**half-height** Of a disk drive that occupies half of the approximate 3-inch space allocated in most personal computers for the installation of a full-height disk in the computer's system unit.

**half-modular cable** A cable that has a modular plug on one end and spade lugs on the other end.

**half-session** In IBM's SNA, a component that provides FMD services, data flow control, and transmission control for one of the sessions of a network addressable unit (NAU).

**halon** A fire-suppression gas used to protect electrical rooms and systems.

**Hamming code** A forward error correcting code named for its inventor that corrects for as little as a single bit received in error.

**Hamming distance** The Hamming distance between two binary words (of the same length) is the number of the corresponding bit positions in which the two words have different bit values. Also known as "signal distance." The Hamming distance between the words 10100010 and 10101110 is two, since the fifth and sixth bits (counting from the left) are different.

**handler**  A software package designed to run in a processor module in a master network processor.

**hand-off**  The process used in cellular mobile radio telephony to transfer the communications link to a mobile user from one transmitter to another.

**handset**  Another name for any ordinary telephone. As distinguished from a headset. Also used to refer to the part of the telephone containing the mouthpiece and receiver.

**handset switch**  A switch frequently used in local-battery telephone sets. When pressed the switch connects the transmitter in the talking circuit. When released, the transmitter is placed out of the circuit, thereby conserving the battery when the transmitter is not used.

**handshake, handshaking**  1. A preliminary procedure, usually part of a communications protocol, to establish a connection. 2. A data transfer protcol developed by Scitex Corporation Ltd. of Israel which enables graphic arts professionals using desktop publishing systems to obtain direct access to high-resolution, full-color imaging.

**handshake protocol**  In communications, a predefined exchange of signals or control characters between two devices or nodes that sets up the conditions for data transfer or transmission.

**hard copy**  A printed copy of machine output in readable form.

**hard disk drive**  A high-capacity magnetic storage device which allows a user to save, read, or erase data.

**hard error**  A serious error on the network that requires that the network be reconfigured or that the source of the error be removed before the network can resume reliable operation.

**hardware**  Equipment (as opposed to a computer program or a method of use), such as mechanical, electrical, magnetic, or electronic devices. Compare with firmware and software.

**hardware interface**  Physical hardware used to connect electrically two devices to one another.

**hardwired**  1. In communications, of a permanent connection between two nodes, stations or devices. 2. Refers to electronic circuitry that perform fixed logic operations using unalterable circuitry rather than under computer or stored program control.

**hardwired FEP**  Non-programmable FEP, or front end. Also line adapter.

**hardwired logic**  A procedure in which the decision-making elements cannot be altered by external means but only by changing the internal connections.

**harmonic**  A frequency equal to a whole-number multiple of a fundamental frequency present in the resultant frequencies because of a sinusoidal stimulus.

**harmonic distortion**  The resultant presence of harmonic frequencies due to non-linear characteristics of a transmission line. Harmonic distortion occurs when attenuation of the receive level varies with the amplitude of the signal; for example, a 1 V signal may be attenuated by one-half, while a 5 V signal many be attenuated by two-thirds. In this way, the signal may be flattened at the peaks, which is equivalent to adding low-amplitude harmonics; this distortion is measured according to the relative degree of second and third order harmonics (two and three times the standard frequency, respectively). The result is distorted data transmission.

**HARNET**  Hong Kong Academic & Research NETwork.

**hashing**  Automatic conversion of input information (such as a network address) into a table location (offset).

**HASP**  Houston Automatic Spooling Program.

**HASP/MLI**  Houston Automatic Spooling Program/ Multi-Leaving Interface.

**HASP workstation**  A remote batch terminal capable of interfacing with the Houston Automatic Spooling Program by means of a special multi-leaving version of BSC designed for this application. Often a System/360 Model 20 computer.

**Hayes compatible**  A term used to note that the command set of a modem matches the command set of Hayes Microcomputer Products modems.

**HA1-receiving weighting**  A telephone class of noise weighting used in a noise measuring set to measure noise across the HA1 receiver of a 302-type or similar subset.

**HCTDS**  High Capacity Terrestrial Digital Service.

**HCX-5000**  A PBX manufactured by Hitachi America Inc.'s Telecommunications Division that supports both the Primary Rate and Basic Rate ISDN interfaces. The HCX-5000 can support up to a maximum of 32 000 ports.

**HCX-5400**  A voice/data PBX switch marketed by Hitachi America Ltd, capable of supporting up to 1500 lines. The HCX-5400 is compatible with AT&T's ISDN Primary Rate Interface.

**HCX-5500**  A voice/data PBX switch marketed by Hitachi America Ltd, capable of supporting up to 3000 lines. The HCX-5500 is compatible with AT&T's ISDN Primary Rate Interface.

**HD**  1. Half-Duplex. 2. High Day.

**HDB3**  High-Density Binary 3.

**HDDI**  Host-Displaywriter Document Interchange.

**HDLC**  High-Level Data Link Control.

**HDFI**  Host Digital Facilities Interface.

**HDSL**  High-rate Digital Subscriber Line.

**HDX**  Half-Duplex Transmission.

**headend**  Location where cable television systems collect and distribute satellite programming.

**headend unit**  In LAN technology, an item of hardware on a single or dual cable broadband network using split frequency bands to provide multiple services. The headend allows devices on a network to send and receive on a single cable.

**header**  The control information added to the beginning of a message—either a transmission block or a packet—for control, synchronization, routing and sequencing of a transmitted data block or packet.

**HEANET**  Higher Education Authority NETwork.

**headset**  Any configuration of earpiece and speaker which fits over the head to allow hands-free telephone use.

**hearing**  The perception of sound by the brain.

**heat coil**  An electrical protection device used to prevent equipment from overheating as a result of foreign voltages on a conductor that do not operate voltage limiting devices. It typically consists of a coil of fine wire around a brass tube that encloses a pin soldered with a low-melting alloy. When abnormal currents occur, the coil heats the brass to soften the solder allowing the spring-loaded pin to move against a ground plate directing currents to ground.

**Helsinki Telephone Company (HTC)**  The largest member of Finland's association of postal and telecommunications administrations.

**HEP**  Host-End Processor.

**HEPnet**  High Energy Physics network.

**hertz (Hz)**  A measure of frequency or bandwidth equal to one cycle per second. Named after the experimenter, Heinrich Hertz.

**Hertz antenna**  An antenna system in which the ground is not an essential part. Its resonant frequency depends upon its electrical length, which is approximately half the wavelength.

**heterodyning**  The mixing of frequencies which produces the original signals as well as signals representing their sum and differences in the frequency domain. The use of a mixer and filters permits frequency division multiplexing.

**heuristic**  A procedure which uses trial and error or random searching and therefore cannot be certain of its results. In contrast an algorithm is expected always to arrive at a correct or optimal result.

**heuristic routing** A routing method, proposed by Baran, in which delay data produced by normal data routing packets coming in on different links from a given source node are used to guide outgoing packets as to the best link for getting to that node.

**HEX** Hexadecimal.

**hexadecimal** A digital system that has sixteen states, 0 through 9 followed by A through F. Any 8-bit byte can be represented by two hexadecimal digits. The decimal and hexadecimal notation of values 0 through 9 is identical. The equivalent notation of the remaining basic hexadecimal values is as follows:

| Hexadecimal | Decimal |
|---|---|
| A | 10 |
| B | 11 |
| C | 12 |
| D | 13 |
| E | 14 |
| F | 15 |

The basic hexadecimal values are 0 through F.

**HF** High Frequency.

**HFCRSP** High Frequency Communications Replacement System Program.

**HFDF** High Frequency Direction Finding.

**HFDM** High Frequency Digital Modem.

**HFIP** High Frequency Improvement Program.

**hierarchic data base model** Scheme of logical representation of entities and relationships between entities of a data base by use of tree or hierarchic structures.

**hierarchical network** 1. A communications network consisting of one host processing center and one or more satellite processors. 2. A multiplering network configuration providing only one path through intermediate rings between source rings and destination rings.

**hierarchical network structure** Network structure in which functions are broken down into layers, each having a specific role (OSI Reference Model).

**hierarchical office class** The functional ranking of a telephone company network switching center based upon its transmission requirements and its hierarchical relationship to other switching centers.

**hierarchical routing** A routing mechanism based upon a hierarchical addressing scheme. Under TCP/IP most Internet routing is based upon a two-level hierarchy in which an Internet address is subdivided into a network and a host portion.

**hierarchical switching** In LAN technology similar to star switching; the switching is done in stages.

**hierarchically distributed processing** The distribution of functions over components connected in a tree-structured hierarchy. Also called vertical distribution.

**hierarchy** In IBM's Network Problem Determination Application (NPDA), a term frequently used in lieu of the terms resource or display levels (or types).

**high-ASCII** ASCII characters whose values exceed 127. In most bulletin board networks the use of high-ASCII in messages is prohibited since some types of personal computers cannot correctly interpret those characters.

**High Capacity Satellite Digital Service (HCSDS)** See Skynet 1.5.

**high capacity service** A term which generally refers to tariffed, digital data transmission service equal to, or in excess of, T1 data rates (1.544 Mbps).

**High Capacity Terrestrial Digital Service (HCTDS)** AT&T's original name for its T1 service. Replaced in 1983 by "Accunet T1.5 Service."

**High Definition Television (HDTV)** Refers to a process of dividing television pictures into twice as many horizontal lines per frame as the typical television picture, thus increasing the picture's horizontal and vertical definition to the level of 35-mm film.

**High Density Bipolar 3 (HDB3)** A line coding technique used to obtain a minimum "1's density" on CEPT PCM-30 2.048 Mbps circuits. Under

HDB3 a group of four consecutive zeros is replaced with one of two HDB3 codes as indicated.

**High Density Binary 3 (HDB3)** A technique for maintaining ones density by substituting a specific pattern for a string of three consecutive zeros. Similar in concept to B8ZS, HDB3 is used primarily in Europe.

HDB3 Coding

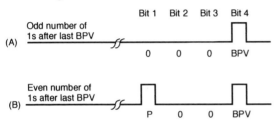

**High Density Bipolar (HDB)** A modified bipolar code which avoids the long absence of pulses and thus eases timing recovery. There are many versions.

**High Energy Physics Network (HEPnet)** A world-wide network used to connect researchers in the field of high energy physics.

**High Frequency (HF)** Portion of the electro-magnetic spectrum typically used in short-wave radio applications. Frequency approximately in the 3 to 30 MHz range.

**High Frequency Direction Finding (HFDF)** Direction finding effort directed against trans-mitters using the frequency range of 1.5 to 30 mega-hertz (MHz).

**high intensity** An attribute of a display field; causes data in that field to be displayed at a brighter level than other data displayed on the screen.

**high order** The port(s) of communications con-trollers, multiplexers and other devices connected to the end of a rotary or hunt group.

**high pass** Frequency level, above which a filter will allow all frequencies to pass.

**high performance option** Same as D1 conditioning.

**highgate module** A hardware-software enhance-ment for AT&T's 5ESS switch which enables tele-phone companies to provide data communications support over existing twisted-pair telephone wiring at speeds up to 1 Mbps.

**High-Level Data Link Control (HDLC)** The international standard communications protocol (similar to SDLC) developed by ISO.

**High-Level Data Link Control Packet Assemb-ler/Disassembler (HPAD)** Supports Synch-ronous Data Link Control (SDLC) devices using two-way alternate transmission (logically half-duplex).

**High-Level Data Link Control (HDLC) Procedure** A set of protocols defined by ISO for carrying data over a link with error and flow control. Versions of HDLC are also being developed for multi-point lines. In spite of its name HDLC is not a "high-level protocol."

**High-Level Data Link Control (HDLC) Station** A process located at one end of the link which sends and receives HDLC frames in accordance with the HDLC procedures.

**High Level Language Application Program Inter-face (HLLAPI)** An IBM standard which defines and simplifies access to mainframe data.

**high-level protocol** A protocol to allow network users to carry out functions at a higher level than merely transporting streams or blocks of data.

**HIgh-Power, COherent Radar (HIPCOR)** A 95-GHz research radar system developed by the Georgia Institue of Technology and the U.S. Army Missile System Command.

**high-precedence message** A term normally applied to messages of priority or higher precedence.

**High-Rate Digital Subscriber Line (HDSL)** A Bellcore project to provide data rates up to 1.54 Mbps over existing copper loops.

**HIM** Host Interface Module.

**Hinsdale Disaster** The Mother's Day 1988 fire that

curtailed telephone service for several days in the area served by the switching center located in suburban Chicago.

**HIPCOR**   HIgh-Power, COherent Radar.

**HISPASAT**   The national communications satellite sytem of Spain.

**hit**   Any random deviation during transmission which causes data to be received in error.

**hit on the line**   General term used to describe errors caused by external interferences such as impulse noise caused by lightning or man-made interference.

**HKG-Phil**   The name of a fiber optic cable route connecting Hong Kong to the Philippines.

**HLF**   High-Level Function.

**HLLAPI**   High-Level Language Application Program Interface.

**HLSC**   High-Level Service Circuit.

**HMINET I, II**   Hahn–Meitner Institut NETwork.

**HN**   Host of Network.

**HNS**   Hospitality Network Service.

**hold**   A PBX feature which enables a station user to remain connected to a line while not off-hook to the line.

**holding time**   The length of time that a communications channel is in use for each transmission.

**Holidex 2000**   The advanced reservation system used by more than 1700 U.S. Holiday Inn Hotels.

**home carrier**   In cellular communications the telephone company with which a user's telephone is registered and from which the user is billed monthly access and usage charges.

**home directory**   In a local area network, the directory to which a user's first network drive is mapped when the user logs in to a file server.

**home loop**   An operation involving only those input and output units associated with the local terminal.

**Home Network Controller**   A device developed at AT&T Network Systems which turns a regular telephone into a digital unit that can access an ISDN service.

**HONOR**   An automated bank teller network in the state of Florida.

**HONTAI-2**   The name of fiber optic cable route connecting Hong Kong to Taiwan.

**hookswitch**   The device on which the telephone receiver hangs or on which a telephone headset hangs or rests when not in use. The weight of the receiver or handset operates a switch which opens the telephone circuit, leaving only the bell connected to the line.

**hop**   The transmission of a radio wave from the earth to the ionosphere and back to the earth.

**hop count**   1. A measure of distance between two points in a network. A hop count of X means that X gateways separate a source from its destination. 2. The number of bridges through which a frame has passed on the way to its destination.

**hop count limit**   The maximum number of bridges through which a frame may pass on the way to its destination.

**Horizon**   A hybrid key system introduced by AT&T in 1979.

**Horizontal Redundancy Check (HRC)**   A validity method to check blocks in which redundant information is included with the information to be checked.

**horizontal wiring subsystem**   That part of a premises distribution system installed on one floor that includes the cabling and distribution components connecting the backbone subsystem and equipment wiring subsystem to the information outlet via cross-connects, components of the administration subsystem.

**Hospitality Network Service (HNS)**   An AT&T flat-rate, discounted long distance direct dial service. Under HNS calls to any domestic location at any time of the day are billed at a flat rate per minute.

**host** Any computer to which remote terminals are attached and in which the application program resides.

**host access method** The access method that controls communication with a mainframe computer.

**host application program** An application program processed in the host computer.

**host attachment** A mode of SNA communication in which the processor acts as a secondary SNA device.

**host carrier** In cellular communications the telephone company that provides service to nonlocal subscribers.

**host computer** The central computer (or one of a collection of computers) in a data communications system which provides the primary data processing functions such as computation, data base access, or special programs or programming languages; often shortened to "host."

**host interface** Interface between a network and the host computer.

**host LU** In IBM's SNA, a logical unit located in a host processor, for example, a VTAM application program.

**host master key** In IBM's SNA, deprecated term for master cryptography key. The cryptography key is used to encipher operational keys that will be used at the host processor.

**host mode** The operating mode of an HASP main processor communicating with an HASP workstation.

**host node** 1. A node at which a host processor is located. 2. In SNA, a subarea node that contains a system services control point (SSCP); for example, a System/370 computer with OS/VS2 and ACF/TCAM.

**host personnel** The people at the host computer site responsible for the connection of terminals and personal computers to the host computer network.

**host print** A print mode in the IBM PC 3270 Emulation Program in which the host computer sends a file to print on the PC printer.

**host processor** 1. (TC97) A processor that controls all or part of a user application network. 2. In a network, the processing unit in which the data communication access method resides. 3. In an SNA network, the processing unit that contains a system services control point (SSCP).

**host system** 1. A data processing system used to prepare programs and operating environments for use on another computer or controller. 2. The data processing system to which a network is connected and with which the system can communicate. 3. The controlling or highest-level system in a data communication configuration; for example, a System/38 is the host system for the work stations connected to it.

**Host-Displaywriter Document Interchange (HDDI)** The component of IBM's VM/SP that works with that vendor's PROFS and Office Vision to transfer documents between PROFS and Office Vision disks and Displaywriter diskettes. HDDI allows a user to convert Document Composition Facility (DCF) to Displaywriter revisable form using a subset of formatting controls.

**Host-End Processor (HEP)** A high-speed communications processor marketed by BBN Communications Corp. of Cambridge, MA. A HEP connects IBM SNA hosts and front-end processors via a BBN private packet network.

**hot key switch** A keystroke sequence enabling a fast change from the session of one program to another.

**hot line service** A special connection between two stations where the second phone will automatically ring when the first phone is lifted off-hook.

**hot potato routing** Packet routing which sends a packet out from a node as soon as possible even though this may mean a poor choice of an outgoing line when the preferred choices are not available.

**hot standby** Equipment ready to take over an operation immediately if the equipment presently in use should fail.

**Hotfix** A Novell utility incorporated into that vendor's NetWare product which compensates for defective areas on the hard disks of a file server.

**hotline** The facility used to report problems for correction.

**Hotline** The name of a cellular telephone produced by a joint venture between L. M. Ericsson and the General Electric Company.

**house cables** Conductors within a building used to connect communications equipment to outside lines.

**HOUSTON Automatic Spooling Program (HASP)** A mainframe job control program developed by IBM and widely used in the late 1960s and 1970s. HASP was developed to facilitate the flow of remote jobs into and out of mainframes; required a special multi-leaving version of the BSC protocol which took its name from the program.

**High-Performance Option (HPO)** Same as D1 conditioning.

**HPA** High Power Amplifier.

**HPIB** Hewlett-Packard Interface Bus.

**HPO** High Performance Option.

**HRC** Horizontal Redundancy Check.

**HSM** Host Switching Module.

**HSI** High-Speed Technology, referring to a modem (tm) USRobotics.

**HT** Holding Time.

**HTC** Helsinki Telephone Company.

**HTR** Hard-To-Reach.

**HU** Host Unit.

**hub** 1. In LAN technology the center or a star topology network or cabling system. 2. A multi-node network topology that consists of a central multiplexer node with multiple nodes feeding into and through the central node. The outlying nodes do not normally interconnect with one another. 3. A communications center. 4. A bulletin board system that calls another hub or is called by one or more nodes to transfer mail.

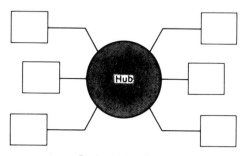

Basic star topology

**hub (DDS)** An office in a Digital Data System that combines the T1 data streams from a number of local offices into signals suitable for transmission.

**hubless DDS** A generic term which references tariffed digital service offerngs of local exchange carriers (LECs) which do not use AT&T Dataphone Digital Service (DDS) hubs. Hubless DDS offerings appear under various names to include:

| LEC | Hubless DDS service |
| --- | --- |
| Ameritech | Basic Digital Service |
| Bell Atlantic | Digital Connect Service |
| BellSouth | Synchronet |
| Nynex | Digipath |
| PacBell | Advanced Digital Network |
| Southwestern Bell | Megalink I |
| U.S. West | Digicom |

**Huffman coding** A compression technique based upon the statistical probability of occurrence of characters in the character set to be compressed. The resulting code assigned to each character is a maximum efficient code and is often referred to as an optimum code.

**hundred call seconds** See CCS.

**hunt group**   The telephone numbers assigned to a rotary such that a person dialing a single listed number is automatically connected by telephone company switching equipment to an available line in the group.

**hunting**   The movement of a call as it progresses through a series of related lines. If the call cannot connect on the first line of the group it will move to the second, then the third, etc.

**hybrid**   1. In transmission systems, a passive elecical component used for bridging between legs of a circuit, such as interfacing a backbone multidrop circuit to individual drop circuits. 2. Combination of two or more technologies. 3. Hybrid teleone circuit, which interfaces 2-wire local loop.

**hybrid coil**   A two- to four-wire converter used to connect subscriber telephone lines to a common carrier transmission facility. The hybrid coil suppresses echoes caused by the impedance mismatch of the two- to four-wire conversion.

**hybrid interface structure**   An interface that has a mixture of labeled and positioned channels.

**Hybridge**   A local area network bridge marketed by Cisco Systems, Inc., of Menlo Park, CA.

**Hyperaccess**   A communications software program from Hilgraeve, Inc. of Monroe, MI, designed to operate on IBM PC and compatible computers. The program is noted for its powerful script language, informative tutorial and reliable and efficient file transfer capability.

**Hytelnet**   A hypertext system which contains information about the Internet, such as accessable library catalogs, Freenets, Gophers, bulletin boards, etc.

**Hz**   See hertz.

**H0 channel**   In ISDN, a digital channel of 384 Kbps, available in Primary Rate service to the end user for high-speed data. A North American ISDN Primary Rate circuit can carry up to four H0 channels; a European ISDN Primary Rate circuit can carry up to five H0 channels.

**H1 channel**   In North American ISDN, digital channel of 1.536 Mbps, available in Primary Rate service to the end user for high-speed data, image, or video. An H1 channel occupies the entire available bandwidth of the Primary Rate channel. (1.544 Mbps aggregate—8 Kbps Framing and 1.536 Mbps data).

**H.120**   A CCITT recommendation which defines the operation of codecs at rates from 768 Kbps to 2 Mbps.

**H.261**   A CCITT standard for video compression called "Codes for Audiovisual Services at NX384 Kilobits per Second." This standard establishes a common algorithm for converting analog video signals to digital signals operating at or above 384 Kbps.

# I

**I** Current.

**IA** International Alphabet.

**I-A** A communications satellite operated by Arabsat.

**IAB** 1. Internet Activities Board. 2. Internet Architecture Board.

**IAC** Integrated Access Controller.

**IACS** Integrated Access and Cross-connect System.

**IAM** Initial Address Message.

**IAS-Net** A Russian packet network which has electronic mail and international gateways operated by the Institute of Advanced Systems in Moscow.

**IAT** Integrated Access Terminal.

**IATA** International Air Transport Association.

**IA5** International Alphabet number 5.

**I-B** A communications satellite operated by Arabsat.

**Ibertex** The videotex service marketed by Spain's national carrier Telefonica.

**IBM** International Business Machines.

**IBM Cabling System (IBMCS)** A cabling system originally introduced by IBM in 1984 which is designed to support most types of cabling requirements to include the vendor's Token-Ring Network. The cabling system consists of specified cables, faceplates, distribution panels, and a special connector.

**IBM Type 1 (indoor) cable** A cable used to connect terminal workstations and other equipment to computers. This cable is for indoor use and consists of two twisted pairs, 22 AWG solid conductors in a foil braid shield. Type 1 cable is also available in plenum.

**IBM Type 1 (outdoor) cable** A cable used for installation in aerial or in underground conduits. The cable consists of two twisted pairs, 22 AWG solid conductors in a corrugated metallic shield.

**IBM Type 2 cable** A cable which provides the same interconnections as a Type 1 indoor cable as well as supporting PBX requirements. The cable consists of two twisted pairs, 22 AWG solid conductors in a foil, braid shield, and four voice-grade twisted pair 22 AWG for telephone use. Type 2 is also available in plenum.

**IBM Type 3 cable** A PVC media cable which consists of four pairs 24 AWG wire and which is used for telephone cable. Type 3 is also available in plenum.

**IBM Type 5 cable** A cable which consists of two optical fiber conductors. This cable can be used for indoor installations, aerial installations, or placed in an underground conduit.

**IBM Type 6 cable** This cable is designed for patching applications within a wiring closet or from a faceplate in the back of a workstation or computer terminal. Type 6 cable consists of two twisted pairs, 26 AWG stranded conductors which provides flexibility.

**IBM Type 8 cable** This cable is designed for use under carpets and is useful for an open office environment. Type 8 cable consists of two parallel pairs of 26 AWG solid conductors arranged flat.

**IBM Type 9 cable**  This cable is considered as an economy version of Type 1 plenum and has a maximum transmission distance approximately two-thirds of that of Type 1 cable. Type 9 cable consists of two twisted pairs of 26 AWG stranded conductors and is only available in plenum.

**IBM Information Network**  A commercial value added network operated by IBM. This network supports both IBM and non-IBM terminal devices and offers such services as electronic mail, access to information services and electronic data interchange capability.

**IBM TSO**  Timesharing System (IBM Network).

**IBM VNET**  Virtual NETwork.

**IBM 3270**  The term "3270" originally referred to an IBM product—a dumb terminal that was interactive with its host. The term now refers to a whole group of hardware, manufactured by IBM as well as other vendors that emulate or imitate the functions of the 3270. The group includes:

— IBM'S 3270 family of display stations.

— 3270 emulators produced by plug and system compatible vendor's.

— Proprietary terminals produced by the other major mainframe, manufacturers (e.g. Unisys, NCR, Control Data, and Honeywell). These terminals use data communications equipment configurations similar to the 3270's, but are not compatible with the 3270. Proprietary terminals are compatible only with their own hosts.

— Multifunction terminal controllers that allow terminals to do 3270 applications, such as data entry, word processing and electronic mail.

**IBM 3600 System**  A loop network configuration developed by IBM to exchange messages between a controller and several terminals. Data flow around the loop is unidirectional.

**IBMCS**  IBM Cabling System.

**IBMLink**  An electronic service of IBM which provides customers with direct access to technical, marketing and service information.

**IBS**  International Business Services.

**IC**  1. Integrated Circuit. 2. Information Center.

**ICA**  1. Integrated Communications Adapter. 2. International Communications Association.

**ICD**  Imitative Communications Deception.

**ICI**  Incoming Call Identification.

**ICL**  International Computers Limited.

**ICLID**  Individual Calling Line IDentification.

**ICMP**  Internet Control Message Protocol.

**ICMS**  Intelligent Chassis Management System.

**I-CNOS**  International-Customer Network Optimization System.

**ICS**  1. Integrated Communications System. 2. Interdigital Communications Subsystem.

**ICSU**  Intelligent Channel Service Unit.

**ICV**  Initial Chaining Value (IBM's SNA).

**ID**  1. Identification. 2. Individualized Dialing. 3. Insulation Displacing.

**ID CODE**  On the RIME network an ID CODE is used to identify each node in the network. The ID CODE can be up to 12 alphanumeric characters in length and appears at the bottom of every message relayed throughout the network.

**IDA**  Integrated Digital Access.

**IDC**  1. Insulation Displacement Connector. 2. Integrated Digital Capability.

**IDCMA**  Independent Data Communications Manufacturers Association.

**IDentified Outward Dialing (IDOD)**  A service used to aid in billing which lists all toll calls made from a specific station.

**IDF**  Intermediate Distribution Frame.

**IDHS**  Intelligent Data Handling System.

**idle character**  A transmitted character which indicates a "no information" status. It does not manifest itself as part of the received message at the destination.

**idle condition** A non-busy state for equipment or circuits, indicating that they are not in operation and/or ready for use.

**idle device** Equipment that is not engaged or busy resulting in it being available for use.

**IDN** Integrated Digital Network.

**IDNID** Immediate Diversion Network Invalid Dialing.

**IDI** Initial Domain Identifier.

**IDP** Initial Domain Part.

**IDT** 1. Integrated Digital Terminal. 2. Integrated Display Terminal.

**IDTN** Interim Data Transmission Network.

**IDTS** Interim Data Transmission System.

**IEC** 1. InterExchange Carrier. 2. International Electrotechnical Commission.

**IEEE** Institute of Electrical and Electronic Engineers.

**IEEE PROJECT 802** In LAN technology an IEEE team that developed the IEEE 802 family of LAN standards.

**IEEE 488** An IEEE standard parallel interface often used to connect test instruments to computers.

**IEEE 696** The IEEE specification for Data Communications Interface Devices.

**IEEE 796** The IEEE specification for Microcomputer System Bus.

**IEEE 802.2** In LAN technology a Data Link layer standard used with IEEE 802.3, IEEE 802.4, and IEEE 802.5.

**IEEE 802.3** A specification published by the Institute of Electrical and Electronic Engineers which defines a physical cabling standard for local area networks as well as the means of transmitting data and controlling access to the cable. The Physical layer standard uses the CSMA/CD access method on a bus topology LAN. Similar to Ethernet.

**IEEE 802.4** In LAN technology a Physical layer standard that uses the token-passing access method on a bus topology LAN. Nearly identical to MAP.

**IEEE 802.5** In LAN technology, a physical layer standard that uses the token-passing access method on a ring topology LAN.

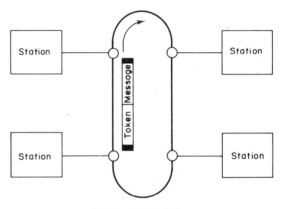

Token-passing ring

**IEEE 802.6** The IEEE specification for Metropolitan Area Network.

**IEEE 802.9** A multimedia LAN standard that supports three types of channels—an ISDN basic rate channel containing two 64 Kbps B-channels and one 16 or 64 Kbps D-channel, an 802 LAN packet service, and one or more wideband isochronous channels.

**IETF** Internet Engineering Task Force.

**IF** Intermediate Frequency.

**IFIPS** International Federation of Information Processing Societies.

**IFL** InterFacility Link.

**IFRB** International Frequency Registration Board.

**IGRP** Interior Gateway Routing Protocol.

**ILAN** A loal area network bridge/router marketed by CrossComm Corp. of Marlborough, MA.

**ILD** Injection Laser Diode.

**IllumiNET**  A trademark of Hughes Network Systems of Germantown, MD, as well as a network management system from that vendor's Personal Earth Station satellite offering.

**IM**  Interface Module.

**ImagePlus**  An IBM image management package of hardware and software which reduces paper documents to digitized images and provides for the storage and distribution of those images.

**IMINT**  Imagery Intelligence.

**IML**  Initial Microcode Load.

**IMLT**  Integrated Mechanized Loop Testing.

**immediate-request mode**  In IBM's SNA, an operational mode in which the sender stops sending request units (RUs) on a given flow (normal or expedited) after sending a definite-response request chain on that flow until that chain has been responded to.

**immediate-response mode**  In IBM's SNA, an operational mode in which the receiver responds to request units (RUs) on a given normal flow in the order it receives them; that is, in a first-in, first-out sequence.

**IMP**  Interface Message Processor.

**impact printer**  A printer in which printing is the result of mechanical impact.

**impedance (Z)**  The total opposition (resistance, inductance, and capacitance) to the flow of current in an electrical circuit.

**impulse**  A high-amplitude, short-duration, pulse. A rapid change in current flow or intensity of a magnetic field.

**impulse noise**  A type of interference on communication lines characterized by high amplitude and short duration. This type of interference may be caused by lightning, by electrical sparking action, or by the make–break or switching devices, etc.

**impulse hits**  Cause of errors in telephone line data transmission. AT&T suggests no more than 15 impulse hits per 15-minute period (spikes).

**IMR**  Internet Monthly Reports.

**IMS**  1. Information Management System. 2. Interprocess Message Switch.

**IMS/VS**  Information Management System/Virtual Storage.

**in**  inch (or inches).

**inactive**  In IBM's VTAM, pertaining to a major or minor node that has not been activated or for which the VARY INACT command has been issued.

**in-band signaling**  Call control signalling transmitted within the bandwidth of the call it controls; also called channel-associated signalling. In T1 transmission, in-band signalling is performed by bit-robbing.

**Inbox**  The name of an electronic mail system marketed by the TOPS division of Sun Microsystems, Inc.

**INCA**  Intelligent Communications Architecture.

**Incoming Call Identification (ICI)**  A switching system feature of a PBX which enables an attendant to identify visually the type of service or trunk group associated with a call directed to the attendant's position.

**Incoming Call Line Identification**  An AT&T term for the ISDN call number identification feature.

**incoming message**  A received message.

**Independent Company (ICO)**  A generic acronym for any provider of local telephone service in the U.S. that is not part of a Regional Bell Operating Company.

**independent logical unit**  In IBM's SNA, a logical unit that does not rely on the SSCP to establish its sessions. To be independent, a logical unit must meet the following criteria:

(1) It must be a Logical Unit Type 6.2;

(2) the Physical Unit node it resides on must be Type 2.1;

(3) it must be capable of sending a "Bind" request;

(4) it must support parallel sessions;

(5) its node must be able to send an XID, which includes a Control Point name vector.

**Independent Telephone Company (ITC)** A non-Bell Telephone operating company that furnishes telecommunications service within a geographic area. There are approximately 1400 independent telephone companies across the United States.

**index information** Information in Revisable-Form Text (RFT) and Document Composition Facility (DCF) documents that IBM's PROFS and Office Vision uses to find documents in storage.

**index matching material** A material, often a liquid or cement, whose refractive index is nearly equal to the core index, used to reduce Fresnel reflections from a fiber end face.

**index tag** An IBM PROFS and Office Vision command that marks index information in the text of a Revisable-Form Text (RFT) document. Each tag has a control associated with it. The controls are inserted into the text of a document.

**indexed file** A file whose records are organized to be accessed sequentially in key sequence or directly by key value.

**indirect activation** In IBM's VTAM, the activation of a lower-level resource of the resource hierarchy as a result of SCOPE or ISTATUS specifications related to an activation command naming a higher-level resource.

**indirect deactivation** In IBM's VTAM, the deactivation of a lower-level resource of the resource hierarchy as a result of a deactivation command naming a higher-level resource.

**Indium–Gallium–Arsenide (InGaAs)** A semiconductor material used in fiber optic system light detectors and emitters.

**Individual Calling Line Identification (ICLI)** A feature which provides calling party information on calls originated from both analog and digital lines on the same AT&T 5ESS switch for ISDN subscribers.

**induction** The process by which an electrically charged object induces an electrical current in a nearby conductor.

**induction coil** A transformer in which the common connection between the windings does not affect the mutual inductance between the coils.

**inductive reactance** The opposition caused by inductance.

**IND$FILE** A file transfer program included in IBM's PC3270 emulation program.

**INEWS** Integrated Electronic Warfare System.

**inflection** Variation in the pitch or loudness which a speaker uses for emphasis or special meaning.

**INFNET** Istituto Nazionale Fisica Nucleare.

**Infolan** A service provided by Infonet of El Segundo, CA, which provides connectivity between internationally dispersed local area networks.

**Info-Look** A videotext gateway service provided by NYNEX.

**Infomaster** An on-line data base and retrieval service marketed by Western Union which provides subscribers access to over 300 data bases located on computer systems around the world.

**Infonet** The public data network operated by Computer Sciences Corporation of El Segundo, CA.

**Infonet 2000** A corporate network for voice and data proposed, designed and installed by NYNEX for Grumman Corporation.

**Infopath Service** The packet switched network operated by New York Telephone.

**Infoplex** An electronic mail service marketed by CompuServe to corporations.

**information** 1. The meaning assigned to data by people. 2. The organizational content of a signal.

209

**information bit**   A data bit, as opposed to an overhead bit.

**Information Management Data Base**   In IBM's NPDA, a system management tool that helps collect, organize, and keep track of problems and their resolutions.

**Information Management System**   A major IBM mainframe software system responsible for controlling extensive batch and transaction processing activities.

**Information Management Systems/Virtual Storage (IMS/VS)**   An IBM host operating system, oriented toward batch processing, and communications-based transaction processing.

**Information Network System (INS)**   The name used for the Japanese pre-ISDN trial.

**information outlet**   A connecting device designed for a fixed location (usually a wall in an office) on which horizontal-wiring-subsystem cable pairs terminate and which receives an inserted plug; it is an administration point located between the horizontal wiring subsystem and work location wiring subsystem. Although such devices are also referred to as "jacks," the term "information outlets" encompasses the integration of voice, data, and other communication services that can be supported via a premises' distribution system.

**information path**   The route by which information is transferred.

**Information Separator (IS)**   A category of control characters used to separate and qualify data logically; its specific meaning has to be defined for each application. Samples of IS characters are: IS1(US), IS2(RS), IS3(GS), IS4(FS).

**information service**   Commercial dial-up service that provides many and various services to its subscribers, to include large on-line data bases, travel booking, banking, bulletin board services, special interest groups, electronic mail and other features.

**Information Systems Engineering Command (ISEC)**   A U.S. Army command tasked with system testing of computer and communications systems and providing technical support for the Army's network of computers and strategic communications equipment.

**Information Technology Management (ITM)**   A term which is used to encompass telecommunications and data management. As telecommunications technology and computer technology advance and merge, the distinctions between the two blur. ITM is used to describe the merging of these technologies.

**Information Technology Requirements Council (ITRC)**   A vendor and user organization that provides vendors with user requirements for a range of information technologies. The MAP/TOP users group is a division of the ITRC.

**Infostream**   A registered trademark of Infotron Systems (now Gandalf) for equipment that provides data PBX functionality.

**Info-2**   An AT&T automatic number identification facility (an ISDN service feature) which forwards the calling party's telephone number to the receiving party (ISDN customer). Info-2 is available with AT&T's Megacom 800 service.

**infrared**   Portion of the electromagnetic spectrum used for optical fiber transmission as well as for short-haul, open-air transmission. Infrared transmission wavelengths are approximately 0.7 micrometres or longer.

**infrared light**   Light with a frequency just below that of visible light which moves through many types of lightguides with little loss of power.

**InGaAs**   Indium–Gallium–Arsenide.

**inhibited**   In IBM's VTAM, pertaining to an LU that has indicated to its SSCP that it is not ready to establish LU–LU sessions. An initiate request for a session with an inhibited LU will be rejected by the SSCP. The LU can separately indicate whether this applies to its ability to act as a primary logical unit (PLU) or as a secondary logical unit (SLU).

**INIT**   1. INITialization Resource. 2. INITialize.

**Initial Chaining Value (ICV)**   In IBM's SNA, an eight-byte pseudo-random number used to verify that both ends of a session with cryptography have the same session cryptography key. The initial chaining value is also used as input to the Data Encryption Standard (DES) algorithm to encipher or decipher data in a session with cryptography. Synonymous with session seed.

**Initial Domain Identifier (IDI)**   In Digital Network Corporation Network Architecture (DECnet), the part of an NSAP address which identifies the authority responsible for the assignment of the DSP.

**Initial Domain Part (IDP)**   In Digital Network Corporation Network Architecture (DECnet), the part of an NSAP address assigned by the first level addressing authority.

**initial microcode load (IML)**   The action of loading the operational microcode.

**Initial Program Load (IPL)**   The set of programs loaded into a computer's active memory areas at startup.

**initialization**   Concerns specific preliminary steps required before execution of iterative cycles to establish or determine efficient start procedures. Usually a single, non-repetitive operation after a cycle has begun and/or until a full cycle is again begun.

**initialization parameter**   An installation-specified parameter that controls the initialization and ultimate operation of JES2.

**Initialization Resource (INIT)**   A system resource that is loaded into part of the Apple Macintosh computer's random access memory (RAM) called the System heap and run as part of a normal startup procedure.

**initialization statement**   An installation-specified statement that controls the initialization and ultimate operation of JES2.

**initialize**   1. The process of setting hardware, such as a protocol analyzer, to a known operating state.

2. To set counters, switches, addresses, or contents of storage to starting values.

**initiate**   In IBM's SNA, a network services request, sent from an LU to an SSCP, requesting that an LU–LU session be established.

**Injection Laser Diode (ILD)**   A laser-based fiber optic light source that is also known as a semiconductor diode.

**inkjet printer**   A printer that produces the image by controlled spraying of droplets of ink at the paper.

**inline exit routine**   In IBM's VTAM, a SYNAD or LERAD exit routine.

**INM**   Integrated Network Manager.

**INMARSAT**   International Maritime Satellite Organization.

**INN**   1. Integrated Network Node. 2. In IBM's SNA, a deprecated term for Intermediate Routing Node (IRN).

**InPath**   An intelligent T1 multiplexer manufactured by Telecommunications Technology, Inc., of Milpitas, CA.

**in-plant system**   A system whose parts, including remote terminals, are all situated in one building or localized area. The term is also used for communication systems spanning several buildings and sometimes covering a large distance, but in which no common carrier facilities are used.

**input**   1. The data to be processed. 2. The state or sequence of states occurring on a specified input channel. 3. The device or collective set of devices. 4. A channel for impressing a state on a device or logic element. 5. The process of transferring data from an external storage to an internal storage.

**input devices**   Devices which allow the users to "talk" to the computer system. For example, a keyboard on a CRT terminal.

**input service processing**   In JES2, the process of performing the following for each job: reading the input data, building the system input data set, and building control table entries.

**Input/Output (I/O)** The process of moving information between the central processing unit and peripheral devices. May refer to the particular peripheral hardware used.

**input/output device** A peripheral or communications device.

**inquiry** A request or interrogation regarding the stored information at the computer system.

**inquiry/response** An exchange of messages and responses with one exchange usually requesting information (inquiry) and the response that provides information.

**INS** 1. Information Network System. 2. INSert.

**INSCOM** International Satellite Communication.

**insert** In a token-ring local area network, to make an attaching device an active part of the ring.

**insertion loss** Signal power loss due to connecting communications equipment units with dissimilar impedance values.

**Inside Wiring (IW)** Customer premises wiring; a simple twisted pair wire which might be used for a service which goes from one floor in the building to another.

**Insight** A network management system marketed by British Telecom for Packet Switch-Stream (PSS) users.

**Insite** The first commercial network management system marketed by Sprint Communications Co.

**INS-Net** NTT's domestic ISDN service in Japan.

**ISN-Net 1500** The name for NTT's ISDN primary rate interface services in Japan.

**installation exit routine** In IBM's VTAM, a user-written exit routine that can perform functions related to initiation and termination of sessions and is run as part of VTAM rather than as part of an application program. Examples are the accounting, authorization, logon-interpret, and virtual route selection exit routines.

**Institute of Electrical and Electronic Engineers (IEEE)** An international professional society that issues its own standards and is a member of ANSI and ISO; created IEEE Project 802.

**insulation** Material that provides a high resistance path for current flow between wires, preventing the current from flowing from one wire to another.

**Insulation Displacement Connector (IDC)** A block with double metal connectors that strips wire as you punch it down.

**Integrated Access and Cross-connect System (IACS)** An AT&T wideband packet product based upon open protocols for international transmission. IACS can be used for compression of voice, fax and data. IACS consists of transmission terminals called Integrated Access Terminals (IATs) and a network management system called the Integrated Access Controller (IAC).

**Integrated Access Controller (IAC)** The network management system of AT&T's Integrated Access and Cross-Connect System (IACS).

**Integrated Access Terminal (IAT)** The terminal used with AT&T's Integrated Access and Cross-connect System (IACS).

**Integrated Circuits (IC)** A complete complex electronic circuit that is capable of performing all the functions of a conventional circuit containing numerous transistors, diodes, capacitors, and or resistors.

**Integrated Communication Adapter (ICA)** In IBM's SNA, an integrated adapter that allows connection of multiple telecommunication lines to a processing unit.

**Integrated Digital Access (IDA)** British Telecom's ISDN-type network based on System X digital exchanges that does not support the CCITT I.420 interface for basic rate ISDN service. IDA provides one 80 Kbps and one 16 Kbps signalling control channel while ISDN offers two 64 Kbps and one 16 Kbps signalling control channel.

**Integrated Digital Network (IDN)** Digital telephony network extending to, but excluding, the subscriber's line and terminal apparatus.

**Integrated Electronic Warfare System (INEWS)** A suite of radar warning and defense electronic countermeasures equipment.

**integrated modem** A modem designed into an information product, such as a terminal or computer, rather than used externally to it and connected to it by means of a cable.

**Integrated Network Management Services (INMS)** A service of MCI Communications Corporation which enables subscribers to monitor and reconfigure their networks.

**Integrated Network Manager (INM)** A trademark of Infotron Systems (now merged with Gandalf) as well as a full-color graphics-driven network management and control system from that vendor. INM enables the configuration, event monitoring, diagnostic testing of Infotron equipment.

**integrated optical circuit** The optical equivalent of a microelectronic circuit, it acts on the light in a lightwave system to carry out communications functions. Generates, detects, switches and transmits light.

**Integrated Service Unit (ISU)** A single device that combines the functions of both a CSU and a DSU.

**Integrated Services Digital Network (ISDN)** A CCITT (now ITU) standard, currently under development, that will cover a wide range of data communications issues but primarily the total integration of voice and data. Access channels include basic (144 Kbps) and primary rate (1.544 and 2.048 Mbps). The ISDN is described in the CCITT's I-series of recommendations (1984 Red Book). Already having major effects on exchange and multiplexer design.

**Integrated Services Line Unit (ISLU)** The AT&T 5ESS switch line unit under development to provide ISDN standard access to the 5ESS switch.

**Integrated Services PBX (ISPBX)** A PBX that employs ISDN standards.

**integrated system** 1. A computer system which extends the concept of integrated software to all end users. Integrated software is groups of programs that perform different tasks but which share data. A major advantage results from different applications all updating the one common data base, keeping it current. 2. A system that transfers analog and digital traffic over the same switched network.

**Integrated Telecommunications Network (ITN)** The American Express network used for credit card authorization.

**Integrated Voice/Data Terminal (IVDT)** A family of devices that feature a terminal keyboard/display and voice telephone. May work with a specific vendor's data PBX.

**integration** The process of merging separate environments or the ability to connect diverse resources into one functional system.

**integrity of data** The status of information after being processed by software.

**Intel** A semiconductor (chip) manufacturer, one of the sponsors of Ethernet.

**Intelli-flex** A communications service marketed by Cable & Wireless Communications which offers subscribers fractional T1 capabilities, such as 4, 6, 8, 10 or more 64 Kbps channels.

**intelligence, intelligent** A term for equipment (or a system or network) which has a built-in processing power (often furnished by a microprocessor) that allows it to perform sophisticated tasks in accordance with its firmware.

**Intelligent Channel Service Unit (ICSU)** A combination Channel Service Unit (CSU) and Data Service Unit (DSU) marketed by Timeplex of Woodcliff Lake, NJ, which permits the user to view diagnostic Extended Superframe (ESF) data.

**Intelligent Chassis Management System (ICMS)** A personal computer based network management system marketed by NEC America of San Jose, CA, for use with that vendor's I series of modems, controllers, data service units and channel service units.

213

**intelligent communications network**  A network which usually permits alternate routing and other com-plex control functions.

**intelligent port selector**  Same as data PABX.

**Intelligent Synchronous (I-SYNC)**  A General DataComm, Inc., proprietary feature found on some GDC modems. With I-SYNC a special "y" cable is used to provide simultaneous connection to both the asynchronous and synchronous boards on a PC. Once the asynchronous auto dialer has established a connection to the remote end modem, GDC communications software automatically switches the modem into synchronous mode.

**intelligent TDM**  Same as concentrator.

**intelligent terminal**  A terminal with some level of programmability; contains processor and memory. Used for pre-transmissions or post-transmission communications processing as well as for offloading of data processing.

**Intelligent Time Division Multiplexer (ITDM)**  A device which assigns time slots on demand rather than fixed subchannel basis.

**intelligent workstation**  A workstation that contains storage which allows it to either process data before sending it to the computer or perform some stand-alone processing.

**Intellipath**  A trademark and service of NYNEX Corporation. Intellipath is a digital Centrex service.

**Intelsat**  INTErnationaL SATellite Service.

**InteMail**  A voice-mail system marketed by Inte-Com, Inc., of Allen, TX.

**intensity**  The relative strength of electrical, magnetic, or vibrational energy.

**intensive mode**  An IBM Network Control Program (NCP) function that forces recording of temporary errors for a specified resource. Data so recorded is not included in error counts on Network Problem Determination Application (NPDA) displays. Intensive mode data appears on the Most Recent Events displays when retrieved by CTRL SEC and MRECENT commands, or when the Most Recent Event display is retrieved by an EV request from a statistical display, or by a SEL# M request from an Alert display.

**interactive**  A time-dependent (real-time) data communications operation where a user typically enters data and awaits a response message from the destination prior to continuing; contrast with batch processing.

**Interactive System Productivity Facility (ISPF)**  An IBM program product that serves as a full screen editor and dialogue manager. Used for writing application programs, it provides a means of generating standard screen panels and interactive dialogs between the application programmer and terminal user.

**Interactive System Productivity Facility/Program Development Facility (ISPF/PDF)**  An IBM dialog manager that provides application developers with services to shortcut the coding required to produce interactive applications.

**Interactive Terminal Interface (ITI)**  In packet-switched networks, a PAD that supports network access by asynchronous terminals.

**interactive testing**  Testing performed through user and system communication. The system acknowledges and reacts to requests entered from the keyboard or to conditions detected on the data lines.

**interbuilding cable**  The communications cable that is part of the campus subsystem and runs between buildings. There are four methods of installing interbuilding cable: in-conduit (in underground conduit); direct-buried (in trenches); aerial (on poles); and in-tunnel (in steam tunnels).

**interbuilding cable entrance**  The point at which --campus subsystem cabling enters a building.

**intercept**  1. The process of rerouting a call from an invalid or out-of-service number to an alternative number, or to an operator position. 2. A PBX feature which stops an in-house call from going out to an unauthorized area code or exchange and redirects that call to an operator or a recording.

**interchange** In U.S. military communications, the transmission of message between different operating elements of the Defense Communications System; e.g., from automatic digital network to Defense Communications System tape relay network.

**interchange circuit** In any interface, a circuit with an associated pin assignment on the interface connector that is assigned a data, timing, testing, or control function.

**intercom** An option on many business phone systems which allows you to use an abbreviated dialing sequence to reach another telephone connected to the same communications system.

**interconnect** 1. The electrical and functional association of two different services, often provided by different suppliers. 2. A supplier of non-telephone-company alternative equipment.

**interconnect company** A provider of communications terminal equipment for connection of telephone lines.

**interconnect industry** The industry to manufacture, market, and service equipment and services that connects to telephone lines.

**interconnected networks** In IBM's SNA, networks connected by gateways.

**interconnection** See SNA network interconnection (IBM's SNA).

**Interdigital Communications Subsystem (ICS)** A system in JSTARS aircraft which controls access to the voice networks aboard the aircraft and to external communications equipment.

**Interexchange Carrier (IEC)** Any carrier registered with the Federal Communications Commission (FCC) that is authorized to carry customer transmissions between LATA's interstate or, if approved by a state public utility commission (PUC), intrastate.

**interface** 1. A shared boundary defined by common physical interconnection characteristics, signal characteristics, and meanings of interchanged signals. 2. The equipment which provides this shared boundary.

**Interface-CCITT** The present mandatory method in Europe (and a possible international standard) for the interface requirements between data terminal equipment and data circuit-terminating equipment (modems). The CCITT (now ITU) V-24 modem interface standard resembles the American EIA RS-232 standard.

**Interface-EIA RS-232** A standardized method adopted by the Electronic Industries Association, in order to ensure uniformity of interface between data communication equipment and data circuit-terminating equipment. This standard interface has been generally accepted by a majority of the manufacturers of data transmission (modems) and data terminal equipment in the United States. The latest version of the RS-232 standard is version D.

**interface computer** Part of a packet switching network which mediates between network subscriber and high-level or trunk network. It can be regarded as containing a local area switching terminal processors.

**Interface Message Processor (IMP)** A computer used to interface terminals and other computers to a network; commonly used in packet switching networks to interface incompatible devices with the network.

**interface processor** Communications processor specialized for handling the function of interfacing computers and terminals to a network.

**interference** The addition of unwanted signals through a communications facility.

**inteframe coding** A video compression technique in which the differences between frames are transmitted.

**Interior Gateway Routing Protocol (IGRP)** A multi-parameter technique for automatic packet routing and flow optimization developed by Cicso Systems, Inc.

**interLATA**  A call between local access and transport areas.

**interleave**  The process of putting bits or characters alternately into time slots in a TDM (Time Division Multiplexer).

**interleaving**  The process of alternately transmitting characters, blocks, ..., messages.

**interlock code**  In packet switching, a numerical value indicating Closed User Group (CUG) membership. The interlock code is sent in a Call Request Packet.

**Intermediate Block Character (ITB)**  A transmission control character that terminates an intermediate block. A Block Check Character (BCC) usually follows. Use of ITBs allows for error checking of smaller transmission blocks.

**Intermediate Distribution Frame (IDF)**  An equipment unit used to connect communications equipment by use of connection blocks.

**intermediate node**  A node that is capable of routing path information units to another subarea and that contains neither the origin NAU, nor the destination NAU, nor any associated boundary function for these NAUs.

**intermediate routing function**  In IBM's SNA, a path control capability in a subarea node that receives and routes path information units (PIUs) that neither originate in nor are destined for network addressable units (NAUs) in the subarea node.

**Intermediate Routing Node (IRN)**  In IBM's SNA, a subarea node with intermediate routing function. A subarea node may be a boundary node, an intermediate routing node, both, or neither, depending on how it is used in the network.

**intermediate SSCP**  In IBM's SNA, an SSCP along a session initiation path that owns neither of the LUs involved in a cross-network LU–LU session.

**intermediate system**  A system which relays data between other systems.

**intermodulation distortion**  An analog line impairment where two frequencies create an erroneous frequency which in turn distorts the data signal.

**internal clock**  Clock signals generated by the modem.

**internal drive**  A disk drive designed to be mounted inside the system unit of a computer.

**International Access Code**  The code that must prefix the country code, city code, and local number of a directly dialed international telephone call. If international dialing is available, dial:

International   Country   City   Local
  Access    +  Code  + Code + Number
  Code

International Access Codes vary, based upon the country you are located in, and are listed in the table on the opposite page.

**International Alphabet (IA)**  An internationally-defined code for communication. Baudot code is IA2 and ASCII code is IA5.

**International Business Machines Corporation (IBM)**  A very large computer company known more recently for its Personal Computer (IBM PC) and 3270 Information Display System products.

**International Business Service (IBS)**  An all digital private line transmission facility offered by Comsat World Systems.

**International Business Service**  High-speed digital satellite
circuit transmission marketed by the International Telecommunications Satellite Organization.

**International Communications Association**  The trade association representing the largest users of communication and telecommunication products and services.

**International-Customer Network Optimization System (I-CNOS)**  An AT&T software product used by that vendor to plan, design, price and track cost-optimized networks based upon customer requirements.

*International Access Codes*

| | | | | | | | |
|---|---|---|---|---|---|---|---|
| Australia | 0011 | Iran | 00 | Panama | 00 |
| Austria—Linz | 00 | Iraq | 00 | Philippines | 00 |
| —Vienna | 900 | Ireland, Rep. of | 16 | Portugal (Lisbon only) | 097 |
| Bahrain | 0 | Israel | 00 | Qatar | 0 |
| Belgium | 00 | Italy | 00 | Saudi Arabia | 00 |
| Brazil | 00 | Ivory Coast | 00 | Senegal | 12 |
| Colombia | 90 | Japan | 001 | Singapore | 005 |
| Costa Rica | 00 | Korea | 001 | South Africa/Namibia | 09 |
| Cyprus | 00 | Kuwait | 00 | Spain | 07* |
| Czechoslovakia | 00 | Lebanon | 00 | Sweden | 009 |
| Denmark | 009 | Libya | 00 | Switzerland | 00 |
| El Salvador | 0 | Liechtenstein | 00 | Taiwan | 002 |
| Finland | 990 | Luxembourg | 00 | Thailand | 001 |
| France | 19* | Malaysia (Kuala Lumpur & | | Tunisia | 00 |
| French Antilles | 19 | Penang only) | 00 | Turkey | 99 |
| Germany | 00 | Monaco | 19 | United Arab Emirates | 00 |
| Greece | 00 | Morocco | 00* | United Kingdom | 010 |
| Guam | 001 | Netherlands | 09* | United States | 011 |
| Guatemala | 00 | Netherlands Antilles | 00 | Vatican City | 00 |
| Honduras | 00 | New Zealand | 00 | Venezuela | 00 |
| Hong Kong | 106 | Nicaragua | 00 | | |
| Hungary | 00 | Norway | 095 | *Await dial tone | |

**International Dialing Access Codes**  The codes you
must use to direct-dial different countries from the
United States.

| | | | | | |
|---|---|---|---|---|---|
| Algeria | 213 | Bulgariqa | 359 | Finland | 358 |
| American Samoa | 684 | Burkina Faso | 226 | France | 33 |
| Andorra | 33 | Burma | 95 | French Antilles | 596 |
| Anguilla | 809 | Cameroon | 237 | French Guiana | 594 |
| Antigua | 809 | Canada (area code) | | French Polynesia | 689 |
| Argentina | 54 | Cape Verde Islands | 238 | Gabon | 241 |
| Aruba | 297 | Cayman Islands | 809 | Gambia | 220 |
| Ascension Island | 247 | Chile | 56 | Germany | 49 |
| Australia | 61 | China | 86 | Ghana | 233 |
| Austria | 43 | Colombia | 57 | Gibraltar | 350 |
| Bahamas | 809 | Costa Rica | 506 | Great Britain | 44 |
| Bahrain | 973 | Cyprus | 357 | Greece | 30 |
| Bangladesh | 880 | Czechoslovakia | 42 | Greenland | 299 |
| Barbados | 809 | Denmark | 45 | Grenada | 809 |
| Belgium | 32 | Djibouti | 253 | Guadeloupe | 590 |
| Belize | 501 | Dominica | 809 | Guam | 671 |
| Benin | 229 | Dominican Republic | 809 | Guantanamo Bay | 5399 |
| Bermuda | 809 | Ecuador | 593 | Guatemala | 502 |
| Bolivia | 591 | Egypt | 20 | Guinea | 224 |
| Botswana | 267 | El Salvador | 503 | Guyana | 592 |
| Brazil | 55 | Ethiopia | 251 | Haiti | 509 |
| British Virgin Islands | 809 | Faeroe Islands | 298 | Honduras | 504 |
| Brunei | 673 | Fiji | 679 | Hong Kong | 852 |

217

*International Dialing Access Codes (continued)*

| | | | | | |
|---|---|---|---|---|---|
| Hungary | 36 | Morocco | 212 | Senegal | 221 |
| Iceland | 354 | Namibia | 264 | Seychelles Island | 248 |
| India | 91 | Nauru Islands | 674 | Sierra Leone | 232 |
| Indian Ocean Region | 873 | Nepal | 977 | Singapore | 65 |
| Indonesia | 62 | Netherlands | 31 | Solomon Islands | 677 |
| Iran | 98 | Netherlands Antilles | 599 | South Africa | 27 |
| Iraq | 964 | Nevis | 809 | Soviet Union | 7 |
| Ireland | 353 | New Caledonia | 687 | Spain | 34 |
| Israel | 972 | New Zealand | 64 | Sri Lanka | 94 |
| Italy | 39 | Nicaragua | 505 | Surinam | 597 |
| Ivory Coast | 225 | Niger Republic | 227 | Swaziland | 268 |
| Jamaica | 809 | Nigeria | 234 | Sweden | 46 |
| Japan | 81 | North Yemen | 967 | Switzerland | 41 |
| Jordan | 962 | Norway | 47 | Taiwan | 886 |
| Kenya | 254 | Oman | 968 | Tanzania | 255 |
| Kiribati | 686 | Pacific Ocean Region | 872 | Thailand | 66 |
| Korea | 82 | Pakistan | 92 | Togo | 228 |
| Kuwait | 965 | Panama | 507 | Tonga | 676 |
| Lesotho | 266 | Papua New Guinea | 675 | Trinidad & Tobago | 809 |
| Liberia | 231 | Paraguay | 595 | Tunisia | 216 |
| Libya | 218 | Peru | 51 | Turkey | 90 |
| Liechtenstein | 41 | Philippines | 63 | Turks & Caicos Islands | 809 |
| Luxembourg | 352 | Poland | 48 | Uganda | 256 |
| Macao | 853 | Portugal | 351 | Union Island | 809 |
| Malawi | 265 | Qatar | 974 | United Arab Emirates | 971 |
| Malaysia | 60 | Reunion Island | 262 | United Kingdom | 44 |
| Maldives | 960 | Romania | 40 | United States | 01 |
| Mali | 223 | Rwanda | 250 | Uruguay | 598 |
| Malta | 356 | St. Kitts | 809 | Vatican City | 39 |
| Marshall Islands | 692 | St. Lucia | 809 | Venezuela | 58 |
| Mauritius | 230 | St. Pierre & Miquelon | 508 | Wales | 44 |
| Mayotte Island | 269 | St. Vincent | 809 | Western Samoa | 685 |
| Mexico | 52 | Saipan | 670 | Zaire | 243 |
| Micronesia | 691 | San Marino | 39 | Zambia | 260 |
| Monaco | 33 | Saudi Arabia | 966 | Zimbabwe | 263 |
| Montserrat | 809 | Scotland | 44 | | |

**International Directory Assistance** At&T provides free overseas directory assistance to persons dialing 10288-0. For assistance with overseas calls you can call the following carriers at the toll free numbers indicated below:

| | |
|---|---|
| AT&T | 1-800-874-4000 |
| MCI | 1-800-624-6240 |
| Sprint | 1-800-531-4646 |

**International Federation for Information Processing (IFIP)** A federation of professional and technical societies concerned with information processing. One society is admitted from each participating nation. IFIP has established a number of technical committees and these have formed working groups. The technical committee for data communication is TC 6 and was formed in 1971. Among its working groups is WG 6.1 which is also known as the International Network Working Group. Working Group WG 6.3 deals with human–computer communications.

**International Federation of Information Processing Societies (IFIPS)** An organization comprising trade and professional organizations related to the computer industry.

**International Kilostream** A private leased circuit facility from British Telecom International which provides point-to-point digital communications between the UK and other countries.

**International Marine Satellite (Inmarsat)** A 55-nation cooperative which operates eight satellite systems which provide mobile communications services to shipping and aircraft.

**International MultiStream Synchronous Service** A service marketed by British Telecom International in conjunction with Telenet Communications Corporation which enables users to send SDLC traffic between sites in the U.S. and the UK via packet networks operated by both companies. Telenet Communications Corporation markets this service under the name Global Multidrop Service.

**International Network Working Group (INWG)** INWG was formed in 1972 to be a forum for discussion of network standards and protocols. In 1973 it was adopted as working group 1 of IFIP TC 6 with the title "International Packet Switching for Computer Sharing". It has about 100 members organized into four study groups. This working group distributes *INWG Notes* to exchange ideas on protocols, on interworking and on network research generally.

**International Record Carrier (IRC)** A common carrier originally dedicated to carrying telecommunication traffic from various gateway cities in the U.S. to foreign locations; now, any common carrier who performs these services, whether on a dedicated basis or along with providing domestic services.

**International Record Carrier Selection (IRC Select)** An optional X.25 facility that allows the caller to select the gateway service for an international call.

**INternational Satellite COMmunication (INSCOM)** A joint venture between the Brazilian rocket manufacturer Avibras Aerospacial and the China Great Wall Industry Corporation to market satellite launch services.

**INTErnationaL SATellite Services (INTELSAT)** The International Telecommunications Satellite organization which operates a network of satellites that are used for international transmission.

**International Standard** An ISO standard document that has been approved in final balloting.

**International Standards Organization (ISO)** An international and voluntary standards organization, closely aligned with the ITU (formerly known as CCITT); its OSI model is widely quoted, and its OSI communications protocols are widely accepted. Membership includes standards organizations from participating nations (ANSI is the USA representative). ISO standards listed under ISO.

**International Telecommunications Union (ITU)** The telecommunications agency of the United Nations, established to provide standardized communications procedures and practices, including frequency allocation and radio regulations, on a worldwide basis. The International Telecommunications Union regulates telecommunications at the international level. It was established in 1863 and served as one of the founding agencies in the United Nations. Membership now stands at over 150 nations.

**International Telecommunications Users Group (INTUG)** A London, UK based international association comprising representatives from multinational corporations and 20 national users groups. The association represents users' interests before such international groups as the ITU.

**INTERNET** 1. When written in lowercase, internet refers to a collection of packet switching networks interconnected by gateways as well as the protocols that enable them to function as a single, large virtual network. 2. When written in uppercase, Internet refers to the set of interconnected networks that share the same network address scheme and use the TCP/IP protocol.

**Internet Activities Board (IAB)**   The group which sets the technical direction and decides which protocols are a required parts of the TCP/IP suite.

**Internet address**   A 32-bit address which identifies the network to which each computer on a TCP/IP network is attached as well as the computer's unique identification. The three major types of addresses are illustrated below and defined as follows:

A Class A address is for networks that have more than $2^{16}$ (65 536) hosts.

A Class B addressed is used for networks that have $2^8$ (256) to $2^{16}$ hosts.

A Class C address is for networks that have fewer than $2^8$ hosts.

*Internet (IP) addresses*

```
     0 1 2 3 4 5 6 7 8      16       24        31

Class A 0 ---netid----  hostid ----------------------

Class B 1 0 -netid------------- hostid--------------

Class C 1 1 0netid--------------------- hostid-----
```

An address with hostid zero refers to the network itself. A broadcast address has a hostid with all bits set to one. IP addresses are written as four decimal integers separated by decimal points where each integer is the value of one octet.

**Internet Architect**   The chairman of the Internet Activities Board (IAB).

**Internet Architecture Board (IAB)**   A group tasked with addressing Internet technical problems and developing solutions to those problems.

**Internet Control Message Protocol (ICMP)**   The portion of the Internet protocol which handles error and control messages. ICMP is used by gateways and hosts on an Internet to send reports of problems

about datagrams back to the originator. ICMP is a series of nine types of messages carried by the Internet Protocol (IP) which defines an unreachable destination, echo, echo reply, parameter problem, redirect, source quelch, time exceeded, timestamp, and timestamp reply. Each of these messages has five fields as indicated below:

| Field | Size (bytes) | Explanation |
|---|---|---|
| Type | 1 | The type of ICMP message |
| Code | 1 | Parameters |
| Checksum | 2 | Checksum of entire message |
| Parameters | 4 | Any necessary parameters |
| Information | N | Additional message information |

**Internet datagram**   The basic information transfer unit on Internet.

**Internet Engineering Task Force (IETF)**   The group in the Internet Activities Board (IAB) which concentrates on short- and medium-term engineering problems.

**Internet Network Information Center (InterNIC)**   A joint NSFnet and NREN information center which provides voice and electronic assistance concerning the Internet.

**Internet Packet Exchange Protocol (IPX)**   A derivative of the Xerox Network Services (XNS) Internet Transport Protocol marketed by Novell which provides applications with the ability to send and receive messages across a network.

**Internet Protocol**   The envelope coding developed by the Department of Defense (DOD) to enable Transmission Control Protocol (TCP) segments to travel between different networks. A collection of IP networks is called an internet.

**Internet Research Task Force (IRTF)**   The group in the Internet Activities Board which coordinates research activities related to TCP/IP protocols and Internet architecture.

**Internet Zone**   An Appletalk cable segment that can

theoretically support up to 255 nodes; however, Apple Computer Corporation recommends a limit of 32 per cluster when LocalTalk cabling is used. When Ethernet or broadband cabling is used, a maximum of 255 clusters or individual users and file servers can be connected to create a complex internetwork.

**internetwork** Two or more networks of the same or different hardware types connected by means of special bridge hardware and software.

**Internet Packet Exchange/Sequenced Packet Exchange (IPX/SPX)** One of two parts of the Novell NetWare shell that operates on a workstation. IPX/SPX provides the hardware-specific routines which enables a workstation to communicate with its installed network card which in turn communicates with other devices on the network.

**Internetwork Protocol Exchange (IPX)** A subset of the Xerox Network Services (XNS) protocol used by Novell Corporation in its NetWare software.

**internetwork router** In LAN technology, a device used for communications between subnetworks; only messages for the corrected subnetwork are transmitted by this device. Internetwork routers function at the network layer of the OSI model.

**Internetworking** Communication between two or more different networks.

**InterNic** Internet Network Information Center.

**internodal** In a telecommunications network, pertaining to any matter, such as an event, transmission, or message exchange that takes place between any two or more nodes of the network.

**internodal delay** The delay of customer data that results from the time it takes to process a bypass channel through a multiplexer node.

**interoffice trunk** A direct circuit between telephone company central offices.

**interoperability** The ability of hardware and software on multiple computers manufactured by different vendors to communicate with one another.

**interpacket gap** The time between the conclusion of transmission or reception of one packet and the transmission or reception of the next packet.

**Interpersonal Message (IPM)** In the CCITT Message Handling System the IPM is a defined user agent (UA) class.

**InterPoll** A set of network management tools marketed by Apple Computer which lets network administrators isolate faults, test device links and receive reports on the systems softre operating on each Macintosh computer connected to an AppleTalk local area network.

**interposition trunk** A connection between two positions of a large switchboard so that a line on one position can be connected to a line on another position.

**interpret table** In IBM's VTAM, an installation-defined correlation list that translates an argument into a string of eight characters. Interpret tables can be used to translate logon data into the name of an application program for which the logon is intended.

**Interprocess Message Switch (IMS)** An element of the AT&T Autoplex third generation cellular telecommunications system which provides interfaces between cellular components, cell sites, and other networked Autoplex systems.

**interrupt** 1. The initiation, by hardware, of a routine intended to respond to an external (device-orignated) or internal (software-originated) event that is either unrelated or asynchronous with the executing program. 2. A break in the normal flow of a system or routine that allows the flow to be resumed from that point at a later time. An interrupt is usually caused by a signal from an external source.

**interrupt driven** An interactive system whose operations and transmissions are determined by user interrupts (e.g., requests, transmission).

**Interrupt Request (IRQ)** A computer activity which interrupts a program.

**interstate** Any connection made between two states.

**inter-symbol interference** An (analogue) waveform which carries binary data is generally transmitted as a number of separate signal elements. When received, the signal elements may be influenced by neighboring elements. This is called inter-symbol interference.

**intertoll trunk** A trunk between toll offices in different telephone exchanges.

**interval timer** A timer can be set by a program to produce an interrupt after a specified interval.

**Interview** A trademark used by Atlantic Research Corporation of Alexandria, VA, for a series of protocol analyzers.

**Interview 7000 Series** A series of high-speed, high-performance protocol analyzers from Atlantic Research Corration.

**interworking unit** Another name for a gateway.

**intrabuilding network cable** Cable (formerly referred to as house or riser cable) that extends to the outside plant distribution from the building entrance point to equipment rooms, cross-connection points, or other distribution points within the building.

**intraLATA** A call within a local access and transport area.

**intraoffice trunk** A trunk connection within a central office.

**intrastate** Any connection made that remains within the boundaries of a single state.

**intrasystem wiring** Wire which is used to connect system components, e.g. a PBX to its stations.

**INTTAB** Interpret Table (VTAM).

**INTUG** International Telecommunications User's Group.

**inventory control** Logistics functions that include management, cataloging, requirements determination, procurement, inspection, storage, distribution, overhaul and disposal of material.

**invoke** To start a command, procedure, or program.

**in-WATS** The use of a WATS service permits toll-free calling (by means of an "800" number) into a subscriber's facility, from various extended geographical areas. Billing to the subscriber is based on zones and time blocks rather than metered usage.

**INWG** International Network Working Group.

**I/O** Input/Output.

**I–O channel** An equipment forming part of the input–output system of a computer. Under the control of I–O commands the "channel" transfers blocks of data between the main store and peripherals.

**I/O port** An input/output (I/O) port is a plug-in location which enables different peripheral devices to be connected to a computer. Examples of I/O ports include serial, parallel, and mouse ports.

**I/O nest** An enclosure that contains input/output logic modules as well as other equipment that enables the nest to be attached to a controller, multiplexer or other type of communications device.

**IOMI** Input–Output Microprocessor Interface.

**IOP** 1. Independent Company. 2. Input–Output Processor.

**IP** Internal Protocol.

**IPARS** International Passenger Airline Reservation System (IBM).

**IPL** Initial Progam Load.

**IPM** InterPersonal Message.

**IPN** Instant Private Network.

**IPX** 1. Internetwork Protocol Exchange. 2. The name of Novell Corporation's network protocol.

**IRAC** Interdepartment Radio Advisory Committee, OTP.

**IRC**  1. International Record Carrier. 2. Internet Relay Chat.

**IRC SELECT**  International Record Carrier Selection.

**Irma**  The name used by Digital Communications Associates of Alpharetta, GA for a series of communications products.

**Irmalan**  A family of hardware and software products marketed by Digital Communications Associates designed to provide personal computers on a local area network with access to mainframe applications.

**IRMAprint**  A device manufactured by Digital Communications Associates of Alpharetta, GA, which converts data from an IBM 3270 mainframe protocol to an ASCII format, enabling personal computer printers to emulate IBM 4224 OR 3287 printers.

**IRN**  Intermediate Routing Node (IBM's SNA).

**IRQ**  Interrupt Request.

**IRTF**  Internet Research Task Force.

**IR&D**  Independent Research & Development.

**IS**  Information Separator.

**isarithmic control**  The control of flow in a packet switching in such a way that the number of packets in transit is held constant.

**ISDN**  Integrated Services Digital Network.

**ISDN access line**  A digital transmission facility between a customer's premises and a local exchange which provides a mechanism to transmit data via ISDN. The ISDN interfaces are illustrated in the diagram.

**ISDN**  Access Line Interface Relationships

**ISDN Gateway (IG)**  A one-way AT&T PBX-to-host interface which complements the ISDN network capabilities of that vendor's Definity, System 85 and System 75 PBXs.

**ISDN multi-button key system set**  A digital voice telephone equipped with an array of buttons, many

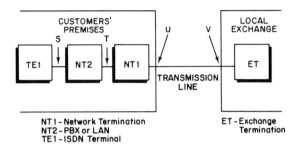

NT1 – Network Termination
NT2 – PBX or LAN
TE1 – ISDN Terminal

ET – Exchange Termination

*ISDN*

of which can be programmed by the customer for one-touch activation of selected features to include call hold, call transfer, and conferencing. It includes a display for information such as call progress, incoming call ID, time, and messages.

**ISDN terminal**  A digital voice and/or data communications device that enables a person to use the ISDN capabilities of a network. There are two major types of ISDN terminals—one operates with a two-wire "U" interface and the other uses a four-wire "T" interface, both of which are part of the Basic Rate Interface (BRI).

**ISDN user part**  The higher layer protocol in Common Channel Signalling System No. 7 that deals with end-user signalling on the ISDN.

**ISDN2**  The name used by British Telecom for their Integrated Services Digital Network Basic Rate Interface (BRI) service marketed in the United Kingdom.

**ISEC**  Information Systems Engineering Command.

**I-Series Recommendations**  ITU (formerly CCITT) recommendations on standards for ISDN services, ISDN networks, user–network interfaces, and internetwork and maintenance principles.

I.100 SERIES RECOMMENDATIONS

**I.100 Series**  The standards recommendations that cover ISDN general concepts and issues.

*RECOMMENDATION*
*Number Title*

I.110  General structure of the I-Series Recommendations.

I.111    Relationship with other recommendations relevant to ISDNs.

I.112    Vocabulary of terms for ISDN.

I.113    Vocabulary of terms for broadband aspects of ISDNs.

I.120    Integrated Services Digital Networks (ISDNs).

I.121    Broadband aspects of ISDNs.

I.122    Framework for providing additional packet mode services.

I.130    Method of characterization of telecommunication services supported by an ISDN and network capabilities of an ISDN.

I.140    Attributes for the characterization of telecommunication services supported by an ISDN and network capabilites of an ISDN.

I.141    ISDN charging capability attributes.

## I.200 SERIES RECOMMENDATIONS

**I.200 Series**   The standards recommendations that cover ISDN services.

*RECOMMENDATION*
*Number Title*

I.200    Guidance to the I.200 Series.

I.210    Principles of telecommunication services supported by an ISDN and network capabilities of an ISDN.

I.211    Bearer services supported by an ISDN.

I.212    Teleservices supported by an ISDN.

I.220    Common dynamic description of basic telecommunication services.

I.221    Common specific characteristics of services.

I.230    Definition of bearer services.

I.231    Circuit mode bearer services categories.

I.232    Packet mode bearer services categories.

I.240    Definition of teleservices.

I.241    Teleservices supported by an ISDN.

I.250    Definition of supplementary services.

I.251    Number of identification services.

I.252    Call offering services.

I.253    Call completion services.

I.254    Multiparty services.

I.255    "Community of interest" services.

I.256    Charging services.

I.257    Additional information transfer services.

**I.300 Series**   The standards recommendations that covers ISDN networking.

*RECOMMENDATION*
*Number Title*

I.310    ISDN—network functional principles.

I.320    ISDN protocol reference model.

I.324    ISDN network architecture.

I.325    Reference configurations for ISDN connection types.

I.326    Reference configurations for relative network resource requirements.

I.330    ISDN numbering and addressing principles.

I.331    Numbering plan for the ISDN era.

I.332    Numbering principles for interwork between ISDNs and dedicated networks with different numbering plans.

I.333    Terminal selection in ISDN.

I.334    Principles relating ISDN numbers/subaddress to OSI network layer addresses.

I.335    ISDN routing principles.

I.340    ISDN connection types.

I.350    General aspects of quality of service and network performance in digital networks, including ISDNs.

I.351    Recommendations in other Series including network performance objectives that apply at T reference points of an ISDN.

I.352    Network performance objectives for call processing delays.

## I.400 SERIES RECOMMENDATIONS

**I.400 Series**   The standards recommendations that covers ISDN user-network interface aspects.

*RECOMMENDATION*
*Number Title*

I.410    General aspects and principles relating to Recommendations on ISDN user–network interface.

I.411    ISDN user–network interfaces—reference configurations.

I.412  ISDN user–network interfaces—interface structures and access capabilities.

I.420  Basic user-network interface.

I.421  Primary rate user–network interface.

I.430  Basic user–network interface—Layer 1 specification.

I.431  Primary rate user–network interface—Layer 1 specification.

I.440  ISDN user–network interface data link layer— general aspects.

I.441  ISDN user–network interface data link layer specification.

I.450  ISDN user–network interface Layer 3— general aspects.

I.451  ISDN user–network interface Layer 3 specification for basic call.

I.452  ISDN user–network interface layer 3 specification—generic procedures for the control of ISDN supplementary services.

I.460  Multiplexing, rate adaptation and support of existing interfaces.

I.461  Support of X.21 and X.21bis and X.20bis based Data Terminal Equipment (DTEs) by an Integrated Services Digital Network (ISDN).

I.462  Support of packet mode terminal equipment by an ISDN.

I.463  Support of Data Terminal Equipment (DTEs) with V-Series type interfaces by an ISDN.

I.464  Multiplexing, rate adaptation and support of existing interfaces for restricted 64 Kbps data transfer capability.

I.465  Support by an ISDN of data terminal equipment with V-Series type interfaces with provision for statistical multiplexing.

I.470  Relationship of terminal functions to ISDN.

## I.500 SERIES RECOMMENDATIONS

**I.500 Series**  The standards recommendations that cover internetwork interfaces to include how the I-Series standards relate to the E-Series, V-Series and other standards.

*RECOMMENDATION*
*Number Title*

I.500  General structure of ISDN interworking recommendations.

I.510  Definitions and general principles for ISDN interworking.

I.511  ISDN-to-ISDN Layer 1 internetwork interface.

I.515  Parameter exchange for ISDN interworking.

I.520  General arrangements for network interworking between ISDNs.

I.530  Network interworking between an ISDN and a public switched telephone network (PSTN).

I.540  General arrangements for interworking between Circuit Switched Public Data Networks (CSPDNS) and Integrated Services Digital Networks (ISDNs) for the provision of data transmisssion services.

I.560  Requirements to be met in providing the telex service within the ISDN.

## I.600 SERIES RECOMMENDATIONS

**I.600 Series**  The standards recommendations that cover ISDN maintenance.

*RECOMMENDATION*
*Number Title*

I.601  General maintenance principles of ISDN subscriber access and subscriber installation.

I.602  Application of maintenance principles to ISDN subscriber installation.

I.603  Application of maintenance principles to ISDN basic accesses.

I.604  Application of maintenance principles to ISDN primary rate accesses.

I.605  Application of maintenance principles to static multiplexed ISDN basic accesses.

**ISLU**  Integrated Services Line Unit.

**ISO**  International Standards Organization.

**ISO**  Standards

ISO 646  7-bit character set for information interchange.

ISO 1155  Defines Longitudinal Redundancy Checking (LRC).

ISO 1177  Asynchronous and synchronous character structure.

ISO 1745  Character-oriented protocol operating procedures.

ISO 2022  Code extension technique for use with 7-bit code.

ISO 2110  25-pin DTE/DCE interface connector with pin assignments (RS232/V.24).

ISO 2111  Code independent information transfer (transparency).

ISO 2593  Connector pin allocations for high-speed data terminal equipment (DTE).

ISO 2628  ISO 1745 complements.

ISO 2629  ISO 1745 Conversational message transfer.

ISO 3309  High-level data link procedures; frame structure (HDLC).

ISO 4335  HDLC elements of procedure.

ISO 4902  HDLC unbalanced classes of procedures.

ISO 4903  15-pin DTE/DCE interface connector and pin assignments.

ISO 6159  HDLC unbalanced procedures.

ISO 6256  HDLC balanced procedures.

ISO 6429  ISO 646 control functions for character imaging devices.

ISO 6936  Conversion between ISO 646 and CCITT ITA2.

ISO 6937  Character sets—page image format.

ISO 7478  Data communications—multilink procedures.

ISO 7942  Graphical Kernel System (GKS).

ISO 8072  OSI transport service definition.

ISO 8073  OSI transport protocol.

ISO 8481  X.24 DTE to DTE connection with DTE provided timing.

**isochronous**  A form of data transmission in which individual characters are only separated by a whole number of bit-length intervals.

**isochronous distortion**  Caused by clock jitter in synchronous modems. The jitter can be early (positive jitter), in which the bits are too short, or the jitter can be late (negative jitter), in which the bits are too long. The isochronous distortion is the percentage of the sum of the absolute peak values of the two jitters compared to the ideal pulse width.

Isochronous distortion = ((Positive jitter + Negative jitter)/Total ideal pulse width) × 100%

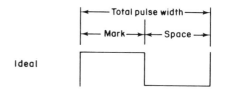

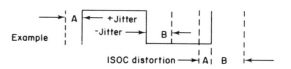

**isolated adaptive routing**  A method of routing in which the decisions are made solely on the basis of information available in each node.

**isolation transformer**  A device used to bond the neutral and ground wires of a transformer together which shorts out common mode noise. Also called a power line condition.

**ISO–OSI**  International Organization for Standardization–Open Systems Interconnect.

**ISPBX**  Integrated Services PBX.

**ISPF**  Interactive System Productivity Facility (IBM's SNA).

**ISPF/PDF**  Interactive System Productivity Facility/Program Development Facility.

**ISSN**  Integrated Special Services Network.

**IST**  InterSwitch Trunk.

**ISTF**  Integrated Services Test Facility.

**ISU**  Integrated Service Unit.

**ISUP**  ISDN User Part.

**I-SYNC**  Intelligent Synchronous (General DataComm, Inc.).

**Italcable**  A subsidiary of IRI-Stet, the Italian state-owned holdings company under the control of the Ministry of State Participation. Italcable is responsible for most non-European international telephone traffic in Italy.

**italic**  A typestyle in which the characters are slanted.

**item** In IBM's CCP, any of the components, such as communication controllers, lines, cluster controllers, and terminals, that comprise an IBM 3710 Network Controller configuration.

**IT** Intelligent Terminal.

**ITD** Intermediate Block Character.

**ITDM** Intelligent Time Division Multiplexer.

**ITI** Interactive Terminal Interface.

**ITRC** Information Technology Requirements Council.

**ITS** Invitation To Send.

**ITT Paynet** A service of ITT World Communications, Inc., of Seacaucus, NJ, which entitles owners and operators of public pay telephones to a commission on certain interstate calls from their telephones.

**ITU** International Telecommunication Union.

**IUP** ISDN User Part.

**IVS** Interactive Video Service.

**IW** Inside Wiring.

# J

**J** Jack.

**jabber control** In local area networking, the ability of a station to automatically interrupt transmission to inhibit an abnormally long data stream.

**jabber frame** A frame on a local area network whose length exceeds the maximum allowable frame length. A loose connector is the most prominent cause of jabber frames.

**jabbering** In LAN technology, the continuous sending of random data (garbage); normally used to describe the action of a station (whose circuitry or logic has failed) that locks up the network with its incessant transmission.

**Jack (J)** 1. A device used for terminating the permanent wiring of a circuit to which access is obtained by inserting a plug. 2. The name used for the plug portion of a manual board plug and socket.

**jam** In local area networking, a collision enforcement technique that ensures detection of a collision by all stations.

**jam signal** In a local area network, the signal generated by a card to ensure that other cards know that a packet collision has taken place.

**jamming** Intentional interference of open-air radio transmission to prevent reception of signals in a specific frequency band.

**JANET** Joint Academic NETwork.

**JCL** Job Control Language.

**JES** Job Entry System.

**JESSI** Joint European Submicron SIlicon.

**JES2** A functional extension of the HASP II program that receives jobs into the system and processes all output data produced by the job.

**JINTACCS** Joint INteroperability of TActical Command and Control Systems.

**jitter** Slight movement of a transmitted signal in time or phase that can cause errors or loss of synchronization; an error condition on a digital circuit caused by short-term deviations of signal pulses from phase (i.e. the drifting of the centers of pulses from the center of the specified bit time slots).

**Jitter Buster** A trademark of Proteon, Inc., as well as the name used to refer to technology involving circuitry and software that allows users to build large-scale, IBM-compatible 16 Mbps token-ring networks over unshielded twisted pair wire.

**job** One or more related programs; a set of data including programs, files, and instructions to a computer.

**job class** Any one of a number of job categories that can be defined. With the classification of jobs and direction of initiator/terminators to initiate specific classes of jobs, it is possible to control the mixture of jobs that are performed concurrently.

**job deck** A sequence of job control and data cards used by an operating system to activate a job.

**Job Entry Subsystem (JES)**   A system facility for spooling, job queueing, and managing the scheduler work area.

**Job Entry System (JES)**   An IBM mainframe control protocol and procedure designed to handle the execution of specific jobs and tasks input to the computer.

**Job Output Element (JOE)**   Information that describes a unit of work for the HASP or JES2 output processor and represents that unit of work for queueing purposes.

**Job Queue Element (JQE)**   A control block that represents an element of work for the system (job) and is moved from queue to queue as that work moves through each successive stage of JES2 processing.

**Job Separator Page Data Area (JSPA)**   A data area that contains job-level information for a data set. This information is used to generate job header, job trailer or data set header pages. The JSPA can be used by an installation-defined JES2 exit routine to duplicate the information currently in the JES2 separator page exit routine.

**job separator pages**   Those pages of printed output that delimit jobs.

**Joint Academic Network (JANET)**   A British networking organization.

**Joint Electronics Type Designation System**   A system used to identify U.S. military electronics equipment. The format of the designation system is: AN/XYZ-DDDL (A,B, or C) (V), where:

| | | |
|---|---|---|
| AN | = | Indicates system |
| X | = | Installation |
| Y | = | Type of equipment |
| Z | = | Purpose |
| DDD | = | Model number |
| L | = | Modification letter |
| A,B or C | = | Changes in voltage, phase or frequency |
| V | = | Variable grouping |

*Installation codes*

A—Airborne
B—Underwater

C—Air transportable
D—Pilotless carrier
F—Fixed
G—Ground, general
K—Amphibious
M—Ground, mobile
P—Pack, portable
S—Water surface craft
T—Ground, transportable
U—General, utility
V—Ground, vehicular
W—Water, surface, and underwater

*Type of equipment codes*

A—Invisible light, heat radiation
B—Pigeon
C—Carrier
D—Radiac
E—Nupac
F—Photograph
G—Telegraph or teletypewriter
I—Interphone and PA
J—Electromechanical
K—Telemetering
L—Countermeasures
M—Meterological
N—Sound in air
P—Radar
Q—Sonar
R—Radio
S—Special types
T—Telephone (wire)
V—Visual

*Purpose codes*

A—Auxiliary assemblies
B—Bombing
C—Communications
D—Direction finding
E—Ejection release
G—Fire control
H—Recording
L—Searchlight control
M—Maintenance and test assemblies
N—Navigational aids
P—Reproduction
Q—Special or combination of purposes
R—Receiving
S—Detecting range bearing

T—Transmitting
W—Control

**European Submicron SIlicon (JESSI)** A collective European venture designed to develop the technology to produce 64-Mbit memory chips.

**Joint STARS (JSTARS)** The joint Surveillance and Target Attack Radar System which connects airborne radars aboard a U.S. Air Force E-8 (modified Boeing 707) aircraft with a U.S. Army ground station which processes the data for use in corps and divisions.

**Joint Tactical Information Distribution System (JTIDS)** A U.S. military jam-resistant, secure digital information system design for managing the tactical air environment.

**joule** A unit of energy expended when a force of one newton (a unit of measurement) moves the point of application one meter in the direction of the force (1 watt = 1 joule/second).

**JSTARS** Joint Surveillance and Target Attack Radar System.

**JTIDS** Joint Tactical Information Distribution System.

**Jughead** A data base of Gopher links which accepts word searches and allows search results to be used on many remots Gophers.

**JUGHEAD** Jonzy's Universal Gopher Hierarchy Excavation and Display.

**jumbo group** A FDM carrier system multiplexing level containing 3600 voice frequency (VF) or telephone channels (six master groups). Also hyper group.

**jump** Also known as a control jump or control transfer. It is a change in the normal sequence of obeying instructions in a computer. A conditional jump is a jump which happens only if a certain criterion is met.

**jumper** A cable (wire) used to establish a circuit, often on a temporary basis, for testing or diagnostics.

**junction** The term used in the UK to describe a local trunk between two local exchanges.

**junctor** Part of a circuit switching exchange. The switching matrix brings to the junctor the two lines which are to be connected. In the junctor there is the common equipment needed in the circuit during the call.

**JUNET** Japanese Unix Network.

**junk fax** Unwanted facsimile transmission.

**JvNCnet** Princeton University's John von Neumann Computer Network which connects universities in the eastern USA and Europe.

# K

**k**  Kilo; notation for one thousand (e.g., kHz).

**K**  Abbreviation for kilo, which means 1000. Used as a measurement of computer memory capacity; however, 1K in these terms means 1024 bits of memory. Thus, a 4K memory chip really carries 4096 bits, not 4000.

**Katakana**  A character set of symbols used in one of the two common Japanese phonetic alphabets.

**K-band**  Portion of the electromagnetic spectrum in the 10 GHz to 12 GHz range increasingly used for satellite communications.

**Ka band**  A portion of the electromagnetic spectrum; frequencies approximately in the 18 to 30 GHz range.

**KAK**  Key-Auto-Key.

**KAU**  Keystation Adapter Unit.

**K-bit**  1024 bits (128 bytes).

**Kbps**  Thousands of bits per second (bps).

**Kbytes (kilobytes)**  1024 BYTES.

**KDS**  Keyboard Display Station.

**KDU**  Keyboard display unit.

**Kermit**  Asynchronous file transfer protocol originally developed at Columbia University.

**Kermit 370**  A version of the Kermit protocol which operates on all IBM 370 series mainframes to include the 43XX, 308X, 309X, and 937X models.

**key**  In encryption systems, the digital code used with a standard combination algorithm to render an encrypted data stream unique.

**key generator**  A device used to generate keys used in encryption systems for communication security.

**key management**  The process of generating, distributing, monitoring, and destroying keys used in a security system. Usually, there are levels of keys. A root key (or base key) protects the generation or transport of another key. It is changed relatively rarely (yearly) and is transported and loaded manually for security and accountability reasons. A session key (sometimes called a data encrypting key) is used to encrypt your data. It is changed relatively frequently (weekly or daily) and, since it is protected by root keys, can be transported securely over the network and changed electronically without visits to remote sites.

**key pad**  12 pushbuttons or "keys" on the face of a touch-tone telephone set, used by a human to dial a circuit to another telephone or sometimes to control certain computer functions.

**Key Service Unit (KSU)**  The main operating unit of a key telephone system.

**key set**  A telephone set where each telephone line is accessed by pushing a button or key directly on the set.

**key system**  Telephone equipment originally designed to obtain additional service from a telephone company central office. Telephones are equipped with pushbuttons representing line appearances that permit users to select appropriate incoming or outgoing calls for their own use. Each telephone line is accessed by a pushbutton telephone for incoming or outgoing calls. Incoming calls are routed by an operator to the requested extension. In a pure key system, an outgoing line is similarly accessed by pushing a button.

**key system features** PBX switching equipment which contains all the necessary circuitry and control elements for servicing standard key telephone stations.

**key telephone set** Instrument with pushbuttons used to select a line for incoming or outgoing traffic.

**Key Telephone System (KTS)** Limited PBX type equipment typically with up to 12 trunk lines and up to 40 extensions.

**key value** The IBM Network Problem Determination Application (NPDA) data record key structure that contains resource type and label identification as well as other pertinent identifying information.

**keyboard** A manual coding device in which pressing keys is the means of generating the desired code elements.

**keyboard definition** A customizing procedure for defining a modified keyboard layout for modifiable keyboards. Most characters, symbols, and functions can be relocated, duplicated, or deleted from almost any keyboard position.

**keyboard lockout** An interlock feature used with teleprinters that prevent sending from the keyboard while the tap transmitter or another station is sending on the same circuit. This feature is used to avoid breaking up a transmission by simultaneous sending.

**keyboard perforator** A perforator provided with a bank of keys, the manual depression of any one of which will cause the code of the corresponding character or function to be punched in a tape.

**Keyboard Select Routing (KSR)** The ability of a calling channel to select its destination channel or channel group in a switching system, such as a data PBX or switching multiplexer.

**Keyboard Send/Receive (KSR)** A combination teleprinter transmitter and receiver with transmission capability from the keyboard only.

**Keyboard Send/Receive (KSR) terminal** A terminal which consists of a printer and keyboard, enabling it to receive and print as well as transmit data.

**key-encrypting key** In IBM's SNA, a key used in sessions with cryptography to encipher and decipher other keys.

**keying** 1. Modulation of a carrier, usually by frequency or phase, to encode data. 2. The interruption of a DC circuit for the purpose of signaling information. 3. The breaking or interrupting of a radio carrier wave (either manually or automatically).

**keyword** One of the predefined words of a programming language; a reserved word.

**KG** Key Generator.

**kilo (k)** 1. A prefix for one thousand times a specific unit. 2. A measure of computer memory size equal to 1024 units.

**kilobit per second (Kbps)** One thousand bits per second. A measurement of the rate of data transmission.

**kilohertz** A term used to state 1000 (kilo) cycles per second (hertz).

**kilopacket (kpkt)** One thousand packets.

**kilosegment** A unit of data traffic equal to 64 000 characters. Several packet switching carriers include a usage charge based upon the number of kilosegments transmitted and received.

**Kilostream** A digital transmission service provided by British Telecom in the United Kingdom. A Kilostream customer is provided with an interface device which is called a Network Terminating

Unit (NTU). The NTU provides a CCITT (now ITU) interface for customer data at 2.4, 4.8, 9.6 or 48 Kbps to include performing data control and supervision, which is known as structured data. At 64 Kbps, the NTU provides a CCITT interface for customer data without performing data control and supervision, which is known as unstructured data. The NTU controls the interface via CCITT recommendation X.21, which is the standard interface for synchronous operation on public data networks. An optional V.24 interface is available at 2.4, 4.8, and 9.6 bps while an optional V.35 interface can be obtained at 48 Kbps.

**Kjoebenhauns Telefon Aktieselskab (KTAS)** A regional Danish telephone company.

**Kluge** Hardware or software assembled from a variety of parts or modules, usually in haste. A Rube Goldberg device.

**KMS** Key Management System.

**K-Network International** A value-added network provider based in Osaka, Japan which provides international service to the U.S. and the UK.

**Knowbots** A program which functions as a front end to all of the white pages on the Internet, searching for the information a user enters. Knowbot Information Services of Reston, VA, developed the Knowbot concept.

**Knowledge Index** An information service which provides on-line access to approximately 50 of Dialog's most popular data bases at night and on weekends. For additional information telephone 1-800-334-2564.

**Korean Telecommunication Authority (KTA)** The sole supplier of all telecommunications services in South Korea.

**kpkt** Kilopacket.

**KPSI** Tensile strength in thousands of pounds per square inch.

**KRTN Photo Service** A satellite-based news photo service marketed by the Knight-Ridder Tribune News Information Service. KRTN Photo Service uses 64 Kbps Ku-band satellite channels provided by Independent Network Systems Inc.

**K-Share** An implementation of the AppleTalk Filing Protocol (AFP) for Sun Microsystem's UNIX or Digital Equipment Corporation's ULTRIX from Kinetics of Walnut Creek, CA.

**KSR** 1. Keyboard Select Routing. 2. Keyboard Send/Receive.

**KTA** Korean Telecommunications Authority.

**KTAS** Kjoebenhauns Telefon Aktieselskab.

**KTS** Key Telephone System.

**Ku band** Portion of the electromagnetic spectrum being used increasingly for satellite communications; frequencies approximately in the 14/12 GHz range. A table listing Ku-band geostationary satellite information to include the satellite name, operator, orbit, use and designation is included under the entry "satellite" in this book.

**Kyutai Cable** A fiber optic undersea cable connecting Miyazaki on Japan's southernmost main island of Kyushi and Tou Cheng in Taiwan.

**K12NET** A loosely organized network of school-based electronic bulletin board systems located throughout North America, Australia and Europe that share curriculum-related information, classroom projects, and other education-related information.

# L

**L** Inductance.

**L** Left.

**label** The portion of a message, usually a data message, which indicates its nature.

**LADC** Local Area Data Channel.

**LADT** Local Area Data Transport.

**LAMA** Local Automatic Message Accounting.

**lambda** A symbol used in queueing systems to represent the average number of arrivals per second.

**LAN** Local Area Network.

**LAN Manager** A local area network operating system originally developed by Microsoft Corporation and Hewlett-Packard Corporation which enables UNIX machines to be used as servers for MS-DOS and OS/2 clients.

**LAN Manager for UNIX (LMU)** The revised name for LAN Manager/X developed by AT&T and Microsoft Corporation.

**LAN Manager/X** Software developed by AT&T and Microsoft Corporation which allows personal computer access to UNIX computer file servers to include native UNIX files under UNIX security mechanisms.

**LAN segment** A portion of a local area network (LAN) separated from other portions of the LAN by one or more bridges.

**LAN Span** A local area network bridge marketed by Infotron Systems Corp. (now Gandalf) of Cherry Hill, NJ.

**LANbridge** A dual-port, high-speed bridge marketed by Digital Equipment Corp. which supports the IEEE 802.1 spanning tree technique for complex internetworking of Ethernet or fiber optic local area networks.

**LANCE** A trademark of Micro Technology of Anaheim, CA, as well as an SNMP-based multi-segment network management system from that vendor.

**landscape printing** The process of printing across the length of a page in comparison to portrait printing where printing occurs across the width of a page. The term "landscape" originated from landscape pictures, which are usually horizontal in format.

**Lanexpress** A local area network marketed by Lanex Corp. of Beltsville, MD.

**language** A set of terms or symbols used according to very precise rules to write instructions, or programs, for computers.

**LAN–PDL** Local Area Network–Program Design Language.

**LANSight** A trademark of LANSystems, Inc., of New York City, NY, as well as a workstation management program from that vendor which operates on Novell networks and which provides real-time diagnostic data about remote work-stations.

**LANtern** A trademark of Novell as well as the name of that vendor's LAN network management system.

**LANVista** A local area network diagnostic and management system marketed by Digilog, Inc., of Montgomery, PA.

**LanWORKS** Software developed by Digital Equipment Corporation and Apple Computer, Inc., which lets DEC VAX computers run native Apple Macintosh computer protocols. This software allows Macintosh computers to operate as a client on a VAX network.

**LAN2LAN** The name of a router manufactured by Newport Systems Solutions, Inc., of Newport Beach, CA.

**LAP** Link Access Procedure.

**LAP M** Link Access Procedure (LAP) M protocol is the primary error control technique specified in the CCITT V.42 standard.

**LAPB** Link Access Procedure-Balanced.

**LAPD** Link Access Protocol for the D-channel.

**LAPM** Link Access Procedure Modem.

**Large Message Performance Enhancement Outbound (LMPEO)** In IBM's VTAM, a facility in which VTAM reformats function management (FM) data that exceeds the maximum request unit (RU) size (as specified in the BIND) into a chain or partial chain of RUs.

**Large-Scale Integration (LSI)** A term used to describe a multi-function semiconductor device, such as a microprocessor, with a high density (up to 1000 circuits) of electronic circuitry contained on a single silicon chip. See following table for comparison of circuit density ranges.

| Scale | Circuit range |
| --- | --- |
| Small (SSI) | 2 to 10 circuits |
| Medium (MSI) | 10 to 100 circuits |
| Large (LSI) | 100 to 1,000 circuits |
| Very large | (VLSI)1000 to 10 000 circuits |
| Ultra large | (ULSI)Over 10 000 circuits |

**LARS** Learning and Recognition System.

**LASER (Light Amplification by Stimulated Emission of Radiation)** A device that transmits a very narrow and coherent (separate waves in phase with one another rather than jumbled as in natural light) single-frequency beam of electromagnetic energy in the visible light spectrum.

**LaserBridge** A local area network bridge marketed by Simple Net Systems, Inc., of Brea, CA.

**LASS** Local Area Signaling Services.

**Last-In-Chain (LIC)** In IBM's VTAM, a request unit (RU) whose request header (RH) end chain indicator is on and whose RH begin chain indicator is off.

**Last In, First Out (LIFO)** A method used to process the latest item to a queue first, without regard to the age of the oldest item.

**Last number dialed** A PBX or telephone set feature which allows a user to quickly redial the previously dialed number.

**LAT** Local Area Transport.

**LATA** Local Access and Transport Area.

**latency** In local area networking, the time (measured in bits at the transmission rate) for a signal to propagate around or throughout the network.

**Lattisnet** A 10-Mbps Ethernet local area network marketed by Synoptics Communications of Mountain View, CA. Lattisnet supports a cabling distance of 330 feet between workstations and a wiring center using unshielded twisted pair wiring.

**Laverne** The name of a voice-mail system marketed by Granite Telecom Corporation of Manchester, NH.

**layer** Refers to a collection of related network-processing functions, which comprise one level of hierarchy of functions. Functions in the same layer coordinate their actions via protocols, while functions in adjacent layers cooordinate their actions via interfaces.

**LAYER** 1. One of the divisions of the OSI model (see following table). 2. One of the division of SNA and other communications protocols.

LAYERS OF THE OSI MODEL

| Layer | Description |
| --- | --- |
| 7. Application | Provides interface with network users. |
| 8. Presentation | Performs format and code conversion. |
| 5. Session | Manages connections for application programs. |
| 4. Transport | Ensures error-free, end-to-end delivery. |
| 3. Network | Handles internetwork addressing and routing. |
| 2. Data Link | Performs local addressing and error detection. |
| 1. Physical | Includes physical signaling and interfaces. |

**Layer Management (LM)** That part of a given layer (1 to 7) which manages the resources and parameters residing in its layer protocl entity.

**layered protocol** A protocol designed to obtain services from, and deliver services to other protocols in the manner described by the ISO layered model.

**lays** The twists in twisted pair cable. Two single wires are twisted together to form a pair; by varying the length of the twists, or lays, the potential for signal interference between pairs is reduced.

**L-band** A portion of the electromagnetic spectrum commonly used in satellite and microwave applications with frequencies approximately in the 1 GHz region. The frequency spectrum from 1.53 to 1.66 GHz is planned for mobile satellite services.

**LBM** Load Balance Module.

**LBO** Line Build Out.

**LCAMOS** Loop Cable Administration Maintenance Operations System.

**LCB** Line Control Block.

**LCC** Life Cycle Cost.

**LCD** 1. Liquid Crystal Display. 2. Line Current Disconnect.

**LCN** Logical Channel Number.

**LCP** Link Control Procedure/Protocol.

**LCT** Low Cost Terminal.

**LDC** Link Data Channel.

**LDDS** Local Digital Data Service.

**LDM** Limited Distance Modem.

**LDN** Listed Directory Number.

**LDSU** Local Digital Service Unit.

**LDSU2** Local Digital Service Unit—Model 2.

**learning bridge** In a local area network, a bridge which adaptively creates its own tables of network topology by performing an analysis of the traffic it processes.

**leased line** A telephone line reserved for the exclusive use of a leasing customer without interexchange switching arrangements. A leased line may be point-to-point or multipoint.

**least-cost routing** A method used to select the least expensive, long-distance carrier to send out a call.

**LEC** Local Exchange Carrier.

**LED** Light Emitting Diode.

**leakage** The current flowing through insulation. The symbol for leakage is the capital letter *G*.

**LEN** 1. Line Equipment Number. 2. Low Entry Networking.

**LEO** Low Earth Orbit.

**LEQ** Line Equalizer.

**lerad exit routine** In IBM's VTAM, a synchronous EXLST exit routine that is entered automatically when a logic error is detected.

**Letter Quality (LQ) print mode** A method of printing on a dot-matrix printer in which the number of dots used to form each character is increased to increase the print quality; however, the extra dots decrease the print speed.

**Letters Shift (LTRS)** 1. A physical shift in a terminal using Baudot Code that enables the printing of alphabetic characters. 2. The character that causes the shift.

**level** 1. Magnitude, as in signal level or power level. 2. Used as a synonym for layer.

**Level 1 router** In Digital Equipment Corporation Network Architecture (DECnet), a router which performs routing within a single area. Messages for destinations in other areas are routed to the nearest Level 2 router.

**Level 2 router** In Digital Equipment Corporation Network Architecture (DECnet), a router which acts as a Level 1 router within its own area, but in addition routes messages between areas.

**Level Of Repair (LOR)** The locations and facilities where items are to be repaired. Typical levels are operator, field technician, bench, and factory.

**LF** 1. Line Feed. 2. Low Frequency.

**LFACS** Loop Facilities Assignment And Control System.

**LHT** Long Holding Time.

**liaison** A virtual connection (like a virtual circuit) which can be set up between two transport stations. A concept used in the end-to-end transport protocol defined by International Network Working Group (INWG).

**LIC** Last-In-Chain.

**LIDB** Line Information Data Base.

**LIFO** Last In, First Out.

**Light Emitting Diode (LED)** A semiconductor which emits incoherent light for a p–n junction (when biased with an electrical current in the forward direction). Light may exit from the junction strip edge or from its surface (depending on the device's structure).

**LightFax** A CCITT Group 3 facsimile and 2400 bps modem in one housing which is marketed by Computer Friends, Inc., of Portland, OR.

**lightguide** Optical waveguide.

**lightguide cable** An optical fiber, multiple fiber, or fiber bundle which includes a cable jacket and strength members, fabricated to meet optical, mechanical, and environmental specifications.

**Lightnet** A fiber-optic communications carrier based in Rockville, Md.

**lightwave** Electromagnetic wavelengths of approximately 0.8 to 1.6 micrometres in the region of visible light.

**LIM** 1. Line Interface Module. 2. Lotus/Intel/Microsoft expanded memory specification.

**Limited Distance Modem (LDM)** A signal converter which conditions and boosts a digital signal so that it may be transmitted much further than a standard RS-232 signal.

**limiter** The part of a frequency modulation receiver that eliminates all variations in carrier amplitude, thus removing all noise present in the carrier as amplitude modulation.

**line** 1. A communication medium connecting two or more points. 2. A physical path which provides direct communications among some number of stations.

**Line Access Procedure (LAP)** In packet switched networks, superseded by LAPB.

**line adapter** A device used for switching between switched (DDD) backup and dedicated lines. This adapter compensates for differences in receive and transmit power levels.

**line analysis** The process of measuring telecommunication circuit (line) parameters and analyzing the condition and quality of the circuit.

**Line Build Out** 1. An attenuator which stimulates cable loss with a frequency rolloff. The amount of attenuation is specified in units of dB. 2. An electronic simulation of a length of wire line that reduces the signal power so that it falls within certain defined limits. On a T1 circuit a LBO is used to reduce the potential for one T1 transmitter to "crosstalk" into the receiver of other services within the same cable binder and to accommodate the first span line repeater's receiver. The communications carrier establishes the appropriate LBO setting and instructs the enduser to select an appropriate Channel Service Unit setting.

**line conditioning**    Conditioning is a procedure that is used to make the levels of transmission impairments fall within specified limits. Line conditioning can be used on a leased line circuit to improve transmission quality. The level of conditioning required is a function of both speed and distance. Several different types are offered which result in a higher transmission rate and/or a reduction in data errors. Conditioning cannot ensure a "clean" line, but that distortion will fall within the limits prescribed.

**line control**    In IBM's SNA, the scheme of operating procedures and control signals by which a data link is controlled. For example, sychronous data link control (SDLC). Synonym for data link control protocol.

**Line Control Block (LCB)**    An area of main storage containing control data for operations on a line. The LCB can be divided into several groups of fields; most of these groups can be identified as generalized control blocks.

**Line Control Unit (LCU)**    A communications controller.

**line discipline**    Archaic term for communications protocol, e.g., the sequence of operations involving the actual transmission and reception of data.

**line driver**    A signal converter which conditions the digital signal transmitted by an RS-232 interface to ensure reliable transmission beyond the 50-foot RS-232 limit and often up to several miles; it is a baseband transmission device.

**Line Equalizer (LEQ)**    A fixed equalizer used to improve transmission and the effectiveness of a modem. LEQ offsets the sloping high frequency rolloff of a metallic local balanced pair cable, which is a function of cable length. LEQ may improve transmission on long cable runs which can experience greater losses and rolloff.

**Line Feed (LF)**    A control character used to move to the next line on a printer or display terminal.

**line finder**    A switch designed to find a calling line among a group of lines and then connect it to another device.

**line folding**    The procedure which is necessary when a text message has a line longer than the maximum allowed by a printer. The excess characters are printed on the next line by generating a local new line signal. The appearance of the message is marred, but its sense is preserved.

**line group**    In IBM's SNA, one or more communication lines of the same type.

**line hits**    An incident of electrical interference causing unwanted signals to be introduced onto a transmission circuit. There are four different types of line hits (or momentary electrical disturbances) on the line caused by atmospheric conditions, telephone company switching equipment, and radio/microwave transmission.

**Line Information Data Base (LIDB)**    AT&T software which provides the message formats and equipment interface for an operator service position system. The LIDB is managed and administrated by the telephone company.

**line load control**    A PBX or central office service feature permitting the selective denial of call origination to certain lines when excessive demands for service are required of a switching center.

**line monitor**    A device capable of passively intercepting a data transmission and, based on a prior knowledge of the protocol in use and other factors, providing its operator with a display and analysis of the transmission.

**line of sight transmission**    A characteristic of some

open-air transmission technologies where the area between a transmitter and a receiver must be unobstructed. Examples of line of sight transmission include microwave, laser, and infrared.

**line printer**  A device that prints all the characters of a line as a unit.

**line probe**  A generic term for the IBM 3867 Link Diagnostic Unit, a device that provides the Network Problem Determination Application (NPDA) user with line quality data and other link information.

**Line Problem Determination Aid (LPDA)**  IBM modem resident software that provides line and device status information to that vendor's Network Problem Determination Application (NPDA).

**line processing**  An ordered series of transactions that are processed as they are received.

**Line Processing Unit (LPU)**  A card in the Telenet Processor (TP) from which lines (links) to the network or users emanate. LPUs can contain four or eight lines and can run at low, medium, or high speed.

**line protocol**  A set of rules used to organize and control the flow of information between two or more stations connected by a common transmission facility.

**Line Protocol Handler (LPH)**  A communications program that processes messages, interrupts, and timeouts; handles protocol acknowledgements, error recovery and other communications functions.

**Line Quality Analysis (LQA)**  Diagnostic software that operates on a network management system in conjunction with modems to provide the measurement and display of analog line parameters.

**line segment scrambling**  A method of video signal scrambling in which segments of lines are moved to other lines, totally obscuring the picture.

**line sharing**  A form of X.21 switched line sharing in which many clients have access to a line, but only one client has access to any single call.

**line shuffle scrambling**  A method of video signal scrambling in which lines are randomly interchanged within the image.

**line speed**  The transmission rate of signals over a circuit, usually expressed in bits per second.

**line splitter**  A device which splits a single line among a cluster of terminals; a modem sharing unit.

**line switching**  Switching in which a current path is set up between the incoming and outgoing lines. Contrast with message switching in which no such physical path is established.

**line test set**  Analog test equipment that measures characteristics of a circuit: level, frequency, noise, signal-to-noise ratio, and equalization.

**Line TurnAround (LTA)**  On a two-wire circuit, the time required for one end to stop transmitting and then start receiving from the other end.

**Linear Predictive Coding (LPC)**  A voice digitization technique in which speech parameters to include pitch, voice, and unvoiced sound parameters are used to develop an algorithm that can be encoded at a low data rate, typically 2400 or 4800 bps.

**linearity**  The property of a transmission medium or of an item of equipment that allows it to carry signals without introducing distortion.

**Linelock**  A communications program from Crystal Point Inc. of Kirkland, WA, which monitors and controls modem usage. Linelock operates on Ungermann–Bass networks and allows users to access a network modem only after they enter their log-in identification and password, and identify the destination of their call.

**line-mode data**  A type of data that is formatted on a physical page by a printer only as a single line.

**link**  1. Communications circuit or transmission path connecting two points. 2. In IBM's SNA, the combination of the link connection and the link stations joining network nodes; for example: (1) a System/370 channel and its associated protocols, (2) a serial-by-bit connection under the control of Sychronous Data Link Control (SDLC). Synonymous with data link. *Note.* A link connection is the physical medium of transmission; for example, a telephone wire or a microwave beam. A link includes the physical medium of transmission, the protocol, and associated communication devices and programming; it is both logical and physical.

**Link Access Procedure (LAP)**   The data link-level protocol specified in the CCITT X.25 interface standard; supplemented by LAPB (LAP-Balanced) and LAPD.

**Link Access Procedure-Balanced (LAPB)**   In X.25 packet switched networks, a link initialization procedure which establishes and maintains communications between the DTE and DCE; LAPB involves the T1 timer and N2 count parameters. All public data networks (PDNs) now support LAPB.

**Link Access Protocol for the D-Channel (LAPD)** The standard Link layer protocol which defines the transmission and reception of information frames, the detection of errors and their correction by retransmission. LAPD is synonymous with the CCITT ISDN international standard Q.921.

**link budget**   In a fiber optic system link budget is a term used to define the maximum allowable light loss for a path from an optical transmitter to an optical receiver. The link budget is the difference, in dB, between the transmitter output and the maximum usable input level of the photodetector.

**link communication**   The physical means of connecting one location to another for the purpose of transmitting and receiving information.

**link connection**   In IBM's SNA, the physical equipment providing two-way communication between one link station and one or more other link stations; for example, a communication line and data circuit terminating equipment (DCE).

**Link Control Procedure/Protocol (LCP)**   A standard procedure by which data is transferred over any communications link to ensure order and accuracy.

**Link Data Channel (LDC)**   A channel of 4 Kbps carried in the framing bits of the 4th, 8th, 12th, 16th, 20th, and 24th frames of a T1 channel using the Extended Superframe format. AT&T has specified that the Link Data Channel on its circuits will carry circuit performance information in a proprietary protocol. CCITT (now ITU) Recommendation G.703 makes no provision for the contents of the LDC.

**link exchange**   A technique for improving network topology by trying the effect of substituting one link for another.

**Link layer**   The logical entity in the OSI model concerned with transmission of data between adjacent network nodes; the second layer processing in the OSI model, between the Physical and the Network layers.

**link level 2 test**   Same as link test.

**link loopback**   A diagnostic technique in which a signal transmitted at the aggregate bit rate of a digital link (e.g. T1) is returned to the sending device in the opposite direction on the same link. Link loopback tests can be performed by some BERT testers, and by some Channel Service Units equipped for diagnostic testing.

**link margin**   In fiber optic systems link margin is a term used to define the margin of available optical power above the minimum usable input level at the photodetector after all path losses are subtracted. The link margin defines the safety margin in a fiber optic system for emergency splices and other alternatives.

**Link Node**   A local area network bridge marketed by Wellfleet Communiactions, Inc., of Bedford, MA.

**Link Problem Determination Aids (LPDA)**   A series of testing procedures initiated by an IBM Network Control Program (NCP) that provide modem status, attached device status, and the overall quality of a communications link.

**Link State PDU (LSP)** A protocol data unit (PDU) used by the routing algorithm to exchange information about each system's neighbors.

**Link State Protocol** A routing protocol in which routing information is transmitted to reflect changes to network connections.

**link station** 1. In IBM's SNA, the combination of hardware and software that allows a node to attach to and provide control for a link. 2. In IBM's ACF/VTAM, a named resource within a subarea node representing another subarea node directly attached by a cross-subarea link. In the resource hierarchy, the link station is subordinate to the cross-subarea link. *Note.* An SDLC link station is defined in an NCP subarea node with a PU PUTYPE=4 macro in the NCP definition. A channel link station is defined dynamically by ACF/VTAM when a channel-attached NCP is activated.

**Link Telecommunications** A subsidiary of Bell-South Enterprises, the holding company for all unregulated BellSouth companies which provides paging and telephone answering service in Australia.

**link test** In IBM's SNA, a test in which one link station returns data received from another link station without changing the data in order to test the operation of the link. *Note.* Three tests can be made; they differ in the resources that are dedicated during the test. A link test, level 0 requires a dedicated subarea node, link, and secondary link station. A link test, level 1 requires a dedicated link and secondary link station. A link test, level 2 requires only the dedicated link station.

**link-attached** A term used to describe devices that are connected to a communications link or telecommunications circuit. Compare with channel-attached.

**link-attached communications controller** In IBM's SNA, an IBM communication controller that is attached to another communication controller by means of a link.

**Link/Design** A support service of Timeplex, Inc., of Woodcliff Lake, NJ, to both existing and prospective customers that can be used to reconfigure and expand an existing Timeplex network or to create a new network from scratch.

**Linkletter** The name of a bimonthly NSFNET newsletter published by Merit Network Information Center.

**Liquid Crystal Display (LCD)** A type of display screen used primarily in laptop computers. A reflective LCD screen reflects existing light while a backlit LCD screen has its own source of light.

**list** A data structure in which each item of data can contain pointers to other items. Any data structure can be represented in this way, which allows the structure to be independent of the storage of the items.

**Listserv** An automated system that contains BITNET discussion lists.

**LIU** Line Interface Unit.

**Liv Zempel** A data compression algorithm developed in 1977 in which strings of characters are replaced by fixed-length code words.

**LL** Longest Lobe.

**LLC** Logical Link Control.

**LLF** Low Layer Functions.

**LLN** Line Link Network.

**LM** Layer Management.

**LMOS** Loop Management Operations System.

**LMOS** Line Maintenance Operation System.

**LMP** Loopback Mirror Protocol.

**LMPEO** Large Message Performance Enhancement Outbound.

**LMSS** Land Mobile Satellite Service.

**LMU** LAN Manager for UNIX.

**LNA** Low Noise Amplifier.

**LNC** Low Noise Converter.

**LNI**  Local Network Interface.

**LO**  Line Occupancy.

**load**  1. To move data or programs into memory. 2. To place a diskette into a diskette drive. 3. To insert paper into a printer. 4. Usage.

**load coils**  Coils used to reduce the attenuation distortion by making the attenuation nearly constant across the frequency band of 300 to 3000 Hz.

**load module**  In ISO, a program unit that is suitable for loading into main storage for execution; it is usually the output of a linkage editor.

**loaded line**  A telephone line equipped with coils (called load coils or loading coils) which minimize voice frequency amplitude distortion by restoring the response at the higher frequencies within the voice bandwidth. To be used only with analog voice. Digital data cannot be used on these lines without experiencing severe attenuation of the data signal.

**loaded wire**  Local loops installed between a telephone company central office and a subscriber which have loading coils installed.

**loading**  Adding inductance by the use of load coils to a transmission line to minimize amplitude distortion.

**loading coil**  An induction device used in telephone local loops, generally exceeding 18 000 feet in length, that compensates for the wire capacitance and serves to boost voice-grade frequencies.

**load sharing**  A multiple-computer system that shares the load during peak hours. During nonpeak periods or standard operation, one computer can handle the entire load with the others acting as fallback units.

**lobe**  In the IBM Token-Ring Network, the section of cable (which may consist of several segments) that connects an attaching device to an access unit.

**lobe receptacle**  An outlet on an access unit for connecting a lobe.

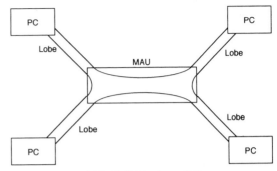

MAU = Multistation Access Unit

*lobe*

**LOC**  Lines of Communication.

**LOC**  Local Operating Company.

**LOC CD**  Local Carrier Detect (Control Character).

**local**  In IBM's SNA: 1. Synonymous with channel-attached. 2. Pertaining to a device that is attached to a controlling unit by cables, rather than by a communication line.

**Local Access and Transport Area (LATA)**  One of 161 USA geographical subdivisions used to define local (as opposed to long distance) telephone service.

**local address**  In IBM's SNA, an address used in a peripheral node in place of a network address and transformed to or from a network address by the boundary function in a subarea node.

**local analog loopback**  An analog loopback test that forms the loop at the line side (analog output) of the local modem.

**Local Area Data (LAD) channel**  Same as Bell 43401 circuit.

**Local Area Data Transport (LADT)**  A method by which AT&T customers can send and receive digital data over existing wires between their premises and a telephone company office. Dial-up LADT lets customers use their lines for occasional data services. Direct Access LADT transmits simultaneous voice and data traffic on the same lines.

245

**Local Area Transport (LAT)** A proprietary Digital Equipment Corp. protocol used by that vendor's terminal servers which accept characters from terminals and multiplexers them into a single packet for transmission to a VAX computer.

**Local Area Network (LAN)** A data communications network confined to a limited geographic area (up to 6 miles or about 10 kilometers) with moderate to high data rates (100 Kbps to 50 Mbps). The area served may consist of a single building, a cluster of buildings, or a campus-type arrangement. It is owned by its user, includes some type of switching technology, and does not use common carrier circuits—although it may have gateways or bridges to other public or private networks. Because it uses physical media (wires or coaxial cables) owned by the operator and does not normally cross public roads, it is not regulated by a body such as the FCC.

**Local Area Signaling Services (LASS)** Intra-LATA voice and data services using the AT&T 1A ESS switch and data services using the 1A ESS switch and signal transfer point (STP).

**Local Automatic Message Accounting (LAMA)** A PBX or central office service which results in the recording of accounting messages for dialed toll or long distance calls.

**local battery telephone system** A telephone system in which batteries are installed at each substation.

**local attachment** In IBM environments, the connection of a peripheral device or control unit directly to a host channel.

**local bridge** A bridge is a combination of hardware and software used to connect two Local Area Networks (LANs).

**local call** 1. Any call utilizing a single switching facility. 2. Any call within a local charging area.

**local central office** A common carrier switching office in which users' lines terminate. Also called local exchange or end office.

**local channel** A cable pair within the cable that goes from the building complex to the telecommunications carrier's office. The local channel can be a major source of trouble for data services.

**local channel loopback** A channel loopback test that forms the loop at the input (channel side) to the local multiplexer.

**local composite loopback** A composite loopback that forms the loop at the output (composite side) of the local multiplexer.

**local dataset** A signal converter which conditions the digital signal transmitted by a RS-232 interface to ensure reliable transmission over a DC continuous metallic circuit without interfering with adjacent pairs in the same telephone cable. Normally conforms with Bell Publication 43401. Also called baseband modem, limited distance modem, local modem, or short haul modem.

**local digital loopback** A digital loopback test that forms the loop at the DTE side (digital input) of the local modem.

**local echo** The ability of a device to echo, or send a copy of the incoming data stream back to the sending device.

**local echoplex** A method of checking data transmission accuracy whereby data characters are returned to the sending terminal's display screen for comparison with the original transmitted data. Some host computers can provide echoplex to the terminal; however, local echoplex is especially desirable when operating over links where long time delays are encountered, e.g., satellite communications.

**local exchange, local central office** The exchange or central office in which the subscriber's lines terminate.

**local exchange company/local operating company** The local telephone company responsible for service within LATAs.

**local line** A term equivalent to local loop.

**local loop** A channel connecting the subscriber's

equipment to the line terminating equipment in the central office, usually a metallic circuit, either 2-wire or 4-wire.

**Local Management Interface (LMI)** A protocol which operates in one dedicated permanent virtual circuit (PVC) of a frame relay link and enables the subscriber and network to exchange information about the link and the status of the other PVCs.

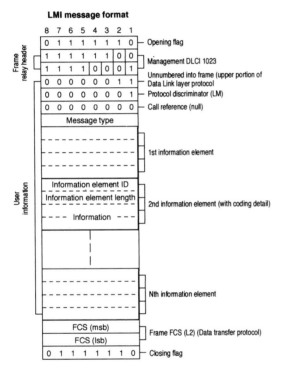

**LMI message format**

local name That part of the entity name which identifies an entity within a node.

**Local Network Interface (LNI)** A device that attaches to a transceiver on a LAN cable and branches out to multiple ports; both transmits and receives signals.

local non-SNA major node In IBM's ACF/VTAM, a major node whose minor nodes are channel-attached non-SNA terminals.

local operating company The telephone company supplying local services to a customer.

local oscillator An oscillator that is a part of the receiver and is used to generate an RF output which is combined with the incoming RF signal to produce an intermediate frequency.

local service area The entire area within which a customer may call at the local rates without incurring toll charges.

**Local Session Identification (LSID)** In IBM's SNA, a field in a FID3 transmission header that contains an indication of the type of session (SSCP–PU, SSCP–LU, or LU–LU) and the local address of the peripheral logical unit (LU) or physical unit (PU).

local SNA major node In IBM's ACF/VTAM, a major node whose minor nodes are channel-attached peripheral nodes.

local 3270 major node See local non-SNA major node (IBM's SNA).

local-attached In IBM's SNA, pertaining to a device that is attached to a controlling unit by cables, rather than by a telecommunication line. Synonym for channel-attached.

locally administered address A temporary address assigned to a token-ring adapter which overrides a universally administered address.

**LocalTalk** The cabling scheme used in an AppleTalk local area network. LocalTalk uses RS-422 signaling and has a rated transmission speed of 230.4 Kbps. LocalTalk cabling uses shielded twisted-pair wire and is limited to a maximum of 32 connections per network.

location sharing A Software-Defined Network (SDN) service feature offered by AT&T which permits a user to share access circuits to another customer's SDN location.

lockout The inability of one or both telephone users to get through due to excessive local circuit noise or because of continuous speech from either or both users.

lock-up 1. An unwanted state of a system from

which it cannot escape, such as a "deadly embrace" in the claiming of common resources. 2. A condition in which a steady mark is received when a device should be receiving an input signal.

**LOCUS**   Library of Congress Information System.

**logfile**   An exact duplicate of a program session recorded in a file.

**logging**   The act of recording something for future reference such as error events or transactions. Also log, tape, logrec.

**logic**   In computer programming, the procedure used to perform a task or solve a problem. In computer hardware, the circuits that carry out logical operations and do arithmetic.

**logic channel, logical connection**   See virtual circuit.

**logic error**   In IBM's ACF/VTAM, an error condition that results from an invalid request; a program logic error.

**logical channel group number and logical channel number**   1. Specifies the route a packet will take through an X.25 packet switched network; contained in the first and second octet of the packet header (12 bits). 2. In packet switched networks, a number assigned when a virtual call is placed; up to 4095 independent logical channels may exist on a single link.

**logical connection**   1. A call following a physical connection to a packet network, which establishes communication with another device in the network. A logical call continues until the user initiates a disconnect request. 2. Link established by a virtual circuit.

**logical group, logical group number**   In packet switched networks, logical channels are divided into one of 16 logical groups.

**Logical Link Control (LLC)**   A data link protocol based on HDLC developed for local area networks. The protocol was developed by the IEEE 802 committee and is common to all of its LAN standards for data link-level transmission control.

**logical record**   A collection of items independent of their physical environment. Portions of the same logical record may be located on other physical records.

**logical screen buffer**   An area to which applications may write keystrokes and copy data to be communicated to a System/370 program in the PC 3270 emulation environment.

**Logical Terminal (LT)**   In Multiple Logical Terminals (MLT), one of five sessions available to share one display station.

**Logical Unit (LU)**   The combination of programming and hardware of a teleprocessing subsystem that comprises a terminal. 2. On an SNA network, a type of network-addressable unit that represents end users (application programs or operators at a device) to the network; a collection of programs that provide the interface through which end users access network resources and then manage information transmission between end users. Current IBM SNA LU session types are listed in the following table.

SNA LU SESSION TYPES

| LU type | Session type |
| --- | --- |
| LU1 | Host application and a remote batch terminal |
| LU2 | Host application and a 3270 display terminal |
| LU3 | Host application and a 3270 printer |
| LU4 | Host application and SNA word processor or between two terminals via mainframe |
| LU6 | Between applications programs typically residing on different mainframe computers |
| LU6.2 | Peer-to-peer |
| LU7 | Host application and a 5250 terminal |

**logical unit services**   In IBM's SNA, capabilities in a logical unit to: (1) receive requests from an end user and, in turn, issue requests to the system services control point (SSCP) in order to perform the requested functions, typically for session initiation; (2) receive requests from the SSCP, for example to activate LU–LU sessions via bind session requests;

and (3) provide session presentation and other services for LU–LU sessions.

**logical unit 6.2**  A logical unit used to implement program-to-program communications.

**logmode table**  In IBM's ACF/VTAM, a logon mode (logmode) table is a set of macros that repesent the session protocols that are to be used in an SNA session.

**logoff**  1. The procedure by which a user ends a terminal session. 2. In IBM's ACF/VTAM, an unformatted session-termination request.

**logon**  1. The process of establishing communications with a computer, including identification of the user and verification (by use of a password) of the user's identity. 2. In IBM's VTAM, a request that a terminal be connected to a VTAM application program.

**logon data**  In IBM's ACF/VTAM: 1. The user data portion of a field-formatted or unformatted session-initiation request. 2. The entire logon sequence or message from an LU. Synonymous with logon message.

**logon-interpret routine**  In IBM's ACF/VTAM, an installation exit routine, associated with an interpret table entry, that translates logon information. It may also verify the logon.

**logon message**  Synonym for logon data (IBM's ACF/VTAM).

**logon mode**  In IBM's ACF/VTAM, a subset of session parameters specified in a logon mode table for communication with a logical unit.

**logon mode table**  In IBM's ACT/VTAM, a set of entries for one or more logon modes. Each logon mode is identified by a logon mode name.

**LOG-TIME**  A trademark of Computronics of Addison, IL, as well as an accounting system from that vendor which operates on Prime computers.

**long distance access code**  A code used to gain access to a long distance network.

**long distance call**  A call placed by a subscriber in one area code to another subscriber in a different area code.

**long distance company/carrier**  Those telephone companies which provide services between LATAs and, in some cases, international service.

**long distance service information**  As indicated in the following table, communications carriers in the United States have toll-free telephone numbers that subscribers and potential users can call for information about their services.

| | |
|---|---|
| AT&T | 1-800-222-0300 |
| ITT | 1-800-526-3000 |
| Metromedia Long Distance | 1-800-292-1052 |
| Sprint | 1-800-521-4949 |

**long haul**  Long distance telephone circuits that cross out of the local exchange, generally applied to any inter-LATA circuits.

**long line**  A communication line of a long distance.

**long wavelength**  A term used to reference the spectrum from 1200 to 1600 nanometers.

**longitudinal parity check**  See Longitudinal Redundancy Check.

**Longitudinal Redundancy Check (LRC)**  An error-detection method in which the Block Check Character (BCC) consists of bits calculated on the basis of odd or even parity for all the characters in the transmission block. The first bit of the LRC is set to produce an odd (or even) number of first bits that are set to 1; the second through eighth bits are set similarly. Also called horizontal parity check.

**loop**  1. Instructions in a program that cause a computer to repeat an operation until a task is completed or until a predetermined condition is achieved. 2. Local transmission that connects your telephones to the nearest central office; also known as subscriber loop.

**loop adapter**  In IBM's SNA, a feature of the 4331 Processor that supports the attachment of a variety of SNA and non-SNA devices. To ACF/VTAM, these devices appear as channel-attached devices.

**loop antenna**  An antenna consisting of one or more complete turns (loops) of wire. It is designed for directional transmission or reception.

**Loop Cable Administration Maintenance Operations System (LCAMOS)**  AT&T software designed to receive outside plant-related messages for the purpose of detecting cable problems before they become service-affecting to the customer.

**loop checking**  A method of checking the accuracy of transmission of data in which the received data are returned to the sending end for comparison with the original data, which are stored there for this purpose.

**loop current**  1. The current from a battery at a central office switch formed by closure of the phone hook switch. It is the presence or absence of the current that enables the automatic equipment in the central office to observe the telephone set's operating status. 2. A teletypewriter to line interface and operating technique without modems.

**Loop Facilities Assignment and Control System (LFACS)**  AT&T software which maintains a mechanized inventory of outside plant facilities and provides mechanized assignment capability for these facilities.

**Loop Management Operation System (LMOS)**  An AT&T system which automates the entry of trouble tickets into the telephone company's repair service bureau and provides a mechanism to monitor the status of repairs.

**loop network**  A central network topology that includes a continuous circuit connecting all nodes in which messages are routed around the loop to and through a central controller.

**loop start**  Method of signaling an off-hook condition between an analog telephone set and switch. Picking up the receiver closes a loop, allowing DC current to flow.

**loopback**  Type of diagnostic test in which the transmitted signal is returned to the sending device after passing through all, or a portion of, a data communications link or network. A loopback test permits the comparison of a returned signal with the transmitted signal. In a digital transmission system, the electrical connection of a CSU's receive circuit back to its transmit circuit. With this connection made, all signals received by the CSU are retransmitted back to the sending unit which allows one-ended testing. Loopbacks may be made by sending an Up-Loop code and broken with a Down-Loop Code.

**Loopback Mirror Protocol (LMP)**  A Digital Network Architecture (DNA) protocol which allows a test message to be transmitted over a logical link and looped back to the sending node.

**loopback test**  Type of diagnostic test in which the transmitted signal is returned to the sending device after passing through all, or a portion of, a data communications link or network; this allows a technician (or a built-in diagnostic circuit) to compare the returned signal with the transmitted signal. This comparison provides the basis for evaluating the operational status of the equipment and the transmission paths through which the signal traveled. (See figure p. 241)

**loosely coupled**  term used to describe processors that are connected by means of channel-to-channel adapters (IBM).

**LOR**  Level Of Repair.

**LOS**  1. Line Of Sight. 2. Loss Of Signal.

**Loss**  1. A measure of calls lost due to congestion. Often expressed as "grade of service," e.g. 1 lost call in 100. 2. Loss due to impedance of transmission medium and circuits. Expressed as decibels (dB) with respect to a reference; opposite of gain.

**loss of signal (LOS)**  An error condition on a T1 circuit defined by AT&T as the absence of a DS1 signal on the circuit for more than 150 milliseconds. A Loss-of-Signal condition triggers a Blue Alarm in the receiving hardware.

**loss of synchronization**  An error condition on a framed T1 circuit, detected when two or more consecutive framing bits are received in error.

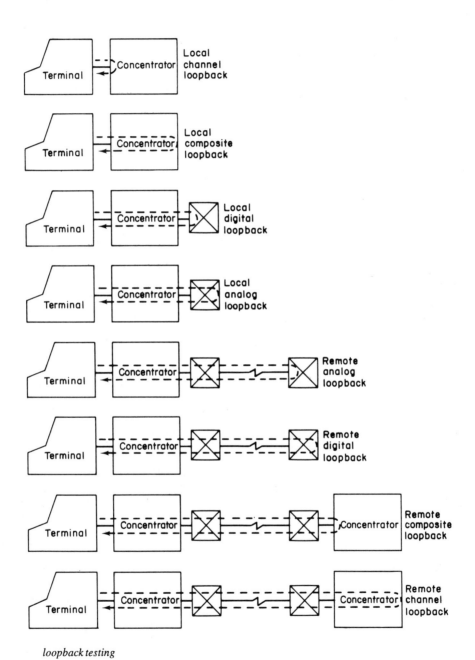

*loopback testing*

**loss set**   In fiber optic systems, an instrument used for measuring the relative attenuation of optical fibers.

**lost call**   A call that has not been completed for any reason other than cases where the called party is busy.

**loudness of sound**   The intensity or strength of the sensation sound causes in the human ear.

**Low Entry Networking (LEN)**   A peer-oriented extension to IBM's SNA first implemented on the System/36. LEN allows networks to be much more easily built and managed due to dynamic route selection, data base exchange, and other features in the extension.

**low earth orbit**   In satellite communications, an orbit often referred to as a parking orbit. It is circular and covers a range from 150 km to 300 km from the earth's surface.

**Low Frequency (LF)**   A portion of the electromagnetic spectrum with frequencies of approximately 30 to 300 KHz.

**Low Noise Amplifier (LNA)**   An amplifier used in conjunction with a satellite receiver to amplify extremely weak satellite signals without the introduction of noise.

**Low Noise Converter (LNC)**   Equipment that combines a low noise amplifier (LNA) and down-converter in one package.

**low pass**   A specific frequency level, below which a filter will allow all frequencies to pass (opposite of high pass).

**low-speed line**   A narrowband line which is usually used for telegraph transmissions and teletypewriter transmissions.

**Low-Speed Line Processing Unit (LSLPU)**   A card of the Telenet Processor (TP) computer. It interfaces asynchronous hosts and terminals to the network using start/stop protocol.

**LPC**   1. Linear Predictive Coding. 2. Longitudinal Parity Check.

**LPDA**   Link Problem Determination Aids (IBM's SNA).

**LPH**   Line Protocol Handler.

**LQ**   Letter Quality (printing).

**LQA**   Line Quality Analysis.

**LRC**   Longitudinal Redundancy Check.

**LRE**   Low Rate Encoding.

**LSD**   Line Signal Detect (control character).

**LSI**   Large Scale Integration; Large Scale Integrated (Circuit).

**LSID**   Local Session IDentification (IBM's SNA).

**LSLPU**   Low-Speed Line Processing Unit.

**LSM**   Local Switching Module.

**LSP**   Link State PDU.

**LT**   Logical Terminal.

**LTA**   Line Turnaround.

**LTP**   Logical Test Port.

**LTRS**   1. Letter Shift (teletypewriters). 2. Letters.

**LTS**   1. Loop Testing System. 2. Line Test Set.

**LU**   1. Line Unit. 2. Logical Unit.

**LU connection test**   In IBM's SNA, a diagnostic aid that permits a terminal operator to check whether the path between a system services control point (SSCP) and logical unit (LU) is operational.

**LU–LU session**   In IBM's SNA, a session between two logical units in an SNA network. It provides communication between two end users, or between an end user and an LU services component.

**LU–LU session type**   In IBM's SNA, the classification of an LU–LU session in terms of the specific subset of SNA protocols and options supported by the logical units (LUs) for that session, namely: (1) the mandatory and optional values allowed in the session activation request, (2) The usage of data stream controls, FM headers,

RU parameters, and sense codes, (3) Presentation services protocols such as those associated with FM header usage. LU–LU session types 0,1,2,3,4,6, and 7 are defined *Note*. At session activation, one LU–LU half-session selects the session type and includes or excludes optional protocols of the session type by sending the session activation request, and the other half-session concurs with the selection by sending a positive response or rejects the selection by sending a negative response. In LU–LU session types 4 and 6, the half-sessions may negotiate the optional parameters to be used. For the other session types, the primary half-session selects the optional protocols without negotiating with the secondary half-session.

**LU type**   In IBM's SNA, deprecated term for LU–LU session type.

**LU 6.2**   Logical Unit 6.2.

**Luma**   A videophone manufactured by Mitsubishi Electric Company aimed at business users.

**LUTS**   Locked-Up Trunk Scan.

**LWIR**   Long Wavelength InfraRed.

# M

**m** 1. meter (or meters). 2. Milli; designation for one thousandth.

**M** Mega; designation for one million.

**M bit** The More Data mark in an X.25 packet which allows the DTE or the DCE to indicate a sequence of more than one packet.

**ma** Milliampere.

**MAC** Media (Medium) Access Control.

**MacAPPC** Apple Computer Corporation's implementation of IBM's Advanced Program-to-Program Communications (APPC) protocol.

**MacBlast** A communications software program designed by Communications Research Group of Baton Rouge, LA, which operates on the Apple Macintosh series of personal computers.

**Macintosh** A series of Apple Computer Corporation microcomputers that represent the first wide-scale use of a window-based display system using icons and a mouse for the selection of specific operations.

**Macintosh Communications Toolbox** A set of standard communications interfaces developed by Apple Computer Company for use by Macintosh programmers.

**MacIRMA** A 3270 emulation board for the Macintosh SE manufactured by Digital Communications Associates (DCA) of Alpharetta, GA. MacIRMA provides a Macintosh computer with the abiliy to access an IBM 3270 network.

**macro instruction** In IBM's SNA: 1. An instruction in a source language that is to be replaced by a defined sequence of instructions in the same source language. The macro instruction may also specify values for paramenters in the instructions that are to replace it. 2. In assembler programming, an assembler language statement that causes the assembler to process a predefined set of statements called a macro definition. The statements normally produced from the macro definition replace the macro instruction in the program.

**Macrolink** The name of Australia's first tariffed ISDN service.

**Macstar end customer management system** A trademark of AT&T for a system which performs customer station rearrangements. The end customer can rearrange and/or move analog, Centrex, and all digital lines as well as perform automatic route selection with Macstar.

**MacTAPS** Software from Total Tec Systems of Edison, NJ, which transforms a Macintosh into a front end workstation to a Digital Equipment Corporation VAX computer. Users can access data and execute applications on a VAX via the Macintosh icon user interface.

**Mac3270** A terminal-emulation and file-transfer communications package marketed by Simware Inc. of Ottawa, Canada. Also a trademark of Simware Inc.

**magic name** A command that a bulletin board

255

system is configured to recognize which then provides the requester with a specific file or set of files.

**Magnetic Ink Character Recognition (MICR)**  A method of character recognition in which printed characters containing particles of magnetic material are read by a scanner and converted into a computer readable digital format.

**magnetic medium**  Any data-storage medium, including disks, diskettes, and tapes.

**magnetic stripe**  A stripe of magnetic material, similar to a piece of magnetic tape, usually affixed to a credit card, badge or other portable item, on which data is recorded and from which data can be read.

**Magyar Posta**  The Hungarian PTT.

**mail log**  In IBM's PROFS and Office Vision, a mail log is a user's personal file of information about documents that are in their PROFS storage.

**mail reader**  Software which enables a user to select unread mail and unread conference messages and have them downloaded for reading off-line. Most mail readers also permit users to create responses off-line and upload them at their convenience.

**mail reflector**  A special type of electronic mailbox which upon receipt of a message resends it to a list of other mailboxes. A mail reflector provides the ability to create a discussion group.

**mail server**  A computer system and associated software which performs the functions analogous to that provided by a post office box accessible by a number of persons. Users may send or forward electronic mail messages to anyone served by the system.

**Mailbus**  A set of software products marketed by Digital Equipment Corporation of Maynard, MA. Based on DEC's Message Router Software, Mailbus acts as a transport system and carries messages between different electronic mail (E-mail) systems.

**mailing list**  A list of persons or organizational ad-

dresses that a copy of a message when it is addressed to a mailing list.

**MailNet 400**  An electronic messaging system operated by P&T-Tele in Finland.

**main (PBX or Centrex)**  Switch into which other PBXs are homed.

**Main Distribution Frame (MDF)**  An equipment unit is used to connect communications equipment to lines, by use of connecting blocks.

**main network address**  In IBM's SNA, the logical unit (LU), network address, within ACF/VTAM that is used for SSCP–LU sessions and for certain LU–LU sessions.

**main program**  1. The highest-level program involved in a job. 2. The first program unit to receive control when a program is run.

**main ring path**  In a token-ring network, the part of the ring made up of access units and the cables connecting them.

**main station**  The telephone set a user employs to answer incoming calls and to send out calls. It is connected to a local loop.

**mainframe, mainframe computer**  A large-scale computer (such as those made by IBM, Unisys, Control Data, and others) normally supplied complete with peripherals and software by a single large vendor, often with a closed architecture. Also called host or CPU.

**mainline program**  In IBM's ACF/VTAM, that part of the application program that issues OPEN and CLOSE macro instructions.

**Mainlink**  A trademark of Quadram Corporation of Norcross, GA, as well as a terminal emulation system. Mainlink provides register-level compatibility with both Digtal Communications Associates IRMA and IBM 3270 emulation and can be obtained for use in IBM PC bus or IBM PS/2 microchannel configurations.

**MainLink**  A trademark of National Semiconductor Corporation as well as an adapter card which provides IBM 3270 access for MS-DOS and PC-DOS personal computers.

**Mainstreet**  A trademark of Newbridge Networks, Inc., of Herndon, VA, as well as that vendor's name for a series of networking products.

**Maintain System History Program (MSHP)**  In IBM's SNA, a program that facilitates the process of installing and servicing a VSE system. For NPDA, MSHP is required for installation in a VSE system.

**maintenance**  An activity intended to eliminate faults or to keep hardware or programs in satisfactory working condition.

**Maintenance Analysis Procedure (MAP)**  A maintenance document that gives an IBM service representative a step-by-step procedure for tracing a symptom to the cause of failure.

**Maintenance and Operator Subsystem (MOSS)**  In IBM's SNA, a subsystem of the 3725 Communication Controller that contains a processor and operates independently of the rest of the controller. It loads and supervises the 3725, runs problem determination procedures, and assists in maintaining both hardware and software.

**maintenance mode**  In DDCMP, the mode of operation used by Maintenance Operations Protocol (MOP).

**maintenance operations protocol**  In DDCMP, a management protocol used for low-level communications with a system which is not fully operational or which is being tested.

**maintenance services**  In IBM's SNA, network services performed between a host SSCP and remote physical units (PUs) that test links and collect and record error information. Related facilities include configuration services, management services, and session services.

**major alarm**  Any alarm condition that causes loss of two or more channels of customer data.

**major node**  In IBM's ACF/VTAM, a set of minor nodes that can be activated and deactivated as a group. This collection of minor nodes are all within one member of the VTAMLST file.

**male-to-female connector**  A connector typically used to connect a DTE interface to a modem or multiplexer, to connect a multiplexer's network port to its trunk modem, to or connect a modem to a digital patch device.

**male-to-male connector**  A connector typically used to connect a modem to a digital patch panel or to connect a DCE interface to a digital patch panel.

**MAN**  1. MANual. 2. Metropolitan Area Network.

**MAN SWBD**  MANual SWitchBoarD.

**Managed Data Network Services (MDNS)**  A proposed joint European Post, Telegraph and Telephone offering which will enable companies to have a "one-stop communications shopping facility." MDNS will provide a set of management tools for monitoring, billing and maintaining intercompany and intracompany communications through national packet networks in Western Europe.

**Management Domain (MD)**  In the CCITT (now ITU) Message Handling System (MHS) MDs are responsible for the routing of messages within its domain.

**Management Event Notification Protocol (MEN)**  An Application layer management protocol used in DNA Phase V for communication between an event dispatcher and an event sink.

**Management Information Control and Exchange (MICE)**  An Application layer management protol used in DNA Phase V.

**management services**  In IBM's SNA, management services are network services performed between a host SSCP and remote physical units (PUs) that include the request and retrieval or network statistics.

**Manbridge**  A local area network bridge marketed by Artel Communications Corp. of Hudson, MA.

**Manchester encoding**  A binary signaling mechanism that combines data and clock pulses. Each bit period in Manchester encoding is divided into two complementary halves; a negative-to-positive voltage transition in the middle of the bit period designates a binary "1" while a positive-to-negative transition represents a binary "0."

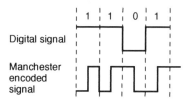

**mandatory cryptographic session**  In IBM's SNA, a synonym for "required cryptographic session."

**Manor System**  The name of a central hotel reservation system marketed by General Electric Information Services.

**manual answering**  In data communications, manual answering is performed by a person who hears the telephone ring, lifts the receiver, causes the called modem to send an answer tone to the calling modem and places the called modem into its data transmission mode of operation.

**manual calling**  In data communications, calling performed by a person who dials a number, waits for an answer, then places the calling modem into its data transmission mode of operation.

**manual teletypewriter network**  A manual teletypewriter network consists of two or more teletypewriter stations capable of direct communications on a common channel or frequency.

**Manufacturing Automation Protocol (MAP)**  In LAN technology, a token-passing bus designed for factory environments by General Motors; the IEEE 802.4 standard is nearly identical to MAP.

**Manufacturing Message Format Specification (MMFS)**  The Manufacturing Automation Protocol (MAP) 2.X equivalent of the Manufacturing Message Specification (MMS). MMFS was MAP 2.X specific, and was not an ISO standard.

**Manufacturing Message Specification (MMS)**  An International Standards Organization (ISO) standard Application layer protocol for communications among factory floor devices.

**MAP**  1. Maintenance Analysis Procedure. 2. Manufacturing Automation Protocol.

**MAP 2.1**  A broadband version of the Manufacturing Automation Protocol (MAP) which supports simultaneous multiple data channels.

**MAP 2.2**  A carrier band version of the Manufacturing Automation Protocol (MAP) which supports the physical, data, and application layers of the OSI model. MAP 2.2 is sometimes called Minimap or Enhanced Performance Architecture (EPA) because messages do not flow through all seven layers and are delivered faster with this protocol.

**MAP 3.0**  A version of the Manufacturing Automation Protocol (MAP) which offers functionality at all protocol layers of the OSI model.

**mapper**  An IBM Network Problem Determination Application (NPDA) function that executes as an NCCF subtask to record errors from resources attached to a 370X-EP or from certain channel-attached devices.

**mapping**  1. The establishment of one-to-one correspondence between two sets of data. 2. The translation of one group of data into a new set of data.

**Marathon**  The name used by Micom Communications Corporation of Simi Valley, CA, for a series of fast packet based servers and multiplexers.

**marconi antenna**  An antenna system in which the ground is an essential part.

**margin**  In an optical system, the amount of signal in excess of that required for a given level of performance.

**marine telephone**  Marine telephones operate on assigned radiotelephone frequencies much as a radio broadcast does. Marine telephones can be used to contact other marine telephones or to reach landbased telephones through an operator.

**mark** 1. In single-current telegraph communications, a mark represents the closed, current-flowing condition. 2. In data communications, a mark represents a binary 1; the steady-state, no-traffic state for asynchronous transmission. 3. The idle condition. 4. In the context of the virtual terminal, a mark is a signal inserted into an output data stream by the virtual terminal, to acknowledge that an attention or interrupt input signal has been received.

**marker** The common control device for a crossbar switch.

**mark-hold** The normal no-traffic line condition whereby a steady mark is transmitted.

**Markov constraint** A constraint on the routing method according to which the future route of a packet is independent of its past history, such as its source or its route so far. This constraint is implied by directory routing.

**mark-to-space transition** The transition, or switching from a marking impulse to a spacing impulse.

**MARS** Military Affiliate Radio System.

**MASER (Microwave Amplification by Simulated Emission of Radiation)** A device that generates signals in the microwave range, with low-noise characteristics.

**mask** Pattern of bits (1s or 0s) specified by the user that can be used with the trap mode of a communications test set.

**masking** A method of transforming one set of data into another while blocking or excluding some data from this process on the basis of code patterns or position.

**master clock** The source of timing signals, or the signals themselves, which all network stations use for synchronization.

**master cryptography key** In IBM's SNA, a

cryptographic key used to encipher operational keys that will be used at a node.

**master group** In Frequency Division Multiplexing (FDM), an assembly of 10 supergroups occupying adjacent bands in the transmission spectrum for purposes of simultaneous modulation and demodulation.

**master modem** In a multipoint system, the modem that transmits constantly in the outbound direction. Usually the modem at the central site. In a multitier system, the term represents a remote master, or a master modem that is not located at the central site.

**master station** 1. In multipoint circuits, the unit which controls/polls the nodes. 2. In point-to-point circuits, the unit which controls the slave station. 3. In LAN technology, the unit on a token-passing ring that allows recovery from error conditions, such as lost, busy, or duplicate tokens; a monitor station.

**MATE** An asynchronous communications software program from Concept Automation that operates on Data General MV computers.

**mathematical model** A mathematical description or approximation of some real event.

**matrix** In switch technology, that portion of the switch architecture where any input leads and any output leads meet.

**matrix switch** A device that allows any input to be cross-connected to any output.

**Matrix Switch Host Facility 2 (MSHF2)** Software from Bytex Corporation of Southborough, MA, which operates on IBM mainframes, enabling Bytex and IBM matrix switches to send network management data directly to IBM's Netview operating on the host.

**MAU** Multistation Access Unit.

**maximum SSCP rerouting count** In IBM's SNA, the maximum number of times a session initiation request will be rerouted to intermediate SSCPs before the request reaches the destination SSCP. This count is used to prevent endless rerouting of session initiation requests.

259

**MAX1**  A trademark of Teltrend of St. Charles, IL, as well as the name of a byte-interleaved M1 multiplexer marketed by that vendor.

**MB**  1. Megabyte; 1 048 576 bytes. 2. Optical fiber Mounting Bracket.

**M-BIT**  1 024 000 bits.

**MBKS**  MultiButton Key Service.

**Mbps**  Millions of bits per second (bps).

**Mb/s**  Stands for megabits per second. Refers to transmission speed of 1 million bits per second.

**MBTA**  Multiple Beam Tows Antenna.

**Mbytes (megabytes)**  1 024 000 bytes (1000 Kbytes).

**MCC**  1. Master Control Center. 2. Mission Control Center. 3. Mobile Control Center.

**MCEB**  Military Communications Electronics Board.

**MCG**  Memory Controller Group.

**MCH**  Machine Check Handler.

**MCHAN**  MultiCHANnel.

**MCI CALL USA**  Toll-free telephone numbers in countries outside the United States that persons can dial to reach an MCI operator in the U.S. The use of MCI CALL USA can significantly reduce the cost of international calls in comparison to making those calls through the switchboard of a hotel. MCI CALL USA access numbers are listed in the following table.

| Country | USA Access Number |
|---|---|
| Australia | 0014-881-100 |
| Bahrain | 800-002 |
| Belgium | 11-00-12 |
| Brazil | 000-8012 |
| Chile | 00*-0316 |
| Colombia | 980-16-0001 |
| Denmark | 8001-0022 |
| France | 19*00-19 |
| Greece | 00-800-1211 |
| Guam | 950-1022 |
| Hong Kong | 008-1121 |
| Italy | 172-1022 |
| Japan | 0039-121 |
| Netherlands | 06*-022-91-22 |
| Singapore | 800-0012 |
| Sweden | 020-795-922 |
| Switzerland | 046-05-0222 |
| United Kingdom | 0800-89-0222 |

*Note*: * means wait for second dial tone.

**MCI FAX**  A facsimile transmission service marketed by MCI Communications Corporation that uses that vendor's digital network. MCI provides MCI FAX users with management reports that denote facsimile usage.

**MCI Mail**  An electronic mail service provided by MCI Communications Corporation of Washington, DC.

**MCI Vision**  A volume discount rate plan marketed by MCI Communications Corporation to individual businesses and aimed at small users.

**MCI800**  A WATS service offered by MCI Communications Corporation.

**MCP**  Multi-location Calling Plan.

**MCS**  Maritime Communications Subsystem.

**MCU**  Monitor and Control Unit.

**MCVF**  Multi-Channel Voice Frequency.

**MCW**  Modulated Continuous Wave.

**MD**  1. Management Domain. 2. Multiple Dissemination.

**MDF**  Main Distributing Frame.

**MDM**  Modular Data Module.

**MDNS**  Managed Data Network Services.

**MDNS BV**  A company set up by 22 European telecommunications administrations to market data communications services throughout Europe. The company is named for the Managed Data Network Services it will provide.

**MDR** 1. Message Detail Recording. 2. Miscellaneous Data Record.

**MDS** Multiple Dataset System.

**Mean Recovery Time (MRT)** Relates to the normal repair time over a given period. Is sometimes used as a measure of assessing equipment reliability.

**Mean Time Between Failures (MTBF)** A figure of merit for electronic equipment or systems that indicates the average duration of periods of fault-free operation. Used in conjunction with Mean Time To Repair (MTTR) to derive availability figures.

**Mean Time to Failure (MTF)** The average length of time for which the system, or a component of the system, works without fault.

**Mean Time To Repair (MTTR)** A figure of merit for electronic equipment or systems that indicates the average time required to fix the equipment or system. Used in conjunction with Mean Time Before Failure (MTBF) to derive availability figures.

**means of signal communications** A medium by which a message is conveyed from one person or place to another.

**measured local service** Telephone service where a charge is made in accordance with measured usage (measured units).

**measured rate** A rate structure or tariff in which the rental includes payment for a specified number of calls within a defined area, plus a charge for additional calls.

**MED** Manipulate Electronic Deception.

**media** The paths along which the signal is propagated, such as wire pair, coaxial cable, waveguide, optical fiber, or radio path. Synonymous with medium.

**Media Access Control (MAC)** A local network control protocol that governs station access to a shared transmission medium. Examples are token-passing and Carrier Sense, Multiple Access (CSMA).

**media conversion** Transformation of electrical signals used to transmit information to or from human usable form, e.g., words, or numbers on a printed page, characters on a digital display.

**medium** Any material substance by which signals are conducted from point to point; includes wire, coaxial cable, fiber optics, water, air, or free space.

**medium band** A term used for voice telephone transmissions and data transmissions linked to visual display terminals and similar devices. A medium band is also called a voice-band.

**Medium-Scale Integration (MSI)** A term used to describe a multi-function semiconductor device with a medium density (up to 100 circuits) of electronic circuitry contained on a single silicon chip.

**MEECN** Minimum Essential Emergency Communications Network.

**mega (M)** A prefix for one million times (10*E6) a specific unit.

**megabit** One million binary digits, or bits.

**Megabit** A trademark of Megabit Communications, Inc., of St. Paul, MN, as well as the name of a channel extender that enables local peripherals to be used at a remote distance from an IBM mainframe.

**megabyte (Mbyte or M)** 1 048 576 bytes; equal to 1024 K bytes.

**megahertz (MHz)** A unit of analog frequency equal to 1 000 000 hertz or cycles per second.

**memo** An electronic mail system developed by Volvo Data and marketed by Verimation of Northvale, NJ. MEMO runs directly under VTAM and can be operated under MVS or DOS/VE. The MEMO system can be accessed from any IBM 3270 or compatible display terminal.

Channel extender usage

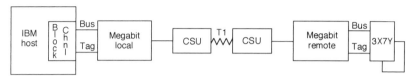

CSU = Channel Service Unit

*Megabit*

**memory** The part of a computer system or terminal where information is stored.

**memory manager** Software which controls dynamic requests for memory or returns unused memory to a memory pool.

**MEN protocol** Management Event Notification protocol.

**menu** An organized collection of fixed captions (field headers) and fields to accept variable data associated with each caption.

**Mercury Communications** A wholly owned subsidiary of Cables & Wireless which operates communications facilities within the United Kingdom.

**Mercury Communications Ltd** The second major UK network provider after British Telecom.

**Meridian Mail** A voice processing system marketed by Northern Telecom, Inc.

**Meridian Norstar** A sophisticated but easy-to-use phone system manufactured by Northern Telecom which is equipped with a visual display.

**MERIT** The statewide computer network run out of the University of Michigan in Ann Arbor, MI.

**Merlin** A trademark used by British Telecom for a series of modems.

**mesh** A multi-node network topology that consists of more than three multiplexer nodes each interconnected with another node such that more than one aggregate path exists for each channel circuit.

**mesh network** A network that has at least two pathways to each node.

**message** 1. Any thought or idea expressed briefly in intelligible or secret language, prepared in a form suitable for transmission by any means of communication. 2. A complete transmission; used as a synonym for packet, but a message is often made up of several packets. 3. In IBM's VTAM, the amount of FM data transferred to VTAM by the application program with one SEND request.

**message address** The information contained in the message header that indicates the destination of the message.

**message center section** In U.S. military communications, that portion of a communications center charged with accepting and routing originating and refile messages and with preparing terminating messages for delivery.

**Message Express** A trademark of Metrocast as well as the name of a packet size, long distance messaging unit manufactured by that vendor.

**message format** Rules for the placement of message elements such as the header, text, and closing.

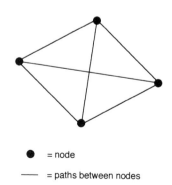

● = node

—— = paths between nodes

*mesh network*

**Message Handling Service (MHS)** A personal computer based software system whose primary function is the transportation of messages between application programs. MHS was developed by Action Technologies and Novell acquired full marketing and development rights to MHS for NetWare-based LANs. Under MHS each application sends messages to the server\mhs\mail\snd directory. MHS delivers messages to an application by placing them in the application's assigned directory.

**Message Handling System (MHS)** The standard defined by the CCITT (now ITU) X.400 Recommendation and by International Standards Organization (ISO) as the Message-Oriented Text Interchange Standard (MOTIS). It is the standard Application layer protocol which defines a framework for the distribution of data from one network to other networks. MHS transfers relatively small messages in a store-and-forward manner.

**message header** That portion of a message which contains the control, routing, and identification information for the message.

**message heading** That part of a message containing all components preceding the text.

**message numbering** The identification of each message in a communications system by the assignment of a sequential number. This numbering is frequently used to facilitate message tracking and accounting.

**message oriented signaling** A common channel signaling protocol for AT&T's Digital Multiplexed Interface; based on the CCITT's Common Channel Signaling System No. 7.

**Message-Oriented Text Interchange Systems (MOTIS)** The term used by the ISO to refer to standard international electronic mail system.

**message reference block** An area of storage allocated to a message until it is processed.

**message routing** The selection of a path or a channel for sending a message.

**message store (MS)** In the CCITT (now ITU) Message Handling System (MHS) a message store holds messages from the Message Transfer Systems (MTS) and makes them available to the user agent (UA) at the discretion of the user.

**message switch** A device used to receive a message, store it until the proper outgoing line is available, and then retransmit it.

**message switching** A data communications technique in which a complete message is stored and then forwarded to one or more destinations when the required destination(s) are free to receive traffic. Because of the high transit time through the network, message switching is not used much for computer data but rather for administrative messages.

**message switching network** A public communications network over which subscribers send primarily textual messages to one another. TWX and Telex are examples of message switching networks.

**Message Telecommunications Service (MTS)** The official designation for long distance dialed (switched), tariffed telephone service.

**message text** That portion of a message which contains the data or information content of the message.

**Message Transfer Agent (MTA)** The functional component that with other MTAs constitutes the Message Transfer System (MTS). The MTA provides Message Transfer Service by: (1) interacting with originating User Agents (UAs); (2) relaying messages to other MTAs based on recipient designations; and (3) interacting with recipient UA(s) (X.400 specific).

**Message Transfer Part (MTP)** Collectively, the lower layers of CCITT Signaling System No. 7 (SS7), consisting of:

Level 1—Signaling data link
Level 2—Link control
Level 3—Common transfer functions

**Message Transfer Protocol (P1)** The protocol that

defines the relaying of messages between Message Transfer Agents (MTAs) (X.400 specific).

**message transfer service** The set of optional service elements provided by the Message Transfer System (MTS) (X.400 specific). *Note.* Message Transfer Service is not referred to by the acronym MTS. This acronym is used only for Message Transfer System.

**Message Transfer System (MTS)** The collection of Message Transfer Agents (MTAs) that provide the message transfer service elements (X.400 specific).

**Message Transmission System (MTS)** The UK version of CCITT (now ITU signaling system) #7 used between and within System X exchanges.

**Message Unit (MU)** 1. In IBM's SNA, a message unit is that portion of data within a message that is passed to, and processed by a particular Network layer (e.g., path information unit (PIU) or request/response unit (RU)). 2. The measurement used for charging local calls.

**Message-Edge** Software marketed by IBS Corporation of La Jolla, CA, which is used to broadcast system-wide messages on IBM mainframes.

**Messavia** The name of an electronic messaging system operated by KDD in Japan.

**Metal Oxide Semiconductor (MOS)** Technology describing a transistor composed of a semiconductor layer including "source" and "drain" regions separated by a channel. Above the channel is a thin layer of oxide and over that a metal electrode called a gate. A voltage applied to this gate controls the current between the source and drain regions or, in another format, stops a flow between the two areas.

**metallic circuits** Circuits which use metal wire (copper) from end to end. Their use implies that no amplifiers or any other devices are interposed between the ends of the circuit. Metallic circuits have electrical (DC) continuity from end to end.

**Meteosat PX** A European Space Agency satellite launched on June 15, 1988. This satellite is designed to perform earth imaging, dissemination of meteorological data, and the collection of environmental data. The satellite is in a geostationary orbit over the gulf of Guinea.

**metered usage** A type of telecommunications service offering in which the amount of usage of the service is monitored and billed accordingly; opposite of a full period service in which usage is not a factor in billing.

**metric prefixes** A series of terms and their associated abbreviations used in the metric system to indicate multiples or portions (sub-multiples) of quantities which can be expressed as positive or negative powers of 10.

| Prefix | Symbol | Multiple |
|--------|--------|----------|
| nano- | n | 0.000 000 001 |
| micro- | $\mu$ | 0.000 001 |
| milli- | m | 0.001 |
| centi- | c | 0.01 |
| deci- | d | 0.1 |
| deka- | da | 10 |
| hecto- | h | 100 |
| kilo- | k | 1 000 |
| mega- | M | 1 000 000 |
| giga- | G | 1 000 000 000 |

**MetroBridge** A local area network bridge marketed by T3 Technologies, Inc., of Research Triangle Park, NC.

**MetroCarrier Express** A service marketed by Metropolitan Fiber Systems, Inc., (MFS) in which the vendor provides a T1 line and multiplexing equipment to the customer to link their site to an interexchange carrier's point of presence, bypassing the local exchange carrier.

**Metro-Digital Transition Plan** An AT&T plan to enable telephone companies to migrate customers needing digital services using ISDN without replacing existing analog stored program control

(SPC) switches. This plan involves the colocation of AT&T 5ESS switches or its remote entities in telephone company SPC wire centers.

**MetroExpress I**   A T1 service marketed by Metropolitan Fiber Systems, Inc., (MFS) which can be used to connect a customer's T1 equipment to a long haul carrier's T1 service, bypassing the local exchange carrier.

**MetroExpress II**   A T2 (6.312 Mbps) service marketed by Metropolitan Fiber Systems, Inc., (MFS) which can be used to connect a customer's equipment to a long haul carrier's T2 service, bypassing the local exchange carrier.

**MetroExpress III**   A T3 (45 Mbps) service marketed by Metropolitan Fiber Systems, Inc., which can be used to connect a customer's equipment to a long haul carrier's T3 service, bypassing the local exchange carrier.

**MetroHub I**   A service marketed by Metropolitan Fiber Systems, Inc., (MFS) under which the vendor will combine circuits from multiple sites onto a single T1 circuit which will be routed to a long haul carrier's point of presence, bypassing the local exchange carrier.

**MetroHub III**   A service marketed by Metropolitan Fiber Systems, Inc., (MFS) under which the vendor will combine T1 circuits from multiple sites onto a T3 circuit routed to an interexchange carrier's point of presence, bypassing the local exchange carrier.

**Metropolitan Area Network (MAN)**   A communications network that spans geographical areas whose size is between that of a local area network and a wide area or long distance network. Other characteristics are a high data rate, moderate delay and moderate error rate.

**MEU**   Memory Expansion Unit.

**MEWSG**   Multi-service Electronic Warfare Support Group.

**MEWSS**   Mobile Electronic Warfare Support Group.

**MF**   Medium Frequency.

**MFC**   Modular Feature Construction.

**MFJ**   Modified Final Judgment.

**MFOS**   MultiFunction Operations System.

**MFTDMA**   Multiple Frequency Time Division Multiple Access.

**MG**   Master Group.

**MHD**   Moving Head Disk.

**MHP**   Message Handling Processor.

**MHS**   1. Message Handling Service. 2. Message Handling System.

**MHz**   Megahertz.

**MIC**   Middle-In-Chain (IBM's SNA).

**MICE protocol**   Management Information Control and Exchange Protocol.

**Micom, Micom Systems, Inc.**   A supplier of data communications equipment.

**MICR**   Magnetic Ink Character Recognition.

**micro ($\mu$)**   A prefix for one millionth of a specific unit.

**micro channel**   The 16-bit and 32-bit bus structure used in most members of the IBM PS/2 family of personal computers. The micro channel architecture supports bus arbitration which allows multiple devices to operate concurrently.

**Micro TAC**   A Motorola packet-sized cellular phone that weighs 10.7 ounces and measures 13.5 cubic inches.

**microbend**   Local discontinuities on a microscopic scale which are caused by mechanical stress on a fiber. Microbends result in additional attenuation.

**microbend loss**   The leakage of light caused by very tiny, sharp curves in a lightguide that may result from imperfections where the glass fiber meets the sheathing that covers it.

**microcode**   A set of software instructions which execute a macro instruction.

**Microcom Networking Protocol (MNP)** An error-correction and data compression protocol developed by the Microcom Corporation of Norwood, MA. The Microcom Networking Protocol (MNP) communications protocol supports interactive and file-transfer applications, divided into six classes, or performance levels. MNP performance levels include the following classes:

Class 1—the lowest performance level, uses an asynchronous byte-oriented half-duplex method of exchanging data. The protocol efficiency of a Class 1 implementation is about 70 percent; (a 2400-bps modem using MNP Class 1 will have a 1690-bit-per-second (bps) throughput).

Class 2—uses asynchronous byte-oriented full-duplex data exchange. The protocol efficiency of a Class 2 modem is about 84 percent (a 2400-bps modem will realize 2000-bps throughput).

Class 3—uses synchronous bit-oriented full-duplex data exchange. This approach is more efficient than the asynchronous, byte-oriented approach, which takes 10 bits to represent eight data bits because of the "start" and "stop" framing bits. The synchronous data format eliminates the need for start and stop bits. Users still send data asynchronously to a Class 3 modem, but the modems communicate with each other synchronously. The protocol efficiency of a Class 3 implementation is about 108 percent (a 2400-bps modem will actually run at a 2600-bps throughput).

Class 4—adds two techniques: Adaptive Packet Assembly and Data Phase Optimization. In the former technique, if the data channel is relatively error-free, MNP assembles larger data packets to increase throughput. If the data channel is introducing many errors, then MNP assembles smaller data packets for transmission. Although smaller data packets increase protocol overhead, they concurrently decrease the throughput penalty of data retransmissions—more data is successfully transmitted on the first try. Data Phase Optimization is a technique for eliminating some of the administrative information in the data packets, which further reduces protocol overhead. The protocol efficiency of a Class 4 implementation is about 120 percent (a 2400-bps modem will effectively yield a throughput of 2900 bps).

Class 5—adds data compression, which uses a real-time adaptive algorithm to compress data. The real-time capabilities of the algorithm allow the data compression to operate on interactive terminal data as well as file-

transfer data. The adaptive nature of the algorithm refers its ability to continuously analyze user data and adjust the compression parameters to maximize data throughput. The effectiveness of data compression algorithms depends on the data pattern being processed. Most data patterns will benefit from data compression, with performance advantages typically ranging from 1.3 to 1.0 and 2.0 to 1.0, although some files may be compressed at an even higher ratio. Based on a 1.6 to 1 compression ratio, Microcom gives Class 5 MNP a 200 percent protocol efficiency, or 4800-bps throughput in a 2400-bps modem installation.

Class 6—adds 9600-bps V.29 modulation, universal link negotiation and statistical duplexing to MNP Class 5 features. Universal link negotiation allows two unlike MNP Class 6 modems to find the highest operating speed (between 300 and 9600 bps) at which both can operate. The modems begin to talk at a common lower speed, and automatically "negotiate" the use of progressively higher speeds. Statistical duplexing is a technique for simulating full-duplex service over half-duplex high-speed carriers. Once the modem link has been established using full-duplex V.22 modulation, user data streams move via the carrier's faster half-duplex mode. However, the modems monitor the data streams, and allocate each modem's use of the line to best approximate a full-duplex exchange. Microcom claims that a 9600-bps V.29 modem using MNP Class 6 (and Class 5 data compression) can achieve 19.2 Kbps throughput over dial circuits.

Class 7—uses an advanced form of Huffman encoding called Enhanced Data Compression. Enchanced Data Compression has all of the characteristics of Class 5 compression but, in addition, predicts the probability of repetitive characters in the datastream. Class 7 compression on the average reduces data by 42 percent.

Class 8—adds CCITT (now ITU) V.29 Fast-Train modem technology to Class 7 Enhanced Data Compression, enabling half-duplex devices to emulate full-duplex transmission.

Class 9—combines CCITT (now ITU) V.32 modem modulation technology with Class 7 Enhanced Data Compression, resulting in a full-duplex throughput that can exceed that obtainable with a V.32 modem by 300 percent. Class 9 also employs selective retransmission, in which error packets are retransmitted, as well as piggybacking, in which acknowledgment information is added to the data.

Class 10—adds adverse channel enhancements to MNP which allow a modem to compensate for and improve the quality of dial-up telephone calls. Adverse channel enhancements include Robust Auto Reliable, Aggressive Adaptive Packet Assembly and Dynamic and Negotiated Shift Speeds.

— Robust Auto Reliable results in up to 12 attempts, without redialing, to establish a connection when there is interference on a line.

— Aggressive Adaptive Packet Assembly adapts the size of the data packets between modems to accommodate varying levels of interference. When line quality is poor the size of the packets decreases to minimize retransmission time.

— Dynamic and Negotiated Shift Speeds raise the speed and packet size as line quality improves.

**microcomputer** 1. A desktop (or knee-top) computer; a personal computer. 2. A microprocessor system.

**Micro-Fone II** A data terminal which is a microprocessor-controlled telephone unit for both data and voice communications. It communicates with host computers either through a packet network or through the dial telephone network. (Micro-Fone II is designed primarily for transaction [credit card] communication.)

**MicroLink 1** A switched digital 56 Kbps service marketed by Southwestern Bell Telephone Company of St. Louis, MO.

**micron** One-millionth of a meter (one micrometer). Commonly used to describe the core diameter of an optical fiber.

**Microphone II** A full-featured asynchronous communications program from Software Ventures Corporation of Berkeley, CA, that operates on the Apple Macintosh.

**microprocessor** An electronic circuit on the surface of a small silicon chip which can be programmed to perform a wide variety of functions within the computer system or terminal. The Intel 8088, 80286, 80386, and 80486, Motorola 68000, and Zilog Z80 are examples of popular microprocessors.

**microprocessor chip** A single device cut from a wafer of silicon used in microcircuitry.

**micro-programming** The process of building a sequence of instructions into read-only memory to carry out functions that would otherwise be directed by stored program instructions at a much lower speed.

**microsecond** One millionth of a second.

**Microsoft Disk Operating System (MS-DOS)** A microcomputer operating system developed by Microsoft Corporation for the IBM PC and compatible personal computers. Also known as PC-DOS by IBM.

**microwave** 1. An electromagnetic wave between 1 centimeter (10*E10 Hz) and 100 centimeters (10*E8 Hz) in length. 2. Those frequencies in the super high frequency (SHF) band above 890 megahertz (MHz) used for data and voice communications.

**microwave communications** A line-of-sight communications system which transmits on microwave frequencies.

**Microwave Pulse Generator (MPG)** A device that generates electrical pulses at microwave frequencies.

**microwave relay system** A microwave relay system consists of towers spaced as much as thirty miles apart. Using microwave signals, transmissions can criss-cross the country. Microwave transmissions travel in straight lines, so the towers must be in "line-of-sight" of each other. This is due to curvature of the earth, mountains, and other obstacles that can block transmission between land-based microwave towers. Repeaters at each tower amplify and retransmit the signals until they reach their destination.

Microwave

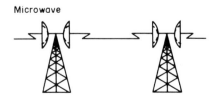

267

**microwave transmission** Relating to data communications systems in which ultra-high frequency waveforms are used to transmit voice or data messages. Line-of-site transmission between antennas, with repeaters every 20–30 miles.

**MIDAS** Multiple Indexed Data Access System.

**Middle-in-chain (MIC)** In IBM's VTAM, a request unit (RU) whose request header (RH) begin chain indicator and RH end chain indicator are both off.

**midicomputer** Sometimes called a super mini. Has a 24- or 32-bit word size.

**MIDnet** Midwestern States Network.

**midsplit** A method of allocating the available bandwidth in a single-cable broadband system. Transmissions from the headend to users are in the 168 to 300 MHz frequency range; transmissions from the users to the headend are in the 5 to 116 MHz sequence range.

**MIF** Minimum Internetworking Functionality.

**MIJI** Meaconing, Intrusion, Jamming and Interference.

**Military Affiliate Radio System (MARS)** A communications system which permits operators at a station nearest a U.S. soldier to place a radio call to a ham radio operator as near to the caller's destination as possible. The ham operator calls the soldier's party, reversing the charges and transmits the call from the telephone onto the radio airwaves, and back to the calling station. MARS is primarily designed to assist U.S. soldiers in making long distance calls from remote locations and depends upon volunteer ham radio operators.

**Military Network (MILNET)** A U.S. Department of Defense network which was splintered off from ARPANET to provide a separate military communications facility.

**military standards (United States)** Two major U.S. military standards affecting the communications industry are MIL-STD-188C and MIL-STD-188-114. MIL-STD-188C defines the technical design of U.S. military communications systems such as permissible signal deterioration, modulation schemes, and hardware operation. MIL-STD-188-114 defines the electrical characteristics of military data communications systems.

**Military Strategic/Tactical And Relay System (MILSTAR)** A planned replacement for the US Air Force's defense satellite communications system (DSCS) satellites. It will use a 44 GHz uplink and a 20 GHz downlink.

**milli (m)** A prefix for one thousandth of a specific unit.

**milliampere (mA)** A measurement unit of electric current.

**millions of instructions per second (Mips)** One measure of processing power.

**millisecond** One thousandth of one second.

**milliwatt** One thousandth of one watt.

**MILNET** MILitary NETwork.; the military network within the Internet.

**MILSATCOM** MILitary SATellite COMmunications System.

**MILSTAR** MILitary Strategic/Tactical And Relay system.

**MIL-STD-1777** The U.S. Department of Defense Internet Protocol (IP).

**MIL-STD-1778** The U.S. Department of Defense Transmission Control Protocol (TCP).

**MIL-STD-1780** The U.S. Department of Defense File-Transfer Protocol (FTP).

**MIL-STD-1781** The U.S. Department of Defense Simple Mail Transfer Protocol (SMTP).

**MIL-STD-1782** The U.S. Department of Defense TELNET Protocol.

**MIL-188C** A shielded, military standard interface equivalent to RS-232C with the exception that the data and clocks are inverted and signal levels are +6, −6 volts.

**MIMI**   Multipurpose Internet Mail Extension.

**MIN**   Mobile Identification Number.

**Mind**   A network design tool of Contel Business Networks of Great Neck, NY.

**minibased**   Any device containing and operating with a minicomputer.

**minicall**   In packet-switched networks, the process of sending a datagram.

**minicomputer**   A small-scale or medium-scale computer (such as those made by DEC, Data General, Hewlett-Packard, and others) usually operated with interactive, dumb terminals and often having an open architecture. Also called mini for short.

**Mini-Manufacturing Automation Protocol (MINI-MAP)**   A version of the Manufacturing Automation Protocol (MAP) consisting of only Physical, Link, and Application layers that is intended for use in lower-cost process-control networks. Under MINI-MAP a device with a token can request a response from an addressed device; however, unlike a standard MAP protocol, the addressed MINI-MAP device does not have to wait for the token to respond.

**Minimap**   Manufacturing Automation Protocol (MAP) version 2.2, also referred to as Enhanced Performance Architecture (EPA).

**MINI-MAP**   Mini-Manufacturing Automation Protocol.

**minimize**   A condition that results in normal message and telephone traffic being drastically reduced so that messages connected with an actual or simulated emergency are not delayed.

**Minimum Internetworking Functionality (MIF)**   A general principle within the ISO that calls for minimum local area n complexity when interconnecting with resources outside the local area network.

**Minimum Point Of Penetration (MPOP)**   A convenient point within a building or on a multibuilding facility where the communications carrier may choose to terminate its entrance cable. Beyond the MPOP all cable and wire responsibility falls on the customer.

**minimum weight routing**   A routing scheme which minimizes the sum of the weights of the links employed route. These "weights" could be link delays, cost or error rate—anything which adds together in transit and should be minimized.

**Miniset 270**   A trademark of Siemens for that firm's quality telephones designed for home and business use.

**Minitel**   A terminal developed in France for videotex usage.

**Minitel 1**   The first Minitel terminal which had an alphamosaic screen display of 25 rows by 40 characters.

**Minitel 1B**   The basic Minitel terminal since 1986. This terminal is a dual-standard device, which supports both the Teletel standard and the 80-column ASCII standard.

**minor node**   In IBM's VTAM, a uniquely defined resource within a major node.

**MINSES.EXE**   An Apple Computer Company's AppleShare PC executable program that handles communications between nodes on an AppleTalk network.

**Mips**   Millions of Instructions Per Second.

**MIRROR**   A communications software program for personal computers marketed by Softklone of Tallahassee, FL.

**MIS**   Management Information System.

**Miscellaneous Data Record (MDR)**   A record originated by communications resources attached to a communications controller to describe permanent errors.

**misrouted message**   A message bearing an incorrect routing instruction.

**missent message**   A message that bears the correct routing instruction but has been transmitted to a station other than that indicated.

**MIU**   Multistation Interface Unit.

**MJU**   Multipoint Junction Unit.

**MLCP**   MultiLine Communications Processor.

**MLHG**   MultiLine Hunt Group.

**MLPP**   MultiLevel Precedence and Preemption.

**MLS**   MultiLan Switch.

**MLT**   1. Mechanized Loop Testing. 2. Multiple Logical Terminals.

**mm**   millimeter (or millimeters).

**MMC**   Man–Machine Communication.

**MMFS**   Manufacturing Message Format Specification.

**MMIC**   Monolithic Microwave Integrated Circuit.

**MMP**   Module Message Processor.

**MMS**   1. Manufacturing Message Specification. 2. Manufacturing Messaging Service.

**MMSU**   Modular Metallic Service Unit.

**MNCS**   Multipoint Network Control System.

**mnemonic code**   Instructions for the computer written in a form that is easy for the programmer to remember. A program written in mnemonics must be converted to machine code prior to execution.

**MNP**   Microcom Networking Protocol.

**Mobile Identification Number (MIN)**   In cellular telephone communications a binary encoded (34 bits) version of a mobile unit's telephone number.

**mobile phone**   Any telephone which can be operated without a physical line.

**mobile radio telephone**   A voice telephone service set in a mobile station.

**Mobile Satellite Service (MSS)**   Mobile Satellite Service refers to the provision of mobile voice, global positioning and alphanumeric data services via satellite.

**mobile telephone service**   Telephone service for moving vehicles that use both the telephone network and radio.

**Mobile Telephone Switching Office (MTSO)**   1. An office which controls individual call switching for all traffic emanating from or terminating on cellular radios within a certain area. 2. That portion of an AT&T Autoplex System that controls the overall system.

**Mobilenet**   A mobile cellular telephone service operated by Telecom Australia.

**Mobitex**   A mobile data system using the cellular communications concept of reusing frequencies. Mobitex was developed in Sweden and systems are operational in several countries.

**MOD**   MODulator.

**mode**   1. The path a light ray follows through a fiber. 2. The electromagnetic field pattern allowed within an optical fiber.

**mode indicator**   On-screen symbols or abbreviations that indicate the operating mode of an IBM Personal Computer using the IBM PC 3270 Emulation Program.

**mode scrambler**   A device used for multimode fiber bandwidth measurements. A mode scrambler mixes the modes for a uniform modal distribution.

**model**   A representation of a system that is frequently constructed from the mathematical expressions which characterize the behavior of the system components.

**modem**   A contraction of the term MOdulator–DEModulator. A modem is an electronic device used to convert digital signals to analog form for transmission over the telephone network. Since the telephone network was designed for analog voice transmission, it is not possible to transmit digital information from a terminal or a computer in its binary form. Since the telephone network has a bandwidth of approximately 3000 Hz, modems using the telephone network must condition signals to fit within this band. Also known as a data set.

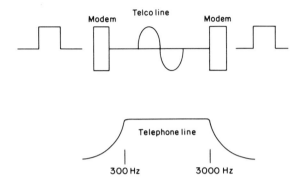

*modem*

**modem, multiport** A device that combines a multiplexer and a modem allowing two or more DTEs to be connected to the same line. Also split stream modem.

**modem, quick turnaround** A modem with minimal turnaround time when the line is used in a half-duplex mode. Also Quick Poll (QP), Fast Poll (FP).

**modem, short haul** Description of both line drivers and limited distance modems.

**modem, wideband** A modem designed to operate at speeds greater than those used with high-speed modems, such as 19.2 or 56 Kbps. Wideband modems will not operate over voice-grade circuits but require a wideband circuit.

**Modem-7** Communications software program supporting the public-domain, X-modem, error-correcting file transfer protocol. This version of the X-modem has multifile transfer capability.

**modem connect** The name used in DNA for that class of communications links governed by industry standards for modem connection.

**modem eliminator** A device used to connect a local terminal and a computer port in lieu of the pair of modems that they would expect to connect to: allows DTE-to-DTE data and control signal connections otherwise not easily achieved by standard cables or connectors. Modified cables (crossover cables) or connectors (adapters) can also perform this function.

**modem operation characteristics** See page 270.

**modem pooling** A feature of a PABX and other communications products that permits subscribers to be automatically or manually connected to a group of shared or "pooled" modems.

**modem sharing unit** A device that splits a signal among a cluster of terminals and allows them to share one modem.

**modem substitution switch** An external option that allows you to reroute your data through a "hot" spare (a modem that is already powered up) in the event the original modem fails.

**moderator** A participant who is in charge of a conference. A moderator is responsible for keeping the discussion on track, for alleviating fights, and for similar functions.

**MODES** Discrete optical waves that can propagate in optical waveguides. Whereas, in a single-mode fiber, only one mode, the fundamental mode, can propagate. There are several hundred modes in a multimode fiber which differ in field pattern and propagation velocity (multimode dispersion). The upper limit to the number of modes is determined by the core diameter and numerical aperture of the waveguide.

**Modified Chemical Vapor Deposition** An AT&T Bell Laboratories-patented process that uses high temperatures to speed the manufacture of large quantities of fiber lightguide. The glass is made by allowing hot vapors to form a coating inside a tube of heated silica, which is later drawn into fiber. Temperatures reach 4000 degrees F. (The melting point of steel is 2800 degrees F.)

**Modified Final Judgment (MFJ)** The 1982 Federal Court ruling that determined the rules governing the divestiture of the Bell Operating Companies from AT&T and other antitrust and deregulation issues. Presided over by Judge Harold Greene, as was the AT&T Antitrust settlement which the MFJ modified. Judge Greene continues his involvement in enforcing and interpreting the provisions of this settlement.

**modular distribution accessories** A term used to reference splitters, modular adapters and modular

271

| Modem type | Data rate | Transmission technique | Modulation technique | Transmission mode | Line use |
|---|---|---|---|---|---|
| **Bell System** | | | | | |
| 103A,E | 300 | asynchronous | FSK | Half, Full | Switched |
| 103F | 300 | asynchronous | FSK | Half, Full | Leased |
| 201B | 2400 | synchronous | PSK | Half, Full | Leased |
| 201C | 2400 | synchronous | PSK | Half, Full | Switched |
| 202C | 1200 | asynchronous | FSK | Half | Switched |
| 202S | 1200 | asynchronous | FSK | Half | Switched |
| 202D/R | 1800 | asynchronous | FSK | Half, Full | Leased |
| 202T | 1800 | asynchronous | FSK | Half, Full | Leased |
| 208A | 4800 | synchronous | PSK | Half, Full | Leased |
| 208B | 4800 | synchronous | PSK | Half | Switched |
| 209A | 9600 | synchronous | QAM | Full | Leased |
| 212 | 0–300 | asynchronous | FSK | Half, Full | Switched |
| | 1200 | asynchronous/ synchronous | PSK | Half, Full | Switched |
| **CCIT** | | | | | |
| V.21 | 300 | asynchronous | FSK | Half, Full | Switched |
| V.22 | 600 | asynchronous | PSK | Half, Full | Switched/ Leased |
| | 1200 | asynchronous/ synchronous | PSK | Half, Full | Switched/ Leased |
| V.22 bis | 2400 | asynchronous | QAM | Half, Full | Switched |
| V.23 | 600 | asynchronous/ synchronous | FSK | Half, Full | Switched |
| | 1200 | asynchronous synchronous | FSK | Half, Full | Switched |
| V.26 | 2400 | synchronous | PSK | Half, Full | Leased |
| | 1200 | synchronous | PSK | Half | Switched |
| V.26 bis | 2400 | synchronous | PSK | Half | Switched |
| V.26 ter | 2400 | synchronous | PSK | Half, Full | Switched |
| V.27 | 4800 | synchronous | PSK | | |
| V.29 | 9600 | synchronous | QAM | Half, Full | Leased |
| V.32 | 9600 | synchronous | QAM | Half, Full | Switched |
| V.33 | 14400 | synchronous | TCM | Full | Leased |

*modem operation characteristics*

cables that are used to convert modular jacks to RS-232 DB-25 connectors.

**Modular Feature Construction (MFC)**   The unique capability on the AT&T 5ESS switch that permits operating companies to actually create new features by varying the way in which existing features work and interwork.

**modular jack cable**   A cable designed to connect modems or telephones to modular telephone jacks; or connect telephones to modems.

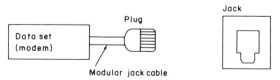

**modular splitter**  A device which converts 50-pin (25 pair) PBX connectors to multiple module jacks. Modular splitters are used with modular cables and modular adapters to convert modular jacks to RS-232 DB-25 connectors.

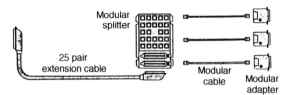

*modular splitter*

**modulation**  The process by which a carrier is varied to represent an information-carrying signal. See amplitude modulation, frequency modulation and phase modulation.

**modulation frequency**  The frequency of the modulating wave.

**modulation with a fixed reference**  A type of modulation in which the choice of the significant condition for any signal element is based on a fixed reference.

**modulator**  An electronic circuit that combines digital data to be transmitted with a carrier signal. The combination of the digital data and the carrier form a composite signal that is suitable for transmission over data communications lines.

**module**  1. (Hardware) Short for card module. 2. (Software) A program unit or subdivision that performs one or more functions.

**modulo**  A term used to express the maximum number of states for a counter; this term is used to describe several packet-switched network parameters, such as packet number (usually set to modulo 8—counted from 0 to 7). When the maximum count is exceeded, the counter is reset to 0.

**modulo-N**  In communications modulo-$N$ refers to a quantity, such as the number of messages or frames that can be counted before the counter resets to zero, or the number of messages $(N-1)$ that can be outstanding before an acknowledgment is required from the receiver. An example is modulo 8 or modulo 128.

**MODULO 2 ADDITION**  A method of adding binary digits which gives: 0+0 = 0; 0+1 = 1; 1+0 = 1; 1+1 = 0 (binary addition without carries). Other names for the operation are "not equivalent" and "exclusive OR".

**Molina blocking**  A formula used to compute the probability that a call will be blocked based upon the assumption that blocked traffic is held for a duration equal to the time of the call had it not been blocked.

**Molniya**  The name for a series of Russian communications satellites used to form a domestic communications network.

**Molniya 1-1**  The first Russian operational communications satellite that was launched on April 23, 1965.

**MOMCOMS**  Man-On-the-Move COMmunications System.

**monitor**  A function or device that involves the observation of activity on a data communication line without interference.

**monitor station**  In LAN technology on ring networks, the unit responsible for removing damaged packets and for making sure that the ring is intact.

**monitored bulletin board**  An information system or electronic mail service bulletin board that remembers a user's last access and reminds the user when new mail has been sent to that board.

**monitoring**  A testing function in which a protocol analyzer displays, records, or gathers statistics on the data transmitted over a circuit without interrupting the circuit or transmitting test data.

**monochromator**  A wavelength tunable, optical filter based on a diffraction grating which provides a narrow linewidth source when used with a white light source.

**monomode**  In optical systems a term commonly used in place of singlemode.

**MOP**  Maintenance Operations Protocol.

**Morse Code**  The series of short (dot) and relatively long (dash) key depressions first used in telegraph systems to represent characters. The following table lists the International Morse Code telegraph characters where a ( . ) indicates a short (dot) key depression and a ( __ ) indicates a relatively long (dash) key depression.

### INTERNATIONAL MORSE CODE

*Telegraph characters*

| | |
|---|---|
| A | . — |
| B | — . . . |
| C | — . — . |
| D | — . . |
| E | . |
| F | . . — . |
| G | — — . |
| H | . . . . |
| I | . . |
| J | . — — — |
| K | — . — |
| L | . — . . |
| M | — — |
| N | — . |
| O | — — — |
| P | . — — . |
| Q | — — . — |
| R | . — . |
| S | . . . |
| T | — |
| U | . . — |
| V | . . . — |
| W | . — — |
| X | — . . — |
| Y | — . — — |
| Z | — — . . |
| , | — — . . — — |
| . | . — . — . — |
| 1 | . — — — — |
| 2 | . . — — — |
| 3 | . . . — — |
| 4 | . . . . — |
| 5 | . . . . . |
| 6 | — . . . . |
| 7 | — — . . . |
| 8 | — — — . . |
| 9 | — — — — . |
| 0 | — — — — — |

**Morsviazsputnik**  The agency which is responsible for Soviet navigation satellites.

**MOS**  A common channel signaling protocol for AT&T's Digital Multiplexed Interface: based on the CCITT's Common Channel Signaling System No. 7.

**MOS**  Metal Oxide Semiconductor.

**mosaic coding**  A method of displaying data in which the display area is addressed as a fixed matrix of cells with each cell composed of a matrix of pixels.

**MOSFET**  MOS Field-Effect Transistor.

**MOSS**  Maintenance and Operator SubSystem (IBM's VTAM).

**MOST**  An automated bank teller network in Maryland, Virginia, Tennessee and the District of Columbia.

**MOS-1A**  A Japanese remote sensing satellite.

**MOTIS**  Message-Oriented Text Interchange Standard.

**mouse**  Electronic device that controls movement of a cursor on a video display terminal or monitor, when the user rolls the device along a flat surface by hand.

**move**  A change in the physical location of facilities or equipment provided by a common carrier, either within the same customer or authorized user premises or to different premises, when made at the request of the customer without discontinuance of service.

**MP**  Modem Port.

**MPCC**  Multi-Protocol Communication Controller.

**MPG**  Microwave Pulse Generator.

**MPL**  Multi-Schedule Private Line.

**MPOP**  Minimum Point Of Penetration.

**MPS400**  The name of an electronic messaging system operated by OTC in Australia.

**MPT**  Ministry of Posts and Telecommunications.

**MQ**  Modular Queueing.

**MRT**  1. Mean Recovery Time. 2. Miniature Receiver Terminal.

**MRVT**  Multiple Rate Voice Terminal.

**MS**  Message Store.

**MS NET**  A networking software package by Microsoft, for use with the IBM Personal Computer.

**MSAU**  MultiStation Access Unit.

**MS-DOS**  Microsoft Disk Operating System.

**MSE**  Mobile Subscriber Equipment.

**Msg**  Message.

**MSG**  MeSsaGe switch.

**MSG CEN**  MeSsaGe CENter.

**MSHF2**  Matrix Switch Host Facility 2.

**MSHP**  Maintain System History Program (IBM's VTAM).

**MSI**  1. Medium Scale Integrated (Circuit). 2. Medium Scale Integration.

**MSLPU**  Medium-Speed Line Processing Unit.

**MSM**  Microwave Switch Matrix.

**MSNF**  Multi-System Networking Facility.

**MSRT**  Mobile Subscriber Radio Terminal.

**MSS**  Mobile Satellite Service.

**MSU**  1. Message Signal Unit (CCITT #7 terminology). 2. Message Switching Unit. 3. Multipoint Signaling Unit.

**MT**  Measured Time.

**MTA**  Message Transfer Agent.

**MTAU**  Metallic Test Access Unit.

**MTBF**  Mean Time Between Failures.

**MTCC**  Modular Tactical Communications Center.

**MTF**  Mean Time to Failure.

**MTL**  Master Test Line.

**MTM**  Magnetic Tape System.

**MTP**  1. Message Transfer Part. 2. Message Transfer Point.

**MTS**  1. Message Telecommunications Services. 2. Message Telephone Service. 3. Message Transfer System. 4. Message Transmission System.

**MTSO**  Mobile Telephone Switching Office.

**MTTR**  Mean Time to Repair.

**MTTS**  Magnetic Tape Terminal System.

**Mu**  In queueing systems, the average number of customers served per second. Mu represents the mean of the service distribution.

**MU**  Message Unit.

**MUD**  Multi-User Dungeon.

**MU-LAW**  The North American standard algorithm for the PCM encoding of PAM samples for digital voice communications.

**multi-access**  A system which can be independently accessed by several users simultaneously.

**multi-access spool configuration**  Two to seven systems sharing the JES2 input, job, and output queues through the use of shared DASD.

**multi-aggregate**  Where more than one aggregate or transmission facility is interfaced to a single multiplexer. This term does not imply multi-node.

**Multibutton Key Service (MBKS)**  A service which allows single button access to timed switchhook flash and to business customer features otherwise accessible only via dialed codes.

**multicast bit**  In LAN technology, a bit in the Ethernet addressing structure used to indicate a broadcast message (a message to be sent to all stations). (See figure below.)

**multicast message**  In a LAN, a message intended for a subset of the stations on a network in comparison to an individual station or all stations.

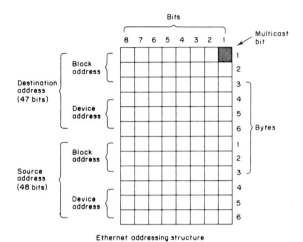

Ethernet addressing structure

*multicast message*

**Multicasting** Transmitting a message to more than one site.

**multichannel** The use of a common channel to derive two or more channels through frequency or time division multiplex.

**Multicom 3270** A trademark of MultiTech Systems of New Brighton, MN, as well as a terminal emulation system consisting of a coaxial card designed for insertion into a PC and 3278/79 terminal emulation and file transfer software.

**multidomain network** In IBM's SNA, a network consisting of two or more host-based system services control points. Usually multi-mainframes.

**multidrop** Describes a telephone line configuration in which a single transmission facility is shared by several end stations.

**multidrop line** Line or circuit interconnecting several stations. Also called multipoint line.

**multiframe** An ordered, functional sequence of frames on a multiplexed digital circuit. "Multiframe" is the CCITT standard term for a "super-frame." D4 framing uses a 12-frame multiframe; ESF uses a 24-frame multiframe; the European format uses a 16-frame multiframe.

**multi-homed hosts** A computer with two or more physical network connections.

**Multilan Switch (MLS)** A product marketed by Alantec of Fremont, CA, which connects up to 10 Ethernet LANs of various media and which can be used with other bridging products and routers.

**multileaving** A characteristic of some character-oriented protocols in which more than one job's set of data can be transmitted, intermixed, at the same time.

**Multiline Communications Processor (MLCP)** A programmable interface between a central processor and one or more communications devices. The MLCP can be programmed to service specific communications devices.

**Multi-Line IDA** A 2.048 Mbps primary rate ISDN service marketed by British Telecom in the UK. This service is based upon British Telecom's proprietary signaling system for private to public exchange, and not on the CCITT Q.931 international standard.

**Multi-location Calling Plan (MCP)** An AT&T bulk-rated multi-location discount service that provides separate bills to each participating location instead of billing a single customer of record.

**multimode** In optical systems, fibers and passive devices which guide many modes. Large core step-index and graded-index fibers are multimode. A fiber-optic waveguide capable of propagating light signals of two or more frequencies or phases.

**multimode fiber** The relatively large core of this lightguide allows light pulses to zigzag along many different paths. It's ideal for light sources larger than lasers, such as LEDs.

**multimode optical fiber** A fiber that will allow more than one bound mode to propagate. May be either graded index or step index fiber.

**MultiNet** The name of an intelligent modular hub manufactured by Lannet Data Communications Ltd of Huntington Beach, CA.

**multi-node** The term implies a network with three or more nodes where networking functionality is employed, such as channel pass-through and alternate routing.

**multi-node network** In telecommunications systems, a network in which users (customers, subscribers) may be interconnected through more than one concentration or switching node.

**multiple access** A network connecting many devices or terminals, all of which use the same channel for broadcasting messages.

**multiple connection port** A dedicated packet network central office access port that supports multiple, simultaneous, virtual connections. Stations connected to the network through such ports must use the X.25 protocol.

**multiple gateways** In IBM's SNA, more than one gateway serving to connect the same two SNA networks for cross-network sessions.

**Multiple Indexed Data Access System (MIDAS)** A Prime database.

**Multiple Logical Terminals (MLT)** In an IBM 3174 control unit, a function that provides a CUT-attached fixed-function display station with the ability to interact with as many as five host sessions. Each session is processed as though it were a separate display station.

**multiple routing** The process of sending a message to more than one recipient, usually when all destinations are specified in the header of the message.

**multiple speed selection** A modem feature which permits continued transmission, even while the quality of the phone line is degrading, by falling back to a lower transmission speed to reduce error rate until line quality is restored.

**Multiple Virtual Storage (MVS)** An IBM operating system whose full name is the Operating System/Virtual Storage (OS/VS) with Multiple Virtual Storage/System Product for the System/370. It controls the execution of programs.

**Multiple Virtual Storage for Extended Architecture (MVS/XA)** An IBM program product whose full name is the Operating System/Virtual Storage (OS/VS) with Multiple Virtual Storage/System Product for Extended Architecture. Extended architecture allows 31-bit storage addressing. MVS/XA is a software operating system controlling the execution of programs.

**multiple-address message** In U.S. military communications, a multiple-address message is one that is destined for two or more addresses, each of whom is informed to all the addressees. Each addressee must be indicated as action or information.

**multiple-call section** In U.S. military communications, that portion of a communications center, normally an integral part of the tape relay station, responsible for multiple reproduction of message tapes at stations of segregation.

**multiple-domain network** In IBM's SNA, a network with more than one system services control point (SSCP).

**multiplex** To interleave or simultaneously transmit two or more messages on a single channel.

**multiplex systems** Used to increase the call-carrying capacities of the different types of carrier systems. Through the use of bundled-coax, microwave, radio systems, and fiber optics. It is possible to stack 100 000 individual voice/data channels on one system.

**Multiplexed Analog Component (MAC)** A direct broadcast satellite television signal standard used in Europe. Under MAC each line of a TV signal is transmitted as a data burst consisting of digital sound, up to 8 data channels, compressed chrominance and compressed luminance.

| Digitally encoded sound | Compressed chrominance 3:1 compression | Compressed luminance 1.5:1 compression | 625 lines |
|---|---|---|---|

MAC signal

**multiplexed channel** A communications channel capable of servicing a number of devices or users at one time.

277

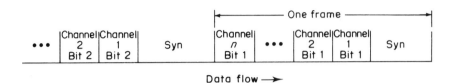

*multiplexing, time division*

**multiplexer (MUX)**  A data communications device that receives signals over several low-speed circuits and combines them in an orderly fashion on a single high-speed circuit. A T1 multiplexer (North American) receives data from 24 circuits at 64 Kbps (or from some number of lower-speed circuits at an equivalent combined bit rate) and combines them according to D4 or ESF framing onto a single circuit at 1.544 Mbps.

**multiplexer channel**  A channel designed to operate with a number of devices simultaneously; may perform its multiplexing function by interleaving either bits or bytes.

**multiplexing**  The process of combining two or more digital signals of low bandwidth into a single signal of higher bandwidth, where the aggregate (combined) bandwidth of the low-speed signals is less than or equal to the bandwidth of the higher-speed signal.

**multiplexing, frequency division**  Multiplexing by splitting the bandwidth of the transmission facility into some number of lower-speed channels or sub-bands.

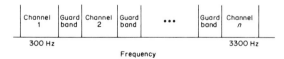

**multiplexing, statistical**  Multiplexing by providing bandwidth on the transmission facility only for those data streams that have data available for transmission. No bandwidth is wasted on terminals not sending data.

**multiplexing, time division**  Multiplexing by assigning each data stream its own time slot during which it transmits data over the transmission facility.

**multipoint**  A term used to describe a data channel which connects terminals at more than two points. Sometimes called a multidrop channel.

**Multipoint Junction Unit (MJU)**  A device which permits one location to serve multiple tributaries.

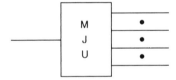

**multipoint line, multipoint connection**  A single communications line or circuit interconnecting several stations supporting terminals in several different locations. Use of this type of line usually requires some kind of polling mechanism, with each terminal having a unique address. Also called multidrop line.

**multipoint network**  A multipoint network is a communications line in which three or more stations are connected. This results in saved lines and saved computer ports. A multipoint network should be connected by a leased line, as opposed to a switched line. There is a master/slave relationship between the modems that organize the communication network. Terminals must possess some intelligence so they can be addressed and polled.

**multiport** Describes a network configuration in which several transmission facilities connect several end stations to a master station.

**Multipoint Signaling Unit (MSU)** A device used in conjunction with a Digital Data test equipment to isolate and test various segments of a digital service multipoint circuit.

**Multiprocessing** Strictly, this term refers to the simultaneous application of more than one process in a multi-CPU computer system to the execution of a single "user job" which is only processed if the job can be effectively defined in terms of a number of independently executable components. The term is more often used to denote multiprogramming operation of multi-CPU computer systems.

**multiprogramming** A method of operation of a computer system whereby a number of independent "user jobs" are processed together. Rather than allow each job to run to completion in turn, the computer switches between them so as to improve the utilization of the system hardware components.

**Multipurpose Internet Mail Extension (MIME)** An extension to Internet mail which adds support for the exchange of multipart messages, including binary files.

**MultiQuest** An AT&T 900-type service which permits businesses to provide callers with information via two-way communications. Unlike 800 service, the call originator pays for the cost of a MultiQuest call.

**Multi-schedule Private Line (MPL)** AT&T's tariff for a voice-grade leased line.

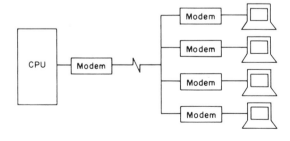

*multipoint network*

**multiserver network** A single network which has two or more file servers. On a multiserver network, users may access files from any server to which they are attached.

**Multistation Access Unit (MAU)** A data concentrator/distributor used in a token-ring local area networks that permits the attachment of up to eight devices. By connecting MAUs to one another, a larger network can be formed.

**Multisystem Networking Facility (MSNF)** An optional software feature that can be added to certain IBM telecommunications access methods which permits more than one host entity running ACF/TCAM or ACF/VTAM to jointly control an ACF/NCP network.

**multitasking** A term used to refer to the concurrent execution of two or more tasks; can also be the concurrent execution of a single program that is used by many tasks.

**multiterminal controller** A terminal controller having more than one terminal device connected to it for subsequent access to the communications line.

**multithread application program** In IBM's VTAM, an application program that processes requests for more than one session concurrently.

**multithreading** Concurrent processing of more than one message (or similar service-request) by an application program.

**multiuser** Used to describe operating systems that allow many users at separate terminals to share a system's processing power, and possibly to also share data and peripherals.

**Multi-User Dungeon (MUD)** A program which enables persons to interact with each other in a simulated environment.

**multiway comparator** An electronic circuit that can simultaneously compare one combination of bits with several other combinations of bits and determine if one combination is equal to any of the others.

**MUM** Multi-Unit Message (CCITT #c6 terminology).

**MUPH** MUltiple Position Hunt.

**Muse** A proposed Japanese High Definition TeleVision (HDTV) transmission standard which would operate at 60 hertz and broadcast 1125 lines.

**MUX** Multiplexer.

**mV** Millivolt.

**MVR** ManeuVeR.

**MVS** Multiple Virtual Storage.

**MVS router** A program running under IBM's TSO/E that uses the Server-Requester Programming Interface (SRPI) to route requests from the PC to the corresponding server on the host. The MVS router is part of the MVSSERV command processor in IBM's TSO/E Release 3.

**MVSSERV** 1. A program that provides the Server-Requester Programming Interface (SRPI) and a service request manager on an IBM System/370 using the TSO/E (time sharing option) on MVS/XA. 2. A command processor in TSO/E Release 3. It initializes, terminates, and provides recovery for an enhanced connectivity facility session between a PC and a host system. It also establishes communication and routes requests from the PC user to the corresponding server on the host.

**MVS/XA** Multiple Virtual Storage for Extended Architecture operating system.

**mW** milliwatt.

**M/W** Microwave.

**MWC** MultiWay calling.

**MX-30GA/G** A U.S. military wire dispenser that holds approximately one-half mile (0.8 km) of WD-1/TT wire.

**M.1020** The CCITT 4-wire international leased line recommendation. M.1020 specifies the permissible group delay distortion and permissible signal loss over the voice frequency range as illustrated in the following two figures.

Group delay distortion
(CCITT M.1020)

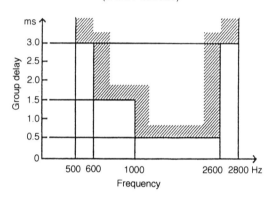

Attenuation distortion
(CCITT M.1020)

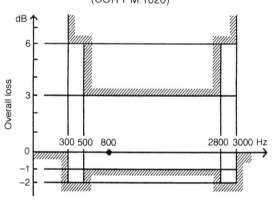

Note: Below 300 Hz and above 3000 Hz the loss shall not be less than 0.0 dB but is otherwise unspecified

*CCITT (now ITU) M.1020 group delay and attenuation distortion*

**M24** A feature offered on AT&T T1 circuits that allows the customer to specify separate destinations for individual channels on a T1 circuit that terminates at a DACS-equipped central office.

**M44** A multiplexing service offered by common carriers that uses ADPCM to compress voice channels and combines up to 44 such channels with signaling onto a single DS1. To access this service, the DS1 format must adhere to M44 specifications as outlined in AT&T Technical Publication 54070.

# N

**NA** Numerical Aperture.

**NAC** Network Administrative Center.

**NACK** Negative ACKnowledgment.

**NADF** North American Directory Forum.

**NAK** Negative AcKnowledgment.

**NAM** Network Access Method.

**name translation** In IBM's SNA network interconnection, converting logical unit names, logon mode table names, and class of service names used in one network into equivalent names to be used in another network. This function can be provided through NCCF and invoked by a gateway SSCP when necessary.

**Named Pipe** A facility included in Microsoft's OS/2 LAN manager which enables processes on separate computers to communicate with each other across a network. The use of Named Pipes provides a simple method for application program developers to write complex distributed network applications.

**namespace** A set of unique names and attributes of network accessible objects. Digital Network Architecture (DNA) Naming Service (DNS) stores objects and attributes in a namespace.

**nameserver** In Digital Equipment Corporation Network Architecture (DECnet), a system with at least one active clearinghouse.

**Nano (n)** A prefix for one billionth of a specific unit.

**nanometer (nm)** One billionth of a meter. The nanometer is often used as a unit of wavelength. Common optical test and operational wavelengths are 850, 1300, and 1550 nm.

**nanosecond** One billionth of a second.

**NAPLPS** North American Presentation Level Protocol Syntax.

**NARC** Non-Automatic Relay Center.

**narrowband channel** Sub-voice-grade channel with a speed ranging from 100 to 200 bits per second.

**NARUC** National Association of Regulatory Utility Commissioners.

**NASI** NetWare Asynchronous Services Interface.

**National Association of Regulatory Utility Commissioners (NARUC)** The trade association for state utility commissioners, representing their interests and concerns before Congress, the FCC, and federal courts.

**National Bureau of Standards (NBS)** A USA government agency that produces Federal Information Processing Standards (FIPS) for all USA government agencies, except the DOD. Membership includes other USA government agencies and network users. NBS was renamed the National Institute of Standards and Technology (NIST).

**National Cable Television Association** Trade organization for cable television carriers.

**National Electrical Code (NEC)** A nationally recognized safety standard for the design, construction, and maintenance of electrical circuits. The NEC, sponsored by the National Fire Protection Association, generally covers electrical power wiring in buildings.

**National Exchange Carrier Association** Trade association for interexchange carriers mandated by the FCC upon the divestiture of AT&T.

**national facilities** In packet switched networks, nonstandard facilities selected for a given (national) network—which may or may not be found on other networks.

**National Institute of Standards and Technology (NIST)** The new name of the U.S. National Bureau of Standards.

**National Research and Education Network (NREN)** A proposed U.S. nationwide multigigabit fiber optic network which will link regional networks, federal research networks and federally sponsored supercomputing centers.

**National Science Foundation Network (NSFnet)** A network operated by the US National Science Foundation which ties together several supercomputer centers and which is designed to support academia.

**National Switch Network** A consortium formed by Telecom*USA, Advanced Telecommunications Corp., Litel Telecommunications Corp. and RCI Corp. to provide each vendor joint access to over 10 000 miles of fiber network.

**National Telecommunications and Information Administration (NTIA)** An agency of the US Department of Commerce concerned with telecommunication standards.

**National Telecommunications Network (NTN)** A consortium of six communications carriers to include Advanced Telecommunications Corp., Consolidated Network Inc., Litel Telecommunications Corp., RCI Long Distance, Telecom*USA, Inc., and Williams Telecommunications Group, Inc.

**native** In IBM's Network Problem Determination Application (NPDA), a term used to describe the network domain controlled by a System Service Control Point (SSCP) that controls the user's terminal.

**native mode** The use of a protocol on the type of network for which that protocol was developed.

**native network** In IBM's SNA, the network attached to a gateway NCP in which that NCP's resources reside.

**NAU** Network Addressable Unit.

**NAUN** Nearest Active Upstream Neighbor.

**NBDVS** NarrowBand Digital Voice System.

**NBH** Network Busy Hour.

**NBS** National Bureau of Standards.

**NC** 1. Network Concept. 2. Network Control (IBM's SNA).

**NCA** National Command Authorization.

**NCC** 1. National Computing Centre, UK. 2. Network Control Center.

**NCCF** Network Communications Control Facility.

**NCCS** Network Control Center System.

**NCL** Network Control Language.

**N-connector** The large-diameter threaded connector used with thick Ethernet cable. N is after Paul Neill.

**NCP** Network Control Program.

**NCP major node** In ACF/VTAM, a set of minor nodes representing resources, such as lines and peripheral nodes, controlled by a network control program.

**NCP PU ADDR** Network Control Program Physcial Unit ADDress.

**NCP PU IDNUM** Network Control Program Physical Unit ID NUMber.

**NCP Token-Ring Interconnection (NTRI)** IBM software for use with a communications controller token-ring gateway. NTRI is a basic feature of IBM's Network Control Program (NCP), beginning with version 4.2.

**NCP/EP Definition Facility (NDF)** In IBM's SNA, a program that is part of System Support Programs (SSP) and is used to generate a partitioned emulation programming (PEP) load module or a load module for a Network Control Program (NCP) or for an Emulation Program (EP).

**NCPGEN** The process on the host computer that creates a network control program (NCP) load module that defines the network configuration and the operating parameters of lines and control units on the data communication network.

**NCR-DNA** NCR Corp.-Distributed Network Architecture.

**NCS** 1. National Communications Systems. 2. Network Control System. 3. Node Center Switch.

**NCSI** Network Communications Services Interface.

**NCT** Network, Control and Timing.

**NCTA** National Cable Television Association.

**NCTE** Network Channel Terminating Equipment.

**NDF** NCP/EP Definition Facility.

**NDIS** Network Driver Interface Specification.

**N/DMI BOS** Network/Digital Multiplex Interface Bit Ordered Signaling.

**NDT** Net Data Throughput.

**NE** Network Element.

**Near-End CrossTalk (NEXT)** Undesired energy transferred from one circuit to an adjoining circuit. Occurs at the end of the transmission link where the signal source is located.

**near-end echo** The reflection of a signal at the two-wire to four-wire interface. This is where the signal leaves the local loop and enters the carriers backbone network. Contrast with far-end echo.

**near-end noise** A noise measurement performed by the Master Test Line unit.

**Near-instantaneous Companding (NIC)** The real-time process of quantizing an analog signal into digital data.

**Nearest Active Upstream Neighbor (NAUN)** For any attaching device on a ring, the device that is sending frames or tokens directly to it.

**NEARNET** New England Academic and Research NETwork.

**NEAT** Network Expert Advisory Tool.

**NEAX-2400** A high-end PBX marketed by Siemens that supports more than 20 000 lines.

**NEC** National Electrical Code.

**NECA** National Exchange Carrier Association.

**Negative Acknowledgment (NAK or NACK)** 1. In BSC communications protocol, a control character used to indicate that the previous transmission block was in error and the receiver is ready to accept retransmission of the erroneous transmission block. 2. In multipoint systems, the not-ready reply to a poll.

**negative polling limit** For a start–stop or BSC terminal in IBM's SNA, the maximum number of consecutive negative responses to polling that the communication controller accepts before suspending polling operations.

**negative response** In IBM's SNA, a response indicating that a request did not arrive successfully or was not processed successfully by the receiver.

**negotiable bind** In IBM's SNA, a capability that allows two logical unit half-sessions to negotiate the parameters of a session when the session is being activated.

**negotiation** When a connection is established between the virtual terminals in different systems, an initial dialogue is needed to establish the parameters that each will use during the transaction. This is known as the negotiation phase.

**neighbor** A system reachable by traversal of a single subnetwork.

285

**nest**  Control or Expander Mainframe of a LINK Family Facilities Management System. Each mainframe accepts various plug-in modules.

**net**  Used as an abbreviated form of network. Within the Telemail system, NET is used to indicate a device connected to a Telenet network.

**Net Partner**  A trademark and network management system of AT&T. Developed by AT&T Bell Laboratories, it allows Integrated Services Digital Network (ISDN) and Centrex customers to work directly with most of the key operations systems controlled by their local exchange carrier for testing, monitoring and maintaining voice and data networks.

**Net 25**  A programmable communications processor marketed by Telematics International of Fort Lauderdale, FL. Net 25 is used mainly in packet switching applications as a packet switch or a network management center.

**NET 1000**  Advanced Information System (ATT-IS).

**NETBIOS**  NETwork Basic Input Output System.

**NetBoss**  A color graphics VTAM network monitor from Allen Systems Group of Naples, FL.

**NetBridge**  A local area network bridge marketed by Shiva Corp. of Cambridge, MA.

**Netcenter**  An IBM graphic user interface that permits users to manage SNA networks. Netcenter also includes a generic interface called the Service Point Interface which enables information from non-SNA equipment to be received and displayed from NetView/PC.

**Netcom**  A consortium of three companies that applied for a license to operate a digital cellular network in Norway. Members of Netcom include Kinevik, Orkla Bouregaard and Nora.

**NetDirector**  A trademark of Ungermann-Bass, Inc., as well as the name of that vendor's network management system.

**netFUSION**  The name of a network management system marketed by BT Tymnet to dedicated service customers.

**Net.God**  A person very visible on a network who may have played an important role in its development.

**NetHub**  The home of the RIME network which is located in Bethesda, MD. All Super-Regional Hubs in the RIME network call the NetHub to exchange mail packets.

**Netiquette**  A term used to indicate proper behavior on a network.

**netline**  An X.25 Data Terminal Equipment (DTE) link from a host or concentrator into the network.

**NetMail**  1. An electronic mail software package marketed by Bramalea Software Systems of Brampton, Ontario, Canada, for Prime Computer systems. 2. A private message transmission capability on FidoNet in which nodes directly communicate with one another on a point-to-point transmission basis. NetMail was originally developed for use by SYSOPS to communicate with one another and is available on some BBSs for regular users.

**Net/Master**  A network management system developed by Cincom Systems Inc. of Cincinnati, OH, which rivals IBM's Netview network Management system.

**NETremote**  A software program marketed by Brightwork Development of Red Bank, NJ, that allows users to view the screen and control the keyboard of another personal computer on the same LAN, a different LAN or across a wide area network.

**Net-Search**  A placement service of Timeplex, Inc., of Woodcliff Lake, NJ, which helps customers find and train communications personnel.

**NetStream**  A British Telecom service that represents a range of local area network products. Under NetStream, users can interconnect PABX's, computers, Kilostream packet services and other transmission facilities.

**Netsys**  The name given to a long-term network and system architecture at Pacific Bell which evolved from a Bellcore prototype called Information Network Architecture (INA).

**NetView** In IBM's System Network Architecture, a mainframe product which provides an unified approach to SNA and non-SNA, IBM and non-IBM network management.

**NetView Status Monitor (STATMON)** A NetView subsystem which provides the network operator with a summary status of network resources.

**NetView/PC** An adaptation of the NetView approach for use in personal computers.

**NetVisualyzer** A set of network management tools from Silicon Graphics, Inc., for distributed, multivendor networks. NetVisualyzer monitors communications on TCP/IP networks and on Ethernet LANs.

**NetWare** LAN networking products produced by Novell, Inc.

**NetWare Asynchronous Service Interface (NASI)** A modem pooling software product that Novell, Inc., licensed from Network Products Inc. Network Products markets its own version as Network Communications Services Interface (NCSI).

**NetWare Operating System** LAN network operating system produced by Novell, Inc.

**NetWare Shell** Novell LAN software which is loaded into the memory of each workstation and which "shells" DOS to enable the workstation to communicate with the LAN file server. The NetWare shell intercepts workstation requests before they reach DOS and reroutes network requests to a file server. The shell allows different types of workstations to use Novell NetWare.

**NetWare 2.X** The revised name for Novell's LAN operating system designed to operate on 80286-based servers. Formerly the software was called NetWare 286.

**NetWare 3.X** The revised name for Novell's LAN operating system designed to operate on 32-bit 80386-based servers. Formerly the software was called NetWare 386.

**NetWare/S-net** Novell's PC network package for IBM PC XT computers.

**Netwise RPC Tool** An application software development product marketed by Netwise, Inc., of Boulder, CO. Netwise RPC Tool is an applications software development product that lets an applications programmer distribute the execution of a program between a server and a client on a local area network using remote procedure calls.

**network** 1. A series of points connected by communications channels. 2. The switched telephone network is the network of telephone lines normally used for dialed phone calls 3. A private network is a network of communications channels confined to the use of one customer. 4. In IBM's SNA, an interconnected group of nodes; a user application network in data processing. 5. A group of computers connected together to facilitate the transfer of information.

**Network A** The portion of the FTS2000 network operated by AT&T for the General Services Administration of the U.S. Government.

**Network B** The portion of the FTS2000 network operated by Sprint for the General Services Administration of the U.S. Government.

**network address** 1. In packet switching, a unique identifier for every device (data terminal, host computer, switch or concentrator) that identifies the device for connection through the network. The address consists of a 12- or 14-digit number made up of the Data Network Identifier Code (DNIC), the area code, the server (TP or host computer), and the individual port subaddress. 2. In IBM's SNA, an address consisting of subarea and element fields that identifies a link, a link station, or a network addressable unit. Subarea nodes use network addresses; peripheral nodes use local addresses. The boundary function in the subarea node to which a peripheral node is attached transforms local addresses to network addresses and vice versa.

**network address translation** In IBM's SNA network interconnection, conversion of the network address assigned to a logical unit in one network into an address in an adjacent network. This function is provided by the gateway NCP that joins the two networks.

**network administrator** A person who manages the use and maintenance of a network.

**network architecture** A computer vendor's plan for configuring networks by interfacing hardware and software products. e.g., SNA.

**Network Basic Input Output System (NETBIOS)** An interface used by application programs in an IBM Personal Computer to access networks and network resources.

**Network Busy Hour (NBH)** The busy hour for an entire network.

**Network Channel Terminating Equipment (NCTE)** Devices required at the customer premises for interfacing to a telephone circuit; the general name for equipment that provides line transmission termination and layer-1 maintenance and multiplexing, terminating a 2-wire U interface in ISDN.

**Network Communications Control Facility (NCCF)** A host-based IBM program through +-which users can monitor and control network operations.

**Network Communications Services Interface (NCSI)** A modem pooling software package for Novell LANs developed by Network Products, Inc. NCSI is licensed to Novell which markets it as NetWare Asynchronous Services Interface (NASI).

**NETwork Computer ARchitecture (NECTAR)** A project funded by DARPA to design and build a high-bandwidth, low-latency network for interconnecting heterogeneous systems for high-performance distributed processing.

**network configuration tables** In IBM's SNA, the tables through which the system services control point (SSCP) interprets the network configuration.

**Network Control (NC)** In IBM's SNA, an RU category used for requests and responses exchanged between physical units (PUs) for such purposes as activating and deactivating explicit and virtual routes and sending load modules to adjacent peripheral nodes.

**Network Control Center (NCC)** Any centralized network diagnostic station used for the control and monitoring of a telecommunications network.

**Network Control Language (NCL)** A high-level structured programming language built into Cincom Systems' Net/Master network management system.

**network control mode** In IBM's SNA, the functions of a network control program that enable it to direct a communication controller to perform activities such as polling, device addressing, dialing, and answering.

**Network Control Program (NCP)** The software provided by IBM to run in its 3705, 3720, 3725, or 3745 communication controllers, responsible for handling links, controlling terminals, etc.

**network control program generation** In IBM's SNA, the process, performed in a host system, of assembling and link-editing a macro instruction program to produce a network control program.

**Network Control Program Physical Unit Address (NCP PU ADDR)** An address which specifies a secondary station on a telecommunications link.

**Network Control Program Physical Unit ID Number (NCP PU IDNUM)** An identification which specifies the physical unit identification number that identifies the switched line to the host computer.

**network controller** An IBM concentrator/protocol converter used with SDLC links. By converting protocols, which manage the way data is sent and received, the IBM 3710 Network Controller allows the use of non-SNA devices with an SNA host processor.

**network coordinator** In FidoNet the network co-ordinator is the "host" of the network who is responsible for coordinating FidoNet within the network, distributing nodelists and the FidoNet newsletter to members of the network. The network coordinator is appointed by the regional co-ordinator.

**Network Driver Interface Specification (NDIS)** An OS/2 LAN Manager executable device driver created by 3Com and Microsoft Corporation. NDIS supports protocol multiplexing allowing such multiple network protocol stacks as IBM's Net-BIOS Extended User Interface and the Xerox Networking System (XNS) to coexist in memory on the same computer.

**Network Element (NE)** A term employed to identify telephone equipment that has active circuitry, such as switching and transmission devices as well as ancillary support equipment.

**Network Expert Advisory Tool (NEAT)** A network management tool initially developed for Harris Corporation's internal use and now marketed to other organizations.

**Network Eye** Software from Artisoft of Tucson, AZ, that works on any NetBIOS compatible local area network and lets one personal computer control up to 100 other PCs, view the screens, take data from their disks and use their processing power.

**network facilities** In packet switched networks, standard facilities are divided into essential facilities (found on all networks) and additional facilities (selected for a given network but which may or may not be selected for other networks).

**network facility** An option on the IBM PC 3270 Emulation Program that determines whether the modem connected to the IBM Personal Computer is using a switched communication line or unswitched communication line.

**Network File System (NFS)** A distributed file system protocol developed by Sun Microsystems and adopted by other vendors which allows computers on a network to use the files and peripherals

of other computers on the network as if they were local. The NFS protocol allows the transfer of parts of files rather than entire files.

**network identifier (Network ID)** In IBM's SNA, the network name defined to NCPs and hosts to indicate the name of the network in which they reside. It is unique across all communicating SNA networks.

**Network Information Center (NIC)** A group at SRI International funded by the Defense Communications Agency to maintain and distribute information about TCP/IP and the connected Internet. The NIC distributes Requests for Comments (RFCs) and Internet Engineering Notes (IENs).

**Network Information Services Company Ltd (NIS)** A company jointly owned by McDonnell Douglas and Marubeni, a leading Japanese trading company, which provides a public data network services in Japan.

**Network Interface (NI)** The physical point (e.g. connecting block, terminal strip, jack) at which terminal equipment or facilities (including station wiring) can connect to the network.

**Network Interface Card (NIC)** An adapter card installed inside the system unit of a computer which provides a physical connection to a network. Examples of NICs include Token-Ring and Ethernet adapter cards.

**network interface machine** A device used to interface a X.25 packet network with non-packet terminals; more commonly known as a Packet Assembler/Disassembler (PAD).

**Network Interface Processor (NIP)** A term used to identify the function of the Telenet Processor 3000 (TP3) product line.

**network job entry facility (jes2)** In IBM's SNA, a facility which provides for the transmission of selected jobs, operator commands, messages, SYSOUT data, and accounting information between communicating job entry nodes that are connected in a network either by binary synchronous communication (BSC) lines or by channel-to-channel (CTC) adapters.

**Network layer**  The third layer in the OSI model; responsible for addressing and routing between subnetworks.

**Network Logical Data Manager (NLDM)**  An IBM program product that collects and correlates LU-LU session-related data and provides the user with on-line access to the information. It runs as an NCCF communication network management application program and provides network problem determination.

**network management**  1. Functions which permit the operation of a network to be controlled and monitored. 2. The systematic approach to the planning, organizing, controlling and evolving of a communications network while optimizing the cost performance relationship.

**Network Management and Administration (NMA)**  An emerging Bellcore standard for managing T1 network operations.

**Network Management Listener**  A Digital Network Architecture (DNA) Network Information and Control Exchange (NICE) protocol routine which listens for NICE protocol requests from remote nodes and passes the function request to the Local Network Management Function. The Listener can also receive commands from the DNA Network Management Access Routines.

**Network Management Productivity Facility (NMPF)**  An IBM product that adds several features to that vendor's NetView product to include a browse facility, a help desk facility, and a library of command lists (CLISTs).

**Network Management Protocol (NMP)**  The protocol used in AT&T's Unified Network Management Architecture (UNMA) to communicate information. NMP is compliant with the OSI seven-layer model and performs the same function as LU6.2 and SSCP–LU for IBM's NetView/PC.

**Network Management System (NMS)**  A minicomputer system located in a Network Control Center (NCC) and connected to a network as a host.

**Network Management Vector Transport (NMVT)**  In IBM's SNA, a solicited or unsolicited record containing alert data and issued by certain SNA resources to the host system.

**network manager**  1. A program or group of programs that is used to monitor, manage, and diagnose the problems of a network. 2. A person using network management.

**network name**  1. In IBM's SNA, the symbolic identifier by which end users refer to a network-addressable unit (NAU), a link station, or a link. 2. In a multiple-domain network, the name of the APPL statement defining an ACF/VTAM application program is its network name and it must be unique across domains.

**network node**  Synonym for node (IBM's SNA).

**network operator**  In IBM's SNA, 1. A person or program responsible for controlling the operation of all or part of a network. 2. The person or program that controls all the domains in a multiple-domain network.

**network operator console**  In IBM's SNA, a system console or terminal in the network from which an operator controls a communication network.

**Network Packet Switching Interface (NPSI)**  An IBM X.25 product which permits that vendor's communications controllers to be interfaced to packet networks.

**Network Performance Monitor (NPM)**  An IBM program host product that uses VTAM. It records performance data collected for various devices in a network.

**network port**  A set of interfaces for receiving data from the outside world and sending data to the outside world.

**Network Problem Determination Application (NPDA)** A host IBM program that aids a network operator in interactively identifying network problems from a central point.

**Network Processing Supervisor (NPS)** A software program resident in Honeywell Datanet 355 or 6600 front-end processors that controls communications.

**Network Processor (NP)** A computer that controls data communications facilities. It performs functions such as routing, switching, concentrations, link/terminal control and, sometimes, information processor interfacing.

**Network Protection Capability (NPC)** An alternate routing feature for T1 and T3 services marketed by AT&T. T1 and T3 services obtained with NPC will have service reestablished within two seconds after an outage occurs in locations where alternate fiber optic transmission facilities are available.

**network relay** Synonym for router.

**Network Routing Facility (NRF)** An IBM program product that resides in the NCP, which provides a path for messages between terminals, and routes messages over this path without going through the host processor.

**network server** A node on a local area network (LAN) that is configured to share its resources with the other PCs on the LAN.

**Network Service Access Point (NSAP)** An addressable point at which the network service is made available.

**Network Services (NS)** In IBM's SNA, the services within the network-addressable units (NAUs) that control network operations via sessions to and from the host system services control point (SSCP).

**network services header** In IBM's SNA, a 3-byte field in an FMD request/response unit (RU) flowing in an SSCP–LU, SSCP–PU, or SSCP–SSCP session. The network services header is used primarily to identify the network services category of the RU (for example, configuration services,

session services) and the particular request code within a category.

**Network Services Procedure Error (NSPE)** In IBM's SNA, a request unit that is sent by an SSCP to an LU when a procedure requested by that LU has failed.

**Network Services Protocol (NSP)** A protocol operating in the DNA Transport layer.

**network station** In the IBM PC 3270 Emulation Program, a node on a local area network (LAN) that must use its gateway to communicate with a System/370 commputer.

**Network Terminal Number (NTN)** Number identifying the logical location of a DTE connected to a network; the NTN may contain a subaddress used by the DTE rather than by the network to identify equipment or circuits attached to it. The NTN can be up to 10 digits long.

**Network Terminal Option (NTO)** IBM program that enables an SNA network to accommodate non-SNA asynchronous and BSC devices via the NC-driven controller.

**network terminating wire** PBX-type service wire used to connect the intrabuilding network cable to the network interface.

**network termination** In ISDN, the device at the end user's premises that receives signals from the provider's network and delivers it to the terminal equipment. The network termination can comprise two separate parts, NT1 and NT2; in this case, NT1 converts the network signal from the 2-wire local loop to the 4-wire end-user interface, and NT2 supplies the necessary demultiplexing, switching, and local signaling intelligence. The interface between NT1 and NT2 is the T-reference point; the interface between NT2 and the terminal equipment is the S-reference point. Outside the United States, the NT can be part of the provider's network; within the United States, the NT is defined as customer premises equipment and cannot be considered part of the provider's network. In this case, the interface between the NT and the network is the U-reference point.

**Network Termination Unit (NTU)** A network termination unit is a piece of equipment forming part of a network. Typically it has a keyboard, to select (dial) calls, and lamps or a display to indicate call progress signals.

**Network Termination 2 (NT2)** A device, such as a PBX, LAN or communications processor that concentrates two or more Terminal Equipment Type 1 (TE1) or Terminal Adapters (TAs) at the ISDN S reference point.

**Network Terminator (NT)** In ISDN the network side of the S reference point.

**network topology** Describing the physical and logical relationship of nodes in a network.

**network transparency** The property of a communications system that makes it independent of the characteristics of the terminals and computers.

**Network User Identification (NUI)** A combination of an ID and password which enables a value added carrier subscriber to obtain access to an overseas network or host computer attached to the carrier's network.

**Network Users Group AT&T (NUGATT)** An independent users group formed to provide recommendations to AT&T for the development of network products and management systems. NUGATT's objective is to encourage unified architecture and integration of products.

**network virtual terminal** A concept where DTEs with different data rates, protocols, codes, and formats are handled on the same network.

**Network Virtual Terminal (NVT) Protocol** A Digital Network Architecture (DNA) protocol which insulates user applications from the network by creating a network virtual terminal.

**Network WATS** A Sprint Communications multi-location pricing plan for WATS users.

**Network ii.5** An analysis tool marketed by CACI of La Jolla, CA, which accepts computer or communications system descriptions and provides measures of hardware utilization, software execution, and conflicts.

**Network 900** A reverse 800 service marketed by Sprint Communications in which the caller pays for the call.

**Network-Addressable Units (NAU)** An SNA term referring to special program code segments that are used to represent software programs and hardware devices to a network; not the actual programs or devices but a portion of a control program that represents the programs or devices.

**Network/Digital Multiplex Interface Bit Ordered Signaling (N/DMIBOS)** A signal scheme in which all signaling bits for 23 data channels on a T1 line are concentrated onto one 64 Kbps signaling channel.

**networking** In an SNA, multiple-domain network, communication among domains.

**networking multiplexer** A T1 multiplexer capable of channel bypass, drop-and-insert multiplexing, and intelligent channel-by-channel switching of T1 circuits; also called a switching multiplexer.

**NET$OS** A Novell file server program.

**neutral current loop** Same as single-current version of current loop; in double-current version, the no-current condition is illegal and indicates a system failure.

**neutral signaling** A method of signaling in which signal levels are referenced to ground. Examples of neutral signaling include the RS-232/V.24 interface.

**neutral transmission** Method of transmitting teletypewriter signals, whereby a mark is represented by current on the line and a space is represented by the absence of current. By extension to tone signaling, neutral transmission is a method of signaling employing two signaling states, one of the states representing both a space condition and also the absence of any signaling. (Also called unipolar transmission.)

**New England Academic and Research Network (NEARNET)** A regional network that links Ethernets at over 40 New England universities, colleges and business and private research laboratories to the National Science Foundation Network (NSFNET).

**Newsnet** An on-line information service which provides full text of over 300 speciality newsletters. For additional information telephone 1-800-345-1301.

**NewWave** A software applications environment for MS-DOS personal computers developed by Hewlett-Packard. NewWave uses object-management technology and automatically updates related files from different applications.

**NEXT** Near End CrossTalk.

**NF** Noise Figure.

**NFS** Network File System.

**NHK** Nippon Hoso Kyokai.

**NI** Network Interface.

**NIB** Node Initialization Block (IBM's SNA).

**NIB list** In IBM's SNA, a series of contiguous node initialization blocks.

**nibble** The first or last half of an 8-bit byte.

**NIC** 1. Near Instantaneous Companding. 2. Network Information Center. 3. Network Interface Card.

**NICE** Network Information and Control Exchange.

**nickname** 1. A locally assigned name used to refer to a global name. 2. In IBM's PROFS and Office Vision a nickname is a name that a user may be able to use in place of a user name and a system name (location). Nicknames are usually easier to remember than user names.

**NICS** NATO Integrated Communications System.

**Nifty-Serve** The Japanese on-line information utility which is modeled after the CompuServe Information Service.

**NIM** Network Interface Machine.

**Nippon Hoso Kyokai (NHK)** The Japan Broadcasting Corporation.

**Nippon Ido Tsushin** A cellular telephone service provider in the Tokyo, Japan, area.

**Nippon Telegraph and Telephone Corporation (NTT)** The dominant Japanese telephone company which became a public company during 1985.

**NIST** National Institute of Standards and Technology.

**NL** New Line.

**NLDM** Network Logical Data Manager (IBM's SNA).

**NLQ** Near Letter Quality (printing).

**NM** Network Management.

**nm** Nanometer.

**NMA** Network Management and Administration.

**NMC** Network Management Center.

**NMCC** 1. National Military Command Center. 2. Network Management Command and Control. 3. National Military Communications Center.

**NMP** Network Management Protocol.

**NMPF** Network Management Productivity Facility.

**NMS** Network Management System.

**NMVT** Network Management Vector Transport (IBM's SNA).

**No.** Number.

**no response** In IBM's SNA, a value in the form-of-response-requested field of the request header (RH) indicating that no response is to be returned to the request, whether or not the request is received and processed successfully.

**nodal processor** A specialized computer system used to control the flow of traffic at a network node, often involving protocol conversion, buffering, multiplexing, and similar network control functions.

**node** 1. In general, a point of interconnection to a network. 2. In multipoint networks, a unit that is polled. 3. In LAN technology, a unit on a ring; often used as a synonym for station. 4. In packet switched networks, one of the switches forming the network's backbone. 5. In IBM's SNA, a junction point in the network that contains a physical unit (PU). A node also contains other network-addressable units, path control components and data link control components, and may contain boundary function. 6. In IBM's ACF/VTAM, a point in a network defined by a symbolic name. Synonymous with network node. 7. A computer that operates as part of a network. On a bulletin board system a node normally calls a hub to exchange mail.

**node entity** The top-level entity in the management hierarchy of a system.

**Node Initialization Block (NIB)** In IBM's ACF/VTAM, a control block associated with a particular node or session that contains information used by the application program to identify the node or session and to indicate how communication requests on a session are to be handled by ACF/VTAM.

**node name** In IBM's ACF/VTAM, the symbolic name assigned to a specific major or minor node during network definition.

**node type** In IBM's SNA, the node type is the classification of a network device based upon the protocols it supports and the network-addressable units (NAUs) it can contain. Currently, Type 1 and Type 2 nodes are peripheral nodes; Type 4 and Type 5 are subarea nodes.

**NODEID** In IBM's PROFS and Office Vision the system name which is also known as a system's location.

**Nodelist** The list of nodes participating in the exchange of messages in a particular bulletin board system network, such as FidoNet or RelayNet. The nodelist defines the topology of the network.

**NOI** Notice Of Inquiry.

**noise** Random electrical signals, generated by circuit components or by natural disturbances, that corrupt the data by introducing errors.

**noise suppressor** Filtering or digital signal-processing circuit that eliminates or reduces noise.

**noise weighting** In telephone communications, a rating system developed to identify the characteristics of the type of receiving element in a telephone handset.

**non blocking** A characteristic of some circuit switching systems, such as PABXs, in which sufficient internal switching paths are provided that, no matter the call demand, no call will be blocked for lack of sufficient switching capacity.

**nondeterministic network** A network in which the access delay cannot be predicted with certainty, because it is based on a probability function. Also called stochastic network.

**nonerasable** Data storage that is unalterable; read-only.

**nonimpact printer** A device that does not use mechanical strikes to print. Also called thermal, laser, and electrostatic.

**noninteractive system** A system where no interaction takes place between the user terminal and the computer during program execution. Usually off-line.

**Non-linear distortion** A type of distortion resulting from attenuation of a signal level. Also clipping.

**non-native network** In IBM's SNA, any network attached to a gateway NCP that does not contain that NCP's resources.

**non-persistent** In LAN technology, a term used to describe a CSMA LAN in which the stations involved in a collision do not try to retransmit immediately—even if the network is quiet.

**Non-Return-to-Zero (NRZ)** A unipolar digital signal which has only two states, High and Low, where Low is typically zero volts and High is a positive voltage. The signal stays at the level (High or Low) for the total clock period.

**Non-Return to Zero Inverted (NRZI)** In SDLC, a binary encoding technique in which a change in state represents a binary 0 and no change in state represents a binary 1. Also known as invert-on-zero coding.

NRZI coding

**non-sharable** A Novell NetWare file attribute which makes an opened file appear to be locked to other users who subsequently attempt to open the file.

**non-SNA terminal** In IBM's SNA, a terminal that does not use SNA protocols.

**nonswitched data link** In IBM's SNA, a connection between a link-attached device and a communication controller that does not have to be established by dialing.

**non-switched line** Leased line; a telecommunications line on which connections do not have to be established by dialing.

**nontransparent mode** A transmission mode of operation in which control characters and control character sequences are recognized through the examination of all transmitted data. Mainly used with byte-oriented protocols, such as IBM's Bisync.

**nonvolatile** A term used to describe a data storage device (memory) that retains its contents when power is lost.

**nonvolatile storage** Any storage medium or circuitry where the contents are not lost when power is turned off or interrupted.

**NORDUNET** NORDic University NETwork.

**NORM** Normal.

**normal contacts** Contacts of an access jack normally closed when a plug is not inserted into the jack and are opened by a plug.

**normal flow** In IBM's SNA, a data flow designated in the transmission header (TH) that is used primarily to carry end-user data. The rate at which requests flow on the normal flow can be regulated by session-level pacing. NOTE: The normal and expedited flows move in both the primary-to-secondary and secondary-to-primary directions. Requests and responses on a given flow (normal or expedited) usually are processed sequentially within the path, but the expedited-flow traffic may be moved ahead of the normal-flow traffic within the path at queueing points in the half-sessions and for half-session support in the boundary functions.

**normal intensity** An attribute of a display field; causes data in that field to be displayed at the normal brightness level.

**normal mode** An HDLC operational mode used in DNA over half-duplex links.

**North American Directory Forum (NADF)** A consortium of electronic messaging service providers that was formed to develop a universal messaging directory based upon the CCITT (now ITU) X.500 recommendation.

**North American Presentation Level Protocol Syntax (NAPLPS)** An ANSI and Canadian standard protocol for videotext graphics and screen formats. NAPLPS is the first standard adopted for describing the presentation of text and graphics on a display. It is used by several videotext services and provides a common language that can be used to develop picture descriptions and text displays.

**NorthWestNet** Northwestern States Network.

**Note Log** In IBM's PROFS and Office Vision a file in which a user keeps their notes.

**Notice** A worldwide electronic mail service marketed by Infonet, Inc., of El Segundo, CA.

**Notice-Fax** A feature added to the Computer Science Corporation Infonet Notice microcomputer software which allows users to create facsimile documents on a personal computer and send them to another PC, a telex terminal or a facsimile machine.

**notify** In IBM's SNA, a network services request unit that is sent by an SSCP to an LU to inform the LU of the status of a procedure requested by the LU.

**NP** Network Processor.

**NPA** Numbering Plan Area.

**NPDA** Network Problem Determination Application.

**NPDA-designated product code** A 12-bit code, similar (and often identical) to the block number in an IBM SNA node-identifier and used to internally identify hardware and software products.

**NPDU** A Network layer Protocol Data Unit.

**NPM** Network Performance Monitor (IBM's SNA).

**NPS** Networking Processing Supervisor.

**NPSI** X.25 NCP Packet Switching Interface (IBM's SNA).

**NREN** National Research and Education Network, an eventual successor to NSFnet.

**NRF** Network Routing Facility (IBM's SNA).

**NRZ** Non-Return to Zero.

**NRZI** Non-Return to Zero Inverted.

**ns** Nanosecond.

**NS** Network Services (IBM's SNA).

**NSA** National Security Agency.

**NSAP** Network Service Access Point.

**NSAP address** The address of a Network Service Access Point.

**NSAP selector** In Digital Equipment Corporation Network Architecture (DECnet), the last byte of an NSAP address, which selects a particular NSAP and hence a Network layer user, within a system identified by the preceding fields of the address.

**NSFnet** National Science Foundation network. The NSFnet forms the national backbone for the Internet, enabling regional networks to interconnect with one another.

**NSK** Narrow Shift Keying.

**NSN** NASA Science Network.

**NSP** 1. Network Service Part. 2. Network Services Protocol.

**NSPE** Network Services Procedure Error (IBM's SNA).

**NT** Network Termination (Terminator).

**NTIA** National Telecommunications and Information Agency.

**NTN** 1. National Telecommunications Network. 2. Network Terminal Number.

**NTO** Network Terminal Option (IBM's SNA).

**NTPF** Number of Terminals Per Failure.

**NTRI** NCP Token-Ring Interconnection.

**NTRI** Network Token-Ring Interface.

**NTSC signal** National Television System specified signalling and display format; *de facto* standard governing the format of television transmission.

**NT1 (Network Termination 1)** In ISDN, the portion of the Network Termination that converts the physical signal from the 2-wire local loop to the 4-wire customer premises circuit. NT1 resides on the network side of the T-reference point. In the United States, NT1 resides on the user side of the U-reference point.

**NT2 (Network Termination 2)** In ISDN, the portion of the Network Termination that provides local multiplexing, switching, and signaling intelligence. NT2 resides on the network side of the S-reference point and the user side of the T-reference point.

**NUA** Network Users Association.

**Nucleus Initialization Program (NIP)** The program that initializes the resident control program; it allows the operator to request last-minute changes to certain options specified during system generation.

**NUGATT**   Network Users Group AT&T.

**NUI**   Network User Identification.

**NUL**   An ASCII code ("nothing") which is used as a fill character in some communications formats.

**null character**   A character (with all bits set to mark) used to allow time for a printer's mechanical actions, such as return of carriage and form feeding, so that the printer will be ready to print the next data character. Same as idle character.

**null modem**   A device that connects two DTE devices directly by emulating the physical connections of a DCE device.

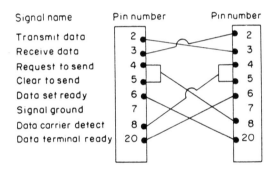

**null point**   The point at which an antenna is positioned to produce a minimum reception of a signal.

**NUM**   Numeric.

**Number 6**   Signaling system CCITT #6. CCITT recommendations Q.251–Q.300.

**Number 7**   Signaling system CCITT #7. CCITT recommendations Q.701–Q.795 and Q.920–Q.931.

**Numbering Plan Area (NPA)**   The geographic subdivision of territory covered by a national or integrated numbering plan. An NPA is identified by a distinctive area code (NPA code).

**numeric lock**   A keyboard condition in which a computer locks the keyboard if a key other than a number, a decimal, or a minus is pressed.

**Numerical Aperture (NA)**   A measure of the angular acceptance of an optical fiber. The numerical aperture is approximately the sine of the half-angle of the acceptance cone.

**Numeris**   The name under which ISDN services are being marketed in France.

**Numero Vert**   France's toll-free service which includes call forwarding during predefined time windows or during traffic overload as well as international access.

**NUTS**   Non-used Trunk Scan.

**NVT**   Network Virtual Terminal.

**NXX**   A code which designates a specific central office. "N" is any number from 2 to 9. "X" is any number from 0 to 9. Also called the exchange code.

**Nyquist sampling rate**   In communication theory, the rate at which a band-limited signal must be sampled in order to accurately reproduce the signal; equal to twice the highest frequency component of the signal.

**Nyquist theorem**   A rule in information theory, Nyquist's theorem states that, in order to reproduce a signal of a given frequency accurately through digital sampling, the sampling frequency must be at least twice the frequency of the signal being sampled; used to derive the bandwidth necessary for PCM signals.

**N2 count**   In X.25 packet switched networks, the count for allowable number of retransmissions.

**N9635**   A V.32 modem manufactured by NEC America Inc. that is downward compatible with 9600 and 4800 bps modems as well as with most 300 and 1200 bps modems. The N9635 supports the MNP Class 5 protocol and features an optional module for diagnostics and network configuration control.

# O

**OA**  Office Automation.

**OAM**  Once-A-Month.

**OA&M**  Operations, Administration, and Maintenance.

**object code**  Executable machine code resulting from programs that were compiled or assembled.

**object entry**  In Digital Equipment Corporation Network Architecture (DECnet), a Naming Service entry which contains the attributes of a network object.

**OBR**  OutBound Record.

**OCC**  Other Common Carriers.

**OCCF**  Operator Communication Control Facility (IBM's SNA).

**OCLID**  Outgoing Called Line IDentification.

**OCR**  Optical Character Recognition.

**OCTAL**  A digital system with eight states, 0 through 7.

| Octal | Decimal |
|-------|---------|
| 0 | 0 |
| 1 | 1 |
| 2 | 2 |
| 3 | 3 |
| 4 | 4 |
| 5 | 5 |
| 6 | 6 |
| 7 | 7 |

| Octal | Decimal |
|-------|---------|
| 10 | 8 |
| 11 | 9 |
| 12 | 10 |

**OCTET**  A logical association of eight consecutive bits. In packet switched networks, a grouping of 8 bits; similar but not identical to byte. In T1 transmission, the octet is the unit of transmission for an end-user channel; each octet in a D4 or ESF frame carries one channel.

**octet interleaving**  A multiplexing technique in which the low-speed channels are combined onto the high-speed channel 8 bits at a time. The DS1 format uses octet interleaving.

**OCU**  Office Channel Unit.

**OC-1**  Optical carrier level 1. Lowest optical rate in the Sonet standard (51.84 Mb/s).

**OC-3**  Optical carrier level 3. Next highest optical rate in the Sonet standard (155.52 Mb/s).

**ODD**  Office Dependent Data.

**ODP**  Originator Detection Pattern.

**OEM**  Original Equipment Manufacturer.

**OFEQ-1**  The first satellite launched by Israel in 1988. OFEQ-1 is the experimental predecessor to Amos, Israel's first communications satellite.

**off-hook**  In a telephone environment, a condition indicating the active state of a subscriber's telephone circuit. A modem automatically answering a call on the dial network is said to go "off-hook."

**Office Automation (OA)** A term used to describe the process of making wide use of the latest data processing and data communications technology—electronic mail, word processing, file and peripheral sharing, and electronic publishing—in the office environment, usually involving the installation of LAN.

**Office Channel Unit (OCU)** A telephone company device used at a central office to reciprocate the functions of a channel service unit or data service unit on a digital circuit.

**office class** The functional ranking of a telephone company network switching center based upon its transmission requirements and its hierarchical relationship to other switching centers.

**Office Facsimile Application (OFA)** IBM's first facsimile software which enables PC or workstation users connected to an IBM mainframe or mid-range computer to communicate with Group III fax machines.

**Office of Telecommunications (OFTEL)** A regulatory agency in the United Kingdom.

**OfficeVision** An OS/2 Presentation Manager application program marketed by IBM Corporation which enables users of PCs, ATs, PS/2s, AS/400s, minicomputers and mainframes to swap and share files.

**OFFLIN** Off-line.

**Off-line** 1. Equipment not directly connected to a central system. 2. A condition in which a user, terminal, or other device is not actively transmitting data. Originally, the term referenced not in the line loop. In telegraph usage, paper tapes frequently are punched "off-line" and then transmitted using a paper tape transmitter.

**Off-line operation** In U.S. military communications, a method of operation in which the processes of either encryption and transmission or reception and decryption are performed as separate steps, rather than automatically and simultaneously.

**off-loading** Process of programming communications processing devices to alleviate processing requirements of other more expensive devices in the network, e.g. a FEP off-loads a host. A terminal may off-load a concentrator.

**off-net (DDS)** A location beyond the primary serving area of Digital Data System.

**Off-Premises Extension (OPX)** Telephone extension located at a distance from the main switch.

**OFP** Optical Fiber Patch cable.

**OFRPTR** IBM 8219 Optical Fiber RePeaTeR.

**OFTEL** Office of Telecommunications.

**OIC** Only-In-Chain (IBM's SNA).

**OJCS** Organization of the Joint Chiefs of Staff.

**OLTP** On-Line Transaction Processing.

**OLU** Origin Logical Unit (IBM's SNA).

**Omnimux** A trademark of Racal-Milgo as well as the name for a series of multiplexers manufactured by that vendor.

**Omninet** A combination of hardware and software from Corvus Systems that uses a dedicated hard disk as a disk server on a local area network. Omninet is based on low-speed transmission using twisted pair cables.

**OMTN** Other Military Teletypewriter Networks.

**ONEMail** An electronic mail system marketed by Honeywell Bull, Inc., of Waltham, MA, for use on that vendor's DPS6 computer systems.

**ones' density** The requirement that a digital signal must carry a certain number of logical "one" or "mark" pulses to maintain timing through repeaters. Digital circuits now in use must maintain a ones' density of 1 "one" bit per 8 bits transmitted.

**one-way function** A function which it is reasonably easy to calculate but for which the inverse is so difficult to compute that, for all practical purposes, it cannot be done.

**one-way trunk** A trunk between a PBX and a central office, or between central offices, where traffic originates from only one end.

**on-hook** Deactivated condition of a subscriber's telephone circuit, in which the telephone circuit is not in use. A modem not in use is said to be "on-hook."

**ONI** Optical Network Interface.

**ONLIN** On-line.

**on-line** In data processing, pertaining to equipment or devices which are directly connected and controlled by a central processor. Originally meant being directly in the line loop. In telegraph usage, transmitting directly onto the line rather than, for example, perforating a tape for later transmission.

**on-line computer** A computer used for on-line processing in which the input data enter the computer directly from their point of origin and/or output data are transmitted directly to where they are used. The intermediate stages such as punching data into cards or paper tape, writing magnetic tape, or off-line printing, are largely avoided.

**on-line cryptographic equipment** A cryptographic device connected electrically to a communication circuit, automatically encrypting or decrypting all transmissions passing through the equipment.

**on-line operation** In U.S. military communications, a method of operation whereby messages are encrypted and simultaneously transmitted from one station to one or more distant stations, where reciprocal equipment is automatically operated to permit reception and simultaneous decryption of the messages. The cryptographic equipment and communications equipment are connected electrically.

**on-line processing** A method of processing data in which data is input directly from its point of origin and output directly to its point of use.

**on-line test** A diagnostic test or data collection program that is run without interrupting the normal operation of a device and any terminals connected to it.

**On-Line Transaction Processing (OLTP)** A computer system that is programmed to transform a data base from one state to another in real time.

**Only-In-Chain (OIC)** In IBM's SNA, a request unit whose request header (RH) begin chain indicator and RH end chain indicator are both on.

**ONM** Open Network Management.

**ONMM** On-Board Microwave Modem.

**on-net (DDS)** A location within the primary serving area of a Digital Data System.

**OnTyme** A registered trademark of McDonnell Douglas's Tymnet public data network as well as an electronic mail system offered by Tymnet (now BT Tymnet).

**OOF** Out Of Frame.

**OOS** Out Of Service.

**OP CEN** Operations CENter.

**OP INIT** Operator INITials.

**OPC** Originating Point Code.

**OPD** Originator Detection Pattern.

**open** 1. To make an adapter ready for use. 2. A break in an electrical circuit.

**open architecture** An architecture that is compatible with hardware and software from any of many vendors.

**open circuit** 1. A circuit that is available for use. 2. A discontinuity in an electrical conductor or piece of equipment.

**Open ID** In UNIX an Open ID or Open Account requires no log-in password.

**Open Look** A mouse-based, windowing, standard user interface with pull-down menus and icons jointly developed by AT&T and Sun Microsystems, Inc.

**open network architecture** Data communications

networks that do not use physically dedicated transmission paths between end-use equipments.

**Open Network Management (ONM)** An IBM network management architecture which allows users and vendors to incorporate non-IBM and non-SNA resource management under SNA.

**open number** In U.S. military communications, a sequential channel number on the received number sheet for which a transmission bearing a corresponding number has not been received.

**Open Shortest Path First (OSPF)** A dynamic routing protocol designed to facilitate interoperability in large heterogeneous networks and provide services to include least-cost routing not available in the TCP/IP Routing Information Protocol (RIP).

**Open Software Foundation (OSF)** A private organization representing computer manufacturers formed to develop a standard, easy to use interface to manipulate Unix programs.

**open system** An open system is a computer and communications system capable of communicating with other computer and communications systems, where each system implements common international standard data communications protocols.

**Open System Interconnection (OSI)** The Open System Interconnection (OSI) is a seven-layered ISO compliant architecture for network operations that enables two or more OSI devices to communicate with each other. The philosophy guiding OSI is that an open system of interconnection is possible if an encompassing set of standards is created. With the implementation of standardized protocols, computers, and related devices can exchange information despite a diversity in makes and models.

**open wire** A conductor separately supported above the surface of the ground—i.e., supported on insulators.

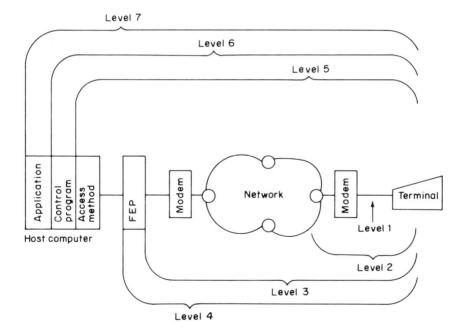

*Open System Interconnection (OSI)*

**open-air transmission** Radio frequency communications technique, including microwave, FM radio, and infrared.

**open-wire line** Parallel bare conductors that are strung on electrical insulators, mounted on the cross arms of telephone poles. Two wires constitute a line.

**operating environment** The combination of computer software to include the operating system, telecommunications access method, data base software and user applications.

**operating signals** In U.S. military communications, an operating signal is a group of either three letters or three letters followed by a digit; the first letter is always either Q or Z. Stations use operating signals as a brevity code during all phases of communications. Operating signals that begin with Q may be used by both military and civilian stations; signals that begin with Z are used only by Allied military stations.

**operating system** A program that manages a computer's hardware and software components. It determines when to run programs, and controls peripheral equipment such as printers.

**Operating Telephone Company (OTC)** The telephone company holding the local franchise to service a defined geographic area.

**operating time** The amount of time required for dialing the call, waiting for the connection to be established, and coordinating the forthcoming transaction with the personnel or equipment at the receiving end.

**Operating Support Systems (OSS)** Software programs and associated hardware designed to help manage specific business functions. They keep records, update inventory, and forecast usage and equipment needs. Many are interlinked.

**operation code** A code which represents specific operations.

**operational mode** In HDLC, the particular operational state or protocol being used.

**operator, computer** The person who operates a computer.

**Operator Communication Control Facility (OCCF)** In NCCF (IBM's network control communications software), a set of transactions that allow communication with and the operation of remote MVS or DOS systems.

**operator information line** A line on a personal computer or workstation screen used to display messages and status information. This information is normally displayed on the last line on the screen.

**operator orientation point** The generic name given to the point in IBM's model 3800 3 printing process at which the data becomes visible to the operator, and is therefore the point at which all operator commands are directed. Also called the transfer station.

**Operator Plus** The name of a voice-mail system marketed by Brooktrout Technology, Inc., of Wellesley Hills, MA.

**Operator Service Providers of America (OSPA)** A trade association consisting of alternative operator service companies which are firms hired by hotels, airports, and other public relations to handle credit card, collect and third-party billed calls.

**operator services** Any of a variety of telephone services which require the assistance of an operator. For example, collect calls, third-party billed calls, person-to-person calls, Calling Card calls, and directory assistance.

**Operator Station Task (OST)** In IBM's Network Communications Control Facility (NCCF), a subtask that establishes and maintains the on-line session with the network operator. There is one operator station task for each network operator who logs on to NCCF.

**OPLAN** Operations PLAN.

**OPM** Operations Per Minute.

**OPORD** OPeration ORDer.

**Optical Carrier (OC) Level** The following table

lists the optical carrier (OC) levels established under Bell Communications Research (Bellcore's) Synchronous Optical Network (SONET).

SONET SIGNAL HIERARCHY

| Level | Line rate |
|---|---|
| OC-1 | 51.84 Mbps |
| OC-3 | 155.52 Mbps |
| OC-9 | 466.56 Mbps |
| OC-12 | 622.08 Mbps |
| OC-18 | 933.12 Mbps |
| OC-24 | 1244.16 Mbps |
| OC-36 | 1866.24 Mbps |
| OC-48 | 2488.32 Mbps |

**optical cavity** The part of a laser where light is amplified by bouncing it between mirrors.

**Optical Character Recognition (OCR)** The machine identification of printed characters through use of light-sensitive devices.

**optical connectors** Connectors designed to terminate and connect either single or multiple optical fibers. Optical connectors are used to connect fiber cable equipment and interconnect cables.

**optical cross-connect panel** A cross-connect unit used for circuit administration and built from modular cabinets. It provides for the connection of individual optical fibers with optical fiber patch cords.

**optical disk** A very high capacity information storage medium that uses light generated by a laser to read information.

**optical fiber** One of the glass strands—each of which is an independent circuit—in a fiber optic cable.

**optical fiber cable** A transmission medium consisting of a core of glass or plastic surrounded by a protective cladding, strengthening material, and outer jacket. Signals are transmitted as light pulses, introduced into the fiber by a light transmitter (either a laser or light-emitting diode). Low data

loss, high-speed transmission, large bandwidth, small physical size, light weight, and freedom from electromagnetic interference and grounding problems are some of the advantages offered by optical fiber cable. Five common types: single, dual, quad, Lightpack, and ribbon.

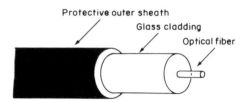

**optical interconnection panel** An interconnection unit used for circuit administration and built from modular cabinets. It provides interconnection for individual optical fibers. Unlike the optical cross-connect panel, the interconnection panel does not use patch cords.

**Optical Isolator** In a fiber optic system a device inserted in series with the fiber to reduce the effect of back reflections.

**optical link budget** A term used to describe the difference, in dB, between the light out of an optical transmitter and the minimum input sensitivity of an optical receiver.

**optical power meter** A light-sensing device used to measure optical power expressed in decibels (dB0 or milliwatts (mW).

**Optical Time Domain Reflectometer (OTDR)** A device used to measure the length of a fiber, the spacing of splices and the loss for each splice. The OTDR can also be used to locate and identify breaks and abnormalities in the fiber.

**optical waveguide** Dielectric waveguide with a core consisting of optically transparent material of low loss (usually silica glass) and with cladding consisting of optically transparent material of lower refractive index than that of the core. It is used for the transmission of signals with lightwaves and is frequently referred to as a fiber. In addition, there are planar dielectric waveguide structures in some optical components, like laser diodes, which also are referred to as optical waveguides.

**optical waveguide fiber**  A transparent filament of high refractive index core and low refractive index cladding that transmits light.

**Optically Remoted Switching Module (ORM)**  An extension of a local switching module into a different georgraphical location. AT&T uses both two-mile and 150-mile ORMs in its fiber network.

**OPX**  Off-Premise Exchange/Extension.

**orderline**  A circuit which connects all of the facilities within a system, usually on a party line basis. An orderline is normally outside the normal band of communications provided by a system for general subscriber use.

**orderly closedown**  In IBM's VTAM, the orderly deactivation of ACF/VTAM and its domain. An orderly closedown does not complete until all application programs have closed their ACBs. Until then, RPL-based operations continue; however, no new sessions can be established and no new ACBs can be opened.

**orders of magnitude**  In order to understand some of the terms used to define measurements, it is necessary to understand the expressions used to delineate very large and very small numbers.

| Terms | Value | Prefix | Example |
|-------|-------|--------|---------|
| Pico | One trillionth (1/1 000 000 000 000) | p | picowatt |
| Nano | One billionth (1/1 000 000 000) | n | nanosecond |
| Micro | One millionth (1/1 000 000) | | microsecond |
| Milli | One thousandth (1/1 000) | m | milliwatt |
| Kilo | One thousand (1 * 1 000) | k | kilohertz |
| Mega | One million times (1 * 1 000 000) | M | megahertz |
| Giga | One billion times (1 * 1 000 000 000) | G | gigahertz |

Examples: 1 000 hertz = 1 kilohertz (1 kHz)
0.000 000 000 001 watt = 1 picowatt (1 pW)

**ORIEN**  ORIENtation.

**Origin Logical Unit (OLU)**  In IBM's SNA, the logical unit from which data is sent.

**Original Equipment Manufacturer (OEM)**  Manufacturer of equipment components built by another manufacturer.

**Originating Point Code (OPC)**  The node address of a central office that is setting up a long distance telephone call on a CCITT (now ITU) common channel Signaling System 7 network.

**originator**  In U.S. military communications, the command by whose authority a message is sent. The originator is responsible for the function of the drafter and releasing officer.

**Originator Detection Pattern (OPD)**  A string of characters sent by a V.42 error-correcting protocol modem which permits the modems communicating with one another to determine if they have similar error-correcting protocols when they first connect.

**Originator/Recipient Name (ORName)**  The specified form used to send messages to users who are part of the X.400 community.

**originators' reference number**  In U.S. military communications, the number assigned to a message by an originator to provide a means of reference.

**ORM**  Optically Remote Module.

**ORName**  ORiginator/Recipient Name.

**OS**  Operating System.

**OSCILLATOR**  An electronic device used to generate repeating signals of a given amplitude or frequency.

**OSDS**  Operating System for Distributed Switching.

**OSDS-M**  Operating System for Distributed Switching in the switching Module.

**OSI**  Open System Interconnection.

**OSinet**  A test network sponsored by the National

Bureau of Standards (NBS) to facilitate the testing of vendor products based on the OSI model.

**Osiscope**   A data analyzer marketed by Artematique, the French national communications authority.

**OSITEL**   A trademark of GSI-Danet of Reston, VA, as well as a family of products from that vendor comprising a full message handling system and directory service according to CCITT standards X.400 and X.500.

**OSITEST**   A trademark of GSI-Danet of Reston, VA, as well as a family of software products from that vendor used for conformance, interoperability and quality assurance testing in the areas of X.25, transport X.224, message handling X.400, teletex and FTAM.

**OSITOOL**   A trademark of GSI-Danet of Reston, VA, as well as a set of software tools from that vendor for supporting the OSI development environment.

**OS/MFT**   IBM Operating System/Multi-programming with Fixed number of Tasks.

**OS/MVT**   IBM Operating System/Multi-programming with Variable number of Tasks.

**OSN**   Operations System Network.

**OSP**   Operator Service Provider.

**OSPA**   Operator Service Providers of America.

**OSPF**   Open Shortest Path First.

**OSPS**   Operator Services Position System.

**OSRI**   Originating Station Routing Indicator.

**OSS**   Operations Support System.

**OST**   1. Office of Science and Technology. 2. Operator Station Task.

**OS/VS1**   IBM Operating System/Virtual Storage 1; used on 370 machines.

**OS/VS2**   IBM Operating System/Virtual Storage 2; used on 370 machines.

**OSS**   Operating Support Systems.

**OSWS**   Operating System WorkStation.

**OS/2**   An operating system jointly developed by Microsoft and IBM which allows programs to use memory in excess of 640 Kbytes.

**OTA**   Office of Technology Assessment.

**OTC**   Operating Telephone Company.

**OTDR**   Optical Time Domain Reflectometer.

**Other Common Carrier (OCC)** Includes specialized common carriers (SCCs), domestic and international record carriers (IRCs), and domestic satellite carriers which are authorized to provide private line services in competition with the established telephone common carriers.

**other field**   Any formatted field that has attributes other than the following: protected, unprotected, normal intensity, or high intensity.

**OTP**   Office of Telecommunications Policy.

**OUSDRE**   Office of Under Secretary of Defense for Research and Engineering.

**outage**   A condition in which a user is completely deprived of service due to any cause within the communication system.

**outage duration**   That period of time between the onset of an outage and the restoration of service.

**outboard primary logical unit**   In IBM's SNA a peripheral node that can start a session by sending a "Bind" request. Outboard PLUs allow two non-host computers to hold sessions over an SNA backbone network.

**Outboard Record (OBR)**   A record originated by I/O and communications components and supported by the access methods, to describe permanent errors or report statistical data.

**outgoing access**   The capability of a user in one network to communicate with a user in another network.

**Outgoing Called Line Identification (OCLI)**   A telephone switch feature which provides a user who is originating a call with information about the called party and the facility or destination.

**outgoing message**  A message addressed to another destination.

**outgoing trunk queueing**  A trunk used to make calls but having no identifiable phone number that can be dialed by others.

**out-of-band (or outband)**  Channel associated signaling which uses signal frequencies within the voice channel bandwidth but outside the voice channel itself, e.g. signals of 0–300 Hz or of 3400–4000 Hz.

**out-of-band signaling**  A technique in which call control signaling for one or more channels is carried in a separate channel dedicated only to signaling, also called common channel signaling. Common channel signaling allows the creation of clear channels. The current international standard for out-of-band signaling is Common Channel Signaling System No. 7.

**Out-Of-Frame (OOF)**  An error condition on a framed, digital circuit, invoked when two of four framing bits are received in error.

**out-of-service testing**  Testing that requires a halt in data transmission to enable a known data pattern to be transmitted and compared. Examples of out-of-service testing include bit error rate (BER), error-free seconds (EFS), severely errored seconds (SES) and percent availability.

**output**  1. Data that has been processed. 2. The state or sequence of states occurring on a specified output channel. 3. The device or collective set of devices used for taking data out of a device. 4. A channel for expressing the state of a device or logic element. 5. The process of transferring data from an internal storage to an external storage device.

**output device**  Device which allows the user to receive data output from a computer system. For example, the screen on a CRT terminal.

**Out-WATS**  Out(ward)-Wide Area Telephone Service.

**Out(ward)-Wide Area Telephone Service (Out-WATS)**  A WATS service that enables customers to call out on a line at a fixed charge, but does not allow incoming fixed charge calls.

**overbooking ratio**  The ratio of the sum of the inputs to the link speed capacity. Usually associated with STDMs and is a result of the concentrating ability of an STDM.

**overflow**  1. Switching equipment which operates when the traffic load exceeds the capacity of the regular equipment. 2. Excess traffic diverted to an alternate route.

**overhead**  Control characters, error control information, synchronization data, etc., sent along with user information in a data communication exchange.

**overhead bit**  A non-data bit used in addressing control, error detection, error control, or synchronization.

**overlay segment**  A portion of a program that can be loaded during execution to overlay another section of the program. An overlay segment is normally used when there is insufficient memory to accommodate all the code of a program.

**override**  An operation that allows certain stations to break into a line already in the busy condition.

**overrun**  Loss of data because a receiving device is unable to accept data at the rate it is transmitted.

**oversampling**  A TDM technique where each bit from each channel is sampled more than once.

**overseas service conditioning**  AT&T conditioning for a two-point overseas service to meet CCITT (now ITU) recommended standards M.1020 or M.1025.

**overspeed**  Condition in which the transmitting device runs slightly faster than the data presented for transmission; overspeeds of 0.1 percent for modems and 0.5 percent for data PABXs are typical.

**over-voltage protection**  Electrical protection provided by devices such as air gap discharge protectors and gas tube protectors and appropriate construction methods (fusing, bonding, grounding, shielding, etc.). Used where served by exposed outside plant.

**owner**   In IBM's PROFS and Office Vision, a document owner is any person who revises a draft document.

# P

**P** 1. Power. 2. Printer. 3. Protected.

**PA** 1. Program Access (key). 2. Program Attention.

**PAB** Process Anchor Block (IBM's SNA).

**PABX** Private Automatic Branch Exchange.

**Pacific Link** The first trans-Pacific fiber optic cable which links the United States and Japan via Guam. Pacific Link consists of three fiber pairs, with each pair capable of supporting transmission speeds up to 240 Mbps. The fiber cable can handle the equivalent of 40 000 simultaneous telephone calls.

**pacing** A method for regulating traffic rates to ensure orderly handling of messages. In IBM's System Network Architecture, it refers to the number of path information units (PIUs) that can be sent before a response is received.

**pacing group** The number of data units (Path Information Units, PIUs) that can be sent before a response is received in IBM's SNA. Also IBM's term for window.

**pacing group size** In IBM's SNA: 1. The number of path information units (PIUs) in a virtual route pacing group. The pacing group size varies according to traffic congestion along the virtual route. Synonymous with window size. 2. The number of requests in a session-level pacing group.

**pacing response** In IBM's SNA, an indicator that signifies a receiving component's readiness to accept another pacing group; the indicator is carried in a response header (RH) for session-level pacing, and in a transmission header (TH) for virtual route pacing.

**packet** A group of bits—including information bits and overhead bits—transmitted as a complete package on a packet switched network. Usually smaller than a transmission block. Often called a message.

**Packet Access Line (PAL)** A dedicated data circuit routed from Southern New England Telephone's ConnNet packet network to a subscribers line.

**Packet Assembler/Disassembler (PAD)** Equipment in a packet network used to convert the format of data from asynchronous or synchronous terminals for transmission over a packet network. A PAD permits terminals which do not have an interface suitable for direct connection to a packet switched network to access such a network as well as convert the terminal's usual data flow to and from packets. The PAD handles all aspects of call setup and addressing.

**packet assembly unit** A user facility which permits non-packet-mode terminals to exchange data on the packet system. Also PCU, NIM.

**packet buffer** Memory space reserved for storing a packet awaiting transmission or for storing a received packet.

**packet data network** Often used to mean packet switched network.

**packet disassembly unit** A user facility which

enables packets to be delivered from the packet network to non packet mode terminals. Combined with a Packet Assembly Unit to form a PAD—Packet Assembler/Disassembler.

**packet header**   In packet switched networks, the first three octets of an X.25 packet.

**Packet Inter Net Groper (PING)**   The name of a program used in the Internet to test the reachability of destinations. This is accomplished by sending them an ICMP echo request and waiting for a reply.

**packet interleaving**   A form of multiplexing in which packets from various subchannels are interleaved on the line. The X.25 interface is an example.

**packet level**   Level 3 of the CCITT X.25 Recommendation which defines the packet format and control procedures for the exchange of packets over a PSDN.

**Packet Level Procedure (PLP)**   A procedure which defines protocols for the transfer of packets between an X.25 DTE and an X.25 DCE. PLP is a full-duplex protocol that supports accountability, data sequencing, flow control and error detection and correction.

**packet mode terminal**   DTE that can control and format, transmit, and receive packets.

**packet radio**   Packet switching in which the transmission paths are radio links and a transmitted packet may be picked up by more than one station. May be used with mobile stations.

**Packet Switch Exchange (PSE)**   A unit which performs packet switching in a network.

**Packet Switch Stream (PSS)**   A value added carrier based upon packet switching operated by British Telecom.

**packet switched data transmission service**   A service that transmits, assembles, disassembles, and controls data in the form of packets, e.g. Tymnet, Telenet.

**packet switched network**   A data communications network that transmits packets. Packets from different sources are interleaved and sent to their destination over virtual circuits. The term includes PDNs and cable-based LANs.

**packet switching**   A technique of switching data in a network whereby individual data blocks or "packets" of controlled size and format are accepted by the networks and routed to their destination. The sequence of packets is maintained and the destination is determined by the exchange of control information (also in packets) between the originating terminal and the network before data transfer starts. The equipment making up the network is shared by all users at all times, packets from different terminals being interleaved throughout the network.

**Packet Switching Exchange (PSE)**   The top level in a network where each PSE is identified by a unique number and contains one or more packet switches. PSEs correspond to the major metropolitan centers within the USA and are interconnected with high-bandwidth, long haul trunks. Area codes are assigned uniquely to the nearest PSE. The lower level in the hierarchy is the node level where each node is identified by a node number unique within its PSE. A X.25 Data Terminal Equipment (DTE) number is known only in the PSE serving that DTE.

**Packet Switching Network (PSN)**   A network designed to assemble data into packets. Each packet is then routed to its destination by the most efficient means possible.

**Packet Switching Node (PSN)**   A minicomputer dedicated to the task of switching packets.

**Packet Switching Service (PSS)**   The U.K. public packet switched data network.

**packet terminal**   A data terminal which can transmit and receive packets without a PAD.

**packet type identifier**   In X.25 packet switched networks, the third octet in the packet header that identifies the packet's function and, if applicable, its sequence number.

**PACKET/74** An SNA Controller/Host PAD designed by PACKET/PC, Inc. of Farmington, CT, which permits communications between an IBM communications controller and a public or private X.25 network. PACKET/74 functions in conjunction with PACKET/3270 software operating on an IBM PC or compatible computer and provides such functions as DFT file transfer, CRC error checking, an Application Program Interface (API) at the PC and bit-compressed communications.

**Packetized Ensemble Protocol (PEP)** A modem operation protocol in which the bandwidth of a dial-up line is split into up to 512 carriers, each of which is capable of supporting data transmission. PEP was developed by Telebit Corporation of Mountain View, CA, and is used in that vendor's series of Trailblazer modems.

**packetized voice** The formation of digitized voice into data packets.

**packet-mode device** Data Circuit-Terminating Equipment (DCE) or Data Terminal Equipment (DTE) that can communicate using packets.

**PACNET** PACific NETwork.

**PAD** Packet Assembler/Disassembler.

**pad characters** 1. In asynchronous transmission, characters added to the end of each line to enable the printhead to return to the beginning of next line before the next line of data is received. 2. In synchronous transmission, characters that are inserted to ensure that the first and last characters of a packet or block are received correctly as they aid in clock synchronization at the receiver. Also called fill characters.

**page** 1. A specified portion of main memory capacity used when allocating memory and for partitioning programs into units or control sections. 2. In IBM's SNA, the portion of a panel that is shown on a display surface at one time. 3. The unit of output from an IBM 3800 model 3 running with full function capability or an IBM 3820 printer.

**page copy** A message in page form that is the result of a transmission.

**page-description language** A special computer language that precisely controls the placement of text and graphics on a page and is able to generate high-quality fonts in a variety of sizes and styles.

**page-mode data** A type of data that can be formatted anywhere on a physical page. This data requires specialized processing such as provided by the Print Services Facility for the IBM 3800-3 and 3820 printers.

**page-mode environment checkpointing** That process which preserves the information necessary to resume page-mode printing.

**page-mode printer** A printer (such as the IBM 3800 model 3 and 3820) that can print page-mode data.

**paging** The ability of the user to move through extended displays (pages) of data by means of commands or the use of Page Up (PgUp) and Page Down (PgDn) or other keys.

**paging/paging system** An audible, one-way communication through an on-site communications system for on-premises paging, or, through special equipment, a means of reaching someone remotely within a specified geographic area. With an internal paging system, you can be notified of important incoming calls even when you are not at your own desk or telephone; external paging systems provide a way to reach you when you are not in the office or at a previously specified telephone number.

**Pair-Selected Ternary (PST)** A pseudo-ternary code in which pairs of binary digits are coded together in such a way that the resultant signal has no long strings of zeros.

**PAL** 1. Packet Access Line. 2. Phase Alternate Line.

**Palcrypt** An encryption system for pay television jointly developed by Thomson S.A. of Paris, France, and News Data Security Products Ltd of the United Kingdom. Palcrypt uses a video scrambling technique based upon line cutting and rotation.

**PAM** 1. Printer Authorization Matrix. 2. Pulse Amplitude Modulation.

**panel** In IBM's SNA computer graphics, a predefined display image that defines the locations and characteristics of display fields on a display surface.

**paper** A document that is on paper, rather than in electronic form. Also known as hardcopy.

**paper bail** A part of a printer which holds the paper against the platen.

**paper jam** The process of paper becoming stuck along the paper path.

**paper tape** An input/output medium, on which data can be recorded as a pattern of punched holes (five or eight channel).

**paper tension unit** A part of a printer which fits on top of its platen to provide proper paper-feed tension.

**paper-out sensor** A small switch behind the platen of a printer that sends a signal when it is not in contact with paper. This signal usually disables the printer.

**PAR** Peak to Average Ratio.

**parallel interface** An interface where an entire group of bits is transmitted at one time by sending each bit over a separate wire. The Centronics parallel interface is a 36-pin, TTL level, byte-wide interface used for computer to parallel printer communications. The transmission of these data bits is controlled by the computer supplied strobe pulse. Flow control is achieved by asserting or deasserting either the ACKNLG or BUSY leads or both. Pins 12, 13, 14, 15, 18, 31, 32, 34, 35, and 36 vary in function depending upon implementation by the manufacturer of the printer. Pins 16 and 17 are commonly used for logic ground and chassis ground respectively.

**parallel interface extender** A subsystem capable of remoting a parallel interface, such as a computer I/O channel. The parallel interface extender functions as a parallel to serial and serial to parallel converter, enabling such devices as line printers and

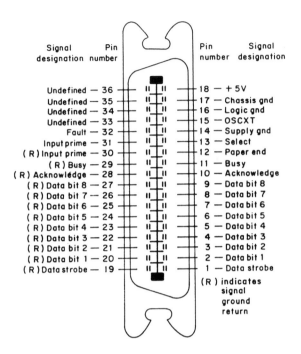

*parallel interface*

card readers to be located remotely from a computer system.

**parallel links** In IBM's SNA, two or more links between adjacent subarea nodes.

**parallel port** An input/output (I/O) port used to connect parallel devices, such as printers and plotters, to a computer. Data transfers occur eight bits at a time over eight wires in parallel via a parallel port.

**parallel processing** Concurrent execution of two or more programs, within the same processor.

**parallel sessions** In IBM's SNA, a parallel session is two or more concurrently active sessions between the same two logical units (LUs) using different network addresses. Each parallel session can have different transmission parameters.

**parallel transmission** A technique that sends each bit simultaneously over a separate line; normally used to send data a byte (eight bits over eight lines) at a time to a high-speed printer or other locally attached peripheral.

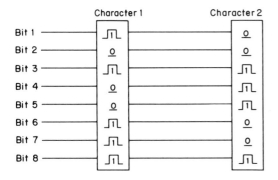

| | Character 1 | Character 2 |
|---|---|---|
| Bit 1 | ⊓ | o |
| Bit 2 | o | o |
| Bit 3 | ⊓ | ⊓ |
| Bit 4 | o | ⊓ |
| Bit 5 | o | ⊓ |
| Bit 6 | ⊓ | o |
| Bit 7 | ⊓ | o |
| Bit 8 | ⊓ | ⊓ |

*parallel transmission*

**parameter**  A variable that takes on any value within a predefined range of values.

**PARC Universal Packet (PUP)**  The Palo Alto Research Center (PARC) protocol developed by Xerox Corporation.

**Parental Right**  An option in Novell NetWare software which enables a user to create or delete subdirectories under the subject directory.

**parity**  A method of checking for errors in data communications. An extra bit (either a "1" or "0"), called the parity bit, is added to the end of each ASCII character to make the final count of "1" bits in the character an even or odd number, according to a prearranged format. Some systems always use even parity, some always use odd parity, and some do not check for parity. Both terminal and system must be set for the same parity.

**parity bit**  The bit which is set to 1 or 0 in a character to ensure that the total number of 1 bits in the data field is even or odd. Or may be fixed at 1 (mark parity), fixed at 0 (space parity), or ignored (no parity). Parity bits trap errors in the following manner: when the transmitting device frames a character, it tallies the number of 0s or 1s within the frame and attaches a parity bit. (The parity bit will vary according to whether the total is even or odd.) Then, the receiving end will count the 0s or 1s and compare the total to the "odd" or "even" recorded on the parity bit. If the receiving end finds a discrepancy, it can flag the data and request a retransmission.

| Parity type | Description |
|---|---|
| odd | Eighth data bit is logical zero if total number of logical 1s in first seven data bits is odd. |
| even | Eighth data bit is logical zero if total number of logical 1s in first seven data bits is even. |
| mark | Eighth data bit always logical 1 (high/mark). |
| space | Eighth data bit always logical 0 (low/space). |
| none/off | Eighth data bit ignored. |

**parity check, horizontal**  A parity check applied to the group of certain bits from every character in a block. Also called longitudinal redundancy check (LRC).

**parity check, longitudinal**  A parity check performed on a group of binary digits in a longitudinal direction for each track. Also called longitudinal redundancy check (LRC).

| | | Character parity bit |
|---|---|---|
| Character | 1 | 10110110 |
| Character | 2 | 01001010 |
| | 3 | 01101000 |
| | 4 | 10010010 |
| | 5 | 01111010 |
| | 6 | 10100001 |
| | 7 | 01011101 |
| | . | 01110011 |
| | . | 10001100 |
| | . | 01101011 |
| Block parity character (LRC) | | 11101011 |

**parity check, transverse**  A parity check performed on a group of binary digits in a transverse direction for each frame. Synonymous with transverse redundancy check.

**parity check, vertical**  A parity check applied to the group which is all bits in one character. Also called vertical redundancy check.

**parity checking**  A technique of error detection in which one bit is added to each data character so that the number of one bits per character is always even (or always odd).

313

**parity error**  An error which occurs in data where an extra or missing bit is detected. Character parity cannot detect an even number of bits.

| | |
|---|---|
| ASCII character R | 1 0 1 0 0 1 0 |
| Adding an even parity bit | 1 0 1 0 0 1 0 1 |
| 1 bit in error | 1 Ø 1 0 0 1 0 1 |
| 2 bits in error | 1 Ø 1 Ø 0 1 0 1 |

**PARS**  The name of the airline reservation system jointly owned by Trans World Airlines and Northwest Airlines.

**parse**  In systems with time sharing, to analyze the operands entered with a command and build up a parameter list for the command processor from the information.

**Part 68**  FCC rules that allow registration of communications equipment provided they meet FCC requirements designed to ensure no harm to the telephone network.

**Partitioned Data Base Management**  A Software-Defined Network (SDN) service feature offered by AT&T which permits a user to group locations into independent subnetworks.

**Partitioned Emulation Programming Extension (PEP)**  IBM software package used with Network Control Program (NCP), allows a communications controller to operate in a split mode, controlling an SNA network while managing a number of non-SNA communications lines.

**party line**  A line where several subscribers' stations are connected with discriminatory signaling, possibly for selective calling.

**Pass Change**  A network security service offered by Tymnet which enables a user to change his/her own password.

**PASS 25**  The name of a multiprotocol concentrator marketed by OST, Inc., of Chantilly, VA, that supports X.25, PAD, Videotex, Honeywell VIP 7700 and SDLC protocols.

**passband**  A contiguous portion of an area in the frequency spectrum which permits a predefined range of frequencies to pass. The passband of a telephone channel permits frequencies between 300 and 3300 Hz to pass.

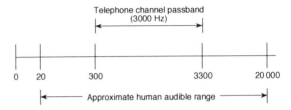

**passband filters**  Filters used to allow only the frequencies within the communications channel to pass while rejecting all frequencies outside the passband.

**Passlife**  A network security service offered by Tymnet which enables an organization using this value added carrier to set a time limit on a user's password.

**passmode**  A terminal state that allows more than one terminal to share the same slot on an IBM 3600 System Loop.

**passthrough**  In general, a term used to describe the ability to gain access to one network element through another. It is also another name for channel bypass. A T1 multiplexing technique similar to drop-and-insert, used when some channels in a DS1 stream must be demultiplexed at an intermediate node. With passthrough, only those channels destined for the intermediate node are demultiplexed—ongoing traffic remains in the T1 signal, and new traffic bound from the intermediate node to the final destination takes the place of the dropped traffic. With drop-and-insert, all channels are demultiplexed, those destined for the intermediate node are dropped, traffic from the intermediate node is added, and a new T1 stream is created to carry all channels to the final destination.

**password**  A combination of letters, or letters and numbers, that are used to gain access to a network and/or computer system. A password is a security measure that helps to prevent unauthorized people from working with another person's computer resources or obtain access to a network.

**password protection**  A method of limiting log-in access to a network by requiring users to enter a password. Unless the password is entered correctly, access will be denied.

**Password Expiration**  A trademark of Computronics of Addison, IL, as well as a program from that vendor which operates on Prime computers to keep users from reusing old passwords, to force users to change passwords periodically and to perform other password functions.

**patch**  1. A temporary change applied to make a software program work correctly. 2. To connect circuits together temporarily by means of a cord (cable) known as a patch cord.

**patch cable**  In the IBM Cabling System, a length of type 6 cable with data connectors on both ends.

**patch cord**  A short length of wire or fiber cable with connectors on each end used to join communication circuits at a cross-connect.

**patch panel**  A system of terminal blocks, patch cords, and backboards which facilitates administration of cross-connect fields for moves and re-arrangements by non-technical end-user personnel, thus enhancing and expanding desirability of PBX and Centrex systems.

**patching jacks**  Series-access devices used to patch around faulty equipment by using spare units.

**path**  In IBM's SNA: 1. The series of path control network components (path control and data link control) that are traversed by the information exchanged between two network-addressable units (NAUs). A path consists of a virtual route and its route extension, if any. 2. In defining a switched major node, a potential dial-out port that can be used to reach a physical unit.

**path attenuation**  The loss of a signal in transit between a transmitter and a receiver, usually measured in decibels.

**path control layer**  In IBM's SNA, the network processing layer that handles routing of data units as they travel through the network and manages shared link resources.

**path control network**  In IBM's SNA, the part of the SNA network that includes the data link control and path control layers.

**path field**  In packet switching, a set of bits within a Call Request packet that indicates Telenet Processors (TPs) have routed or have attempted routing of the Call Request packet. The Path Field prevents the formation of loops in virtual circuit routes.

**Path Information Unit (PIU)**  In IBM's SNA, a message unit consisting of a transmission header (TH) alone, or of a TH followed by a basic information unit (BIU) or a BIU segment.

**path monitoring**  In ISDN the monitoring of basic rate access as a single entity.

**path test**  In IBM's SNA, a test provided by NLDM Release 2 that enables a network operator to determine whether a path between two LUs that are currently in session is available.

**path trace**  A function that may be requested of a bridge by a received frame. The request is for a record of the bridges through which the frame has passed.

**Pathfinder**  A trademark of Ven-Tel Inc. of San Jose, CA, as well as a high-speed modem designed for operation on the public switched telephone network.

**pathname**  A character string which provides the location of a file or directory on a disk.

**pattern sensitivity**  Harmonic distortion caused by the selection of a BERT pattern whose frequency is the same or close to the data line's carrier frequency or the carrier frequency's first, third, or fifth harmonic. Pattern sensitivity causes high error rate test results even if line is functioning properly.

**PAX**  Private Automatic Exchange.

**Pay Per View (PPV)**   The transmission of an event by a cable company to a subscriber that pays for viewing each event.

**payload mapping**   The mapping of network services into synchronous payload envelopes.

**PBX**   Private Branch Exchange.

**PC FormFax**   A memory-resident software program from Commtech International of Atlanta, GA, that enables a personal computer user to design a variety of business forms and transmit the resulting forms to any fax machine or to another PC that has a fax board.

**PC**   1. Path Control (IBM's SNA). 2. Personal Computer. 3. Phase Corrector. 4. Printed Circuit (board).

**PC COMplete**   A communications software program from Transcend Corporation of Portola Valley, CA, which includes a "personal gateway" feature that allows users to exchange mail over public and private E-mail services without using a gateway or file server.

**PC Connect**   A service of MCI which enables computer modem users to obtain discounted long distance rates when calling distant bulletin board systems.

**PC Network**   An IBM broadband local area network.

**PC Pursuit**   An outdial communications service marketed by Sprint Communications of Reston, VA which allows users to communicate with personal computers, host computers, and information services during off-peak hours for a flat fee per month. Subscribers can log onto PC Pursuit and then request the Sprint network to outdial in any city served. This enables subscribers to connect to a host computer or bulletin board service, avoiding the cost of long distance charges.

**PC router**   A program that is a part of the IBM PC 3270 Emulation Program or IBM 3270 PC Control Program that uses the Server–Requester Programming Interface (SRPI), to route requests from the PC requester programs to the corresponding router on the host.

**pcANYWHERE**   A trademark of DMA, Inc., of New York, NY, as well as a PC-based remote computing software program marketed by that vendor.

**PCBoard**   A series of bulletin board software marketed by Clark Development Co., Inc., of Salt Lake City, UT.

**PCI**   Protocol Control Information.

**PCM**   1. Plug Compatible Machine. 2. Pulse Code Modulation.

**PCNET**   A local area network (LAN) developed by Orchid Technology based on coaxial cable running at 1 Mbps transmission rate.

**PC-NFS**   A trademark of Sun Microsystems, Inc., as well as a networking product from that vendor which enables IBM PC and compatible computers to access a variety of small and large computer systems to include DEC VAX and IBM SNA networks.

**PCS**   Plastic Clad Silica.

**PC-SYNC**   An internal synchronous modem and communications software marketed by Barr Systems, Inc., of Gainesville, FL, for the IBM PC series and PS/2 Models 25 and 30.

**PC-SYNC/2**   An internal synchronous modem and communications software marketed by Barr Systems, Inc., of Gainesville, FL, for IBM PS/2 micro-channel personal computers.

**PC-Term**   An asychronous communications and terminal emulation program from Crystal Point of Kirkland, WA, that operates on IBM PC and compatible personal computers.

**PCU**   Packet Control Unit.

**pcX**   The name of a software program from Tropical Communications Associates, Inc., of Debray Beach, FL, that performs packet assembler/disassembler functions. pcX operates on IBM PC and compatible personal computers.

**PD**   Pulse Duration.

**PDI**   Picture Description Instruction.

**PDM** Pulse Duration Modulation.

**PDN** 1. Packet Data Network. 2. Public Data Network.

**PDS** Premises Distribution System.

**PDU** Protocol Data Unit.

**peak limiter** A filter used to reduce the effect of noise on a signal by clipping off noise peaks above the desired peak level of a signal; can be used on frequency modulated (FM), frequency shift keyed (FSK), or pulse-code modulated (PCM) signals.

**peak to average ratio test (P/AR)** The P/AR test is a signal fidelity measurement designed to be sensitive to envelope delay distortion and attenuation distortion and less sensitive to the normal steady interferences or impairments on a channel such as non-linear distortion, noise, and phase jitter. The measurement technique employs the signal transmission of a complex frequency spectrum. The receiver of the analog test set performs a calculation based on the charges to the original signal and generates a value of measurement called P/AR units. A P/AR value of 100 would indicate a channel with excellent fidelity; however, 75 is more of a practical reading to expect on a fairly good line.

**PEEK** A trademark of Computronics of Addison, IL, as well as a terminal monitor program from that vendor which operates on Prime computers.

**peg count** The number of calls made or received in a specific time period.

**pending active session** In IBM's VTAM, the state of an LU–LU session recorded by the SSCP when it finds both LUs available and has sent a CINIT request to the primary logical unit (PLU) of the requested session.

**penetration tap** In an Ethernet local area network, a penetration tap is a device used to connect a transceiver to the bus without requiring that the bus transmission be interrupted for the installation of fittings. This is accomplished by the use of a needle-like device which penetrates the insulation of the coaxial cable bus to reach the center of the coax conductor.

**PEP** 1. Packetized Ensemble Protocol. 2. Partitioned Emulation Program. 3. Peak Envelope Power.

**percent break** The ratio of the open circuit time to the sum of the open and closed circuit times allotted to a single dial pulse cycle, expressed as a percentage.

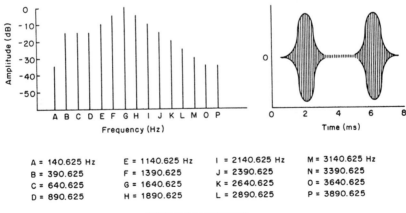

| | | | |
|---|---|---|---|
| A = 140.625 Hz | E = 1140.625 Hz | I = 2140.625 Hz | M = 3140.625 Hz |
| B = 390.625 | F = 1390.625 | J = 2390.625 | N = 3390.625 |
| C = 640.625 | G = 1640.625 | K = 2640.625 | O = 3640.625 |
| D = 890.625 | H = 1890.625 | L = 2890.625 | P = 3890.625 |

P/AR test measurement

*peak limiter*

**Percentage Allocation** The MCI name for a call allocator feature.

**perforator** A device used for the manual preparation of a perforated tape, in which telegraph signals are represented by holes punched in accordance with a predetermined code. The holes correspond to a five-unit code of a character entered from the machine's keyboard. The device is similar to a tape punch with the paper tape being prepared off-line.

**performance classifications** CCITT G.821 has defined four error-rate performance categories as follows:

| | |
|---|---|
| Available and Acceptable | Intervals of test time of at least one minute during which the error rate is less than one-millionth. |
| Available but Degraded | Intervals in the test time of at least one minute during which the error rate is between one-thousandth and one-millionth. |
| Available but Unacceptable | Intervals in the test time of at least one second but less than 10 consecutive seconds during which the error rate is greater than one-thousandth. |
| Unavailable | Intervals in the test time of at least 10 consecutive seconds during which the error rate is less than one-thousandth. |

**performance error** In IBM's NPDA, a resource failure that can be resolved by error recovery programs. Synonym for temporary error.

**performance testing** The process of verifying the implementation of end-to-end service to ensure that it meets overall throughput, delay, and reliability requirements.

**perigee** The lowest point in a satellite's orbit; the point in the orbit of a satellite when it is closest to the object about which it revolves.

**periodic frames** Segments of equal duration that are delineated by incorporation of fixed periodic patterns into the bit stream.

**peripheral device** A device which is connected to a computer to perform an input/output function such as data storage.

**peripheral equipment** Equipment that works in conjunction with a communications system or a computer, but not integral to them. Printers and CRTs, for example.

**peripheral interface** A standard interface used between a computer and its peripherals so that new peripherals may be added or old ones changed without special hardware adaption.

**peripheral LU** In IBM's SNA, a logical unit representing a peripheral node.

**peripheral node** In IBM's SNA, a node that uses local addresses for routing and therefore is not affected by changes in network addresses. A peripheral node requires boundary function assistance from an adjacent subarea node. A peripheral node is a type 1 or type 2 node connected to a subarea node.

**peripheral PU** In IBM's SNA, a physical unit representing a peripheral node.

**permanent error** In IBM's SNA, a resource error that cannot be resolved by error recovery programs.

**Permanent Signal (PS)** An extended off-hook condition not followed by dialing.

**Permanent Virtual Circuit (PVC)** 1. A permanent virtual call existing between two Data Terminal Equipment (DTEs). It is a point-to-point, non-switched circuit over which only data, reset, interrupt, and flow-control packets can flow. 2. A logical channel that is maintained in a data transfer state between two user devices (typically a terminal and host) at all times when the network is operational. Supports only X.25 DTEs.

**permissive (PE) arrangement** A connection arrangement used to connect FCC registered equipment to the DDD network. This arrangement utilizes the type USOC RJ11C jack. The output signal level of the communications equipment is fixed at a maximum of − dBm. An assumption that at least 3 dB signal loss will occur on the local loop insures that the signal won't arrive at the central office at more than the maximum allowable level of −12 dBm.

**permissive device** A classification of registered modems with output limited to −9 dBm.

**persistent** In LAN technology, a term used to describe a CSMA LAN in which the stations involved in a collision try to retransmit almost immediately; $p$-persistent where $p$ (for probability) =1 (hence, also called 1-persistent).

**Personal Computer (PC)** A microcomputer with an end user-oriented application program (used by data processing professionals and non-professionals alike) for an assortment of functions.

**Personal Identification Number (PIN)** A special number assigned to a specific customer or user, to enable him/her to authenticate himself/herself to a system. A PIN is primarily used with automatic bank teller systems to permit a user to withdraw cash from his/her account.

**personal sign** In U.S. military communications, signs composed of one or more letters (normally initials) used when endorsing station records and messages, to indicate responsibility of operating and supervisory personnel.

**Personal System/2 (PS/2)** One of IBM's current family of microcomputers.

**Personal 800** An MCI service that allows customers to set up an 800 toll-free number to receive long distance calls at their home.

**person-to-person call** An operator-assisted call where the calling party specifies the person on the receiving end who must accept the call.

**Perumtel** The state-owned network operator in Indonesia.

**PEST** A program similar to a virus except that it does not attach itself to other software.

**Petroleum Industry Data Exchange** An organization which addresses electronic data exchange issues of the petroleum industry.

**pF** Picofarad.

**PF** Program Function (key).

**PFA** Private Facilities Access.

**PFEP** Programmable Front End Processor.

**phantom** A circuit formed by using the two-wire loop of another simplex channel as one leg of a further channel. The second leg is formed in a similar fashion from another simplex channel.

**phantom telegraph circuit** Telegraph circuit superimposed on two physical circuits reserved for telephony.

**phase** A measure of signal position with respect to a reference signal. The unit of measurement is the angular degree.

**Phase Alternate Line (PAL)** The television format used in the United Kingdom which provides a 625-line picture using a 6 MHz bandwidth. In the UK PAL system the luminance extends to 5.5 MHz, chrominance interleaving is based on a 4.43 MHz carrier and sound is frequently modulated on a 6 MHz carrier. Stereo sound broadcasting uses a digitally encoded signal centered at 6.552 MHz.

**phase conductor** Metallic wires used to carry current.

**Phase Corrector (PC)** A function of synchronous modems which adjusts the local data clocking signal to match the incoming receive signal.

**phase equalizer, delay equalizer** A delay equalizer is a corrective network which is designed to make the phase delay or envelope delay of a circuit or system substantially constant over a desired frequency range.

**phase hit** Undesired shifting in phase of an analog signal. Any case where the phase of a 1004-Hz test signal shifts more than 20 degrees.

**phase inversion modulation** A method of phase modulation, in which the two significant conditions differ in phase by 180 degrees.

**phase jitter** An analog line impairment caused by the shifting of phase of one part of the frequency tone relative to an earlier part of the tone. Line distortion is caused by the variation in phase or frequency of a transmitted signal from its reference timing position. This can cause data transmission errors particularly at high speeds. Phase jitter is caused by primary frequency supplies and power transients. These supplies are prevalent in telephone offices where ringing voltages are generated and from power supplies for frequency division multiplexed carrier systems. Phase jitter rarely occurs above 300 Hz and it is typically measured in two areas. These are called inband and out of band jitter, or high and low frequency jitter. These encompass 20 to 300 Hz for the high band and 4 to 300 Hz for the low or out of band.

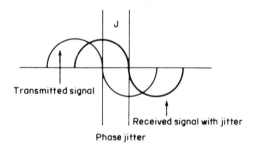

Phase jitter

**phase locked loop** Device that compares the phase of two signals, i.e., a reference signal and a voltage-controlled signal. A phase difference between the two signals produces an error voltage that locks the frequency of the voltage controlled signal to that of the reference signal.

**phase modulation** One of three basic ways (see also amplitude modulation (AM) and frequency modulation (FM)) to add information to a sine wave signal; the phase of the sine wave, or carrier, is modified in accordance with the information to be transmitted. With only discrete changes in phase, this technique is known as phase-shift keying (PSK).

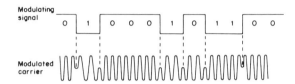

*phase modulation*

**phase shift** A change in the time or amplitude that a signal is delayed with respect to a reference signal.

**Phase-Shift Keying (PSK)** A modulation technique in which the phase of the carrier is modulated by the state of the input signal.

**phasor** Temporary buffer storage that compensates for slight differences in data rate between TDM I/O ports and devices.

**Philippine Long Distance Telephone Company** The company based in Manila that has virtual control over all telecommunications services in the Philippines.

**Phoenix 5500** A trademark of Phoenix Microsystems Corporation. The Phoenix 5500 is a widely used BERT tester, capable of testing high-speed digital circuits such as T1 lines.

**phone mail** A computer-controlled voice processing system which enables people to communicate when both parties are not available at the same time or if one party already has an ongoing conversation.

**phone phreak** A person who tries to steal codes to make free long distance telephone calls.

**Phone/FAX Switch** A switch from Electronic Modules, Inc., of Dallas, TX, which allows a single telephone line to accommodate a telephone, fax, and/or computer modem. In addition, the switch provides a prerecorded message to direct incoming calls.

**PhoneMail** An IBM voice-mail system.

**PhoneNet** A wiring system marketed by Farallon Computing of Berkeley, CA, which integrates with Apple Computer Company's LocalTalk cabling scheme. PhoneNet uses standard telephone wire and permits cable runs up to 3000 feet as well as extensions through the use of bridges and repeaters.

**Phonepoint** A consortium of companies led by British Telecom that offer CT2-based cellular telephone service in the United Kingdom.

**phonetic alphabetic** To help identify spoken characters, a set of easily understood words (phonetic alphabet) was selected by the U.S. Army to avoid confusion between letters. BRAVO, for example, is the phonetic equivalent of the letter B, and DELTA is the phonetic equivalent of the letter D. BRAVO and DELTA are much less likely to be confused in a radiotelephone conversation than B and D. The accented syllables of each word are underlined in the following phonetic alphabet table.

PHONETIC ALPHABET

| Letter | Word | Pronunciation |
|--------|------|---------------|
| A | ALFA | AL FAH |
| B | BRAVO | BRAH VOH |
| C | CHARLIE | CHAR LEE or |
|   |       | SHAR LEE |
| D | DELTA | DELL TAH |
| E | ECHO | ECK OH |
| F | FOXTROT | FOKS TROT |
| G | GOLF | GOLF |
| H | HOTEL | HOH TELL |
| I | INDIA | IN DEE AH |
| J | JULIET | JEW LEE ETT |
| K | KILO | KEY LOH |
| L | LIMA | LEE MAH |
| M | MIKE | MIKE |
| N | NOVEMBER | NO VEM BER |
| O | OSCAR | OSS CAH |
| P | PAPA | PAH PAH |
| Q | QUEBEC | KEH BECK |
| R | ROMEO | ROW ME OH |
| S | SIERRA | SEE AIR RAH |
| T | TANGO | TANG GO |
| U | UNIFORM | YOU NEE FORM OR |
|   |        | OO NEE FORM |
| V | VICTOR | VIK TAH |
| W | WHISKEY | WISS KEY |
| X | XRAY | ECKS RAY |
| Y | YANKEE | YANG KEE |
| Z | ZULU | ZOO LOO |

| Numeral | Word | Pronunciation |
|---------|------|---------------|
| 0 | ZERO | ZE RO |
| 1 | ONE | WUN |
| 2 | TWO | TOO |
| 3 | THREE | TREE |
| 4 | FOUR | FOW ER |
| 5 | FIVE | FIFE |
| 6 | SIX | SIX |
| 7 | SEVEN | SEV EN |
| 8 | EIGHT | AIT |
| 9 | NINE | NIN ER |

**photodetector** In a lightwave system, a device that turns pulses of light into bursts of electricity. A photodetector serves as a light receiver when used in an optical system.

**photoelectric effect** The emission of electrons by a material when it is exposed to light.

**photon** The fundamental unit of light and other forms of electromagnetic energy. Photons are to optical fibers what electrons are to copper wires; like electrons, they have a wave motion.

**photonics** The technology that uses light particles (photons) to carry information streams over hair-thin fibers of pure glass.

**physical connection** 1. The Physical layer communications path between two systems. 2. In IBM's VTAM, a point-to-point connection or multipoint connection.

**Physical layer** The lowest (first) layer in the OSI model; responsible for the physical signaling, including the connectors, timing, voltages, and other related matters.

**physical record** A single block of data transferred between an I/O device and main memory; may contain several logical records.

**physical security** Security concerned with physical measures designed to safeguard personnel, prevent unauthorized access to equipment, facilities, material and documents and safeguard them against espionage, sabotage, damage and theft.

**Physical Unit (PU)** On an IBM SNA network, a type of network-addressable unit that represents hardware devices or nodes to the network; the program that resides in each node and provides the services required to manage and monitor that node's resources. Current IBM SNA PU types arelisted in the following table.

SNA PU SUMMARY

| PU type | Node | Representative hardware |
|---------|------|------------------------|
| PU type 5 | Mainframe | S/370, 43XX, 308X |
| PU type 4 | Communications controller | 3705, 3725, 3720, 3745 |
| PU type 3 | Not currently defined | N/A |
| PU type 2 | Cluster controller | 3274, 3276, 3174 |
| PU type 1 | Terminal | 3180, PC with SNA adapter |

**Physical Unit name (PU name)** In IBM's SNA the mnemonic name given to a node. This name is defined to VTAM and/or NCP, and sometimes to the node itself.

**physical unit services** In IBM's SNA, the components within a physical unit (PU) that provide configuration services and maintenance services for SSCP–PU sessions.

**physical unit type** In IBM's SNA, the classification of a physical unit (PU) according to the type of node in which it resides. The PU type is the same as its node type; that is, a type 1 PU resides in a type 1 node, and so forth.

**PI** Pacing Indicator.

**PIC** Plastic Insulated Conductors.

**pico (p)** A prefix for one trillionth of a unit.

**picosecond** One-millionth of one-millionth of a second. One-trillionth of a second.

**Picture Description Instruction (PDI)** Instructions included in the North American Presentation Level Protocol (NAPLP) which allows the geometric coding of graphic information in a manner that is not dependent upon the resolution of a particular display.

**PIDX** Petroleum Industry Data Exchange.

**PIF** Program Information File.

**piggybacking** 1. A technique in which acknowledgment information is added to data. The Microcom Networking Protocol (MNP) Class 9 uses piggybacking. 2. A term used to refer to the activity during which a perpetrator intercepts a valid user's signon request, fooling the user with a fake screen message that the logoff is complete. The perpetrator then continues to use the active session. Also called spoofing.

**pigtail** 1. In fiber optic systems a short length of fiber, terminated on one end with a source, detector, coupler, or connector and bare fiber on the other end. 2. A corkscrew-shaped antenna attached to a vehicle which enables communications between a car phone and a cell (fixed) site.

**pilot** In U.S. military communcations, instructions appearing in format line 1 relating to the transmission or handling of that message.

**pilot model** A model of the system used for program testing purposes which is less complex than the complete model, e.g., the files used on a pilot model may contain a much smaller number of records than the operational files; there may be few lines and fewer terminals per line.

**pilot tone** A test frequency of controlled amplitude transmitted over a carrier system for monitoring and control purposes.

**PIM** Protocol Insensitive Multiplexing.

**PIN** 1. Personal Identification Number. 2. Positive Intrinsic Negative.

**pin-diode** A photodetector used to convert optical signals to electrical signals. A pin-diode functions as a receiver in a fiber optic system or as a light detector in power meters.

**ping**  A TCP/IP application in which a packed is transmitted and its echoed response indicates the status of a host or network device.

**PING**  Packet Inter Net Groper.

**ping pong**  A method used to emulate full-duplex transmission on a half-duplex circuit where automatic line turnaround occurs when a receiving modem has data to transmit.

**pipeline**  Slang for telephone cables or other telephone circuit bundles.

**pitch**  The number of characters printed per horizontal inch.

**PIU**  Path Information Unit.

**pixel**  An abbreviation for a picture element; the smallest unit into which an image can be divided, and to which can be assigned such characteristics as gray scale, color, and intensity.

**PIXNET-XL**  A trademark of AT&T Paradyne as well as a family of channel extension products from that vendor.

**PK**  PeaK.

**PKS**  Public Key System.

**PKT**  Packet(s).

**Plain Old Telephone Service (POTS)**  A reference to the basic service provided by the public telephone network without any added facilities such as conditioning.

**plaindress**  In U.S. military communications, a type of message in which the originator and addressee designations are indicated externally of the text.

**plaintext**  1. In the context of cryptography, messages in their normal, readable form are called plaintext. 2. Synonym for clear data (IBM's SNA).

**PLANET**  Communications software from Alpha Microsystems of Santa Ana, CA, which allows personal computers to share files and other resources on a network.

**plant**  The physical means by which transmission services are provided. This can include copper wire, conduit, poles, microwave systems, fiber optic trunks, and switching systems.

**PLAR**  Private Line Automatic Ringdown.

**plasma display**  A type of flat visual display in which selected electrodes in a gas-filled panel are energized, causing the gas to be ionized and light to be omitted.

**Plastic Clad Silica (PCA)**  A fiber composed of a glass core and plastic cladding.

**platen**  The roller in a printer which provides a backing for the printing process.

**platform**  A structure in space containing multiple missions.

**PLDC**  Primary Long Distance Carrier.

**plenum, return air**  The utilization of the false ceiling area on each floor to move air (for heating and air conditioning).

**plenum cable**  Cable specifically designed for use in a "plenum" (the space above suspended ceiling used to circulate air back to the heating or cooling system in a building). Plenum cable has insulated conductors often jacketed with PVDF material to give them low flame spread and low smoke-producing properties. Its maximum allowable flame distance is five feet.

**Plesiochronous**  A network with multiple Stratum 1 Primary Reference Sources.

**Plexar**  The name of a switched digital 64 Kbps service marketed by Southwestern Bell Telephone Company of St. Louis, MO.

**plotter**  A type of computer peripheral printer that displays data in a two-dimensional graphics form.

**PLP**  Packet Level Procedure.

**PLR**  Pulse Link Repeater.

**PLRS**  Position Location Reporting System.

**PLRZ**  PoLaRiZation.

**PLS**  Private Line Service.

**PLU** Primary Logical Unit (IBM's SNA).

**plug** A device for connecting wires to a jack. It is typically used on one or both ends of equipment cords, and on wiring for interconnects and cross-connects.

**Plug-Compatible Machine (PCM)** Term used to describe a device which can be directly substituted for an original manufacturer's device; the PCM device is usually an improvement over the original device—less expensive, more fully featured, or both.

**PL/1** A high level programming language oriented toward both business and mathematical programs. PL1 stands for programming Language 1.

**PM** 1. Phase Modulation. 2. Pulse Modulation.

**PMO** Present Method of Operation.

**PMR** Private Mobile Radio.

**PMS** Public Message Service.

**PMX** Packet Multiplexer.

**PND** Present Next Digit.

**PNID** Precedence Network In Dialing.

**PNOD** Precedence Network Out Dialing.

**PO** Pulse Originating.

**Pocket Edition** The name of a miniature modem manufactured by Hayes Microcomputer Products for use with laptop computers. The modem obtains power from the telephone line, eliminating the need for batteries or an external power supply.

**POI** Point Of Interface.

**Point** A computer that is not in a FidoNet nodelist but which communicates in the same manner as a participating nodelist computer. Points are addressed using a boss node nodelist address and the boss node then transfers messages to the point. Points are addressed using the boss node's nodelist address followed by a period (point) and a number.

**point code** The unique address assigned to every node in a CCITT Common Channel Signaling System 7 network.

**point of demarcation** The point where the responsiblity of wire installation and maintenance no longer belongs to the communications carrier, but to the user.

**Point Of Interface (POI)** The physical telecommunications interface between the LATA access and interLATA functions. This point establishes the technical interface, the test points, and the points of operational responsibility.

**Point Of Presence (POP)** The location in each local access transport area (LATA) that the local exchange carrier (LEC) connects to a designated interexchange carrier (IEC). This the point where the local telephone company terminates subscribers' circuits for leased line or long distance dial-up circuits.

**Point Of Sale (POS)** Transaction terminal used in retail.

**Point Of Service (POS)** The point at which a long distance call enters the long distance company's network from the local company's network.

**Point Of Termination (POT)** The point where the entrance cable or regulated riser cable meets the network interface. The POT is the location where the communications carrier's facility responsibility ends. POT is co-located at the Rate Demarcation Point (RDP).

**point size** The height of a character to be printed or displayed.

**Pointel** A project name used by France Telecom to try out Telepoint, the mobile communications system based on digital cordless telephones called CT2 and radio-based stations in public locations.

**pointer** A word in a computer's store containing the address of another item of data. It is saying "point" to that item. The manipulation of pointers may save many operations by avoiding the movement of larger items to which they point.

**point-to-point** A term used to describe a data channel which connects two, and only two, terminals.

**point-to-point circuit** A communication circuit, or system connecting two points through a telephone circuit, or line.

**point-to-point line** In IBM's SNA, a link that connects a single remote link station to a node; it may be either switched or nonswitched.

**point-to-point link** A communications link connecting two stations.

**point-to-point network** A point-to-point network is one in which exactly two stations are connected. It may be a dial connection or a leased line.

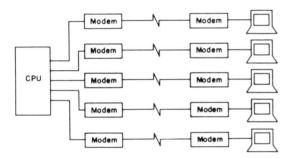

**poisson** A set of tables based on mathematical probabilities that can be used to calculate the number of circuits needed to provide a specified grade of service.

**polar non-return to zero signaling** The polar non-return to zero signal uses positive current to represent a mark and negative current to represent space. As no transmission occurs between two consecutive bits of the same value, the signal must be sampled to determine the value of each received bit.

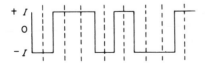

**polar orbit** A highly elliptical orbit that maximizes a satellite's exposure to the northern portion of the earth.

**polar return to zero** The polar return to zero signaling method uses positive current to represent a mark and negative current to represent a space; however, the signal returns to zero after each bit is transmitted. Since there is a pulse that has a discrete value for each bit, sampling of the signal is not required. Thus, the circuitry required to determine whether a mark or space has occurred is reduced.

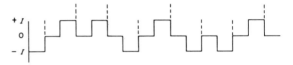

**polar transmission** A method for transmitting teletypewriter signals, whereby marking signal is represented by direct current flowing in one direction and the spacing signal is represented by an equal current flowing in the opposite direction. By extension to tone signaling, polar transmission is a method of transmission employing three distinct states, two to represent a mark and a space and one to represent the absence of a signal. Also called bipolar.

**polarity** Any condition in which there are two opposing voltage levels or charges, such as positive and negative.

**polarization** 1. A characteristic of electromagnetic radiation that occurs when the electric-field vector of the energy wave is perpendicular to the main direction of the electromagnetic beam. 2. The directional aspects of a signal. A signal can have a circular or planar polarization.

**poll** The process by which a computer asks a terminal associated with it if the terminal has any data for the computer.

**poll character** A unique character or sequence sent by the main computer to a device to check for availability for sending and receiving data.

**polling** A means of controlling terminals on a multipoint line. The computer, acting as the master station, sends a message to each terminal in turn saying, "Terminal A: have you anything to send?"

If not, "Terminal B: have you anything to send?" and so on. Each such message is called a poll.

**polling delay**   The elapsed time between successive polls to a given station which becomes the maximum delay after an operator is ready to send before transmission actually takes place.

**polling list**   The order in which stations are polled and maintained in a list associated with each channel. A line polling list can also be used to provide priority in line service.

**Poly-STAR**   Terminal emulation software from Polygon, Inc., of St Louis, MO, which enables an IBM PC or compatible personal computer to communicate with any Digital Equipment Co. VAX or other ASCII host computers.

**Poly-STAR/240**   Terminal emulation software from Polygon, Inc., which make a personal computer operate as a Digital Equipment Corporation (DEC) VT240 terminal.

**PolyVinyl Chloride (PVC)**   A flame-retardant thermoplastic insulation material that is commonly used in the jackets of building cables.

**POP**   Point Of Presence.

**port**   A computer interface capable of attaching to a modem for communicating with a remote terminal; the logical entrance and exit through which data traffic flows into and out of a network.

**port**   PORTable.

**port concentrator**   A device that allows several terminals to share a single computer port; a concentrator link in which the port concentrator simplifies

the software demultiplexing used in lieu of the demultiplexing normally performed by the computer-site concentrator.

**port contention**   The use of switching for incoming data calls to assign ports into the destination equipment on a first- come, first-served basis. This function is performed by a port selector or port concentrator.

**port group**   Group of destination ports defined for a particular exchange.

**port selector**   A switching device that extends the capability of a computer to handle more data traffic without more ports. It eliminates dedicated line-to-port interfaces, so fewer ports may handle more data lines. Also called a port concentrator and data PBX.

**port sharing**   An arrangement by which a port in a concentrator, multiplexer, computer, or controller is used consecutively by two or more devices, such as terminals or printers.

**port sharing device**   Digital device that treats several point-to-point lines as if they were a single multipoint line enabling the circuit capacity of a FEP to be exceeded. Located at FEP end of line. Also port selector.

**Portable NetWare**   A version of Novell LAN software developed for modification by computer and operating system manufacturers to run in their specific environments.

**portrait printing**   Printing across the width of a page. This is the opposite of landscape printing in which printing occurs across the length of a

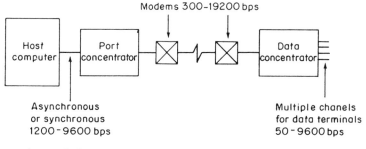

*port concentrator*

page. The term "portrait" originated from portraits of people which are usually vertical in format.

**POS** Point Of Sale.

**Position Location Reporting System (PLRS)** A navigation tool manufactured by Hughes Aircraft which denotes a ground position location within 10 to 15 meters.

**Positive, Intrinsic, Negative (PIN)** A type of photodetector used to sense lightwave energy and convert it into electrical signals.

**positive response** In IBM's SNA, a response indicating that a request was received and processed.

**POST** Power On Self Test.

**Post, Telephone, and Telegraph Authority (PTT)** The governmental agency that functions as the communications common carrier and administrator in many areas of the world.

**posten** The Swedish post office.

**PostLink** A program used by a RIME bulletin board system node in place of mailer software.

**PostMARC** An electronic mail and telephone messaging system marketed by MARC Analysis Research Corporation of Palo Alto, CA, for operation on Prime computers.

**postmaster** A person responsible for administration of electronic messaging at a site.

**postmortem** Pertaining to the analysis of an operation after its completion.

**PostScript** A programming language developed by Adobe Systems which describes a printed page.

**POT** Point Of Termination.

**POTS** Plain Old Telephone Service.

**power level** The ratio of power at a given point to an arbitrary amount of power chosen as a reference. Usually expressed in decibels based on 1 milliwatt (dBm) or 1 watt (dBw).

**power line connector** A technique that uses a radio

frequency carrier transmitted over the AC power line usually in a building.

**Power On Self Test (POST)** A sequence of test procedures, such as keyboard checking and memory parity, performed by a computer when it is powered on.

**power supply** A hardware system that converts line current and voltage to a current and voltage that is suitable for circuit components. Most power supplies convert alternating current to filtered direct current.

**Powerline** A voice mail system marketed by Talking Technology, Inc., of Oakland, CA.

**PP** Precedence and Preemption.

**PPDN** Public Packet Data Network.

*p*-**persistent** In LAN technology, a term used to describe a CSMA LAN in which the stations involved in a collision try to retransmit almost immediately—with a probability of $p$.

**PPM** Pulse Position Modulation.

**PPP** Point-to-Point Protocol.

**PPV** Pay Per View.

**PRBS** Pseudo-Random Binary-pulse Sequence.

**PRC** Primary Rate Center.

**preamble** 1. A sequence of bits sent at the beginning of a transmission to condition the electronics at the receiving device. 2. In U.S. military communications, one of the components contained in the heading of a message, whose elements include the degree of precedence, the date-time group, month and year abbreviated, and message instructions.

**precedence** A designation assigned to a message by the originator to indicate to communications personnel the relative order of handling and, to the addressee, the order in which the message is to be noted.

**predictive dialing** A digital switch which automatically dials telephone numbers contained in a data base. When a connection is established the system forwards the call to a waiting agent while simultaneously transferring information about the called party to the agent's terminal.

**preemption**　The act of interrupting a lower-priority call or message to use the same circuit to transmit an urgent message.

**preferred call forwarding**　A telephone feature offered by some communications carriers which lets a customer forward calls from specified numbers. Also known as select forward and selective call forwarding.

**preform**　A solid glass rod that is heated and drawn out to form several kilometers of fiber lightguide.

**PREMIS**　PREMises Information System.

**premise distribution**　A scheme used to distribute services throughout a building complex. This may be a combination of twisted pairs and coaxial cables.

**premises distribution system**　An AT&T intra-facility wiring scheme. The transmission network inside a building or group of buildings that connects various types of voice and data communications devices, switching equipment, and other information management systems together, as well as to outside communications network. It includes the cabling and distribution hardware components and facilities between the point where building wiring connects to the outside network lines back to the voice and data terminals in your office or other user work location. The system consists of all the transmission media and electronics, administration points, connectors, adapters, plugs, and support hardware between the building's side of the network interface and the terminal equipment required to make the system operational.

**Premises Information System (PREMIS)**　AT&T software which provides such information as street address, modular wiring data, and service data to help telephone companies determine dates for providing service to customers.

**premises lightwave system**　An AT&T fiber optic intra-facility wiring scheme.

**premises network**　Same as cable system.

**premises wiring**　1. Wiring that connects separately housed equipment entities or system components to one another. 2. Wiring that connects an equipment entity or system component located at the customer's premises and not within an equipment housing with a telephone network interface.

**Presentation layer**　The sixth layer in the OSI model; responsible for format and code conversion.

**Presentation Manager**　The graphic interface delivered as part of IBM's OS/2 which features icons and windows that enables users to perform file operations and manipulate information by pointing and clicking a mouse.

**Presentation Services Command Processor (PSCP)**　An IBM Network Communications Control Facility (NCCF) component that processes a user request from a user terminal and formats displays to be presented at the user terminal.

**presentation space**　In the 3270 environment, a region in computer memory (either host, or PC) that can be displayed, in whole or part, in a window on the screen. For example, a spreadsheet consisting of 4096 rows and 4096 columns is a presentation space that cannot be viewed in its entirety on one screen. However, it can be viewed in sequence, part-by-part. Sometimes presentation space is used synonymously with session, although not all presentation spaces are sessions (technically, some internal components are presentation spaces).

**Prestel**　The term for the Videotex service provided by British Telecom in the United Kingdom.

**PRF**　Pulse Repetition Frequency.

**PRI**　1. PRImary. 2. Primary Rate Interface. 3. PRIority. 4. Pulse Repetition Interval.

**primary application program**　In IBM's SNA, an application program acting as the primary end of an LU–LU session.

**primary center**  A control center connecting toll centers; a class 3 office. It can also serve as a toll center for its local end offices.

**primary commands**  IBM ISPF Editor commands that are typed on the command line denoted by three equal signs followed by the greater than sign (===>).

**primary end of a session**  In IBM's SNA, the end of a session that uses primary protocols. The primary end establishes the session. For an LU–LU session, the primary end of the session is the primary logical unit.

**primary group**  The lowest level of a multiplexing hierarchy. The multiplexing of a large group of channels (for example telephone channels) is carried out by stages. The basic signals are first multiplexed into a primary group then a set of prmary groups is multiplexed, and so forth. In the frequency division multiplexing of 4 kHz speech channels the primary group contains 12 channels and occupies 48 kHz. The primary group of PCM channels contains either 24 or 30 speech channels and uses approximately 1.5 or 2.0 Mbit/s of channel capacity, respectively.

**primary half-session**  In IBM's SNA, the half-session that sends the session activation request.

**primary link station**  In IBM's System Network Architecture (SNA), the link station on a link that is responsible for the control of that link. A link has only one primary link station. All traffic over the link is between the primary link station and a secondary link station.

**Primary Logical Unit (PLU)**  In IBM's SNA, the logical unit (LU) that contains the primary half-session for a particular LU–LU session. *Note.* A particular logical unit may contain primary and secondary half-sessions for different active LU–LU sessions.

**Primary Long Distance Carrier (PLDC)**  The interexchange carrier selected to carry long distance telephone calls. You can determine which

PLDC you are currently connected to by dialing 1-700-555-4141 in the United States.

**Primary Rate**  In ISDN, a high-bandwidth service offered to the end user. The Primary Rate signal comprises one H1 channel or any combination of B- and HO-channels possible within its aggregate bandwidth. In North America, the Primary Rate Service is offered at 1.544 Mbps; in Europe, the Primary Rate Service is offered at 2.048 Mbps. *Note.* The terms "23B+D" (North America) and "30B+D" (Europe) are often used interchangeably with "Primary Rate." Each of these terms describes only one possible Primary Rate channel configuration. In North America, there are 89 possible channel arrangements for the Primary Rate service.

**Primary Rate Interface (PRI)**  (North American) Twenty-three 64 Kbps B-channels and one 64 Kbps D-channel. (European) Thirty 64 Kbps B-channels and one 64 Kbps D-channel (23+d/30B+D).

**Primary Reference Source (PRS)**  The precise term for a synchronous master clock.

**primary route**  The path normally used to establish a virtual circuit to a destination. If this path is unavailable, other "secondary" routes are used.

**primary station**  A station responsible for controlling a data link; controls one or more secondary stations.

**primary-detected**  In IBM's Network Problem Determination Application (NPDA), refers to event and statistical data that reports conditions detected by resources at the primary end-of-line.

**Prime Operating System (PRIMOS)**  The operating system for Prime computers providing time-shared access, segmented virtual address space, and a file system with user-implemented passwords and file-protection attributes.

**Primex**  A private network switching service offered by British Telecom which is designed to permit users to benefit from international lease-circuit networks. Circuits interfaced can include lines routed to teleprinters, telex machines, word processors, video display units, personal computers and mainframe computers.

**primitive name**   A name that denotes a single, unique object.

**primitives**   The basic unit of machine instruction.

**PRIMOS**   PRIMe Operating System.

**print density**   A term used to reference the relative darkness of print on a page. Very dense print appears totally black.

**print server**   In local area networks, a computer on the network that makes one or more of its attached printers available to other users. The computer normally uses a hard disk to spool the print jobs while they wait in a queue for the printer.

**print services facility**   The program (code) that operates the IBM 3800 model 3 and 3820 printers. The print services facility operates as an a functional subsystem.

**Printed Circuit Board (PCB)**   A plastic board that supports electronic components interconnected by conductor deposits.

**printer authorization matrix**   A matrix stored in an IBM control unit that establishes printer assignment and classification.

**printer converter**   A coaxial converter that allows an asynchronous printer to emulate an IBM 3287 Printer (as shown in diagram following Type A Coax).

**printhead**   The mechanical part of a printer that moves back and forth across the page to make the image on the paper.

**priority**   A condition denoting a level of relative urgency, importance, or value of a specific message over others, which serves as the basis for invoking special handling to ensure rapid delivery, or a higher level of protection.

**priority indicators**   Such message priorities as urgent, rush, routine, and deferred are typically indicated by a code in the message header to define the sequence of transmission.

**priority level**   A value that is assigned to a task or device to control processing.

**privacy**   The techniques used for preventing access to specific system information from system users.

**privacy feature**   Ports of packet networks to include private dial ports, and single-connection ports associated with Dedicated Access Facilities (DAFs) may be equipped with the privacy feature. This feature uses a privacy list established by the customer at restricted access host port/rotaries to block virtual connections to or from unauthorized Data Terminal Equipment (DTEs). A privacy list must be established by the customer for each access port/rotary for which the privacy feature is requested.

**Private Automatic Branch Exchange (PABX)**   A telephone switch in private ownership connecting extension telephones to each other and to the public network via exchange lines.

**Private Automatic Exchange (PAX)**   A telephone exchange which provides private telephone service within an organization, but not to/from the public switched network.

**Private Branch Exchange (PBX)**   A manual, user-owned telephone exchange. Sometimes used in a general sense to include both PBXs and PABXs.

**private dial ports**   Private dial ports are available on a leased basis for the exclusive use of a particular customer or his authorized users. Such ports provide a means of access via dial telephone networks and Foreign Exchange (FX) access channels. Two types of private dial ports are available: those equipped for inward dialing to the network and those equipped for outward dialing.

**Private Exchange (PX)**   An exchange serving a subscriber's premises without connection to the public switched network.

**private line**   A circuit which is 100 percent dedicated to the user. Unlike the circuits which are continually built up in the message telephone network, the facilities on the private line are permanently assigned to the customer and no one else uses the circuit. Private lines may be both two-wire and four-wire. They are also called leased line and dedicated line. Private lines are physically connected at the central office independent of public switching and signaling equipment. They are leased for a flat monthly rate based on line length and quality, regardless of use. These are the Bell Private Line offerings:

| Series | Examples of service |
| --- | --- |
| 1000 | Low-speed (narrowband) data, for example, private line telegraph, teletypewriter, and remote metering (telemetering) |
| 2000 | Voice |
| 2001 | This was originally developed for private line voice communications |
| 3000 | Medium-speed (voiceband) data |
| 3002 | This is used for applications requiring dedicated lines for voiceband data transmission |
| 4000 | Telephoto/facsimile |
| 6000 | Audio (music transmission) |
| 7000 | Television |
| 8000 | High-speed (broadband) data |

**private line service (pls)** Initially private line service was point-to-point telecommunications service over a channel dedicated to a particular customer's private user. FCC regulation now allows all parts of private line services, except access lines, to be used in common by many customers. Private line services are utilized by customers with high volume or specialized requirements. However, most private lines are connected directly or indirectly (PBX) to the Public Switched Telephone Network.

**Private Management Domain (PRMD)** Under X.400 addressing the PRMD represents a private electronic messaging system that may be connected to an Administrative Management Domain. The PRMD is usually a corporate or government agency E-mail system connected to an ADMD.

**private network** A network established and operated by a private organization for the benefit of members of the organization.

**Private Transatlantic Telecommunications System** The first private transoceanic fiber optic cable system between the United States and Europe.

**private wire** Same as leased circuit.

**PRMD** PRivate Management Domain.

**PRN** Pseudo-Random Noise.

**Pro WATS** A communications service offering of AT&T that has a distance-sensitive pricing structure and is targeted at companies that use between 7 and 200 hours of interstate and international calls each month. Pro WATS was created by the merging of Pro America, Pro America II and Pro American III and covers calls made in the United States and to 45 foreign countries. Pro WATS provides customers with a variety of call-detail reports as well as discounts on calls made during the day, evening, and night/weekend periods.

**Probe** In the CCITT (now ITU) Message Handling System (MHS) a probe is an empty message sent to a recipient to verify the possibility of delivery of a type of message without incurring the expense of sending the message.

**procedure sign (prosign)** One or more letters or characters, or combination thereof, used to facilitate communications by conveying, in a condensed standard form, certain frequently used orders, instructions, requests, and information related to communications. See prosigns.

**process** An activity in a software system organized as a set of self-contained but interacting activities. A process is most simply regarded as a "pseudo-processor" which may possess certain states such as "active" or "dormant" and which may execute a piece of program code.

**process mode** The mode in which sysout data exists

331

and is to be processed by a JES output device. There are two IBM-defined process modes—line mode and page mode.

**processing, batch** A method of computer operation in which a number of similar input items are accumulated and grouped for processing.

**processing, in-line** The processing of transactions as they occur, with no preliminary editing or sorting of them before they enter the system.

**Procomm** A popular shareware communications program from Datastorm Technologies, Inc. of Columbia, MO designed for use on the IBM PC and compatible computers.

**Prodigy** A videotext service provided by a joint venture of Sears, Roebuck and Co. and IBM. The initial name for this service was Trintex.

**Professional Office System (PROFS)** An IBM office automation software package now known as Office Vision.

**profile** In packet-switched networks, refers to a set of parameter values, such as for a terminal, which can be defined and stored; the parameters can then be recalled and used as a group by identifying and selecting the appropriate profile.

**PROFS** PRofessional OFfice System.

**program** A sequence of step-by-step instructions that tell a computer what to do.

**Program Access (PA) key** On a display device keyboard, a key that produces a call to a program that performs display operations.

**Program Function (PF) key** On a display device keyboard, a key that passes a signal to a program to call for a particular display operation.

**Program Information File (PIF)** 1. A file containing information that IBM's "Top View" requires to allow the application to process properly in the "Top View" environment. 2. System information files.

**program operator** In IBM's ACF/VTAM, an

application program that is authorized to issue ACF/VTAM operator commands and receive ACF/VTAM operator awareness messages.

**Program Temporary Fix (PTF)** A patch sent by the vendor to fix a bug, or bugs, in a software package.

**programmable arrangement** A connection arrangement used to connect FCC registered equipment to the DDD network. This arrangement can employ either of two telephone company supplied data jacks—USOC RJ45S (programmable) or USOC RJ41S (universal). With this arrangement, the telephone company measures signal loss over the local loop between the subscriber's site and the central office. A "programming" resistor is selected and installed in the data jack to enable the communication equipment to transmit at a level that delivers the maximum −12 dBm signal at the central office.

**programmable jack** A jack that contains a resistor whose value can be changed to control the output level of a modem to meet a required signal loss level.

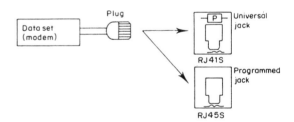

**Programmable Read-Only Memory (PROM)** Permanently stored data in a nonvolatile semiconductor device. The semiconductor is electrically programmed by the equipment manufacturer and can only be changed with special equipment which erases the previous program.

**programmable terminal** A terminal that has processor capability; also intelligent terminal.

**Programmable 800** An 800 service from Cable and Wireless of Vienna, VA, which allows companies to shift calls from one terminating location to another.

**Programmed Symbol Set (PSS)** A set of fonts that can be system-defined or defined by the user and to which a code can be assigned.

**Programmed Symbols (PS)** In an IBM 3270 Information Display System, an optional feature that stores up to six user-definable, program-loadable character sets of 190 characters each in terminal read/write storage for display or printing by the terminal.

**programmer, application** A person who writes or maintains application programs.

**programmer, systems** A person who writes or maintains system programs such as operating systems or data base managers.

**programming plug** An equipment feature which allows the change of the physical presentation of signals to accommodate a physical DTE or physical DCE connection or enable/disable such control signals as ring indicator and data terminal ready and/or enable/disable such features as flow control and physical loopback.

**Project Victoria** A service based upon a technology developed by Pacific Bell which enables a single, twisted copper pair in a telephone company's local loop to simultaneously transmit seven channels—two voice, one 9.6 kbps data channel and four 1.2 kbps data channels.

**PROM** Programmable Read-Only Memory.

**PromiseLAN** A trademark of Moses Computers of Los Gatos, CA, as well as the name of a kit containing hardware and software which enables IBM PC and compatible computers to be networked.

**prompt** A symbol that alerts the user that a device is on-line and connected to a channel for transmission.

**Promulgate** A trademark and an electronic mail gateway system from Management Systems Designers, Inc., of Vienna, VA, that provides electronic mail interchange between 3+Mail, UNIX, DDN and Internet mail users.

**ProNET** A 10 Mbps token-passing LAN network marketed by Proteon Corporation.

**Pronto** A videotex service which was offered by Covidea, a partnership between AT&T and Chemical Banking Corporation.

**proof test** In a fiber optic system, the process of applying constant stress to an entire length of fiber to determine and/or verify its minimum strength.

**propagation delay** The transit time for a signal to travel through a link, network, system, or piece of equipment.

**propagation velocity** The speed at which an electrical or optical signal travels through a transmission medium.

**proportional printing** Printing in which the width of characters varies from character to character.

**proprietary protocol** A network specification owned by an organization which can prohibit any third party use of the protocol.

**prosigns** Abbreviations used in radio teletypewriter operations that are used to expedite the transmission of messages and reduce the number of errors by providing a precise and uniform method of handling traffic. Some of the more commonly used prosigns and their meanings are listed in the following table.

| Prosign | Meaning |
| --- | --- |
| DE | THIS IS |
| IMI | SAY AGAIN |
| AR | OUT |
| EEEEEEEE | CORRECTION |
| K | OVER |
| R | ROGER (can also mean ROUTINE) |
| ZRC2 | TUNE YOUR EQUIPMENT TO MY SIGNAL |
| INT ZBK | HOW DO YOU COPY ME |
| ZBK1 | I COPY YOU CLEAR |
| ZBK2 | I COPY YOU GARBLED |
| ZUJ | STANDBY |
| INT ZNB | AUTHENTICATE |
| ZNB | AUTHENTICATION IS (in teletype, the reply is sent twice) |
| AKJ 1 | CLOSE DOWN |

**protected field** 1. Preset data or an area that cannot be changed or overridden by an operator without altering the program. 2. On a display device, a display field in which a user can't enter, modify, or erase data.

**protected memory** An independently established area in a storage module that protects stored programs from being inadvertently overwritten or destroyed by other programs sharing a core storage area. Other programs cannot write or transfer within the protected area.

**protector** 1. A device which is applied to a telephone line to protect the connected equipment from over-voltage and/or over-current. Excessive voltages and currents are shorted to ground. 2. A base of insulating material equipped with protector units, or carbon protector blocks, and sometimes fuses. Provide protection against over-voltage and over-current.

**protector unit** A small device which screws or plugs into a protector to provide over-voltage or over-current protection.

**protector bypass** Internal shorting or grounding of a protector due to the lack of physical separation between input and output fields, results in an over-current or over-voltage condition being passed into the building wiring without activating the protector unit.

**PROTO** A family of Telenet Processor 400 (TP4) software routines that interface the Packet Assembler/Disassembler (PAD) to the Switch and Link Access Procedure (LAP) software. PROTO provides protocol quality control, privacy screening, facilities processing, and packet reformatting. Other PROTO functions include rotary processing, autoconnects, and table support.

**protocol** 1. A set of rules governing information flow in a communications system. Sometimes called "data link control." 2. The set of rules followed by two computers with they communicate with one another.

**protocol control information** Information sent between communicating protocol modules to coordinate their operation, as distinct from user data.

**protocol conversion** The conversion of data transmissions from one protocol to another, thus enabling compatibility between two dissimilar systems.

**protocol converter** A device that translated from one communications protocol into another, such as IBM SNA/SDLC to ASCII.

**Protocol Data Unit (PDU)** An ISO term which refers to the exchange of packet information between two Network layer entities.

**protocol handler** Programming in a computer adapter that encodes and decodes the protocol used to format signals sent along the network.

**protocol ID** A five-byte field in the header of a SNAP frame on a LAN, used to identify the data link client at the receiving system that is to receive this frame.

**protocol identifier** A character string name for a protocol.

**protocol insensitive** Of a device able to operate upon data without regard to its protocol.

**protocol intelligence** The ability to decode, display, and transmit information about the contents of a bit stream as well as its format, especially about the data communications protocols in use on a circuit.

**protocol sensitive** Of a device able to operate upon data when the protocol is understood by the equipment.

**protocol sequence** An ordered list of Protocol Identifiers.

**protocol translator** A gateway that accepts data for routing in one protocl and forwards it in another protocol.

**protocol type** A two byte field in the header of an Ethernet frame, used to identify the Data Link client at the receiving system that is to receive this frame.

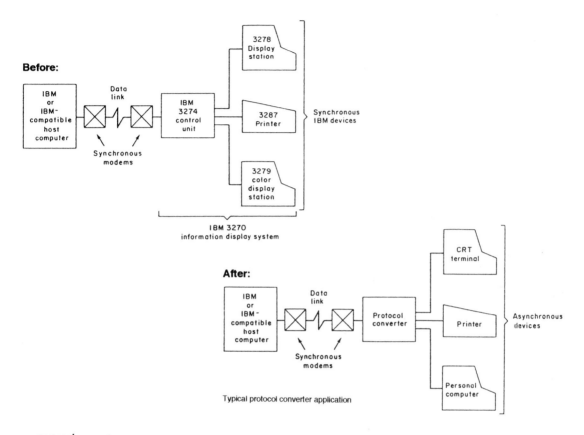

*protocol converter*

**prowords** Procedure words (referred to as "prowords") are words used to convey a specific meaning during radio transmission. They are used in standard phrases between military radio operators to shorten and minimize transmissions. The following table lists commonly used prowords.

| *Proword* | *Meaning* |
|---|---|
| ALL AFTER | The portion of the message to which I have reference is all that which precedes |
| ALL BEFORE | The portion of the message to which I have reference is all that which precedes |
| BREAK | I hereby indicate the separation of the text from other portions of the message. |
| CORRECT | You are correct, or what you have transmitted is correct. |

| Proword | Meaning | Proword | Meaning |
|---------|---------|---------|---------|
| CORRECTION | An error has been made in this transmission. Transmission will continue with the last word correctly transmitted. | I READ BACK | The following is my response to your instruction to read back. |
| | An error has been made in this transmission (or message indicated). The correct version is _____. | I SAY AGAIN | I am repeating transmission or portion indicated. |
| | | I SPELL | I shall spell the next word phonetically. |
| | That which follows is a corrected version in answer to your request for verification. | I VERIFY | That which follows has been verified at your request and is repeated. (To be used only as a reply to VERIFY.) |
| DISREGARD THIS TRANSMISSION—OUT | This transmission is in error. Disregard it. (This proword should not be used to cancel any message that has been completely transmitted and for which receipt or acknowledgment has been received.) | MESSAGE | A message which requires recording is about to follow. (Transmitted immediately after the call.) |
| | | OUT | This is the end of my transmission to you and no answer is required or expected. (Since OVER and OUT have opposite meanings they are never used together.) |
| EXEMPT | The addressees immediately following are exempted from the collective call. | | |
| FIGURE | Numerals or numbers follow. | OVER | This is the end of my transmission to you and a response is necessary. Go ahead; transmit. |
| FLASH | Precedence FLASH. Reserved for initial enemy contact reports or special emergency operational combat traffic originated by specifically designated high commanders of units directly affected. | PRIORITY | Precedence PRIORITY. Reserved for important messages which must have precedence over routine traffic. This is the highest precedence which normally may be assigned to a message of administrative nature. |
| FROM | The originator of this message is indicated by the address designation immediately following. | | |
| GROUPS | This message contains the number of groups indicated. | READ BACK | Repeat this entire transmission back to me exactly as received. |
| INFO | The addressees immediately following are addressed for information. | ROGER | I have received your last transmission satisfactorily. |

| Proword | Meaning | Proword | Meaning |
|---|---|---|---|
| ROUTINE | Precedence ROUTINE. Reserved for all types of messages which are not of sufficient urgency to justify a higher precedence, but must be delivered to the addressee without delay. | UNKNOWN STATION | The identity of the station with whom I am attempting to establish communications is unknown. |
| SAY AGAIN | Repeat all of your last transmission. (Followed by identification data means "Repeat _____ (portion indicated).") | VERIFY | Verify entire message (or portion indicated) with the originator and send correct version. (To be used only at the discretion of the addressee to which the questioned message was directed.) |
| SERVICE | The message that follows is a service message. | WAIT | I must pause for a few seconds. |
| SILENCE | "Cease Transmission Immediately." Silence will be maintained until lifted. (Transmissions imposing silence must be authenticated.) | WAIT OUT | I must pause longer than a few seconds. |
| | | WILCO | I have received your signal, understand it, and will comply. (To be used only by the addressee. Since the meaning of ROGER is included in that of WILCO, the two prowords are never used together.) |
| SILENCE LIFTED | Silence lifted. (When an authentication system is in force the transmission lifting silence is to be authenticated.) | | |
| SPEAK SLOWER | Your transmission is at too fast a speed. Reduce speed of transmission. | WORD AFTER | The word of the message to which I have reference is that which follows _____ |
| THIS IS | This transmission is from the station whose designator immediately follows. | WORD BEFORE | The word of the message to which I have reference is that which precedes _____ |
| TIME | That which immediately follows is the time or date/time group of the message. | WORDS TWICE | Communications is difficult. Transmitting each phrase (or each code group) twice. This proword may be used as an order or a request, or as information. |
| TO | The addressee(s) immediately following is (are) addressed for action. | WRONG | Your last transmission was incorrect. The correct version is _____ |

**PRS**   Primary Reference Source.

**PRTM**   Printing Response Time Monitor.

**PRW**   Pseudo-Random Word.

**PS**   Permanent Signal.

**PSC**   Public Service Commission.

**PSCP**   Presentation Services Command Processor.

**PSDN**   Public Switched Data Network.

**PSDS**   Public Switched Digital Service.

**PSE**   Packet Switch Exchange.

**Pseudo-Random Noise (PRN)**   Another name for a pseudo-random pattern, used in BERT testing.

**pseudo-random pattern**   A repeating bit pattern of a specific length whose bit order appears to be random within the length of the pattern. Used in BERT testing to minimize the effects of regular or repeating sequences in the pattern on transmitting or receiving equipment.

**Pseudo-Random Word (PRW)**   Specific bit pattern that is generated by a special algorithm that simulates random bit patterns associated with normal data transmission.

**PSK**   Phase Shift Keying.

**PSM**   Protocol Sensitive Multiplexing.

**PSN**   1. Packet Switching Network. 2. Packet Switching Node. 3. Private Satellite Network. 4. Public Switched Network.

**PSNs**   Private Switched Networks.

**psophometric weighting**   A type of telephone noise weighting established by the CCITT for use in a psophometer noise measuring set. Similar to F1A line weighting, psophometric weighting is used mostly in Europe.

**PSPDN**   Packet Switched Public Data Network (CCITT terminology).

**PSS**   1. Packet Switch Stream. 2. Programmed Symbol Set.

**PSTN**   Public Switched Telephone Network.

**PSU**   Packet Switch Unit.

**PS/2**   Personal System/2.

**PT**   Pulse Terminating.

**PTAT-1**   Private TransAtlantic Telecommunications System.

**PTR**   Printer.

**PTT**   Post, Telephone, and Telegraph authority.

**PU**   Physical Unit.

**PU type**   Physical Unit type.

**PU 2.1**   In IBM's SNA, a low-level network entry point with limited routing facilities; also referred to as a Single Node Control Point (SNCP).

**PU 5**   In IBM's SNA, ACF/VTAM in the host computer and ACF/NCP in the 37xx communications controller; also referred to as a System Service Control Point (SSCP).

**public**   Provided by a common carrier for use by many customers.

**Public Data Network (PDN)**   A network established and operated by a PTT, common carrier, or private operating company for the specific purpose of providing data communications services to the public. May be a packet switched network or a digital network such as DDS. The Data Network Identification Code (DNIC) of major PDNs are listed in the following table:

| Country | PDN name | DNIC |
|---|---|---|
| AFRICA: | | |
| Egypt | ARENTO | 6023 |
| Gabon | GABONPAC | 6282 |
| Ivory Coast | SYTRANPAC | 6122 |
| Reunion | DOMPAC | 6470 |
| South Africa | SAPONET | 550 |
| Sudan | via Bahrain | |
| Zimbabwe | ZIMNET | 6482 |

| Country | PDN name | DNIC | | Country | PDN name | DNIC |
|---|---|---|---|---|---|---|
| EUROPE | | | | NORTH AMERICA: | | |
| Austria(1) | DATEX-P | 2322 | | Antigua | AGANET | 3443 |
| Belgium | DCS | 2062 | | Bahamas | BATELCO | 4263 |
| Denmark | DATAPAC | 2382 | | Barbados | IDAS | 3420 |
| England(1) | PSS | 2342 | | Bermuda | BERMUDANET | 3503 |
| Finland | DATAPAC | 2442 | | Canada(1) | DATAPAC | 3020 |
| France(1) | TRANSPAC | 2080 | | Cayman Islands | IDAS | 3463 |
| West Germany(1) | DATEX-P | 2624 | | Costa Rica | RACSAPAC | 7120 |
| Greece | HELPAC | 2022 | | Curacao | UDTS | 3620 |
| Hungary | via Austria | | | Dominican Republic | no name | 3701 |
| Iceland | ICEPAC | 2740 | | French Antilles | DOMPAC | 7420 |
| Ireland(1) | EIRPAC | 724 | | Guatemala | GUATEL | 7040 |
| Italy(1) | ITAPAC | 2222 | | Honduras | HONDUTEL | 7080 |
| Luxembourg | LUXPAC | 2703 | | Jamaica | JAMANTEL | 3380 |
| Netherlands | DATANET-1 | 2041 | | Mexico | TELEPAC | 3340 |
| Norway | DATAPAC | 2422 | | Panama | INTELPAQ | 7141 |
| Portugal(1) | TELEPAC | 2680 | | Puerto Rico | PDIA | 3301 |
| Spain(1) | IBERPAC | 2145 | | Trinidad | TEXTEL | 3745 |
| Sweden | DATAPAC | 2402 | | United States(1) | various | various |
| Switzerland | TELEPAC | 2284 | | Virgin Islands | UDTS | 3320 |
| | | | | | | |
| MIDDLE EAST: | | | | SOUTH AMERICA: | | |
| Bahrain(1) | BAHNET | 4263 | | Argentina(1) | ARPAC | 7222 |
| Iraq | via Bahrain | | | Brazil(1) | RENPAC | 7241 |
| Israel | ISRANET | 4251 | | Chile(1) | ENTEL | 7302 |
| Jordan | via Bahrain | | | Columbia | DAPAQ | 7320 |
| Kuwait | via Bahrain | | | French Guiana | DOMPAC | 7420 |
| Qatar | via Bahrain | | | Peru | no name | 7160 |
| Saudi Arabia | via Bahrain | | | | | |
| Turkey | TURPAC | 2862 | | | | |
| United Arab Emirates | EMDAM | 4243 | | | | |

(1)  Indicates that the country has more than one PDN and operates as a gateway PDN to other countries.

| Country | PDN name | DNIC |
|---|---|---|
| ASIA: | | |
| China | no name | 4600 |
| Thailand | IDAR | 5200 |

**public dial-in ports**   Public dial-in ports are available on a network on a continuous basis and may be used by the customer or authorized user upon demand. Local access ports provide a means of access via the local public telephone network. In-WATS access ports provide a toll-free means of access from any point within the continental United States via the telephone network.

| Country | PDN name | DNIC |
|---|---|---|
| PACIFICA: | | |
| Australia(1) | AUSTPAC | 5052 |
| French Polynesia | TOMPAC | 5470 |
| Guam | LSDS | 5350 |
| Hong Kong(1) | DATAPAC | 4545 |
| Indonesia | SKDP | 5101 |
| Japan(1) | DDX-P | 4401 |
| South Korea | DNS | 4501 |
| Malaysia | MAYPAC | 5021 |
| New Caledonia | TOMPAC | 5460 |
| New Zealand | PACNET | 5301 |
| Philippines | EASTNET | 5156 |
| Singapore | TELEPAC | 5250 |
| Taiwan(1) | PACNET | 4872 |

**public exchange**   Central office.

**Public Key Encryption System**   A security system in which the keys required for encryption of data are known to the general public and the decryption keys are only known to the user receiving the encrypted data.

339

**Public Key System (PKS) Data Encryption** An asymmetrical, two-key encryption algorithm that transforms data from plaintext to ciphertext with one key that is made public and converts ciphertext back to plaintext with a different key that remains secret. Also called public key cryptosystem.

**public message** A message available for reading by all network users.

**public network** A network established and operated by communication common carriers or telecommunication administrations for the specific purpose of providing circuit switched, packet switched, and leased-circuit services.

**Public Service Commission (PSC)** An agency charged with regulating communications services at the state level. Also known as the Public Utility Commission (PUC).

**Public Standard Protocol** A network standard specification available to the public. Anyone may write a program that implements a public standard protocol which enables network hosts to communicate.

**Public Switched Digital Service (PSDS)** A service of Regional Bell Operating Companies (RBOC) and AT&T which permits full-duplex dial-up, 56 Kbps transmission on an end-to-end basis.

**Public Switched Network (PSN)** Any switching communications system—such as the Telex, TWX, or public telephone networks—that provides circuit switching to many customers.

**public telephone network** A telephone network which is shared among many users, any one of which can establish communications with any other user by use of a dial or pushbutton telephone; includes DDD service. In the United Kingdom and some other countries, the network is known as the PSTN, public switched telephone network.

**Public Utility Commission (PUC)** See Public Service Commission.

**PUC** Public Utility Commission.

**pullback** The retransmission of the most recently transmitted message or series of messages.

**pulse** A brief change of current or voltage produced in a circuit in order either to operate a relay, or to be detected by a logic circuit.

**Pulse Amplitude Modulation (PAM)** A voice sampling technique, used in digital telephony, in which the amplitude of the voice waveform is sampled 8000 times per second to produce discrete values for pulse code modulation encoding.

**Pulse Code Modulation (PCM)** Modulation in which an analog signal is sampled and the sample

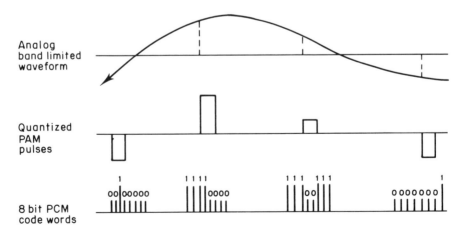

*Pulse Code Modulation (PCM)*

quantized and coded. Standard North American sampling is 8 000 times per second with 8 bits representing each sample pulse, giving a transmission rate of 64 Kbps. The resultant bit stream is sent down the line as interleaved data. The original analog signal, now in digital form, is less susceptible to noise. At the demodulator, the interleaved signals are separated and regenerated into an analog signal.

**pulse dialing**   Older form of phone dialing, utilizing breaks in DC current to indicate the number being dialed.

**Pulse Duration (PD)**   The amount of time a pulse is on. The pulse duration is inversely proportional to the data rate.

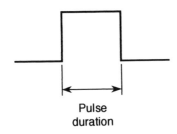

Pulse
duration

**Pulse Duration Modulation (PDM)**   Also called pulse width modulation and pulse length modulation. A form of pulse modulation in which the duration of pulses are varied.

**pulse modulation**   Transmission of information by modulation of a pulsed, or intermittent, carrier. Pulse width, count, position, phase, and/or amplitude may be the varied characteristic.

**Pulse Repetition Interval (PRI)**   The time between consecutive pulses.

**pulse stuffing**   Same as "bit stuffing".

**pulse trap**   A device that monitors any RS-232 lead for changes in logic levels (high to low or low to high).

**PulseLink**   A trademark of BellSouth as well as the name of that vendor's Southeastern regional network designed to interface with interLATA and international carriers.

**Pulsenet** The packet switching network of Cincinnati Bell Telephone.

**Pulse-Position Modulation (PPM)**   A form of pulse modulation in which the positions in time are varied, without modifying their duration.

**punch block**   A punch block permits the distribution of a 25-pair cable to individual terminals. At one end a 25-pair cable (50 wires) is connected to the punch block. The punch block contains rows of pins into which wires can be "punched down," using a special tool, permitting a connection to individual stations. Punch block wire codes are listed in the following table while the connection of a 25-pair telephone cable to a punch block is shown in the illustration.

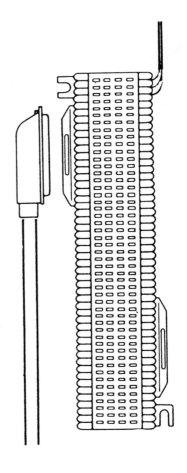

PUNCH BLOCK WIRE COLOR CODES

| Pin no. | Color |
|---|---|
| 1 | WHITE/BLUE |
| 2 | BLUE/WHITE |
| 3 | WHITE/ORANGE |
| 4 | ORANGE/WHITE |
| 5 | WHITE/GREEN |
| 6 | GREEN/WHITE |
| 7 | WHITE/BROWN |
| 8 | BROWN/WHITE |
| 9 | WHITE/SLATE |
| 10 | SLATE/WHITE |
| 11 | RED/BLUE |
| 12 | BLUE/RED |
| 13 | RED/ORANGE |
| 14 | ORANGE/RED |
| 15 | RED/GREEN |
| 16 | GREEN/RED |
| 17 | RED/BROWN |
| 18 | BROWN/RED |
| 19 | RED/SLATE |
| 20 | SLATE/RED |
| 21 | BLACK/BLUE |
| 22 | BLUE/BLACK |
| 23 | BLACK/ORANGE |
| 24 | ORANGE/BLACK |
| 25 | BLACK/GREEN |
| 26 | GREEN/BLACK |
| 27 | BLACK/BROWN |
| 28 | BROWN/BLACK |
| 29 | BLACK/SLATE |
| 30 | SLATE/BLACK |
| 31 | YELLOW/BLUE |
| 32 | BLUE/YELLOW |
| 33 | YELLOW/ORANGE |
| 34 | ORANGE/YELLOW |
| 35 | YELLOW/GREEN |
| 36 | GREEN/YELLOW |
| 37 | YELLOW/BROWN |
| 38 | BROWN/YELLOW |
| 39 | YELLOW/SLATE |
| 40 | SLATE/YELLOW |

| Pin no. | Color |
|---|---|
| 41 | VIOLET/BLUE |
| 42 | BLUE/VIOLET |
| 43 | VIOLET/ORANGE |
| 44 | ORANGE/VIOLET |
| 45 | VIOLET/GREEN |
| 46 | GREEN/VIOLET |
| 47 | VIOLET/BROWN |
| 48 | BROWN/VIOLET |
| 49 | VIOLET/SLATE |
| 50 | SLATE/VIOLET |

**punch-down block** An interface which permits multiconductor twisted pair cable to be connected to standard 50-pin telephone company connectors.

**punch-down tool** A special tool which enables twisted pair wires to be attached to punch-down contacts on punch-down and punch/patch blocks.

**PUP** PARC Universal Packet.

**PU–PU FLOW** In IBM's SNA, the exchange between physical units (PUs) of network control requests and responses.

**pure code** A program written so that none of the program (code) is altered during execution. All data and working storage is outside the program area. A process using the pure code can be stopped at any point and the program re-entered with a different process. Also called re-entrant code.

**PUs** Peripheral Units.

**pushbutton dialing** The use of keys or pushbuttons instead of a rotary dial to generate a sequence of digits to establish a circuit connection. The signal is usually multiple tones. Also called tone dialing, Touch-call, Touch-Tone.

**push-to-talk operation** A method of communication over a voice circuit in which transmission occurs from only one station at a time, the talker being required to keep a switch operated while talking.

**PVC** 1. Permanent Virtual Circuit. 2. PolyVinyl Chloride.

**PWI**   Power Indicator.

**PWM**   Pulse Width Modulation.

**PWR**   PoWeR.

**Px64**   A proposed CCITT standard for video compression which defines the implementation of discrete cosine transforms (DCT) for video compression.

**pyramid configuration**   A communications network in which the data link(s) of 1 or more multiplexers are connected to I/O ports of another multiplexer.

**P1**   A protocol specified by the CCITT X.400 standard which defines how Message Transfer Agents (MTAs) communicate with each other.

**P2**   A protocol specified by the CCITT X.400 standard which defines how messages are structured for transfer, submission, and delivery via Message Transfer Agents (MTAs).

**P3**   A protocol specified by the CCITT X.400 standard which defines how a User Agent (UA) communicates with a Message Transfer Agent (MTA).

**P-300**   A cellular telephone manufactured by NEC Corporation's Mobile Radio Division.

# Q

**QAM**   Quadrature Amplitude Modulation.

**Q-bit**   Qualifier bit.

**QBlazer**   The name of a CCITT V.32 compatible modem manufactured by Telebit Corporation which includes fax transmission capability.

**QCM**   Quad Channel Module.

**QEAM**   Quick Erecting Antenna Mast.

**QLLC**   Qualified Logical Link Control.

**QLLC PAD**   Qualified Logical Link Control Packet Assembler/Disassembler.

**Q-MAIN**   An electronic mail system marketed by Quadratron Systems, Inc., of Westlake Village, CA, that operates on Unix and Xenix based computers.

**QPSK**   Quadrature Phase-Shift Keying.

**QRSS**   Quasi-Random Signal Source.

**QTAM**   Queued Telecommunications Access Method.

**QUAD**   A cable consisting of two twisted pairs of conductors.

**Quadrature Amplitude Modulation (QAM)**   A modulation technique that combines phase modulation and AM techniques to increase the number of bits per baud. It can transmit at rates of 4800, 9600 bps and even higher. It is also capable of transmitting from one to seven bits per baud while keeping within the 3000 kHz limits of the phone line.

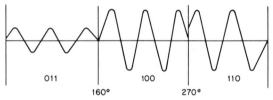

Quadrature amplitude modulation

**quadrature distortion**   Analog signal distortion frequently found in phase modulated modems.

**Qualified Logical Link Control (QLLC)**   Logical link control procedures designed to provide Synchronous Data Link Control (SDLC) station communication through a Packet Switching Network (PSN). The QLLC protocol employs the Qualifier (Q) bit in X.25 data packets to identify unnumbered and supervisory commands and responses.

**Qualified Logical Link Control Packet Assembler/Disassembler (QLLC PAD)**   A PAD that supports inputs from Synchronous Data Link Control (SDLC) devices and allows them to communicate with X.25 devices using QLLC; e.g. IBM's NPS2.

**qualifier**   In IBM's Network Problem Determination Application (NPDA) the string network resource names that fully identify a resource, or a product-specific information transmitted to the host processor that helps explain the data.

**qualifier bit (Q bit)** In X.25 packet switched networks, bit 8 in the first octet of the packet header. It is used to indicate whether the packet contains control information and whether more than one level of data is being sent over a logical channel.

**quality** A measure of the performance level expressed in percent, error-free intervals.

**quality of service** Synonym for "grade of service."

**quantization error** The difference between the signal level encoded as a digital value and its actual analog value.

**quantizer** A device used in digital communication systems to assign one of a discrete set of values to the amplitude of each successive sample of an analog signal.

**Quasi-Random Signal Source (QRSS)** A pseudo-random data pattern with properties similar to Gaussian noise used for out-of-service, bit-error-rate testing of the ones' density requirements of a digital service. Common QRSS data patterns are 2+E20−1 and 2+E15− 1 defined by Bell and CCITT standards respectively.

**quasi-random word** One iteration of the Quasi-Random Signal Source (QRSS) pattern.

**Qube Cable Network** A videotext service offered by Warner Communications and American Express from 1977 to 1984.

**query** The action of searching data for desired information.

**query language** A programming language that makes it relatively simple to engage in a conversational mode with a computer.

**queue** A line or list formed by items waiting for service, such as tasks waiting to be performed, stations waiting for connection, or messages waiting for transmission.

**queue discipline** The method employed by a server mechanism in picking customers from the queue. The most common method is FIFO (first in, first out). Other queue disciplines include customer

selection based on a priority scheme and FILO (first in, last out).

**queued bind** In IBM's VTAM, a BIND request, sent from the primary logical unit (PLU) to the secondary logical unit (SLU), that has not yet been responded to by the SLU. This creates a pending active session at the SLU. When the SLU is a VTAM application program, it responds to a BIND by issuing an OPNSEC or SESSIONC macro instruction.

**queued CINIT** In IBM's VTAM, a CINIT request, sent from an SSCP to an LU, that has not yet been responded to by the LU. This creates a pending active session at the LU. A VTAM application program responds to a CINIT by issuing an OPNDST ACCEPT or a CLSDST macro instruction.

**queued message** A message from the system that is not automatically displayed. This contrasts with a message that is sent directly to the screen for the user to see immediately.

**queued session** In IBM's VTAM, pertaining to a requested LU–LU session that cannot be started because one of the LUs is not available. If the session-initiation request specified queueing, the SSCP(s) will record the request and later continue with the session-establishment procedure when both LUs become available.

**Queued Telecommunications Access Method (QTAM)** IBM teleprocessing access method that provides the capabilities of BTAM plus the capability of queued messages on direct access storage devices. QTAM may be employed for data collection, message switching, and many other teleprocessing uses.

**queueing** A feature that allows transactions to be "held" at the origination point, node, or delivery point, while waiting for a facility to become available.

**quick closedown** In IBM's VTAM, a closedown in which any RPL-based communication macro instruction is terminated (posted complete with an error code) and no new sessions can be established and no new ACBs can be opened.

**Quick-Com**  A trademark of GE Information Services of Rockville, MD, as well as the name of a global electronic messaging service marketed by that vendor.

**quickfix**  A U.S. military division-level heliborne intercept, direction finding and jamming system.

**quicklook**  A U.S. military airborne system designed to provide corps and division commanders with the locations and identification of non-communications emitters.

**QuickMail**  A trademark of CE Software Inc. of West Des Moines, IA, as well as the name of a Macintosh computer electronic mail program marketed by that vendor.

**Quicktran**  A data compression program marketed by Eidolan Technologies of New York City, NY, which compresses files in the background as they are sent and received.

**quiesce protocol**  In IBM's VTAM, a method of communicating in one direction at a time. Either the primary logical unit (PLU) or the secondary logical unit (SLU) assumes the exclusive right to send normal-flow requests, and the other node refrains from sending such requests. When the sender wants to receive, it releases the other node from its quiesced state.

**quit**  A key, command, or action that tells the system to return to a previous state or stop a process.

**Quorum**  A trademark of AT&T as well as an analog conference phone from that vendor. The phone has sound pickup and amplification covering an 8-foot radius.

**QWK-format**  The format of a file which contains bulletin board system mail from selected conferences that will be transmitted for off-line reading. Most BBSs have doors for accessing off-line readers that use the QWK-format.

**Q.920**  The CCITT recommendation which describes the general aspects of the ISDN User–Network Interface Data Link Layer; identical to Recommendation I.440.

**Q.921**  "ISDN User–Network Interface Data Link Layer Specification" is the CCITT recommendation in the Q-Series that describes LAP-D; identical to Recommendation I.441. It is the CCITT layer 2 protocol used in the ISDN D-channel.

**Q.931**  "ISDN User–Network Interface Data Link Layer 3 Specification" is the CCITT recommendation in the Q-Series that describes LAP-D; identical to Recommendation I.451. It is the CCITT layer 3 protocol used on the ISDN D-channel for signaling.

**Q.940**  The CCITT recommendation which describes the ISDN management architecture and provides a general overview of the management services and functions at the user–network interface.

# R

**R**  Resistance.

**R interface**  In ISDN, the 2-wire physical interface which is used for a single customer termination between the TE2 and TA.

**RACF**  Resource Access Control Facility.

**rack**  Same as cabinet.

**rack-mount**  Designed to be installed in a cabinet.

**radial wiring**  Wiring in which all cable runs from a common point to the point requiring service by the most direct means possible.

**radiate**  To send out energy into space, as in the case of radio frequency (RF) waves.

**radio channel**  A band of adjacent frequencies having sufficient width to permit its use for radio communications.

**radio communication**  Communications by means of radio waves.

**Radio Frequency (RF)**  That portion of the electromagnetic spectrum between 10 kHz and 300 MHz where propagation occurs without a guide in free space.

**radio frequency amplification**  The amplification of a radio wave by a radio receiver before detection or by a radio transmitter before radiation.

**Radio Frequency (RF) noise**  Noise caused by an electronic spark developed across relay contacts or electronic motor brush contacts. Usually suppressed by a resistor in series with a capacitor.

**radio frequency spectrum**  The following chart illustrates the radio frequency spectrum.

**radio telephone**  Telephones which operate over radio frequencies.

**radio wave**  Electromagnetic waves of frequencies between 30 kHz and 3 000 000 MHz, propagates without guide in free space.

**radio wave emission classification**  The International Telecommunications and Radio Conference (ITRC) which met in Cairo in 1938 devised the following classification for amplitude-modulated continuous waves:

*Designator*  *Type of emission*

A0  Waves the successive oscillations of which are identical under fixed conditions.

A1  Telegraphy on pure continuous waves. A continuous wave that is keyed according to a telegraph code.

A2  Modulated telegraphy. A carrier wave modulated at one or more audible frequencies, the audible frequencies or their combination with the carrier wave being keyed according to a telegraph code.

A3  Telephony. Waves resulting from the modulation of a carrier wave by frequencies corresponding to the voice to music or to other sounds.

*radio frequency spectrum*

| Frequency | 3 kHz | 30 kHz | 300 kHz | 3 MHz | 30 MHz | 300 MHz | 3 GHz | 30 GHz | 300 GHz |
|---|---|---|---|---|---|---|---|---|---|
| Wavelength | | 30 km | 3 km | 300 m | 30 m | 3 m | 30 cm | 3 cm | 0.3 cm |
| Band designation | VLF | LF | MF | HF | VHF | UHF | SHF | EHF | |
| Band number | 4 | 5 | 6 | 7 | 8 | 9 | 10 | 11 | |

*Legend:*

| | | | | |
|---|---|---|---|---|
| kHz | kilohertz | LF | low frequency |
| MHz | megahertz | MF | medium frequency |
| GHz | gigahertz | VHF | very high frequency |
| km | kilometers | UHF | ultra high frequency |
| m | meters | SHF | super high frequency |
| cm | centimeters | EHF | extremely high frequency |
| VLF | very low frequency | | |

A4    Facsimile. Waves resulting from the modulation of a carrier wave by frequencies produced by the scanning of a fixed image with a view to its reproduction in a permanent form.

A5    Television. Waves resulting from the modulation of a carrier wave by frequencies produced at the time of the scanning of fixed or moving objects.

**radioactive isotopes**   Radioactive gases that are used in some gas tube surge protection devices.

**Radiodetermination Satellite Service (RDSS)**  A satellite-based system that allows two-way communications between moving vehicles and a base station. In addition, the system permits the base station to pinpoint the exact location of a vehicle on a real-time basis.

**Raduga satellites**   A class of Russian geostationary communications satellites that were first launched in December 1975.

**RAID**   Redundant Array of Inexpensive Disks.

**rain loss**   The attenuation of a signal due to rainfall.

**RAM**   Random Access Memory.

**random access**   The process of obtaining information from or placing information in storage, where the time required for such access is independent of the location of the information most recently obtained from or placed in storage.

**Random Access Memory (RAM)**   A storage device into which data can be entered (written) and read; usually but not always a volatile semiconductor memory.

**random noise**   Distortion due to the aggregate of a large number of elementary disturbances that occur at random.

**random retry time**   In an Ethernet local area network, the time a station waits after a collision prior to attempting to transmit again.

**RANGKOM**   RANGkaian KOmputer Malaysia.

**RARP**   Reverse Address Resolution Protocol.

**RAS**   Reliability, Availability, and Serviceability.

**raster**   A scanning pattern used for image representation in which successive horizontal lines are followed in presenting, or detecting, individual picture elements (pixels) of the image. Commonly used in television, facsimile, and electronic (laser) printing.

**raster graphic image**   An image formed by the use of a grid of pixels to form a bit map on a screen.

**rate**   The price charged for telecommunication service or equipment.

**rate adaptation**   The process of converting existing slow-speed data communications to the 64 Kbps synchronous data rate of the ISDN B-channel. The rate adaptation process causes the incoming data stream to be padded with dummy bits and clocked at 64 Kbps.

**rate averaging**   The process telephone companies use to establish toll costs for long distance calls.

**rate center**   A defined geographic point used by telephone companies in distance measurements for interLATA mileage rates.

**Rate Demarcation Point (RDP)**   Also referred to as demarc, this point has been defined by the FCC as: "The point of interconnection between telephone company communications facilities and equipment, protective apparatus or wiring at a subscriber's premises. The network interface or demarcation point shall be located on the subscriber's side of the telephone company's protector, or the equivalent thereof in cases where a protector is not employed" (First Report and Order, CC Docket 81-216, May 18, 1984). The RDP is the point where tariffed charges end and deregulated charges begin.

**RATEPRO**   A trademark of COMDEV of Sarasota, FL, as well as a call accounting system which enables users to verify monthly telephone bills and budget and control other telephone expenses.

**RATG**   RAdioTeleGraph.

**ratio, signal to noise**   A relative measurement of the power of a signal to the power of the noise on a communications channel. Expressed in decibels (dB). Used in measuring channel quality and specifying channel or equipment characteristics.

**RATT**   RAdio TeleTypewriter.

**RATTLER**   A sophisticated battle-proven radar jammer system manufactured by Rafael Armament Development Authority of Haifa, Israel.

**RAU**   Radio Access Unit.

**Rayleigh fading**   A radio impairment encountered in cellular telephone's 800 MHz frequency band that causes an occasional signal interval to be lost.

**Rayleigh scattering**   The random scattering of light along a fiber resulting from submicroscopic differences in material density. An increase in wavelength decreases the amount of Rayleigh scattering.

**RA0**   The initial step in the European Computer Manufacturers Association's 102 Recommendation for rate adaptation. In the RA0 step asynchronous data is converted to synchronous data.

**RA1**   The second step in the European Computer Manufacturers Association's 102 Recommendation for rate adaptation. In the RA1 step a synchronous data stream data rate is adapted to one of three higher intermediate speeds—8 Kbps, 16 Kbps or 32 Kbps.

**RA2**   The third and final step in the European Computer Manufacturers Association's 102 Recommendation for rate adaptation. In the RA2 step the data rate is converted to the ISDN 64 Kbps B-channel rate.

**RB**   Reference Burst.

**RBID**   Reference Burst IDentification.

**RBOC**   Regional Bell Operating Company.

**RBS**   Robbed-Bit-Signaling.

**RBT**   Remote Batch Terminal.

**RCA Global Communications**   An international record carrier that provides international telex, facsimile, leased private-line channels and data transmission services that were purchased by MCI Communications Corporation.

**RCAC**   Remote Computer Access Communications Service.

**RCD**   Receiver Carrier Detect.

**RCI**   Remote Computer Interface.

**RCM**   Reporting and Configuration Module.

**RCOC**   Reserve Communications Operation Center.

**RCV**   ReCeiVer.

**RCVR**   ReCeiVeR.

**RC-292**   A U.S. military general purpose, stationary, ground plane antenna used to increase the distance range of tactical frequency modulation radio sets.

**RD**   1. Receive Data. 2. RingDown.

**RDC**   Remote Data Concentrator.

**RDD**   Receive Data Decapsulation.

**RDF**   Radio Direction Finding.

**RDP**   Rate Demarcation Point.

**RDPS**   Reverse Direction Protection Switch.

**RDR**   ReDirectoR.

**RDS**   Remote Device System.

**RDSS**   Radiodetermination Satellite Service.

**RDT**   Resource Definition Table (IBM's VTAM).

**RDTE**   Research, Development, Test and Evaluation.

**reactance**   Frequency sensitive communications line impairment causing loss of power and phase shifting.

**read only**   A teleprinter receiver without a transmitter.

**readdress**   A method whereby the originator or the original addressee(s) may add new addressee(s), without change in the address or text, to a previously transmitted message.

**read-only bulletin board**   A bulletin board to which messages may be read but no messages can be sent unless the user is the designated owner of the board.

**Read-Only Memory (ROM)**   A nonvolatile memory device manufactured with predefined contents that can only be read.

**read-only replica**   In Digital Equipment Corporation Network Architecture (DECnet), a replica which responds only to lookup requests.

**Read/Write (R/W)**   A memory that allows data to be read from it and written to it. Read/Write memory is generally a Random Access Memory (RAM).

**real name**   1. In IBM's SNA, the name by which a logical unit (LU), logon mode table, or class of service (COS) table is known within the SNA network in which it resides. 2. In IBM's PROFS and other electronic mail systems, a person's given name, such as John M. Doe.

**real network address**   In IBM's SNA, the address by which a logical unit (LU) is known within the SNA network in which it resides.

**real time**   Response to requests for service on demand, in contrast to time sharing, in which all requests for service are responded to on a round-robin basis in a predetermined time sequence.

**real-time system**   An on-line computer that generates output nearly simultaneously with the corresponding inputs. Often, a computer system whose outputs follow by only a very short delay its inputs.

**rearrangeable switch**   A circuit switch which accommodates extra connections by rearrangement of some of the connections it already carries, so that they follow different paths through the switch. By such rearrangement a switch may become non-blocking.

**reasonableness checks**   Tests made on information reaching a real-time system or being transmitted from it to ensure that the data is within a given range. It is a means of protecting a system from data transmission errors. Also limit check.

**reassembly**   The process of reconstructing a complete user data message from the received segments.

**reassignment**   In Digital Equipment Corporation Network Architecture (DECnet), when operating over the CONS, the process of transferring a transport connection to use a new network connection when the original has failed for some reason.

**rebooting** The process of reinitializing an operating system.

**REC** RECeiver.

**receipt** A transmission made by a receiving station to indicate that a message has been satisfactorily received.

**Receive Clock (RxC)** An interface timing signal which synchronizes the transfer of Receive Data (RxD), provided by a DCE device.

**receive only terminal** A terminal which has a printer but lacks a keyboard.

**receive pacing** In IBM's SNA, the pacing of message units that the component is receiving.

**Received Data (RD)** An RS-232 data signal received by DTE from DCE on pin 3.

**Receive Only (RO)** A device usually a printer, which can receive transmissions but cannot transmit.

**Received Line Signal Detector (RLSD)** Modem interface signal, defined in RS-232, that indicates to the attached data terminal equipment that it is receiving a signal from the distant modem.

**receiver** Any device that is capable of receiving a transmitted message.

**RECFMS** RECord Formatted Maintenance Statistics.

**RECMS** Record Maintenance Statistics (IBM's SNA).

**Recommendation X.21 (Geneva 1980)** A Consultative Committee on International Telegraph and Telephone (CCITT) now known as the ITU recommendation for a general purpose interface between data terminal equipment and data circuit equipment for synchronous operations on a public data network.

**Recommendation X.25 (Geneva 1980)** A consultative Committee on International Telegraph and Telephone (CCITT) now known as the ITU recommendation for the interface between data terminal equipment and packet-switched data networks.

**reconfiguration** The process of changing the quantity, types, or arrangement of hardware and/or software.

**record** A collection of logically related fields.

**Record Formatted Maintenance Statistics (RECFMS)** In an IBM SNA network, a RECFMS is sent to a focal point as a solicited reply unit in response to a Request Maintenance Statistic (REQMS) request. A statistical record built by an SNA controller and usually solicited by the host.

**Record Maintenance Statistics (RECMS)** In IBM's NPDA, an SNA error event record built from an NCP or line error and sent unsolicited to the host.

**record mode** In IBM's ACF/VTAM, the mode of data transfer in which the application program can communicate with logical units (LUs).

**Record Separator (RS)** A control character.

**recovery** The necessary actions required to bring a system to a predefined level of operation after a failure.

**recovery procedure** An action performed by the user when an error message appears on the display screen. Usually, this action permits the program to continue or permits the operator to run the next job.

**rectifier** A device built into a power supply which converts AC into DC power for use in operating logic circuits.

**Red Alarm** An alarm condition on a T1 circuit; a Red Alarm is in effect whenever an Out-of-Frame condition persists for 2.5 seconds. A Red Alarm condition at one end of a circuit causes the hardware at the other end to transmit the Yellow Alarm signal.

**Red Book** The 1984 compilation of CCITT now known as the ITU Standards for international telecommunications.

**red box** An illegal device used to simulate the aural system tones that count coins deposited in a pay phone. Its use make operators believe that the correct amount is deposited for a call.

**Red Ryder**   A communications program designed for use on Apple Macintosh personal computers which is marketed by Freesoft of Beaver Falls, PA.

**redirect**   In packet switching, a function that routes a call to an alternate network address if the communications line to the original address is blocked. Redirect occurs at the endpoint switches.

**redirect address**   In packet switching, the 12- or 14-digit X.25 address to which a call is routed if the primary address is unavailable.

**redirect NPDU**   In Digital Equipment Corporation Network Architecture (DECnet), an NPDU issued by a router when it forwards a data NPDU onto the same subnetwork from which it was received. It includes the subnetwork address to which the NPDU was forwarded. This indicates to the sender of the original data NPDU that it can send subsequent NPDUs destined for the same NSAP address directly to the indicated subnetwork address.

**redirector**   A module of DOS 3.1 and later versions of the operating system that allows a user to access the resources of a remote file or print server as if those resources were attached directly to the user's computer.

**Reduced Instruction Set Computing (RISC)**   A computer-designed architecture where the number of processing instructions is reduced to enable most instructions to execute faster.

**redundancy**   In data transmission, the portion of characters and bits that can be eliminated without losing information. Also refers to duplicate facilities.

**redundancy checking**   A technique of error detection involving the transmission of additional data related to the basic data in such a way that the receiving terminal can determine to a certain degree of probability whether an error has occurred in transmission.

**redundant code**   A code using more signal ele-

ments than necessary to represent the intrinsic information. For example, five-unit code using all the characters of International Telegraph Alphabet No. 2 is not redundant; five-unit code using only the figures in International Telegraph Alphabet No. 2 is redundant; seven-unit code using only signals made of four "space" and three "mark" elements is redundant.

**re-entrant code or program**   Same as pure code.

**re-entrant routine**   A software routine that does not alter itself during execution. A re-entrant routine can be entered and reused at any time by any number of callers.

**reference clock**   The master timing source of a synchronous system.

**reference noise (dBrn)**   The level of noise that is equal to 1 picowatt ($-90$ dBm) of power at 1000 Hz.

**reference pilot**   A reference pilot is a different wave from those which transmit the telecommunication signals (telegraphy, telephony). It is used in carrier systems to facilitate the maintenance and adjustment of the carrier transmission system such as automatic level regulation and synchronization of oscillators.

**reference point**   In the CCITT (now ITU) model for ISDN, a reference point is an abstract location between two proposed devices; in other words, a reference point is the potential location of a physical and logical interface. The CCITT identifies reference points by single, capital letters from the end of the alphabet. Thus, the R-reference point resides between a nonstandard terminal and a terminal adapter; the S-reference point stands between a terminal device (standard terminal or terminal adapter) and Network Termination 2; the T-reference point stands between Network Termination 2 and Network Termination 1; and the U-reference point (in the United States) stands between Network Termination 1 and the provider's network. *Note.* In common practice, the physical and logical interface proposed for a given reference point is named for that reference point, thus, the interface at the S-reference point is known as the S-interface.

**refile**  The reprocessing of messages into appropriate format for transfer to another communications system.

**refile message**  An incoming message that is forwarded to another headquarters over a network or facility other than that by which it was received.

**reflecting satellite**  A satellite intended to reflect radiocommunications signals.

**reflection**  1. The abrupt change in direction of a light beam at an interface between two dissimilar media so that the light beam returns into the medium from which it originated. 2. The turning back of a radio wave from an object or the surface of the earth. 3. The name used by Walker Richer & Quinn, Inc., of Seattle, WA, for a series of terminal emulation software packages. Reflection software products operate on IBM PC and comparable computers and the Apple Macintosh family of personal computers. Emulations include:

| Software | IBM | Macintosh |
| --- | --- | --- |
| Reflection 1 Plus | HP 2392A, 700/92 2624B, DEC VT102, VT52 | HP 2392A |
| Reflection 2 Plus | DEC VT220, VT102, VT52, Tektronix 4014 | DEC VT320 |
| Reflection 3 Plus | | HP 2393A, HP 2392A |
| Reflection 4 Plus | DEC VT241, Tektronix 4014, VT220, VT102, VT52 | |
| Reflection 7 Plus | HP 2627A, 2623A, 2392A, 2624B, DEC VT102, VT52, and Tektronix 4010 | |

**refraction**  The bending of a beam of light at an interface between two dissimilar media or in a medium whose refractive index is a continuous function of position (graded-index medium).

**refractive index**  The ratio of the speed of light in a vacuum to its speed in a given material such as glass. The larger the ratio, the more the light entering the material is bent.

**refresh rate**  In CRT displays, the rate per unit of time that a displayed image is renewed in order to appear stable; typically 50 times per second, or 50 Hz, in Europe or 60 times per second, or 60 Hz, in the United States.

**regeneration**  A technique used to restore a digital signal on a cable pair. The digital signal is reconstructed and transmitted to the next regenerator. The regenerator is the device which makes digital transmission superior to analog transmission. Because the signal is reconstructed, the effects of noise and attenuation distortion are minimized.

**regenerative repeater**  Telegraph repeaters which are speed-and code-sensitive, and are used to retime and retransmit a received signal at its original strength.

**regenerator**  Equipment that restores the shape, timing, and amplitude to digital signals that may have been distorted during transmission. In a lightwave system, an electronic tonic for pooped pulses. A photodetector collects the pulses and converts them to electricity. An electronic circuit reconverts the electrical signals to light pulses, and a laser speeds them on their way. A costly part of lightwave systems.

**Regie des Télégraphes et Des Téléphones**  The Belgium PTT.

**Regie van Telegrafie en Telephone**  The Belgian PTT.

**Region**  In a bulletin board system network a region is a geographic area that contains nodes that exchange messages in a defined manner.

**Regional Bell Operating Company (RBOC)**  One of seven holding companies (Ameritech, Nynex, Bell Atlantic, BellSouth, Pacific Telesis, Southwestern Bell, and US West) set up as a result of the AT&T divestiture. Publicly traded, each is responsible for owning various Bell operating companies.

**regional center** A control center (class 1 office) connecting sectional centers of the telephone system together. Every pair of regional centers in the United States has a direct circuit group running from one center to the other.

**register** A storage device having a specified storage capacity, such as a bit, a byte, or a computer word. Usually intended for a special purpose.

**registration program** An FCC program which authorizes the connection of terminal equipment to the telecommunications network without the need for a telephone company-provided protective connecting arrangement known as data access arrangements.

**regulatory agency** An agency which controls common and specialized carrier tariffs, such as the Federal Communications Commission and the State Public Utility Commissions.

**REJ** REJect.

**relative file organization** A file whose records are organized to be accessed sequentially or directly by their record position relative to the beginning of a file.

**relative humidity** The amount of water present in the air as measured at 72 degrees Fahrenheit.

**relative record number** A number representing the position of a record relative to the beginning of a file. The initial record in a file is relative record number 1.

**relative transmission level** The ratio, expressed in decibels, of the test-tone signal power at one point in a circuit to another circuit point selected as a reference.

**relay** 1. An electronic device that utilizes variation in the condition or strength of a circuit to affect the operation of the same or another circuit. 2. A transmission forwarded through an intermediate station. 3. An automated bank teller network in North and South Carolina.

**Relay Baton** A Macintosh to IBM mainframe file transfer software package marketed by Relay Communications, Inc., of Danbury, CT.

**Relay Gold** A full featured communications program from VM Personal Computing of Danbury, CT, for use on the IBM PC and compatible computers. It is marketed by Relay Communications, Inc., of Danbury, CT. The program is noted for its file transfer capabilities, terminal emulation, text editing and script language.

**relay message** An incoming message to be forwarded to another headquarters over the same network or facilities over which it was received.

**RelayNet International Message Exchange** A bulletin board network which provides participating hubs and nodes with a mechanism to exchange messages through common conference areas that are relayed via a central NetHub.

**Relay/3270** A communications software program developed by Relay Communications of Danbury, CT. Relay/3270 enables microcomputers to emulate 3270 terminals and access IBM VM and MVS host computers via the switched telephone network.

**release** In IBM's ACF/VTAM, resource control, to relinquish control of resources (communication controllers or physical units).

**releasing officer** The person who may authorize the transmission of a message for, and in the name of, the originator.

**reliability** A measure of the impact of line failures and the ability to recover. Enhanced by ability to reconfigure or by redundancy.

**relocatable address** A reference to a storage location that has a fixed displacement from the program origin, but whose displacement from absolute memory location zero depends upon the loading address of the program.

**REM CD** REMote Carrier Detect (control signal).

**REM LP**   REMote LooP.

**remote**   1. Physically distant from a local computer, terminal, multiplexers etc. Synonym for link-attached. 2. A communications software program marketed by Crosstalk Communications of Roswell, GA, which enables one personal computer to remotely control the operations of another personal computer.

**remote access**   1. The ability of a transmission station to gain access to a computer from which it is physically removed. 2. A PBX or central office feature that allows a user to operate a phone set at a location other than where the PBX is located.

**remote analog loopback**   An analog loopback test that forms the loop at the line side (analog output) of the remote modem.

**remote batch processing**   A batch process in which the source data is prepared and collected at a remote station. The station then uses remote access to enable the computer to retrieve and process the data.

**Remote Batch Terminal (RBT)**   A collection of input/output devices such as a card reader, line printer, and console that is controlled by a single processing unit which uses a single communications line.

**remote call forwarding**   A function of a PBX or central office switch which forwards calls to a distant exchange but allows the calling party to call as if it were a local call.

**Remote Call Procedure (RCP)**   A command from one computer application that causes another computer or network to carry out a predefined task.

**remote channel loopback**   A channel loopback test that forms the loop at the input (channel side) of the remote multiplexer.

**remote composite loopback**   A composite loopback test that forms the loop at the output (composite side) of the remote multiplexer.

**Remote Computer Interface (RCI)**   The communications procedure used to operate the link between a Honeywell (now Bull) front end processor, and a Level 6 computer. RCI supports one logical stream on each physical line.

**Remote Device System (RDS)**   A trademark of Network Systems of Minneapolis, MN, as well as a combination of hardware and software that supports remotely located, channel-attached IBM peripherals in a locally attached mode. Through the use of RDS, devices to include line printers and control units can be operated at greater distances from a host computer than otherwise possible.

**remote digital loopback**   A digital loopback test that forms the loop at the DTE side (digital input) of the remote modem. No modem operator is required at the looping modem. With RDL a signal is sent down the communications line which instructs the remote modem to place itself in digital loopback mode.

**Remote File Service (RFS)**   A distributed file system network protocol developed by AT&T and adopted by other vendors as part of UNIX V. The protocol permits one computer to use the files and peripherals of another as if they were local.

**Remote Job Entry (RJE)**   Entering batch jobs from remote terminals usually including a card reader and printer. Also remote batch entry.

**remote log-in**   Application software which allows a user directly connected to one computer to be connected to a remote computer and establish an interactive log-in session.

**Remote Memory Administration System (RMAM)**   An AT&T operations support system which provides mechanization of service order input to a Metro-Digital Transition Plan switching system.

**Remote Network Processor (RNP)**   A network that is not directly connected to an information processor.

**Remote Procedure Call (RPC)**   A protocol within Sun's Network File System (NFS) which transmits requests to file servers and services their response.

RPC includes the ability to verify that requesters have authorization to make calls.

**remote processing** Moving part of the main (host) computer's processing responsibility to a computer at a remote location. This remote (mini) computer may be connected to and supervised by the host computer.

**remote station** Any device attached to a controlling unit by a data link.

**Remote Terminal (RT)** 1. The terminating equipment for a digital line furthest from the central office. 2. A terminal attached to a system through a data link. 3. In telephony, a terminal attached through a trunk or tieline.

**Remote Terminal Access Method (RTAM)** A facility that controls operations between the job entry subsystem (JES2 or JES3) and remote terminals.

**remote terminal supervisor** A software program originally developed by General Electric and modified by Honeywell which controls communications on the Honeywell Datanet 355 and 6600 series front end processors.

**remove** In a local area network environment, to make an attaching device inactive on the ring. To stop an adapter from participating in data passing on the network.

**REN** Ringer Equivalence Number.

**RENAN** The name of the French ISDN trial.

**repair** The restoration or replacement of parts or components as necessitated by wear, tear, damage, or failure.

**repairables** Parts or items that are economically and technically repairable.

**Repartee** A voice mail system marketed by Active Voice Corporation of Seattle, WA.

**repeat dialing** A telephone feature offered by many communications carriers and supported by most PBX's which automatically redials the last number

a customer dialed if the first attempt resulted in a busy signal. Also known as call queue.

**repeater** 1. In a lightwave system, an optoelectric device or module that receives an optical signal, converts it to electrical form, amplifies it (or, in the case of a digital signal, reshapes, retimes, or otherwise reconstructs it), and retransmits it in optical form. 2. In digital transmission, a device used to extend transmission ranges/distance by restoring signals to their original size or shape. Repeaters function at the physical layer of the OSI model. 3. In a local area network (LAN), a device that repeats all messages from one LAN segment to another; connects two or more separate LAN segments and logically joins them into a single LAN, increasing the maximum length of the LAN connection.

**repeater, regenerative** Normally, a repeater utilized in telegraph applications. Its function is to retime and retransmit the received signal impulses restored to their original strength.

**repeater, telegraph** A device which receives telegraph signals and automatically retransmits corresponding signals.

**reperforator** A device that can be used to receive teletypewriter signals and convert the signals to typewritten characters which are punched on paper tape or which can be used to prepare tape or transmit characters entered from its keyboard directly to the line.

**reperforator (receiving perforator)** A telegraph instrument in which the received signals cause the code of the corresponding characters or functions to be punched in a tape.

**Reperforator/Transmitter (RT)** A teletypewriter unit consisting of a perforator and a tape transmitter, each independent of the other. It is used as a relaying device and is especially suitable for transforming the incoming speed to a different outgoing speed, and for temporary queuing.

**repertory dialing** The ability of a PABX to dial a

complete external telephone number upon receipt of a predetermined abbreviation code for that number from an internal station.

**replica** In Digital Equipment Corporation Network Architecture (DECnet), a copy of a directory stored in a particular clearinghouse.

**reply** The answer to a service request that came from a device.

**Req** Request.

**reqdiscont.** A term which specifies how the IBM 3270 Emulation Program terminates the host connection when a 3270 task is ended.

**REQMS** REQuest for Maintenance Statistics.

**request** The requirement for service that came from a requester.

**Request For Comments (RFC)** A proposal which when reviewed by the Internet Task Force can be formalized into Official Protocols which are Internet standards. Current Internet RFCs include:

| | |
|---|---|
| FRC-768 | User Datagram Protocol |
| RFC-791 | Internet Protocol |
| RFC-792 | Internetwork Control Message Protocol |
| RFC-793 | Transmission Control Protocol |
| RFC-821 & RFC-822 | Simple Mail Transport Protocol |
| RFC-822 & RFC-823 | Domain Name Service |
| RFC-826 | Address Resolution Protocol |
| RFC-854 & RFC-855 | Telenet Terminal Access Service |
| RFC-959 | File Transfer Protocol |
| RFC-1001 & RFC-1002 | NETBIOS |

**Request of Discussion (RFD)** A period of time during which comments on a particular subject are solicited.

**Request For Information (RFI)** A general notification of an intent of an organization to purchase computer or communications equipment. An RFI is normally sent to potential suppliers to determine their interest and solicit information concerning their products.

**Request for Maintenance Statistics (REQMS)** An IBM host solicitation to an SNA controller for a statistical data record (RECFMS).

**Request for Price Quotation (RPQ)** A document which is used for the solicitation of pricing for a hardware device, software product, service, or system.

**Request For Proposal (RFP)** A document sent to interested vendors which defines the requirements of an organization so the vendor can configure and price their product(s).

**Request Header (RH)** In IBM's SNA, control information preceding a request unit (RU).

**Request Parameter List (RPL)** In IBM's VTAM, a control block that contains the parameters necessary for processing a request for data transfer, for establishing or terminating a session, or for some other operation.

**Request Unit (RU)** In IBM's SNA, a message unit that contains control information such as a request code or FM headers, end-user data, or both.

**Request/Response Header (RH)** In IBM's SNA, control information, preceding a request/resonse unit (RU), that specifies the type of RU (request unit or response unit) and contains control information associated with that RU.

**Request/Response Unit (RU)** In IBM's SNA, a generic term for a request unit or a response unit.

**Request-To-Send (RTS)** An RS-232 modem interface signal (sent from the DTE to the modem on pin 4) which indicates that the DTE has data to transmit.

**required cryptographic session** In IBM's SNA, a cryptographic session in which all outbound data is enciphered and all inbound data is deciphered. Synonymous with mandatory cryptographic session.

**required parameter** A parameter that must have a defined option. The user must provide a value if no default is supplied.

**rerun** The retransmission of a message as a result of a request from a distant station.

**resale carrier** A company that redistributes the services of another carrier on a retail basis.

**reseller** 1. A company that resells products made by another company. 2. A company that buys or leases transmission lines and then resells them to its customers.

**reserved character** A character or symbol that has special (non-literal) meaning unless quoted.

**reserved word** A word that is defined in a programming language for a special purpose, and that must not appear as a user-declared identifier.

**reset** A term referring to system initialization.

**reset packet** A packet used to clear an error condition on an existing switched virtual call or virtual circuit.

**residual error rate, undetected error rate** The ratio of the number of bits, unit elements, characters, or blocks incorrectly received but undetected or uncorrected by the error-control equipment, to the total number of bits, unit elements, characters, or blocks sent.

**resistance** The opposition to the flow of electrons in a conductor measured in ohms. Resistance causes a voltage applied at the transmitting end of a circuit to drop.

**resistor** A circuit component designed to resist electrical current flow. A resistor can be used to provide a desired voltage drop for use in a circuit.

**resolution** A measurement of display or print quality. The number of dots placed in a specified area, usually one square inch.

**resource** 1. Any data processing device in a network. 2. In IBM's SNA, any facility of the computing system or operating system required by a job or task, and including main storage, input/output devices, the processing unit, data sets, and control or processing programs. 3. In IBM's NPDA, any hardware or software component that provides function to the network or to locally attached environments.

**Resource Access Control Facility (RACF)** An IBM mainframe security software product which enables users to secure a mainframe by supporting the use of user identifications and passwords as well as restricting access to defined computer resources and logging detected unauthorized attempts to enter the system and detected accesses to protected data sets.

**resource class** In LAN technology, a collection of computers or computer ports that offer similar facilities, such as the same application program; each can be identified by a symbolic name.

**Resource Definition Table (RDT)** In IBM's VTAM, a table that describes the characteristics of each node available to VTAM and associates each node with a network address. This is the main VTAM network configuration table.

**resource hierarchy** In IBM's VTAM, the relationship among network resources in which some resources are subordinate to others as a result of their position in the network structure and architecture; for example, the LUs of a peripheral PU are subordinate to that PU, which, in turn, is subordinate to the link attaching it to its subarea node.

**resource level** In IBM's NPDA, the hardware (and software contained in the hardware) configuration of a data processing system. For example, a first-level resource would be the communication controller, and the second-level resource would be the line connected to it.

**resource takeover** In IBM's VTAM, action initiated by a network operator to transfer control of resources from one domain to another.

**resource types** In IBM's NPDA, a concept to describe the organization of data displays. Resource types are defined as central processing unit, channel, control unit, and I/O device for one category; and communication controller/adapter, link, cluster controller, and terminal for another category. Resource types are combined with data types and display types to describe NPDA display organization.

**responded output** In IBM's VTAM, a type of output request that is completed when a response is returned.

**response** An answer to an inquiry. In IBM's SNA, the control information sent from a secondary station to the primary station under SDLC.

**Response Header (RH)** In IBM's SNA, a header, optionally followed by a response unit (RU), that indicates whether the response is positive or negative and that may contain a pacing response.

**response time** The elapsed time between the generation of the last character of a message at a terminal and the receipt of first character of the reply (often an echo). It includes all propagation delays.

**Response Time Monitor (RTM)** In IBM's NLDM, a feature available with the 3274 control unit to measure response times.

**Response Unit (RU)** In IBM's SNA, a message unit that acknowledges a request unit; it may contain prefix information received in a request unit. If positive, the response unit may contain additional information (such as session parameters in response to a bind session), or if negative, contains sense data defining the exception condition.

**responsivity** In an optical system, responsivity is the ratio of a detector's output current to the input light power.

**restart packet** A packet whose contents notifies X.25 DTEs that an irrecoverable error exists in an X.25 network. A restart packet clears all existing switched virtual calls and resynchronizes all existing permanent virtual circuits between an X.25 DTE and an X.25 DCE.

**Restoration Priority (RP)** A priority level assigned to a circuit for restoring service during an emergency.

**restorer** A trademark of Atlantic Research Corporation of Sringfield, VA, as well as a system which uses the direct dial network to backup multidrop circuits.

**restricted document** A document in an electronic mail system that only certain people are allowed to work with. In IBM's PROFS, the author of a document may specify who these people are when he or she first puts the document in PROFS storage or sends the document to someone. Only the author of a restricted document can forward it to others and only the people on the document's distribution list can view it.

**restricted resource group** In a switching network, a group of ports that can only be called by other ports within the same group.

**Restructured Extended Executive (REXX)** An interpreter that supports structured coding on an IBM mainframe computer system.

**restricted use cable** Cable designed for use in non-concealed spaces where the exposed cable length does not exceed 10 feet. This type of cable can also be used in an enclosed raceway or noncombustible tubing, or when it is less than 0.25 inches in diameter and installed in one- or multiple-family dwellings.

**RETRANS** RETRANSmission.

**retransmission** The procedure of transmitting a message for a second or subsequent time. Performed when it is assumed that the previous copy of the message was not successfully delivered.

**retransmissive start** An optical fiber component that permits the light signal on an input fiber to be retransmitted on multiple output fibers.

**retry** The process of retransmitting a block of data a prescribed number of times.

**return address** The address of an instruction in a

program to which control is returned after a call to a subroutine occurs.

**return code** 1. A code used to influence the execution of succeeding instructions. 2. A value returned to a program to indicate the results of an operation requested by that program.

**return loss** The measurement of reflected power at various frequencies.

**Return to Zero (RZ)** Encoding information so after each encoded bit, voltage returns to zero level.

**REUNIR** RÉseaux des UNIversités et de la Recherche.

**REV** REVision.

**Revenue Volume Pricing Plan (RVPP)** An AT&T pricing plan for large 800 service customers. Under RVPP the billing for both domestic and international inbound 800 service can be added to obtain a discount whenever the monthly volume exceeds a revenue threshold.

**Reverse Address Resolution Protocol (RARP)** A TCP/IP internet protocol used by a diskless workstation on a LAN to obtain its IP address from a server.

**reverse channel** A feature provided on some modems which provides simultaneous communication from the receiver to the transmitter on a two-wire channel. It may be used for circuit assurance, circuit breaking, error control, and network diagnostics. Also called a backward channel.

**reverse charge call** A telephone call in which the caller specifies that the charge should be paid by the called party. Also called collect call.

**Reverse Direction Protection Switch (RDPS)** A switch used in the U.S. Sprint network to restore service to a Point Of Presence (POP) affected by a route outage by sending transmissions in the opposite direction from a failure. RDPSs are installed along the entire backbone of the U.S. Sprint network.

**Reverse Interrupt (RVI)** In the bisynchronous

protocol, a reverse interrupt is a control character sequence sent by a receiving station to request the premature termination of a transmission in progress.

**reverse video** To display a screen so that the background color is the normal foreground color and the foreground color is the normal background color.

**reverse-battery signaling** A loop signal in which the battery and ground are reversed on the tip and ring of the loop to provide an off-hook signal when the called party answers.

**revertive pulsing** A means of controlling distant switching selections in telephone networks by pulsing in which the near end receives signals from the far end.

**revisable-form document** An electronic document with its formatting information intact, making it readable and modifiable.

**Revisable-Form Text (RFT)** A document form that allows the users of IBM office systems to communicate in a universally understood manner.

**REX** Route EXtension (IBM's SNA).

**REXX** REstructured eXtended eXecutive.

**RF** Radio Frequency.

**RF modem** A modem that converts digital signals to analog signals and analog signals back into digital signals using assigned frequencies. This type of modem is used on a broadband local area network.

**RFA** Radio Frequency Authority.

**RFC** Request For Comments. The set of documents which defines the internal operation of the Internet.

**RFC-822** The standard which defines the format of messages in the TCP/IP Simple Mail Transfer Protocol (SMTP). RFC establishes such fields as "to," "from," "subject," "date," "reply-to" and "CC." It has an escape mechanism that allows new fields to be added by mutual agreement among users.

**RFD**   Request For Discussion.

**RFI**   1. Radio Frequency Interference. 2. Request For Information.

**RFP**   Request For Proposal.

**RFS**   Remote File Service.

**RFSM**   Radio Frequency Spectrum Management.

**RFT**   Revisable-Form Text.

**RFT-D**   A Revisable-Form Text (RFT) document that can be changed in IBM's PROFS.

**RFT-F**   A Revisable-Form Test (RFT) document that can't be changed in IBM's PROFS.

**RH**   1. Request Header. 2. Request/Response Header (IBM's SNA).

**rhosts**   In UNIX rhosts is the home directory which lists systems from which the user can log in without a password.

**RI**   Ring Indicator.

**RIME**   RelayNet International Message Exchange, a multi-tier communications network which exchanges messages among member bulletin board systems.

**ring**   1. In LAN technology, a closed-loop network topology. 2. Audible signal that announces an incoming call. 3. The battery-supplied or "hot" wire in a traditional telephone circuit.

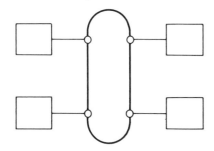

Basic ring topology

**ring around**   An improper routing of a call back through a switching center already involved in attempting to complete the same call.

**ring diagnostic**   In the IBM Token-Ring Network, software to be run in a workstation that provides the user information regarding the performance of the ring.

**ring in (RI)**   In a local area network environment, the receive or input receptacle on an access unit.

**Ring Indicator (RI)**   An RS-232 modem interface signal (sent from the modem to the DTE on pin 22) which indicates that an incoming call is present.

**ring network**   A network in which there is not a central computer, but a series of computers which communicate with one another.

**ring out (RO)**   In a local area network environment, the transmit or output receptacle on an access unit.

**ring sequence**   The order in which devices are attached on a ring network.

**ring status**   The condition of a ring.

**Ring System**   An array of processors distributed along one or more rings or loops. Data is transferred around the ring or loop and used by a particular processor according to a predetermined address.

**ring topology**   A logically circular, unidirectional transmission path without defined ends. Control can be distributed or centralized.

**ringback tone**   The sound heard by the calling party through the handset to indicate that the called telephone is ringing.

**ringdown**   A method of signaling subscribers and operators using either a 20-cycle AC signal, a 135-cycle AC signal, or a 100-cycle signal interrupted 20 times per second.

**Ringer Equivalence Number (REN)**   A number which indicates the quantity of ringers (or products) which may be connected to a single telephone line and still ring. For example, the number 1.0B. The total of all RENs connected to a single line may not exceed the value of 5 or some or all of the ringers may not work. The letter indicates the frequency response of the ringer. The REN is on the "FCC

Registration" label located on the bottom of analog phone products.

**ringing key**   A key that sends the ringing current.

**RingMaster**   A trademark of FiberCom of Roanoke, VA, as well as the name for a high performance FDDI bridge developed by that vendor.

**R-interface**   An ISDN reference point which is the interface for older, non-ISDN equipment accessing an ISDN network.

**RIP**   Routing Information Protocol.

**RIPE**   Reseaux IP Européens.

**RISC**   Reduced Instruction Set Computing (computer).

**riser cable**   A cable designed to be used in vertical shafts (risers) and not for inside plenums unless it is enclosed in noncombustible tubing. Riser cable has fire-resistant characteristics that are designed to prevent it from spreading fire between floors.

**riser closet**   Same as backbone closet.

**RISLU**   Remote Integrated Services Line Unit.

**Rivest, Shamir and Adelman (RSA)**   The developers of a public key encryption system commonly referred to as RSA.

**RJE**   Remote Job Entry.

**RJ11C**   The modular jack connector used on dial

lines (2-wire) and is the connector most commonly used in the office for telephone connections. The RJ11 is an optional connector for 4-wire private lines. For dial line operation, modems are designed to transmit at a level of -9 dBm. The RJ11C is a miniature 6-position surface- or flush-mounting jack. The RJ11C provides a single-line bridged tip and ring electrical network connection and is typically used with single-line non-key telephones and ancillary devices.

**RJ11W**   A wall-mounted version of the RJ11C jack.

**RJ12C**   A miniature 6-position surface- or flush-mounted jack. The RJ12C provides a single-line bridged tip and ring ahead of the line circuit of a key system with A/A1 leads electrical network connection. Typical usage of the RJ12C is for use with single line non-key telephone sets and ancillary devices connected to a key system where registered terminal equipment is not compatible with electrical characteristics of tip and ring behind line circuits.

**Wiring diagram**

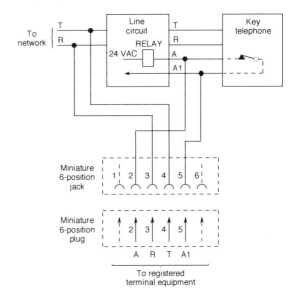

**Wiring diagram**

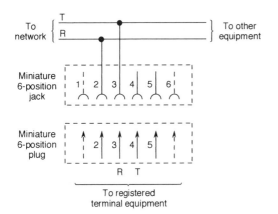

**RJ12W**   A wall-mounted version of the RJ12C.

**RJ13C** A miniature 6-position surface- or flush-mounted jack. The RJ13C provides an electrical network connection for a single-line bridged tip and ring behind the line circuit of a key system with A/A1 leads. Typical usage of the RJ13C is for single-line non-key telephone sets and ancillary devices connected to a key system.

**Wiring diagram**

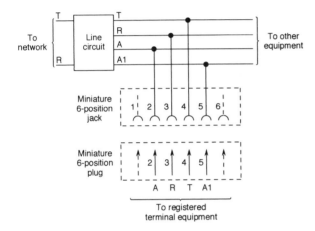

RJ13C–Surface- or flush-mounted jack
RJ13W–Wall-mounted equipment

*RJ13C*

**RJ13W** A wall-mounted version of the RJ13C.

**RJ14C** A miniature 6-position surface- or flush-mounted jack. The RJ14C provides an electrical network connection for two line bridged tip and ring. Its typical usage is for two-line non-key telephone sets and ancillary devices.

**RJ14W** A wall-mounted version of the RJ14C.

**RJ16X** A miniature 6-position surface- or flush-mounted jack which provides an electrical network connection for bridged tip and ring with mode indication (MI) to A series connection ahead of bridged connection. Typical usage of the RJ16X is for permissive data equipment with MI (mode indication) and MIC (mode indication common) leads.

**Wiring diagram**

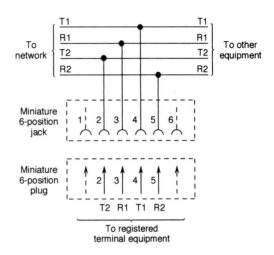

RJ14C–Surface- or flush-mounted jack
RJ14W–Wall-mounted equipment

*RJ13W*

**Wiring diagram**

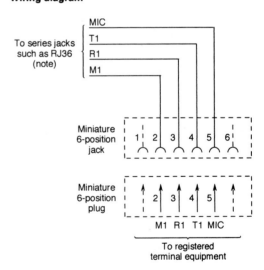

*Note.*
M1 and MIC leads are typically wired to an RJ36X series jack which can be used to connect an exclusion key telephone set ahead of the data equipment.

*RJ16X*

365

**RJ17C** A miniature 6-position surface- or flush-mounted jack which provides an electrical network connection for bridged tip and ring. Typical usage of the RJ17C is for special non-key telephone set or certain hospital ancillary equipment in hospital critical work areas.

**Wiring diagram**

*Subject to effective tariff and/or rulemaking

To network { R T

Miniature 6-position jack | 1 | 2 | 3 | 4 | 5 | 6 |

Miniature 6-position plug | 1 | | | | | 6 |

T          R

To registered terminal equipment

Note.
Equipment must comply with Article 517 of the National Electric Code

**RJ18C** A miniature 6-position surface- or flush-mounted jack which provides an electrical network connection for bridged connection of single-line tip and ring with make-busy leads MB/MB1. Typical usage of the RJ18C is when the registered equipment provides a contact closure between the MB and MB1 leads, a make-busy indication is transmitted to the network equipment busying out the line from further incoming calls.

**RJ19C** A miniature 6-position surface- or flush-mounted jack which provides an electrical network connection for bridged tip and ring behind key telephone line circuit, with A and A1 lead control and make-busy leads. Typical usage of the RJ19C is when the registered equipment provides a contact closure between the MB and MBI leads, a make-busy indication is transmitted to the network equipment busying out the line from further incoming calls.

**Wiring diagram**

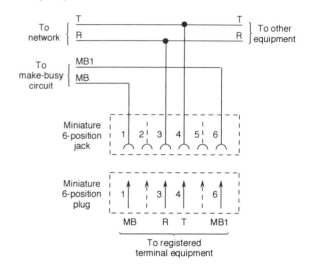

*RJ18C*

**Wiring diagram**

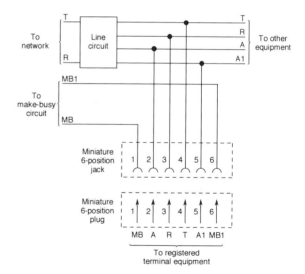

*RJ19C*

**RJ25C** A miniature 6-position surface- or flush-mounted jack which provides an electrical network connection for up to three lines bridged connection. Typical usage of the RJ25C is for

**Wiring diagram**

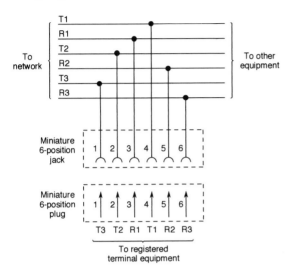

Miniature 6-position jack
| 1 | 2 | 3 | 4 | 5 | 6 |

Miniature 6-position plug
| 1 | 2 | 3 | 4 | 5 | 6 |

T3 T2 R1 T1 R2 R3

To registered terminal equipment

three-line non-key telephone sets, ancillary devices, including message registration, automatic identification, outward dialing, and off-premise station.

**RJ31X**  A miniature 8-position series jack which provides an electrical network connection for series tip and ring ahead of all station equipment. Typical usage is for an alarm-reporting device.

**Wiring diagram**

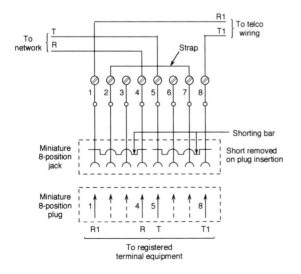

**RJ32X**  A miniature 8-position series jack which provides an electrical network connection for series tip and ring at station. Typical usage is for series ancillary devices such as automatic dialers, single-line sets with exclusion.

**Wiring diagram**

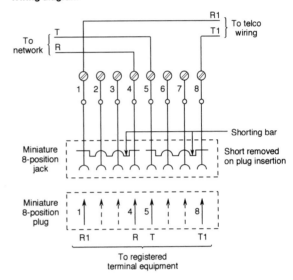

**RJ33X**  A miniature 8-position series jack which provides an electrical network connection for series tip

**Wiring diagram**

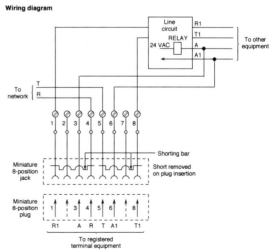

367

and ring ahead of the line circuit of a key telephone system with A/A1 leads. Typical usage is for series ancillary devices connected to a key system where registered terminal equipment is not compatible with electrical characteristics of tip and ring behind line circuit.

**RJ34X** A miniature 8-position series jack which provides an electrical network connection for series tip and ring behind the line circuit of a key telephone system with A/A1 leads. Typical usage is for a single-line set with exclusion or ancillary device connected to a key system.

**Wiring diagram**

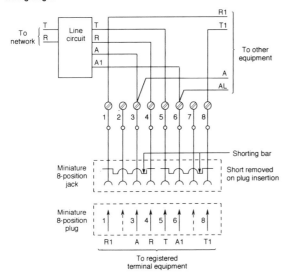

**RJ35X** A miniature 8-position series jack which provides an electrical network connection for series tip and ring at key telephone station, behind station pickup key with A/A1 leads. Typical usage is for series ancillary devices connected to key telephone set.

**RJ36X** A miniature 8-position series jack which provides an electrical network connection for series tip and ring with mode indication signal. Typical usage is for mode indication (exclusion key) telephone set in series with data jack.

**Wiring diagram**

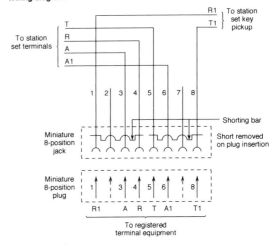

*RJ35X*

**Wiring diagram**

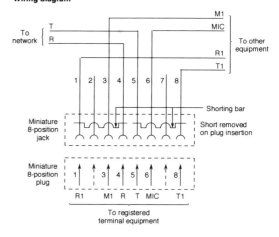

*RJ36X*

**RJ37X** A miniature 8-position series jack which provides an electrical network connection for series tip and ring on first line and bridged tip and ring on second line. Typical usage is for two-line telephones with exclusion on one-line.

368

## Wiring diagram

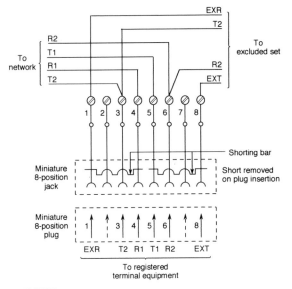

*RJ37X*

**RJ38X**  A miniature 8-position series jack which provides an electrical network connection for series connection identical to RJ31X. Strap provides a continuity circuit between 2 and 7, which is used

## Wiring diagram

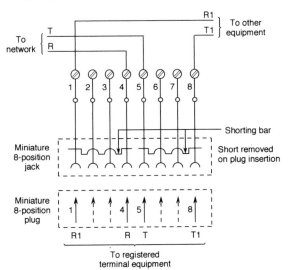

as an indication that the plug of the equipment is engaged with the jack strap provided by the telephone company. Typical usage is for registered alarm dialers.

**RJ41M**  A miniature 8-position keyed data jack; universal which provides an electrical network connection for data, multiple bridged tip and ring. Typical usage is for multiple installations of fixed loss loop or programmed types of data equipment.

## Wiring diagram

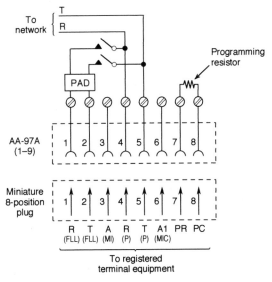

**RJ41S**  The universal "data jack" can connect modems classified either as programmable devices or fixed loss loop devices. This type of jack can function like the RJ45S programmable jack, or allow the modem to transmit at the fixed level of −9 dBm. With a fixed loss loop, the phone company installs the dial line with a "fixed loss" from their central office to your premise. This insures a much closer to optimum level of loss in your dial-up lines. If your modem cannot use the RJ45S programmable jack, but you wish to install a dial data line that is maintained at close tolerances, then the RJ41S and corresponding fixed loss line is a feature to consider. With this type of line and jack, the telephone company is more able to assist you with data communications troubleshooting. This 8-position miniature keyed data jack provides an electrical network connection for data, bridged tip and ring. Typical usage is for a universal jack for fixed loss loop (FLL) or programmed (P) types of data equipment.

**Wiring diagram**

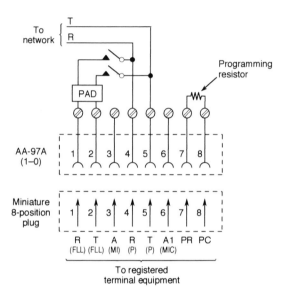

**RJ42S** A miniature 8-position keyed data jack which provides an electrical network connection for data, multiple tip and ring, A and A1 leads, tip and ring ahead of line circuit. Typical usage is for a universal jack for FLL or programmed (P) types of data equipment connected to a key system where the registered terminal equipment is not compatible with electrical characteristics of tip and ring behind the line circuit.

**RJ42M** A miniature 8-position keyed data jack which provides an electrical network connection for data, multiple tip and ring connected ahead of the line circuit. Typical usage is for multiple installations of FLL or P types of data equipment.

**Wiring diagram**

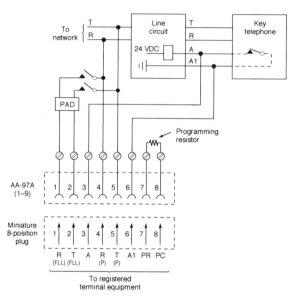

*RJ42S*

**Wiring diagram**

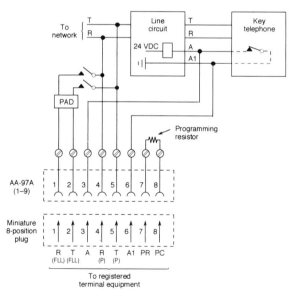

*RJ42M*

370

**RJ43M** A miniature 8-position keyed data jack which provides an electrical network connection for data, multiple tip and ring, A/A1 tip and ring behind line circuit. Typical usage is for multiple installations of FLL and P data equipment.

**Wiring diagram**

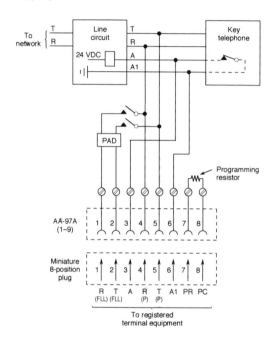

**RJ43S** A miniature 8-position keyed data jack which provides an electrical network connection for data, multiple tip and ring A/A1, tip and ring behind line circuits. Typical usage is for universal jack for FLL or P types of data equipment connected to a key system.

**RJ45M** A miniature 8-position keyed data jack which provides an electrical network connection for data, multiple bridged tip and ring. Typical usage is for multiple installations of programmed types of data equipment.

**RJ45S** This data modular jack can be programmed by inserting a resistor of the proper value. Used with modems classified as "programmable

**Wiring diagram**

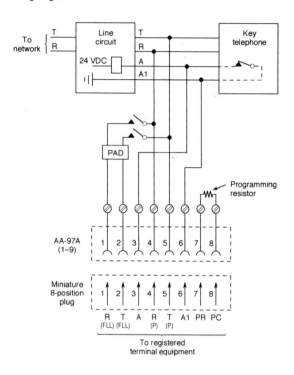

*RJ43S*

**Wiring diagram**

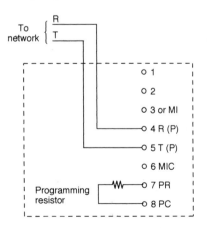

*RJ45M*

371

devices." This jack is used with modems that are engineered to vary their output signal, based on the programmable function of the jack. The jack "programs" the transmit level of the modem, rather than the modem using the default transmit level of −9 dBm. This feature allows the modem to transmit to the central office at a higher level. If you are experiencing problems with dial-up connections, the problem can exist in the link between you and your telephone company central office, as this is the only portion of the dial-up network that is used in every connection. This jack is only used with dial-up lines. This 8-position miniature keyed data jack provides a single-line bridged tip and ring electrical network connection. Typical usage of the RJ45S is for use with programmed data equipment.

**Wiring diagram**

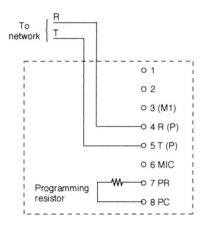

**RJ46M** A miniature 8-position keyed jack similar to the RJ46S but used in multiple mounting arrangements. The RJ46M is typically used with programmed data equipment connected to a key system.

**RJ46S** A miniature 8-position keyed data jack which provides a data, single-line bridged tip and ring, A/A1 tip and ring behind line circuit electrical network connection. Typical usage of the RJ46S is for use with programmed data equipment connected to a key system where the registered terminal

equipment is not compatible with the electrical characteristics of tip and ring behind line circuit.

**Wiring diagram**

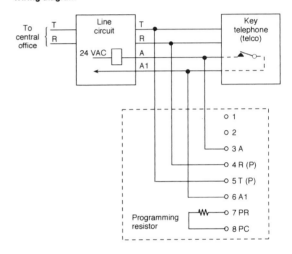

**RJ47M** A miniature 8-position keyed data jack which is the same as an RJ46S but used in multiple mounting arrangements.

**RJ47S** A miniature 8-position keyed data jack which provides an electrical network connection for single-line bridged tip and ring, A/A1 tip

**Wiring diagram**

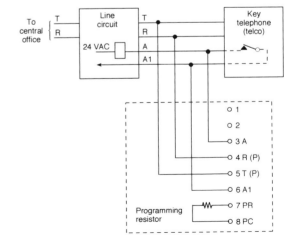

and ring behind line circuit. Typical usage is for programmed data equipment connected to a key system where the registered terminal equipment is not compatible with electrical characteristics of tip and ring behind line circuit.

**RJA1X** A miniature 6-position jack to 4-pin plug adapter which provides a single-line bridged tip and ring electrical network connection. Typical usage of the RJA1X is for use with single-line non-key telephone sets or ancillary devices where a working 4-pin jack exists and the registered terminal equipment is equipped with a miniature 6-position plug.

**Wiring diagram**

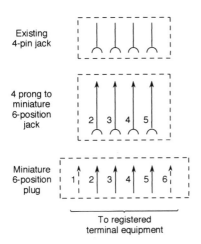

To registered terminal equipment

**RJA2X** A miniature 6-position plug to duplex (2) miniature 6-position jack adapter which provides a single-line bridged tip and ring electrical network connection. Typical usage of the RJA2X is for use with single-line non-key telephone set or an ancillary device where a working miniature 6-position jack exists.

**RJE** Remote Job Entry.

**RJ1DC** A miniature 6-position jack which provides a single-line bridged 4-wire tip/ring and T1/R1 electrical network connection. Typical usage of the RJ1DC is for use with terminal equipment and systems requiring 4-wire exchange access.

**Wiring diagram**

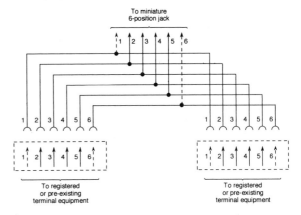

*RJA2X*

**Wiring diagram**

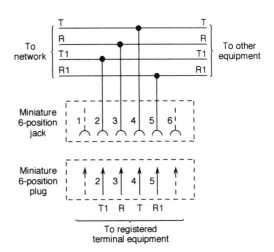

*RJ1DC*

**RLSD** Received Line Signal Detector.

**RL-31-E** A U.S. military lightweight, portable, folding A-frame of steel tubing used for paying-out and recovering field wire and field cable.

**RL-159/U** A U.S. military metal spool-type container used to store, transport, lay or recover field wire. The RL-159/U can hold 1 mile (1.6 km) of field wire.

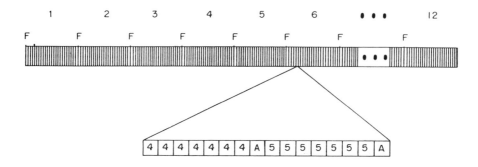

Bit robbing: A 12-frame Superframe. Two octets from the 6th frame have been enlarged to illustrate robbed-bit signaling. The least significant bit of each octet in frames 6 and 12 is "robbed" to provide signaling. Robbed bits in frame 6 are "A" bits; those in frame 12 are "B" bits

*robbed bits*

**RMAS**   Remote Memory Administration System.

**RMS**   Root Mean Square.

**RMT generation**   Generation of remote work stations for remote job entry.

**Rn**   Reference noise.

**RNET**   A trademark of Active Voice Corporation of Seattle, WA, as well as a communications system from that vendor which interconnects multiple voice-mail systems.

**RNP**   Remote Network Processor.

**RNR**   Receive Not Ready.

**RO**   1. Read Only. 2. Receive Only. 3. Ring Out.

**R/O message**   Receiver Only—a message directed to a single individual.

**Roamer Access Number (RAN)**   In cellular communications a ten-digit (area code and telephone number) number that allows users to receive calls when they are located outside their home service area. The call first dials the RAN where the called party is located. Upon receipt of a tone the caller then dials the area code and number assigned to the cellular telephone.

**Roaming Agreement**   In cellular communications an agreement between two cellular carriers to accept the calls of each other's users.

**robbed bits**   The units of AB or ABCD signaling, in which the least significant bit of each channel octet in the 6th and 12th frames of a 12-frame superframe, or the 6th, 12th, 18th, and 24th frames of a 24-frame superframe, are removed from the useful bandwidth for on-hook/off-hook signaling.

**Rockwell Semiconductor Products**   A subsidiary of Rockwell International located in Newport Beach, CA, which manufactures chip sets used in a majority of modems and facsimile machines.

**ROM**   Read-Only Memory.

**ROM bootstrap loader**   A firmware routine in a computer or intelligent terminal that reads the first record from a designated storage device into memory.

**root directory**   The base of the directory structure of a disk.

**rotary**   An arrangement of a group of lines, such as telephone or data PABX lines, that are identified by a single symbolic name or number; upon request, connection is made to the first available (free) line.

**rotary dial**   Calling device that generates pulses for establishing connection in a telephone system.

**rotary feature**   Ports on packet networks associated with Dedicated Access Facilities (DAFs) may be equipped with the rotary feature. The customer must specify which ports are to operate under rotary control. All ports on a rotary list designated by the customer are assigned a single network address, and a virtual connection initiated by a distant Data Terminal Equipment (DTE) to the group is established to the first available port in the list.

**rotary hunt** An arrangement which allows calls placed to seek out an idle circuit in a prearranged multi-circuit group and find the next open line to establish a through circuit.

**ROTL** Remote Office Test Line.

**ROTR** Receive Only Typing Reperforation.

**round robin retraining** A method of training in which the receiving modem asks for a training pattern by sending a training pattern.

**round trip delay** The amount of time it takes for an electrical signal to travel from one end of a transmission medium to the other and back.

**route** 1. (*noun*) The path that a message takes from its source to its destination. 2. (*verb*) The process of directing a message to the appropriate line and terminal, based on information contained in the message header.

**route advance** A feature of U.S. Sprint 800 Service which automatically assigns an alternate communications path when the primary path is busy. The calls are routed to DDD lines.

**Route Extension (REX)** In IBM's SNA, the path control network components, including a peripheral link, that make up the portion of a path between a subarea node and a network-addressable unit (NAU) in an adjacent peripheral node.

**Route Table Generator (RTG)** An IBM-supplied field developed program that assists the user in generating path tables for SNA networks.

**routed message** On a bulletin board system network, a message meant to be delivered to a specific board.

**router** 1. A sophisticated device which divides networks into logical software-oriented subnetworks,

enabling data traffic to be more efficiently routed around a network. 2. In local area networking, an inter-networking device that dynamically routes frames based upon the quality of the service required and the amount of traffic in the network.

**routing** The assignment of the communications path by which a message or telephone call will reach its destination.

**routing, alternate** Assignment of a secondary communications path to a destination when the primary path is unavailable.

**routing code** Third group of digits dialed to place an international call, or the area code for a domestic long distance call. Also area code.

**routing domain** In Digital Equipment Network Architecture (DECnet), a collection of End Systems, Intermediate Systems, and Subnetworks which operate according to the same routing procedures and which is wholly contained within a single administrative domain.

**routing indicator** 1. The bits or characters in a message header that identify the destination of the message to the routing mechanisms. 2. A group of letters assigned to identify a station within a tape relay network to facilitate the routing of traffic.

**Routing Information Protocol (RIP)** A routing protocol supported by Xerox Network Services (XNS), Novell's Internet Packet Exchange Protocol (IPX) and TCP/IP.

**routing line** In U.S. military communications, that format line that contains the routing indicator(s) of the station(s) to which a transmission is routed.

**routing table** A table which contains predefined information concerning the connections or port routing for a network or device.

**Routing Table Maintenance Protocol (RTMP)** A routing protocol incorporated into AppleTalk.

**RP** Restoration Priority.

**RPC** Remote Procedure Call.

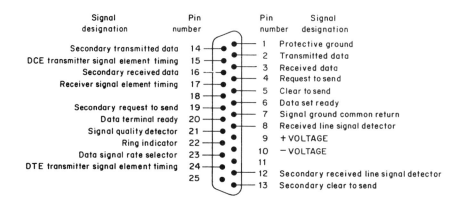

| Signal designation | Pin number | | Pin number | Signal designation |
|---|---|---|---|---|
| Secondary transmitted data | 14 | | 1 | Protective ground |
| DCE transmitter signal element timing | 15 | | 2 | Transmitted data |
| Secondary received data | 16 | | 3 | Received data |
| Receiver signal element timing | 17 | | 4 | Request to send |
| | 18 | | 5 | Clear to send |
| Secondary request to send | 19 | | 6 | Data set ready |
| Data terminal ready | 20 | | 7 | Signal ground common return |
| Signal quality detector | 21 | | 8 | Received line signal detector |
| Ring indicator | 22 | | 9 | + VOLTAGE |
| Data signal rate selector | 23 | | 10 | − VOLTAGE |
| DTE transmitter signal element timing | 24 | | 11 | |
| | 25 | | 12 | Secondary received line signal detector |
| | | | 13 | Secondary clear to send |

*RS-232 interface*

**RPG**   Report Program Generator.

**RPL**   Request Parameter List (IBM's SNA).

**RPL exit routine**   In IBM's VTAM, an application program exit routine whose address has been placed in the EXIT field of a request parameter list (RPL). VTAM invokes the routine to indicate that an asynchronous request has been completed.

**RPL-based macro instruction**   In IBM's VTAM, a macro instruction whose parameters are specified by the user in a request parameter list.

**RPM**   Revolutions Per Minute.

**RPO**   Ringdown Pulse Originating.

**RPOA**   Recognized Private Operating Agency.

**RPQ**   Request to Price Quotation.

**RPT**   Ringdown Pulse Terminating.

**RQMT**   ReQuireMenT.

**RR**   Receive Ready.

**RS**   1. Recommended Standard. 2. Record Separator.

**RSA**   Rivest, Shamir and Adelman.

**RSCS**   Remote Spooling Communications Subsystem.

**RSM**   Remote Switching Module.

**RS-232, RS-232C**   An EIA recommended standard (RS); most common standard for connecting data

| Pin | Interchange circuit | CCITT equivalent | Description | Gnd | Data From DCE | Data To DCE | Control From DCE | Control To DCE | Timing From DCE | Timing To DCE |
|---|---|---|---|---|---|---|---|---|---|---|
| 1 | AA | 101 | Protective ground | X | | | | | | |
| 7 | AB | 102 | Signal ground/common return | X | | | | | | |
| 2 | BA | 103 | Transmitted data | | | X | | | | |
| 3 | BB | 104 | Received data | | X | | | | | |
| 4 | CA | 105 | Request to send | | | | | X | | |
| 5 | CB | 106 | Clear to send | | | | X | | | |
| 6 | CC | 107 | Data set ready | | | | X | | | |
| 20 | CD | 108.2 | Data terminal ready | | | | | X | | |
| 22 | CE | 125 | Ring indicator | | | | X | | | |
| 8 | CF | 109 | Received line signal detector | | | | X | | | |
| 21 | CG | 110 | Signal quality detector | | | | X | | | |
| 23 | CH | 111 | Data signal rate selector (DTE) | | | | | X | | |
| 23 | CI | 112 | Data signal rate selector (DCE) | | | | X | | | |
| 24 | DA | 113 | Transmitter signal element timing (DTE) | | | | | | | X |
| 15 | DB | 114 | Transmitter signal element timing (DCE) | | | | | | X | |
| 17 | DD | 115 | Receiver signal element timing (DCE) | | | | | | X | |
| 14 | SBA | 118 | Secondary transmitted data | | | X | | | | |
| 16 | SBB | 119 | Secondary received data | | X | | | | | |
| 19 | SCA | 120 | Secondary request to send | | | | | X | | |
| 13 | SCB | 121 | Secondary clear to send | | | | X | | | |
| 12 | SCF | 122 | Secondary received signal detector | | | | X | | | |

RS-232 circuit summary with CCITT equivalents

*RS-232, RS-232C*

processing devices. RS-232 defines the electrical characteristics of the signals in the cables that connect DTE with DCE; it specifies a 25-pin connector (the DB-25 connector is almost universally used in RS-232 applications); and it is functionally identical to the CCITT V.24/V.28. Of the 25 pins in the interface, 20 are specified for routine system operating. Of the remaining 5 pins, two (pins 9 and 10) are reserved for modem testing and three (pins 11, 18, and 25) are unassigned. For pin assignments see the diagram and table.

**RS-366-A** The RS-366-A interface is employed to connect terminal devices to automatic calling units. This interface standard uses the same type 25-pin connector as RS-232; however, the pin assignments are different. The RS-366-A interface is illustrated. Note that each actual digit to be dialed is transmitted as parallel binary information over circuits 14 through 17.

**RS-422** An EIA recommended standard for cable lengths that extended the RS-232 50-foot limit. Although introduced as a companion standard with RS-449, RS-422 is most frequently implemented on unused pins of DB-25 (RS-232) connectors. Electrically compatible with CCITT recommendation V.11.

**RS-423** An EIA recommended standard for cable lengths that extended the RS-232 50-foot limit. Although introduced as a companion standard with RS-422, RS-423 is not widely used. Electrically compatible with CCITT recommendation V.10.

**RS-449 interface** This EIA interface specifies the functional and mechanical characteristics of the interconnection between DTE and DCE and compliance to EIA electrical interface standards RS-422 and RS-423. RS-449 specifies a 37-position connector for all interchange circuits with the exception of secondary channel circuits which are accommodated in a separate 9-position connector. This 9-pin connector is only necessary when the secondary channel capability is implemented in the interface. On the 37-pin connector, pins 3 and 21 are undefined. Because RS-449 can support an electrically balanced interface (RS-422), provisions are made on the connector for a return line for the various leads that are balanced. In the illustrations, the RS-449 table shows all return leads under the "B" column and the 37-pin connector shows its respective pin-outs.

**RSA** An encryption algorithm developed by Rivest, Shamir, and Adleman that uses a public key.

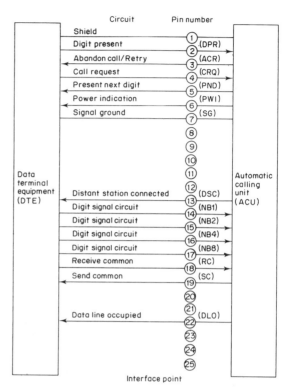

*RS-366A interface*

*RS-449 interface*

| 9 pin | 37 pin A | 37 pin B | RS-449 Circuit | Function | To DTE | To DCE |
|---|---|---|---|---|---|---|
| 1 | 1 | | Shield | Frame ground | | |
| 5 | 19 | | SG | Signal ground | | → |
| 9 | 37 | | SC | Send common | | → |
| 6 | 20 | | RC | Receive common | ← | |
| | 4 | 22 | SD | Send data | | → |
| | 6 | 24 | RD | Receive data | ← | |
| | 7 | 25 | RS | Request to send | | → |
| | 9 | 27 | CS | Clear to send | ← | |
| | 11 | 29 | DM | Data mode | ← | |
| | 12 | 30 | TR | Terminal ready | | → |
| | 15 | | IC | Incoming call | ← | |
| | 13 | 31 | RR | Receiver ready | ← | |
| | 33 | | SQ | Signal quality | ← | |
| | 16 | | SR | Signaling rate | | → |
| | 2 | | SI | Signaling rate indicator | ← | |
| | 17 | 35 | TT | Terminal timing | | → |
| | 5 | 23 | ST | Send timing | ← | |
| | 8 | 26 | RT | Receive timing | ← | |
| 3 | | | SSD | Secondary send data | | |
| 4 | | | SRD | Secondary receive data | | |
| 7 | | | SRS | Secondary request to send | | |
| 8 | | | SCS | Secondary clear to send | | |
| 2 | | | SRR | Secondary receiver ready | | |
| | 10 | | LL | Local loopback | | → |
| | 14 | | RL | Remote loopback | | → |
| | 18 | | TM | Test mode | ← | |
| | 32 | | SS | Select standby | | → |
| | 36 | | SB | Standby indicator | ← | |
| | 16 | | SF | Select frequency | | → |
| | 28 | | IS | Terminal in service | | → |
| | 34 | | NS | New signal | | → |

RS-449 table

**RT** Remote Terminal.

**RT** Reperforator/Transmitter.

**RTA** Routing and Terminal Allocation.

**RTE** The Irish state broadcast company.

**RTG** Route Table Generator (IBM's SNA).

**RTM** Response Time Monitor (IBM's SNA).

**RTMP** Routing Table Maintenance Protocol.

**RTR** Real-Time Reliable.

**RTS** Request To Send.

**RTTU** Remote Trunk Testing Unit.

**RTZ** Return To Zero.

**RT-246/VRC** A U.S. military receiver–transmitter used in radio sets AN/VRC-12 and AN/VRC-43 through -45. The RT-246/VRC supports 920 channels spaced every 50 kHz and uses the 30 to 52.95 and 53 to 75.95 MHz frequency bands.

**RT-524/VRC** A U.S. military receiver–transmitter used in radio sets AN/VRC-46 through AN/VRC-49. The RT-524/VRC supports 920 channels spaced every 50 kHz and uses the 30 to 52.95 and 53 to 75.95 MHz frequency bands.

**RU** Request Unit or Response Unit.

**RU chain** In IBM's SNA, a set of related request/response units (RUs) that are consecutively transmitted on a particular normal or expedited data flow. The request RU chain is the unit of recovery—if one of the RUs in the chain cannot be processed, the entire chain is discarded. *Note.* Each RU belongs to only one chain, which has a beginning and an end indicated via control bits in request/response headers within the RU chain. Each RU can be designated as first-in-chain (FIC), last-in-chain (LIC), middle-in-chain (MIC), or only-in-chain (OIC). Response units and expedited-flow request units are always sent as only-in-chain.

**Rumba** A Microsoft Corporation Windows-based 3270 emulation program marketed by Wall Data of Redmond, WA.

**run book** A tool consisting of three sets of worksheets designed to provide assistance in (a) physically preparing an IBM PC for use as a 3270 Emulation workstation by determining your memory capacity or additional memory requirements, (b) customizing 3270 Emulation to specify the type of workstation and the available 3270 Emulation options, and (c) completing certain host or host-related functions.

**run-length encoding** A data compression technique in which lengths of identical symbols are encoded in terms of the actual symbol value and the number of units of symbols. The "run" of identical characters is replaced by a special character indicating the presence of this compression technique, a character which represents the "run" character, and a count character which indicates the length of the run.

**running closed** A condition in which a machine receives a steady mark from a distant station.

**running open** A condition in which a machine receives a steady space from a distant station.

**runt frame** A frame on a local area network whose length is less than the minimum allowable length of a frame. Runt frames are usually the result of collisions on a network.

**RVI** ReVerse Interrupt.

**RVPP** Revenue Volume Pricing Plan.

**R/W** Radio/Wire.

**RWI** Radio Wire Integration.

**RX** Receive.

**RXC** Receive Clock (control signal).

**RXD** Receive(d) Data (control signal).

**RyBBS** A series of bulletin board systems marketed by the Ryco Company of Shorewood, WI.

**RZ** Return to Zero.

**R&D** Research and Development.

**R1** Line and multi-frequency inter-register signaling system. CCITT recommendations Q.310–Q.331.

**R2** Multi-frequency inter-register signaling system. CCITT recommendations Q.400–Q.490.

**R107** A Russian tactical radio system used at the company level. This FM voice transceiver operates from 20 to 51.5 MHz (HF/VHF) and has a range of six to eight kilometers.

**R-123M** A Russian tactical vehicle FM voice radio which operates from 20 to 51.5 MHz. The R-123M is designed in two bands with a channel spacing of 25 kHz and has a range of 16 to 55 kilometers.

**R-126** A Russian tactical radio system used for communications at low echelons, such as between platoon and company. The R-126 is a manpack system that operates from 48.5 to 51.0 MHz and has a range of two to four kilometers.

**R-148** A Russian tactical radio system designed to replace the R-126. The R-148 operates at 37.0 to 51.95 MHz and has a range of four to five kilometers.

**R2 LAN** A local area network compatible version of Crosstalk Communications software. R2 LAN operates on any IBM Netbios-compatible link to include Digital Communications Associates' 10-Net, Novell's Netware and IBM's PC LAN program.

**R-405** A Russian radio-relay station used for command nets at division level. The R-405 has a UHF frequency range of 320 to 420 MHz and a VHF capability. Its range is 40 to 50 kilometers.

**R-442/VRC** A U.S. military auxiliary, frequency modulation receiver of radio sets AN/VRC-12, AN/VRC-44, AN/VRC-47 and AN/VRC-48. This auxiliary receiver supports 920 channels spaced every 50 kHz in the 30 to 52.95 and 53 to 75.95 MHz frequency bands.

# S

**SAA**   System Application Architecture.

**SABM**   Set Asynchronous Balanced Mode.

**SABRE**   The airline reservation system operated by American Airlines.

**SAC system**   Status Alarm and Control system.

**SAFENET**   Survivable Adaptable Fiber Embedded NETwork.

**SAM**   Status Activity Monitor.

**same-domain LU–LU session**   In IBM's SNA, an LU–LU session between logical units (LUs) in the same domain.

**sampling**   The process of obtaining a group of measurements representative of a universe in order to make inferences about that universe.

**Samsung Data Systems**   A joint venture between IBM Korea and Samsung Electronic Devices.

**SAMT**   State-of-the-Art Medium Terminals.

**San Francisco/Moscow Teleport (SF/MT)**   A San Francisco company which purchases transponder time on a satellite linking Moscow and New York City to provide electronic mail service between the USA and Russia.

**SAP**   Service Access Point.

**SAPI**   Service Access Point Identifier. LAP D terminology.

**SARM**   Set Asynchronous Response Mode.

**SARSAT**   Search And Rescue Satellite Aided Tracking System.

**SARSAT-1**   A United States satellite which is part of a satellite program called COSPAS-SARSAT Spaced-Based Global Search and Rescue System. The SARSAT-1 was launched in March, 1983.

**SATCOMA**   SATellite COMmunications Agency.

**satellite**   Transmission from earth station to antennas in earth orbit and broadcast back to earth stations; signal is delayed approximately 275 milliseconds between earth and satellite.

**C Band Satellite Locations**   The following table is reprinted with the permission of *Satellite Communications* magazine, 6300 S. Syracuse Way, Suite 650, Englewood, CO 80111. This table lists the large number of satellites that transmit in the frequency range of 3.4 GHz and 4.8 GHz. The table also shows the international (COSPAR) designation (for example, 88-123C). The first two digits show the year of launch; the next three, the sequential launch number; and the last letter, the satellite or launch part. Thus 88-123C is the third (C) spacecraft on the 123rd launch that achieved orbit in 1988. A zero in the year column indicates it is a future launch.

ABBREVIATION KEY:

| | | | |
|---|---|---|---|
| FSS | Fixed Satellite Service | DGT | Director General for |
| Int or Int'l | International | | Telecommunications |
| DOM | Domestic | Maritime | Maritime Satellite Service |
| GOVT | Government Service | Aeronautical | Aeronautical Mobile Satellite Service |
| BSS | Broadcasting Satellite Service | M or Mobile | Mobile Satellite Service |
| DIPLOMAT | Diplomatic Service | IO | Indian Ocean Region |
| PT&T | Postal, Telephone & Telegraph | MP | Major Path |
| | Administration | PRI | Primary Path |
| E | East Longitude | PO | Pacific Ocean Region |
| W | West Longitude | AO | Atlantic Ocean Region |

| Orbit | Satellite name | Operator | Year | Use | Designation |
|---|---|---|---|---|---|
| E001-7 | Intelsat Retired | Intelsat | 0 | FSS | |
| E003 | Telecom 1C | France | 88 | DOM,GOVT | 88-018B |
| E007 | F-Sat-1 | France | 0 | FSS | |
| E008 | Statsionar-18 | USSR PT&T | 0 | DOM | |
| E014 | Nigerian Nat Sat-1 | Nigeria | 0 | DOM | |
| E015 | Statsionar-23 | USSR PT&T | 0 | DOM | |
| E015 | AMS-1 | Israel | 90 | FSS | |
| E015 | AMS-2 | Israel | 90 | FSS | |
| E019.2 | Arabsat-1A | Arab Sat Comm | 85 | FSS&BSS | 85-015A |
| E020 | Nigerian Nat Sat-2 | Nigeria | 0 | DOM | |
| E023 | Statsionar-19 | USSR PT&T | 0 | DOM | |
| E026.0 | Arabsat-1B | Arab Sat Comm | 85 | FSS&BSS | 85-048C |
| E034.4 | Raduga-11 | USSR PT&T | 82 | DOM | 82-113A |
| E035 | Statsionar-02 | USSR PT&T | 0 | DOM | |
| E035 | Statsionar-D03 | USSR PT&T | 0 | DIPLOMAT | |
| E035.4 | Raduga-17 | USSR PT&T | 85 | DOM | 85-107A |
| E040 | Statsionar-12 | USSR PT&T | 0 | DOM | |
| E044.7 | Raduga-19 | USSR PT&T | 86 | DOM | 86-082A |
| E045 | Statsionar-09 | USSR PT&T | 0 | DOM | |
| E045 | Statsionar-D04 | USSR PT&T | 0 | DIPLOMAT | |
| E052.7 | Gorizont-11 | USSR | 85 | DOM,GOVT | 85-007A |
| E053 | More-53 | USSR | 0 | Mobile | |
| E053 | Statsionar-05 | USSR PT&T | 0 | DOM | |
| E057 | Intelsat IO-Spare-1 | Intelsat | 0 | Int&Maritime | |
| E057 | Intelsat VI | Intelsat | 0 | Int'l FSS | |
| E058 | Statsionar-24 | USSR PT&T | 0 | DOM | |
| E059.9 | Intelsat VA-F12 | Intelsat | 85 | INT | 85-087A |
| E060 | Intelsat IO-MP/PRI | Intelsat | 0 | Int&Maritime | |
| E060 | Intelsat VI | Intelsat | 0 | Int'l FSS | |
| E062.8 | Intelsat V-F5 | Intelsat | 82 | Int&Maritime | 82-097A |
| E063 | Intelsat IO-PRI/MP | Intelsat | 0 | Int&Maritime | |
| E063 | Intelsat VA Ind 3 | Intelsat | 0 | Int'l FSS | |
| E063* | Avsat 6 | Aeron.Radio Inc. | 0 | Aeronautical | |
| E063.6 | Raduga-14 | USSR PT&T | 84 | COM,GOVT | 84-016A |
| E064.5 | Inmarsat-2 F3 | Inmarsat | 89 | Mobile | |
| E064.5 | Inmarsat-2 F4 | Inmarsat | 91 | Mobile | |
| E064.5 | Marecs C | Inmarsat | 0 | Maritime | |
| E066 | Intelsat IO-Spare-2 | Intelsat | 0 | Int&Maritime | |

| Orbit | Satellite name | Operator | Year | Use | Designation |
|-------|----------------|----------|------|-----|-------------|
| E066 | Intelsat VA | Intelsat | 0 | Int'l FSS | |
| E066.0 | Intelsat V-F7 | Intelsat | 83 | Int&Maritime | 83-105A |
| E069 | Statsionar-20 | USSR PT&T | 0 | DOM | |
| E072.5 | Marisat-Indian-102 | Inmarsat | 76 | Maritime | 76-053A |
| E074 | Insat-IIC | ISRO (India) | 90 | DOM,BSS,M | |
| E079.7 | Gorizont-10 | USSR | 84 | DOM,GOVT | 84-078A |
| E080 | Statsionar-13 | USSR PT&T | 0 | DOM | |
| E080 | Statsionar-01 | USSR PT&T | 0 | DOM | |
| E080 | Potok-2 | USSR PT&T | 0 | DOM | |
| E080.5 | Insat-1B | ISRO (India) | 83 | DOM,BSS,M | 83-089B |
| E081.5 | Foton-2 | USSR PT&T | 0 | FSS | |
| E083 | Insat-IIA | ISRO (India) | 0 | DOM,BSS,M | |
| E083 | Palapa C | Indonesia | 90 | Regional | |
| E084.7 | Raduga-12 | USSR PT&T | 83 | DOM | 83-028A |
| E085 | Statsionar-03 | USSR PT&T | 0 | DOM | |
| E085 | Statsionar-D05 | USSR PT&T | 0 | DIPLOMAT | |
| E085.3 | Raduga-20 | USSR PT&T | 87 | DOM | 87-028A |
| E087.5 | PRC-22 | PR China | 88 | DOM | 88-014A |
| E090 | More-90 | USSR | 0 | Mobile | |
| E090 | Statsionar-06 | USSR PT&T | 0 | DOM,BSS | |
| E090.0 | Gorizont-13 | USSR | 86 | DOM,GOVT | 86-090A |
| E093.5 | Insat-IC | ISRO (India) | 88 | DOM,BSS,M | |
| E093.5 | Insat-ID | ISRO (India) | 89 | DOM,BSS,M | |
| E093.5 | Insat-IIB | ISRO (India) | 91 | DOM,BSS,M | |
| E095 | Statsionar-14 | USSR PT&T | 0 | DOM | |
| E096.3 | Gorizont-05 | USSR PT&T | 82 | DOM,GOVT | 82-020A |
| E098 | Chinasat-3 | PR China | 0 | FSS | |
| E101.8 | PRC-18 | PR China | 86 | DOM | 86-010A |
| E103 | Statsionar-21 | USSR PT&T | 0 | DOM | |
| E106 | Palapa B-1 | Indonesia | 83 | Regional | 83-059C |
| E110.5 | Chinasat-2 | PR China | 0 | FSS | |
| E113.0 | Palapa B-2P | Indonesia | 87 | Regional | 87-029A |
| E118 | Palapa B-3 | Indonesia | 0 | Regional | |
| E123.3 | PRC-15 | PR China | 84 | DOM | 84-035A |
| E128 | Statsionar-15 | USSR PT&T | 0 | DOM | |
| E128 | Statsionar-D06 | USSR PT&T | 0 | DIPLOMAT | |
| E128.3 | Raduga-15 | USSR PT&T | 84 | DOM | 84-063A |
| E128.4 | Raduga-21 | USSR PT&T | 87 | DOM | 87-100A |
| E132* | Avsat 5 | Aeron.Radio Inc. | 0 | Aeronautical | |
| E132.0 | CSJ-2A (Sakura) | NASDA &NTT | 83 | DOM | 83-006A |
| E135.9 | CS-2B (Sakura 2B) | NASDA &NTT | 83 | DOM | 83-081A |
| E139.6 | Gorizont-14 | USSR | 87 | DOM,GOVT | 87-040A |
| E139.6 | Gorizont-06 | USSR PT&T | 82 | DOM,GOVT | 82-103A |
| E140 | More-14j0 | USSR PT&T | 0 | Mobile | |
| E140 | Statsionar-07 | USSR PT&T | 0 | DOM | |
| E145 | Statsionar-16 | USSR PT&T | 0 | DOM | |
| E157.8 | Raduga-13 | USSR PT&T | 83 | DOM | 83-088A |
| E167.05 | Pacstar-1 | Papua N Guinea | 90 | FSS Int'l | |
| E170 | Pac.Area Sat Sys-A | Rite-Japan | 0 | Int'l | |
| E170 | Pac.Area Sat Sys-A | Rite-Japan | 0 | Int'l | |
| E173* | Avsat 4 | Aeron.Radio Inc. | 0 | Aeronautical | |

| Orbit | Satellite name | Operator | Year | Use | Designation |
|-------|----------------|----------|------|-----|-------------|
| E174 | Intelsat V-F3 | Intelsat | 81 | Int'l FSS | 81-119A |
| E174 | Intelsat VA PAC 1 | Intelsat | 0 | Int'l FSS | |
| E174 | Intelsat PO-PRI | Intelsat | 0 | Int'l FSS | |
| E176.1 | Marisat-Pacific-103 | Inmarsat | 76 | Maritime | 76-101A |
| E177 | Intelsat IVA PAC 2 | Intelsat | 0 | Int'l FSS | |
| E177 | Intelsat V PAC 2 | Intelsat | 0 | Int'l FSS | |
| E177 | Intelsat V-F1 | Intelsat | 81 | Int'l FSS | 81-050A |
| E177 | Intelsat VA PAC 2 | Intelsat | 0 | Int'l FSS | |
| E177.0 | Intelsat IVA-F3 | Intelsat | 78 | Int'l FSS | 78-002A |
| E178.0 | Marecs A | Inmarsat | 81 | Maritime | 81-122A |
| E179 | Intelsat PO-Spare | Intelsat | 0 | Int'l FSS | |
| E179.9 | Intelsat V-F8 | Intelsat | 84 | Int&Maritime | 84-023A |
| E180 | Intelsat MCS PAC A | Inmarsat | 0 | Maritime | |
| E180 | Intelsat V PAC 3 | Intelsat | 0 | Int'l FSS | |
| E180 | Intelsat VA PAC 3 | Intelsat | 0 | Int'l FSS | |
| | | | | | |
| W000.2 | Telecom 1B | France Telcom | 85 | DOM,GOVT | 85-035B |
| W001 | Intelsat AO-Spare 4 | Intelsat | 0 | Int'l FSS | |
| W001 | Intelsat VA Cont. 4 | Intelsat | 0 | Int'l FSS | |
| W001.0 | Intelsat V-F2 | Intelsat | 80 | Int'l FSS | 80-098A |
| W003 | Statsionar-22 | USSR PT&T | 0 | DOM | |
| W003 | Telecom 2C | France Telecom | 0 | DOM,GOVT | |
| W004 | Intelsat AO-Spare 5 | Intelsat | 0 | Int'l | |
| W004 | Intelsat V Cont. 3 | Intelsat | 0 | Int'l FSS | |
| W004 | Intelsat VA Cont. 3 | Intelsat | 0 | Int'l FSS | |
| W007.8 | Telecom 1A | France Telecom | 84 | DOM,GOVT | 84-081B |
| W008 | Telecom 2A | France Telecom | 91 | DOM,GOVT | |
| W010.6 | Gorizont-07 | USSR | 83 | DOM,GOVT | 83-066A |
| W010.8 | Gorizont-12 | USSR | 86 | DOM,GOVT | 86-044A |
| W011 | Statsionar-11 | USSR PT&T | 0 | DOM | |
| W013 | Gorizont-15 | USSR | 88 | FSS | 88-028 |
| W013.5 | Potok-1 | USSR PT&T | 0 | DOM | |
| W014 | More-14 | USSR | 0 | Mobile | |
| W014 | Statsionar-04 | USSR PT&T | 0 | DOM | |
| W014.6 | Marisat-Atlantic-101 | Inmarsat | 76 | Maritime | 76-017A |
| W015 | Inmarsat 2 F1 | Inmarsat | 89 | Mobile | |
| W015 | Foton-1 | USSR PT&T | 0 | FSS | |
| W016.5 | Intelsat IVA 343.5E | Intelsat | 0 | Int'l FSS | |
| W016.5 | Intelsat V 343.5E | Intelsat | 0 | Int'l FSS | |
| W016.5 | Intelsat VA 343.5E | Intelsat | 0 | Int'l FSS | |
| W016.5 | Intelsat IBS 343.5E | Intelsat | 0 | Int'l FSS | |
| W018.5 | Intelsat V-F6 | Intelsat | 83 | Int&Maritime | 83-047A |
| W018.5 | Intelsat AO-Maj-2 | Intelsat | 0 | Int&Maritime | |
| W018.5 | Intelsat IBS 341.5E | Intelsat | 0 | Int'l FSS | |
| W021.5 | Intel.AO-Dom&MCS | Intelsat | 0 | Int&Martime | |
| W021.5 | Intelsat IVA-F4 | Intelsat | 77 | Int'l FSS | 77-041A |
| W021.5 | Intelsat VA 338.5E | Intelsat | 0 | Int'l FSS | |
| W022* | Avsat 1 | Aeron.Radio Inc. | 0 | Aeronautical | |
| W023.7 | Raduga-18 | USSR PT&T | 86 | DOM | 86-007A |
| W024.5 | Intelsat VA-F10 | Intelsat | 85 | Int'l FSS | 85-025A |
| W024.5 | Intelsat VI 335.5E | Intelsat | 0 | Int'l FSS | |

| Orbit | Satellite name | Operator | Year | Use | Designation |
|-------|----------------|----------|------|-----|-------------|
| W024.5 | Intelsat AO-PRI | Intelsat | 0 | Int'l FSS | |
| W025 | Statsionar-08 | USSR PT&T | 0 | DOM | |
| W026 | Inmarsat-2 | Inmarsat | 89 | Mobile | |
| W026.0 | Marecs B2 | Inmarsat | 84 | Maritime | 84-114B |
| W026.5 | Statsionar-17 | USSR PT&T | 0 | DOM | |
| W026.5 | Statsionar-D01 | USSR PT&T | 0 | DIPLOMAT | |
| W027.5 | Intelsat AO-Spare 2 | Intelsat | 0 | Int&Maritime | |
| W027.5 | Intelsat VA ATL 2 | Intelsat | 0 | Int'l FSS | |
| W027.5 | Intelsat VI 332.5E | Intelsat | 0 | Int'l FSS | |
| W027.6 | Intelsat VA-F11 | Intelsat | 85 | INT | 85-055A |
| W031 | Intelsat AO-Spare 1 | Intelsat | 0 | Int'l FSS | |
| W031 | Intelsat V ATL 6 | Intelsat | 0 | Int'l FSS | |
| W031 | Intelsat VA ATL 6 | Intelsat | 0 | Int'l FSS | |
| W034.5 | Intelsat AO-MAJ-1 | Intelsat | 0 | Int'l FSS | |
| W034.6 | Intelsat V-F4 | Intelsat | 82 | Int'l FSS | 82-017A |
| W040.5 | Intelsat VA 319.5E | Intelsat | 0 | Int'l FSS | |
| W040.5 | Intelsat IBS 319.5E | Intelsat | 0 | Int'l FSS | |
| W045 | PAS-1 | PanAmSat Corp. | 88 | Int'l | 88-051C |
| W047 | FINANSAT-1 | Finansat | 0 | Int'l | |
| W050 | Intelsat IBS 310E | Intelsat | 0 | Int'l FSS | |
| WO53 | Intelsat AO-Spare 3 | Intelsat | 0 | Int'l | |
| W053 | Intelsat VA Cont. 1 | Intelsat | 0 | Int'l FSS | |
| W053 | Intelsat VA(IBS)-F13 | Intelsat | 88 | INT | 88-040A |
| W053 | Intelsat IBS 307E | Intelsat | 0 | Int'l FSS | |
| W056 | Intelsat VA 304E | Intelsat | 0 | Int'l FSS | |
| W056 | Intelsat IBS 304E | Intelsat | 0 | Int'l FSS | |
| W057 | PAS-2 | PanAmSat Corp. | 0 | Int'l | |
| W058* | Avsat 2 | Aeron.Radio Inc. | 0 | Aeronautical | |
| W060 | Intelsat VA 300E | Intelsat | 0 | Int'l FSS | |
| W060 | Intelsat IBS 300E | Intelsat | 0 | Int'l FSS | |
| W065.5 | SBTS-1 | Brazil | 85 | DOM | 85-015B |
| W069 | Spacenet II | GTE-Spacenet | 84 | DOM | 84-114A |
| W069* | Spacenet IIR | GTE-Spacenet | 93 | DOM | |
| W070.1 | SBTS-2 | Brazil | 86 | DOM | 86-026B |
| W072 | Satcom IIR | GE Americom | 83 | DOM | 83-094A |
| W074.0 | Galaxy II | Hughes Com Gal | 83 | DOM | 83-098A |
| W075 | Satcol II | Colombia | 0 | DOM | |
| W075.4 | Satcol IA | Colombia | 0 | DOM | |
| W075.4 | Satcol IB | Colombia | 0 | DOM | |
| W076.0 | Comstar D2 | Comsat General | 76 | DOM | 76-073A |
| W076.0 | Comstar D4 | Comsat General | 81 | DOM | 81-018A |
| W080 | Nahuel I | Argentina | 0 | DOM | |
| W082 | Satcom IV | GE Americom | 82 | DOM | 82-004A |
| W083 | ASC-2 | Contel-ASC | 90 | DOM | |
| W085 | Telstar 302 | AT&T Comm. | 84 | DOM | 84-093D |
| W085 | Nahuel II | Argentina | 0 | DOM | |
| W087 | Spacenet IIIR | GTE Spacenet | 88 | DOM | 88-018A |
| W091 | Westar VI-S | Hughes Com Gal | 89 | DOM | |
| W091.1 | Westar III | Hughes Com Gal | 79 | DOM | 79-072A |
| W093* | Spotnet 2 | Nat'l Exchange | 93 | DOM | |
| W093* | Telstar 401 | AT&T Comm. | 92 | DOM | |

| Orbit | Satellite name | Operator | Year | Use | Designation |
|-------|---------------|----------|------|-----|-------------|
| W093.5 | Galaxy III | Hughes Com Gal | 84 | DOM | 84-101A |
| W096.0 | Telstar 301 | AT&T Comm. | 83 | DOM | 83-077A |
| W097 | STSC-2 | Cuba | 0 | Regional | |
| W099.0 | Westar IV | Hughes Com Gal | 82 | DOM | 82-014A |
| W101* | Galaxy V | Hughes Com Gal | 92 | DOM | |
| W101* | Contelsat-1 | Contel-ASC | 93 | DOM | |
| W101* | Satcom H-1 | GE Americom | 93 | DOM | |
| W101* | Spotnet 1 | Nat'l Exchange | 93 | DOM | |
| W101* | Telstar 402 | AT&T Comm. | 93 | DOM | |
| W104.5 | Anik D-1 | Telesat Canada | 82 | DOM | 82-082A |
| W107.5 | Anik E1 | Telesat Canada | 90 | DOM | |
| W110.5 | Anik D-2 | Telesat Canada | 84 | DOM | 84-113B |
| W110.5 | Anik E2 | Telesat Canada | 90 | DOM | |
| W113.5 | Morelos-1 | Mexico | 85 | DOM | 85-048B |
| W114* | Avsat 3 | Aeron.Radio Inc. | 0 | Aeronautical | |
| W116.6 | Morelos-2 | Mexico | 85 | DOM | 85-109B |
| W120.0 | Spacenet I | GTE-Spacenet | 84 | DOM | 84-049A |
| W120.0* | Spacenet IR | GTE-Spacenet | 93 | DOM | |
| W122 | Galaxy IV | Hughes Com Gal | 93 | DOM | |
| W122.5 | Westar V | Hughes Com Gal | 82 | DOM | 82-058A |
| W125.0 | Telstar 303 | AT&T Comm. | 85 | DOM | 85-048D |
| W128 | ASC-1 | Contel-ASC | 85 | DOM | 85-076C |
| W128* | Contelsat-2 | Contel-ASC | 93 | DOM | |
| W131 | Satcom IIIR | GE Americom | 81 | DOM | 81-114A |
| W134.0 | Galaxy I | Hughes Com Gal | 83 | DOM | 83-065A |
| W138 | Satcom IR | GE Americom | 83 | DOM | 83-030A |
| W143 | Aurora I | Alascom, Inc. | 82 | DOM | 82-105A |
| W143* | Aurora IR | Alascom, Inc. | 92 | DOM | |
| W168 | Potok-3 | USSR PT&T | 0 | DOM | |
| W168.7 | Raduga-16 | USSR PT&T | 85 | DOM | 85-070A |
| W169.5 | Foton-3 | USSR PT&T | 0 | FSS | |
| W170 | Statsionar-10 | USSR PT&T | 0 | DOM | |
| W170 | Statsionar-D02 | USSR PT&T | 0 | DIPLOMAT | |
| W175 | Pacstar-2 | Papua New Guinea | 91 | FSS Int'l | |
| W178 | Finansat-2 | Finansat | 0 | Int'l | |

*Satellite locations that have been requested, but not assigned.

**Ku-Band Geostationary Satellite Locations**

The following global Ku-band satellite table is reprinted with the permission of *Satellite Communications* of Englewood, CO and lists those satellites which have downlinks in the frequency range of 11 to 13 GHz and operate with uplinks at 14 to 14.5 GHz, with the exception of direct broadcast satellites which have uplinks at 17 GHz. The principal missions of these satellites are communications, research, video and VSAT services. The table shows the international (COSPAR) designation (for example, 88-123C). The first two digits show the year of the launch; the next three, the sequential launch number; and the last letter, the satellite or launch part. Thus, 88-123C is the third (C) spacecraft on the 123rd launch that achieved orbit in 1988. The number "0" in the "year" column signifies a satellite with an undetermined launch date. Some satellites may appear to be colocated. In reality, they generally are separated by 0.1 degrees or more. The locations also are nominal, especially for non-North American satellites. All satellites shown are at their mid-1988 position or the position requested in various satellite applications. The Eutelsat locations listed are the same as those listed by NORAD. Eutelsat will be moving its satellites among these locations (replacing older satellites). The United States Broadcast Satellite Service (BSS) assignments are shown at the locations assigned to the United States by the ITU. Multiple applications have been filed with the FCC by potential operators.

ABBREVIATION KEY:

| | | | |
|---|---|---|---|
| FSS | Fixed Satellite Service | DGT | Director General for Telecommunications |
| Int or Int'l | International | Maritime | Maritime Satellite Service |
| DOM | Domestic | Aeronautical | Aeronautical Mobile Satellite Service |
| GOVT | Government Service | M or Mobile | Mobile Satellite Service |
| BSS | Broadcasting Satellite Service | IO | Indian Ocean Region |
| DIPLOMAT | Diplomatic Service | MP | Major Path |
| PT&T | Postal, Telephone &Telegraph | PRI | Primary Path |
| | Administration | PO | Pacific Ocean Region |
| E | East Longitude | AO | Atlantic Ocean Region |
| W | West Longitude | | |

| Orbit | Satellite name | Operator | Year | Use | Designation |
|---|---|---|---|---|---|
| E003 | Telecom 1C | France | 88 | DOM,GOVT | 88-0188 |
| E005 | Tele-X | Nordic Nations | 89 | Regional | |
| E007.1 | Eutelsat I-F2 | Eutelsat Councl | 84 | Regional | 84081A |
| E010.1 | Telecom 1B | France Telecom | 85 | DOM,GOVT | 85-035B |
| E011.5 | Eutelsat I-F4 | Eutelsat Councl | 87 | Regional | 87-078B |
| E013 | Eutelsat II-F2 | Eutelsat Councl | 90 | Regional | |
| E013.3 | Eutelsat I-F1 | Eutelsat Councl | 83 | Regional | 83-058A |
| E015 | Zenon-B | France | 0 | Aeromobil | |
| E015 | AMS-1 | Israel | 90 | FSS | |
| E015 | AMS-2 | Israel | 90 | FSS | |
| E016 | Sicral 1A | Italy PTT | 0 | GOVT | |
| E016 | Eutelsat I-F5 | Eutelsat Councl | 88 | Regional | 89-063B |
| E017 | SABS | Saudi Arabia | 0 | BSS | |
| E019 | Zenon-C | France | 0 | Aeromobil | |
| E019 | Eutelsat II-F3 | Eutelsat Councl | 90 | Regional | |
| E019.2 | Astra 1 | Soc Euro Sat | 88 | Business | |
| E019.2 | Astra-F2 | Soc Euro Sat | 0 | Video | |
| E022 | Sicral 1B | Italy PTT | 0 | GOVT | |
| E023.5 | DFS-1 | Germany | 89 | DOM,BSS | |
| E028.5 | DFS-2 | Germany | 89 | DOM,BSS | |
| E032 | Videosat-1 | France (DGT) | 0 | Special | |
| E036 | Eutelsat II-F1 | Eutelsat Council | 90 | Regional | |
| E038 | Paksat I | Pakistan | 89 | DOM,MET | |
| E040 | Loutch-07 | USSR PT&T | 0 | DOM | |
| E041 | Paksat-2 | Pakistan | 0 | DOM,MET | |
| E045 | Loutch-P02 | USSR PT&T | 0 | DOM | |

| Orbit | Satellite name | Operator | Year | Use | Designation |
|-------|----------------|----------|------|-----|-------------|
| E053 | Loutch-02 | USSR PT&T | 0 | DOM | |
| E055.5 | Loutch-02 Later | USSR PT&T | 0 | DOM | |
| E057 | Intelsat VI 57E | Intelsat | 0 | Int'l FSS | |
| E060 | Intelsat VI 60E | Intelsat | 0 | Int'l FSS | |
| E060 | Intelsat VI-F5 | Intelsat | 90 | Int'l FSS | |
| E060.0 | Intelsat VA-F12 | Intelsat | 85 | Int'l | 85-087A |
| E062.9 | Intelsat V-F5 | Intelsat | 82 | Int &Marit | 82-097A |
| E063 | Intelsat VA Ind 3 | Intelsat | 0 | Int'l FSS | |
| E063 | Intelsat VI-F4 | Intelsat | 90 | Int'l FSS | |
| E065 | CBSS-1 | China B/C Sat | 0 | BSS | |
| E066 | Intelsat VA 66E | Intelsat | 0 | Int'l FSS | |
| E066.0 | Intelsat V-F7 | Intelsat | 83 | Int &Marit | 83-105A |
| E070 | Celestar II | McCaw | 91 | Int'l | |
| E077 | CSDRN-2 | USSR | 0 | Res Data | |
| E077 | CSSRD-2 | USSR | 0 | Tracking | |
| E080 | Cosmos-1888 | USSR PT&T | 87 | | 87-084A |
| E080 | Loutch-08 | USSR PT&T | 0 | DOM | |
| E080 | CBSS-2 | China B/C Sat | 0 | BSS | |
| E085 | Loutch-P03 | USSR PT&T | 0 | DOM | |
| E090 | Loutch-03 | USSR PT&T | 0 | DOM | |
| E092 | CBSS-3 | China B/C Sat | 0 | BSS | |
| E095 | CSDRN-1 | USSR | 0 | Tracking | |
| E096.3 | Gorizont-05 | USSR PT&T | 82 | DOM,GOVT | 82-020A |
| E103 | Loutch-05 | USSR PT&T | 0 | DOM | |
| E109.7 | Yuri-2A (BS-2A) | MOPT Japan | 84 | BSS | 84-005A |
| E109.9 | Yuri-2B (BS-2B) | Sci&Tech Agency | 86 | BSS | 86-016A |
| E110 | BS 3A (Yuri-3) | Telecomm Sat CP | 90 | BSS | |
| E110 | BS 3B (Yuri-3) | Telecomm Sat CP | 91 | BSS | |
| E124 | Superbird 1A | SCC (Japan) | 89 | DOM | |
| E128 | Superbird 1B | SCC (Japan) | 89 | DOM | |
| E130.0 | Kiku-2 | Japan-RRL | 77 | EXP | 77-014A |
| E139.6 | Gorizont-06 | USSR PT&T | 82 | DOM,GOVT | 82-103A |
| E140 | Loutch-04 | USSR PT&T | 0 | DOM | |
| E150 | JCSat-1 | Japan | 89 | DOM | |
| E154 | JCSat-2 | Japan | 89 | DOM | |
| E155.6 | Aussat-2 | Australia | 85 | DOM,BSS | 85-109C |
| E160.0 | Aussat-1 | Australia | 85 | DOM,BSS | 85-076B |
| E164.0 | Aussat-31 | Australia | 87 | DOM,BSS | 87-078A |
| E167 | VSSRD-2 | USSR | 0 | Tracking | |
| E167 | ESDRN-2 | USSR | 0 | Res Data | |
| E167.05 | Pacstar-1 | Papua New Guinea | 90 | Int'l FSS | |
| E170 | Celestar I | McCaw | 91 | Int'l | |
| E170 | Pac.Area Sat Sys-B | Rite-Japan | 0 | Int'l | |
| E174 | Intelsat V-F3 | Intelsat | 81 | Int'l FSS | 81-119A |
| E174 | Intelsat VA Pac 1 | Intelsat | 0 | Int'l FSS | |
| E174 | Intelsat VI-F1 | Intelsat | 92 | Int'l FSS | |
| E174 | Intelsat VII-F1 | Intelsat | 92 | Int'l FSS | |
| E177 | Intelsat V Pac 2 | Intelsat | 0 | Int'l FSS | |
| E177 | Intelsat V-F1 | Intelsat | 81 | Int'l FSS | 81-050A |
| E177 | Intelsat VA Pac 2 | Intelsat | 0 | Int'l FSS | |
| E177 | Intelsat VII-F2 | Intelsat | 93 | Int'l FSS | |

| Orbit | Satellite name | Operator | Year | Use | Designation |
|-------|----------------|----------|------|-----|-------------|
| E180 | Intelsat V-F8 | Intelsat | 84 | Int &Marit | 84-023A |
| E180 | Intelsat V Pac 3 | Intelsat | 0 | Int'l FSS | |
| E180 | Intelsat VA Pac 3 | Intelsat | 0 | Int'l FSS | |
| | | | | | |
| W001 | Intelsat V-F2 | Intelsat | 80 | Int'l FSS | 80-098A |
| W001 | Intelsat VA Cont. 4 | Intelsat | 0 | Int'l FSS | |
| W001 | Intelsat VA(IBS)-F15 | Intelsat | 88 | Int'l | |
| W003 | Telecom 2C | France Telecom | 0 | DOM,GOVT | |
| W004 | Intelsat V Cont. 3 | Intelsat | 0 | Int'l FSS | |
| W004 | Intelsat VA Cont. 3 | Intelsat | 0 | Int'l FSS | |
| W007.8 | Telecom 1A | France Telecom | 84 | DOM,GOVT | 84-081B |
| W008 | Zenon-A | France | 0 | Aeromobil | |
| W008 | Telecom 2A | French Telecom | 91 | DOM,GOVT | |
| W011 | Loutch-06 | USSR PT&T | 0 | DOM | |
| W011 | F-Sat-2 | France | 0 | FSS | |
| W014 | Loutch-01 | USSR PT&T | 0 | DOM | |
| W016 | WSDRN-1 | USSR | 0 | Tracking | |
| W016 | ZSSRD-2 | USSR | 0 | Tracking | |
| W016 | WSDRN-2 | USSR | 0 | Res Data | |
| W016.5 | Intelsat V 343.5E | Intelsat | 0 | Int'l FSS | |
| W016.5 | Intelsat VA 343.5E | Intelsat | 0 | Int'l FSS | |
| W016.5 | Intelsat IBS 343.5E | Intelsat | 0 | Int'l FSS | |
| W018.5 | Intelsat V-F6 | Intelsat | 83 | Int &Marit | 83-047A |
| W018.5 | Intelsat IBS 341.5E | Intelsat | 0 | Int'l FSS | |
| W019 | Helvesat-1 | Swiss Tel-Tec | 90 | BSS | |
| W019 | Helvesat-2 | Swiss Tel-Tec | 91 | BSS | |
| W019 | TDF-1A | Tele-Dif-France | 88 | BSS | |
| W019 | Olympus 1 | ESA | 89 | EXP,BSS | |
| W019 | Olympus 2 | ESA | 0 | EXP,BSS | |
| W019 | TDF-2 | Tele-Dif-France | 89 | BSS | |
| W019 | TV-Sat2 | Bundepost | 89 | BSS | |
| W019 | Sarit | Italy | 89 | BSS | |
| W021.5 | Intelsat VA 338.5E | Intelsat | 0 | Int'l FSS | |
| W024.4 | Intelsat VA-F10 | Intelsat | 85 | Int'l FSS | 85-025A |
| W024.5 | Intelsat VI 335.5E | Intelsat | 0 | Int'l FSS | |
| W024.5 | Intelsat VI-F2 | Intelsat | 89 | Int'l FSS | |
| W025 | Loutch-P01 | USSR PT&T | 0 | DOM | |
| W027.5 | Intelsat VA ATL 2 | Intelsat | 0 | Int'l FSS | |
| W027.5 | Intelsat VA-F11 | Intelsat | 85 | Int'l | 85-055A |
| WO27.5 | Intelsat VI 332.5E | Intelsat | 0 | Int'l FSS | |
| W027.5 | Intelsat VI-F3 | Intelsat | 90 | Int'l &Marit | |
| W031 | Eiresat-1 | Ireland | 88 | FSS &BSS | |
| W031 | Falklands | UK | 0 | BSS | |
| W031 | Intelsat V ATL 6 | Intelsat | 0 | Int'l FSS | |
| W031 | Intelsat VA ATL 6 | Intelsat | 0 | Int'l FSS | |
| W031 | British Sat B/C-1 | BSB | 89 | BSS | |
| W031 | British Sat B/C-2 | BSB | 90 | BSS | |
| W031 | Atlantic Sat | Ireland/Hughes | 90 | FSS &BSS | |
| W034 | Guyana &Jamaica | Guyana &Jamaica | 0 | BSS | |
| W034.6 | Intelsat V-F4 | Intelsat | 82 | Int'l FSS | 82-017A |
| W037.5 | Orion-1 | Orion Sat Corp | 0 | Int'l | |

| Orbit | Satellite name | Operator | Year | Use | Designation |
|-------|----------------|----------|------|-----|-------------|
| W037.5 | Vidrosat-2 | France (DGT) | 0 | Special | |
| W040.5 | Intelsat VA 319.5E | Intelsat | 0 | Int'l FSS | |
| W040.5 | Intelsat IBS 319.5E | Intelsat | 0 | Int'l FSS | |
| W042 | Grenada | Grenada | 0 | BSS | |
| W043.5 | Videosat-3 | France (DGT) | 0 | Special | |
| W045 | Brazil | Brazil 1-4 | 0 | BSS | |
| W045 | PAS-1 | PanAmSat Corp | 88 | Int'l | 85-051C |
| W047 | Orion-2 | Orion Sat Corp | 0 | Int'l | |
| W050 | Intelsat IBS 310E | Intelsat | 0 | Int'l FSS | |
| W053 | Intelsat VA Cont. 1 | Intelsat | 0 | Int'l FSS | |
| W053 | Intelsat VA(IBS)-F13 | Intelsat | 88 | Int'l | 88-040A |
| W053 | Intelsat IBS 307E | Intelsat | 0 | Int'l FSS | |
| W055 | Argentina | Argentina #2 | 0 | BSS | |
| W056 | ISI-1 | Int'l Sat Inc | 0 | Int'l | |
| W056 | Intelsat VA 304E | Intelsat | 0 | Int'l FSS | |

**Satellite Business Systems (SBS)**  A U.S. satellite carrier originally formed by IBM and now owned by MCI Communications.

**satellite closet**  A walk-in or shallow wall closet that supplements a backbone closet by providing additional facilities for connecting backbone subsystem cables to horizontal wiring subsystem cables from information outlets. Also referred to as "satellite location."

**satellite communications**  The utilization of geo-stationary orbiting satellites to relay the transmission received from one earth station to one or more other earth stations.

**Satellite Master Antenna TV systems (SMATV)**  Refers to the use of earth stations to provide satellite programming to communities via their master antenna systems.

**satellite microwave radio**  Microwave or beam radio system using geosynchronously orbiting communications satellites.

**Satellite PBX**  A PBX or Centrex system not equipped with attendant positions but associated with an attended main PBX or Centrex system.

**Satellite Period of Revolution**  The period of revolution ($T$, hours) is a function of the earth radius ($R = 3913$ km), the satellite altitude ($h$, km) and the universal constant GM = 396 613.52 km$^3$ /s$^2$ such that:

$$T = \pi/4.32 * 10^4 * \sqrt{(R + h)^3/GM}$$

**satellite relay**  An active or passive repeater in geo-synchronous orbit around the earth which amplifies the signal it receives before transmitting it back to earth.

**Satstream**  A private digital communications service using satellites and small dish earth terminals offered by British Telecom International.

**saturation testing**  The testing of systems with large volumes of messages intended to expose errors that occur very infrequently and can be triggered by such rare coincidences as simultaneous arrival of two messages. Also called volume testing.

**Saturn** ®  The registered trademark of Siemens for a series of office communications equipment to include PABXs.

**save and repeat**  A PBX or telephone set feature which allows the caller to store the last number dialed into the phone's memory so that a pushbutton can be pressed later when retrying the same number.

**SBS**  Satellite Business Systems.

**Sbus**  The I/O bus used in the Sun Microsystem Sparcstation.

**SB-22/PT** A U.S. military lightweight, manual switchboard that can support up to 12 local-battery telephone circuits, remote-controlled radio circuits, or voice-frequency teletypewriter circuits.

**SB-86/P** A U.S. military tactical, local-battery switchboard capable of terminating 30 telephone circuits.

**SB-993/GT** A U.S. military lightweight, portable switchboard capable of servicing six local-battery telephone lines.

**SB-3082(V)1/GT** A U.S. military telephone switchboard, capable of terminating 50 telephone circuits, which can be mounted in a $\frac{1}{4}$-ton truck or in a shelter.

**SB-3614/TT** A U.S. military telephone switchboard capable of switching both analog and digital signals from voice, teletypewriter, and data terminal subscribers. The SB-3614/TT is capable of supporting 30 telephone circuits and can be combined to operate as a 60- or a 90-terminal switchboard.

**SC** Session Control (IBM's SNA).

**scale line** A looseness=1 line that is displayed at the top or bottom of an IBM CMS file that divides the line into columns. In the following example, "This" begins in column 1 and ends in column 4. The entire sentence begins in column 1 and ends in column 19.

```
|...+....1....+....2....+....3....+....4....+....5....
```

This is an example.

**Scandinavian Telecommunication Services** A consortium owned by the five Nordic telecommunications administrations—Sweden, Finland, Norway, Denmark and Iceland. STS offers users one-stop shopping for such services as leased lines, telex and X.25 packet switched connections.

**scanner** A program on a bulletin board system which scans the message base for previously entered E-mail and pulls a copy of each message and makes them available to the BBS mailer program.

**Scanner Interface Trace (SIT)** In IBM's SNA, a record of the activity within the communication scanner processor (CSP) for a specified data link between a 3725 Communication Controller and a resource.

**scattering** The cause of lightwave signal loss in optical fiber transmission. It is caused by the glass itself, which bends some of the light so sharply that it leaves the fiber. The second leading cause of scattering is impurities in the glass.

**SCC** 1. Satellite Communications Controller. 2. Specialized Common Carrier.

**SCCP** Signaling Connection and Control Part.

**scheduled output** In IBM's VTAM, a type of output request that is completed, as far as the application program is concerned, when the program's output data area is free.

**SCIP EXIT** Session Control In-bound Processing EXIT (IBM's VTAM).

**SCM** System Controller Module.

**SCN** Strategic Communications Network.

**SCOCS** Selective Class of Call Screening.

**SCP** Service Control Point.

**SCPC** Signal Channel Per Carrier.

**scrambler** A device which transforms data into an apparently random bit sequence. At the receiving locations, the data is unscrambled.

**screen id** Screen identifier.

**screen identifier (screen id)** The number in the upper right corner of an IBM PROFS and Office Vision screen that is used to identify the screen.

**screen print** Screen print allows only the contents of the screen that is currently displaying to be printed. If the operator initiates screen print, then it is called operator-initiated screen print. If screen print is initated by the host, it is called host-initiated screen print.

**scripting** The process of using a programming language to automate functions.

**SCRIPT/VS** The formatter component of IBM's Document Composition Facility. SCRIPT/VS provides capabilities for text formatting and processing.

**scrolling** Moving the contents of a display screen up or down.

**SCS** An SNA character string consisting of EBCDIC control characters optionally mixed with user data that is carried within an SNA request/response unit.

**SCSI** Small Computer System Interface.

**SCTO** Soft Carrier Turn Off.

**SD** Send Data.

**SDCU** Satellite Delay Compensation Unit.

**SDLC** Synchronous Data Link Control.

**SDM** Space-Division Multiplexing.

**SDMA** Space Diversity Multiple Access.

**SDN** 1. Software Defined Network. 2. System Development Network.

**SDNS** 1. Secure Data Network System. 2. Software-Defined Network Services.

**SDP** Service Delivery Point.

**SDPO** Sleeve Control Dial Pulse Originating.

**SDS** Switched Data Service.

**SDU** Service Data Unit.

**SD-80** A Siemens telephone system with features for both small businesses and hotel/motel applications.

**SD 192** A Siemens microprocessor-controlled telephone switching system for small-to-medium-sized businesses.

**SD 232** A Siemens microprocessor-controlled telephone system with special features for hotel/motel industry.

**SEC** SECondary.

**secondary application program** In IBM's SNA, an application program acting as the secondary end of an LU–LU session.

**secondary channel** A data channel derived from the same physical path (telephone lines) as the main data channel but completely independent from it. Running at a low data rate, it carries auxiliary information such as device control, diagnostics, or a simultaneous data stream.

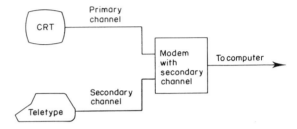

**secondary end of a session** In IBM's SNA, that end of a session that uses secondary protocols. For an LU–LU session, the secondary end of the session is the secondary logical unit (SLU).

**secondary half-session** In IBM's SNA, the half-session that receives the session-activation request.

**secondary link station** In IBM's System Network Architecture (SNA), any link station on a link, using a primary–secondary protocol, that is not the primary link station. A secondary link station can exchange data only with the primary link station; no data traffic flows from one secondary link station to another.

**Secondary Logical Unit (SLU)** In IBM's SNA, the logical unit (LU) that contains the secondary half-session for a particular LU–LU session. *Note.* An LU may contain secondary and primary half-sessions for different active LU–LU sessions.

**Secondary Logical Unit (SLU) KEY** In IBM's SNA, a key-encrypting key used to protect a session cryptography key during its transmission to the secondary half-session.

**secondary route** In packet switching, a path taken to establish a virtual circuit if the "primary" path is not available.

**secondary station**  A remote or tributary station that is under the control of a primary station. In bit-oriented protocols remote stations are secondary stations.

**secondary-detected**  In IBM's Network Problem Determination Application (NPDA), a term that refers to event and statistical data that reports conditions detected by resources at the secondary end-of-line.

**second-party maintenance**  Maintenance performed by an original equipment manufacturer (OEM) owned and operated service department or division, authorized dealer, distributor, or system integrator.

**Secretariat of Communications and Transrtation (SCT)**  The Mexican government related entity responsible for providing telephone service to small rural towns as well as television distribution, telex, time-sharing and data transmission services.

**sectional center**  A control center connecting primary centers; a class 2 office.

**sector**  The smallest unit of data written onto or recieved from a track on a disk.

**Secure Exchange of Keys (SEEK)**  An electronic public key encryption technique created by Cylink, Inc. The SEEK algorithm encodes data based upon the cooperative development of a secret number between two users.

**security**  The techniques used for preventing physical access to information. May involve use of encryption.

**Securities Industry Automation Corporation (SIAC)**  A company formed to provide data processing services to the New York and American Stock Exchanges.

**SEED**  Self-Electro-optic Effect Device.

**SEED**  Specific hex characters placed at the beginning of a pseudo-random word. The SEED is used by test sets in determining the beginning of the pseudo-random word.

**seek**  A mechanical movement involved in locating a record in a random-access file. This can, for example, be the movement of the arm and head mechanism that is necessary before a read instruction can be given to read data in a certain location in the file.

**SEEK**  Secure Exchange of Keys.

**segment**  In a local area network using a bus topology, a segment is an electrically continuous piece of the bus. Segments can be connected to one another via repeaters.

**segmentation**  The process of breaking a large user data message into multiple, smaller messages for transmission.

**segmented monitoring**  In ISDN the monitoring of each transmission system used on a basic rate access as a separate entity.

**segmenting of BIUs**  In IBM's SNA, an optional function of path control that divides a basic information unit (BIU) received from transmission control into two or more path information units (PIUs). The first PIU contains the request header (RH) of the BIU and usually part of the RU; the remaining PIU or PIUs contain the remaining parts of the RU. *Note.* When segmenting is not done, a PIU contains a complete BIU.

**seizure signal**  A signal used by the calling end of a trunk or line to indicate a request for service.

**select**  A command sent from the control station which is an invitation to receive for the remote station.

**selecting port**  A programmable port type which can communicate with any switched port in the network.

**selection**  The process of identifying the destination terminal for a message.

**selection number (SEL#)**  A means by which an IBM Network Problem Determination Application (NPDA) user requests display activity for the data that is described on a single display line. The

393

SEL# is sometimes accompanied by an additional parameter that modifies the request.

**selective call ringing** A service of Signaling System 7 (SS7) which results in calls from important numbers being given a distinctive ring.

**selective call rejection** A service of Signaling System (SS7) which results in calls from a preselected list of telephone numbers being diverted to a recording rather than sent to the subscriber.

**selective calling** The ability of the transmitting station to specify which of several stations on the same line is to receive a message.

**Selective Class of Call Screening (SCOCS)** The coding of telephone lines by a carrier which informs the local exchange carrier and long distance carrier that special billing conditions may exist. As an example, a prison location may only allow collect calls.

**selective cryptographic session** In IBM's VTAM, a cryptographic session in which an application program is allowed to specify the request units to be enciphered.

**selectivity** The degree to which a receiver is capable of discriminating between the desired signal and all other signals.

**selector channel** An input/output (I/O) channel designed to operate with only one I/O device at a time. Once the I/O device is selected, complete records are transferred in one-byte intervals.

**selector lightpen** An instrument that can be attached to the display station as a special feature. When pointed at a portion of the display station's image on the screen and then activated, the selector lightpen identifies that portion of the displayed screen for subsequent processing.

**self-checking numbers** Numbers which contain redundant information so that an error in them, caused, for example, by noise on a transmission line, may be detected.

**Self-Electro-Optic Effect Device (SEED)** A proto-type optical switch developed in 1986 at Bell Laboratories. SEED consists of a six-by-six array of switches controlled by light beams which alter the physics of material so they change from opaque to transparent, or vice versa, based upon light striking or not striking the material.

**self-test** A diagnostic test mode in which the modem is disconnected from the telephone facility and its transmitter's output is connected to its receiver's input to permit the looping of test messages (originated by the modem test circuitry) through the modem to check the performance of the modem.

**SELV** Safety Extra Low Voltage.

**SEL#** Selection Number.

**semaphore** A mechanism for the synchronization of a set of cooperating processes. It is used to prevent two or more processes from entering mutually critical sections at the same time. The process is used to synchronize the exchange of data.

**semiconductor** A material (like silicon, germanium, and gallium arsenide) with properties between those of conductors and insulators. Used to manufacture solid state devices—diodes, transistors, integrated circuits, injection lasers, and light-emitting diodes.

**semiconductor laser** A device made of two or more semiconductors (materials with electrical properties between those of conductors and insulators) that produces an intense, very pure beam of light when stimulated by electricity. Their small size, high speed, and minimal power consumption make these lasers a good choice for lightwave systems.

**send data** Data from DTE to DCE.

**send pacing** In IBM's SNA, pacing of message units that a component is sending.

**sender** The device in a switching system that automatically or manually transfers a message to the next part of the system.

**SENDIT** A telecommunications network for North

Dakota educators and students in the K-12 environment.

**sensitivity** The degree of response of a radio circuit to signals of the frequency to which it is tuned.

**sequence number** A field which indicates the sequential ordering of successive blocks or frames.

**sequencing** The process of dividing a user's message into smaller blocks, frames, or packets for transmission, with each subdivision of data given a sequence number to insure the correct reassembly of the complete message at its destination.

**sequential access** The method of reading or writing a record in a file by requesting the next record in sequence.

**sequential file organization** A File on disk or magnetic tape whose records are organized for access in consecutive order.

**SERCnet** Science Engineering Research Council Network.

**serial communication** The standard method of ASCII character transmission where bits are sent, one at a time, in sequence. Each 7-bit ASCII character is preceded by a start bit and ended with a parity bit and stop bit.

**Serial Input/Output (SIO)** The sequential or consecutive performances of input/output activities from a single channel or device.

**serial interface** Mechanical and electrical components that enable data to be transmitted sequentially bit-by-bit over a transmission medium.

**serial networks** A group of SNA networks connected in series by gateways.

**serial port** An input/output (I/O) port used to connect serial devices, such as a mouse or external modem to a computer. Data transfers occur one bit at a time serially via a serial port.

**serial transmission** A technique in which each bit of information is sent sequentially on a single channel, rather than simultaneously as in parallel transmission. Serial transmission is the normal mode for data communications. Parallel transmission is often used between computers and local peripheral devices.

**server** 1. Any node on a LAN that allows other nodes to access its resources. Dedicated servers are totally dedicated to providing resources to other nodes on the LAN. Non-dedicated servers are used both as servers and as workstations. Common resources shared by a server include files and printing facilities. 2. That part of a queueing system providing the service requested by the customer.

**Server Message Block (SMB)** A distributed file system developed by Microsoft Corporation and adopted by IBM and other vendors. SMB allows one computer on a network to use the files and peripherals of another as if they were local.

**Server–Requester Programming Interface (SRPI)** 1. A protocol between requesters and servers when using IBM's Enhanced Connectivity Facilities. 2. An application programming interface used by requester and server programs to communicate with the PC or host routers.

**service** A set of functions provided by a layer to its client.

**service access point (SAP)** 1. In the OSI model, the destination or source address at a layer boundary. 2. A logical point in a token-ring network made available by an adapter where information can be received and transmitted.

**Service Control Point (SCP)** A centralized node in a communication carrier's network that contains service control logic.

**service data unit** A distinct unit of data passed by a client of a service, for transmission to the remote client; a message.

**Service Delivery Point (SDP)** Under FTS2000 the physical location at which an FTS2000 service is terminated.

395

**service element**  That part of an entity which performs the primary functions of the entity.

**service entrance**  The point at which network communications lines ("telephone company" lines) enter a building.

**Service Information Octet (SIO)**  In CCITT #7 terminology SIO defines the user part and whether the service is national or international.

**Service Initiation Charge (SIC)**  The fixed charge associated with obtaining a circuit or feature.

**service line**  An exchange line that extends from data stations to a telephone company center and is used for monitoring the line for any defects.

**Service Management System (SMS)**  A centralized operations system in a communication carrier's network used for creating and establishing services.

**service message**  A brief message between communications centers pertaining to any phase of traffic handling, communications facilites, or circuit conditions.

**Service Order Analysis and Control System (SOAC)**  The interface between the telephone company service order processor and the Facilities Assignment and Control System (FACS) which manages line assignments for an office.

**Service Oversight Center (SOC)**  A facility established under the FTS2000 contract which enables the General Services Administration (GSA) to monitor and verify every aspect of Sprint contract compliance.

**service point**  In an IBM SNA network, the location which converts native vendor protocol to an SNA format and then transmits the resulting data to the focal point.

**service provider**  In the ISO Model, all the lower-level layers that provide a service to a service user.

**Service Reference Model (SRM)**  A standard which defines the minimum capabilities a monitor must have in order to display an acceptable North American Presentation Level Protocol Syntax

(NAPLPS) image. SRM defines a minimum terminal image of 200 vertical pixels, 256 horizontal pixels and a palette of at least eight colors and eight shades of gray to include black and white.

**service request**  A request for service signal as defined by CCITT Recommendation X.28.

**Service Switching and Control Point (SSCP)**  A centralized node in a communication carrier's network which has integrated service switching point (SSP) and service control point (SCP) functionality.

**Service Switching Point (SSP)**  A distributed switching node in a communication carrier's network that processes calls by interacting with the service control points (SCPs) in the network.

**service terminal**  The equipment needed to terminate a channel and connect to station equipment.

**service user**  In the ISO Model, the recipient of services provided by lower-level functions.

**Service 800**  An international toll-free facsimile service marketed by International 800 Telecom Corporation.

**serving area**  Geographic area handled by a telephone facility; generally a LATA.

**Serving Test Center (STC)**  A test location established to control and maintain circuit layout records (CLR).

**SES**  1. Service Evaluation System. 2. Severely Errored Seconds.

**session**  1. In SNA communications protocol, a session is a logical network connection between two addressable units for the exchange of data. For example, a 3278 Display Station could be a logical unit in a session with a software application. 2. Another name for "connect time".

**session activation request**  In IBM's SNA, a request that activates a session between two network-addressable units (NAUs) and specifies session parameters that control various protocols during session activity; for example, BIND and ACTPU. Synonymous with generic BIND.

RELATIONSHIP OF CARRIER SERVICES PROVIDERS

*Service Switching and Control Point (SSCP)*

**session address space**   In IBM's VTAM, an ACB address space or an associated address space in which an OPNDST or OPNSEC macro instruction is issued to establish a session.

**session awareness data**   In IBM's SNA, data relating to sessions that is collected by NLDM and that includes the session type, the names of session partners, and information about the session activation status. It is collected for LU–LU, SSCP–LU, SSCP–PU, and SSCP–SSCP sessions and for non-SNA terminals not supported by NTO.

**session control**   1. A module providing functions for system-dependent process-to-process communication, name to address mapping, and protocol selection. 2. In IBM's SNA, one of the components of transmission control. Session control is used to purge data flowing in a session after an unrecoverable error occurs, to resynchronize the data flow after such an error, and to perform cryptographic verification. 3. In IBM's SNA, an RU category used for requests and responses exchanged

between the session control components of a session and for session activation/deactivation requests and responses.

**Session Control In-bound Processing Exit (SCIP)**   In IBM's VTAM, a user exit that receives control when certain request units (RUs) are received by VTAM.

**session cryptography key**   In IBM's SNA, a data encrypting key used to encipher and decipher function management data (FMD) requests transmitted in an LU–LU session that uses cryptography.

**session data**   In IBM's SNA, data relating to sessions that is collected by NLDM and that consists of session awareness data and session trace data.

**session deactivation request**   In IBM's SNA, a request that deactivates a session between two network-addressable units (NAUs); for example, UNBIND and DACTPU. Synonymous with generic UNBIND.

**Session Information Retrieval (SIR)** In IBM's SNA, the function allowing the operator to enable/disable session information retrieval for a particular gateway or all gateway sessions. When a gateway session ends, trace information will be passed back to all SSCPs that have enabled session information retrieval (SIR) for that or all sessions. Trace information can also be requested for a particular gateway session to be passed back to the requesting host.

**Session layer** The fifth layer in the OSI model responsible for establishing, managing, and terminating connections for individual application programs.

**session limit** 1. In IBM's SNA, the maximum number of concurrently active LU–LU sessions a particular logical unit can support. *Note.* VTAM application programs acting as logical units have no session limit. Device-type logical units have a session limit of one. 2. In the Network Control Program (NCP), the maximum number of concurrent line-scheduling sessions on a non-SDLC, multipoint line. 3. In an IBM 3174 control unit, the total number of logical terminals or defined Asynchronous Emulation Adapter (AEA) default destinations for an AEA port set.

**session management exit routine** In IBM's SNA, an installation-supplied VTAM exit routine that performs authorization, accounting, and gateway path selection functions.

**session parameters** In IBM's SNA, the parameters that specify or constrain the protocols (such as bracket protocol and pacing) for a session between two network-addressable units.

**session partner** In IBM's SNA, one of the two network-addressable units (NAUs) having an active session.

**session seed** Synonym for initial chaining value in IBM's SNA.

**session sequence number** In IBM's SNA, a sequentially incremented identifier that is assigned by data flow control to each request unit on a particular normal flow of a session, typically an LU–LU session, and is checked by transmission control. The identifier is carried in the transmission header (TH) of the path information unit (PIU) and is returned in the TH of any associated response.

**session services** In IBM's SNA, one of the types of network services in the system services control point (SSCP) and in the logical unit (LU). These services provide facilities for an LU or a network operator to request that the SSCP initiate or terminate sessions between logical units.

**session trace** In IBM's NLDM, the function that collects session trace data for sessions involving specified resource types or involving a specific resource.

**session trace data** In IBM's SNA, data relating to sessions that is collected by NLDM whenever a session trace is started and that consists of session activation parameters, access method PIU data, and NCP data.

**session-establishment macro instructions** In IBM's VTAM, the set of RPL-based macro instructions used to initiate, establish, or terminate LU–LU sessions.

**session-establishment request** In IBM's VTAM, a request to an LU to establish a session. For the primary logical unit (PIU) of the requested session, the session-establishment request is the CINIT sent from the SSCP to the PLU. For the secondary logical unit (SLU) of the requested session, the session-establishment request is the BIND sent from the PLU to the SLU.

**session-initiation request** In IBM's SNA, an Initiate or logon request from a logical unit (LU) to a system services control point (SSCP) that an LU–LU session be activated.

**session-level pacing** In IBM's SNA, a flow control technique that permits a receiving connection point manager to control the data transfer rate (the rate at which it receives request units) on the normal flow. It is used to prevent overloading a receiver with unprocessed requests when the sender can generate requests faster than the receiver can process them.

**session-termination request** In IBM's VTAM, a request that an LU–LU session be terminated.

**Set Asynchronous Balanced Mode (SABM)** A command sent over a communications line to establish two-way, link-level communication. A SABM is generated by Link Access Procedure Balanced Mode (LAPB) software.

**Set Asynchronous Response Mode (SARM)** A command sent over a communications line to establish link-level communication in one direction. SARM, a command generated by the Link Access Procedure (LAP) software, must be sent in both directions of the call.

**setup** The preparation of a computing system to perform a job or job step. Setup is usually performed by an operator and often involves performing routine functions, such as mounting tape reels and loading card decks.

**severely errored second** An error condition on a T1 circuit. A severely errored second is any second of time in which 320 or more Out-Of-Frame conditions or CRC error occur, or which has a bit error rate worse than $1 \times 0.00010$. Ten consecutive Severely Errored Seconds constitute a Failed Signal State.

**SF/MT** San Francisco/Moscow Teleport.

**SG** Signal Ground.

**SGND** Signal GrouND.

**SH** Switch Hook.

**shadow resource** In IBM's VTAM, an alternate representation of a network resource that is retained as a definition for possible future use.

**shadowing** The process of maintaining a duplicate file whereby modifications made to the primary file are also made to the duplicate file as changes occur.

**Shannon equation** A formula for determining the maximum rate of transmission for binary digits based on signal-to-noise ratio and bandwidth.

**Shannon limit** The theoretical maximum data rate capable of being transmitted over a band-limited channel with a given signal-to-noise ratio, for a given error rate.

**share limit** In IBM's SNA, the maximum number of control points that can concurrently control a network resource.

**shared** In IBM's SNA, pertaining to the availability of a resource to more than one user at the same time.

**shared access** In LAN technology, an access method that allows many stations to use the same (shared) transmission medium; contended access and explicit access are two kinds of shared access methods.

**Shared Network Facilities Agreements (SNFA)** Contracts between AT&T and the former Bell Operating Companies that came into use at the time of the AT&T divestiture. All SNFA's were terminated by 1991.

**shared RAM** Random access storage on the adapter that is shared by the computer in which the adapter is installed.

**shared tenant services** A PABX feature which permits multiple tenant organizations to share the use of a communications switch that is normally cost effective for large organizations.

**shared-control gateway** In IBM's SNA, a gateway consisting of one gateway NCP that is controlled by more than one gateway SSCP.

**sheath** The outside, opaque concentric layer in a fiber optic cable.

**SHF** Super High Frequency.

**shield** The metallic layer that surrounds insulated conductors in "shielded" cable. The shield may be the metallic sheath of the cable or the metallic layer inside a nonmetallic sheath.

**shielded pair** Two wires in a cable wrapped tightly to avoid any interference.

**shielding** Protective covering that eliminates electromagnetic and radio frequency interference.

**shift**  In Baudot transmission, the prefix code that defines the interpretation of the following character.

**Shift-In (SI)**  A control character used in conjunction with SO (Shift-Out) and ESC (Escape) to extend the graphic character set of the code.

**Shift-Out (SO)**  A control character used in conjunction with SI (Shift-In) and ESC (Escape) to extend the graphic character set of the code.

**ship-to-shore telephone**  Marine telephones.

**short circuit**  An electrical system in which current flows directly from one conductor to another without passing through the device that is supposed to receive the current.

**short haul**  Referring to transmission distance typically less than 50 miles.

**short haul modem**  A signal converter which conditions a digital signal to ensure reliable transmission over DC continuous private line metallic circuits without interfering with adjacent pairs in the same telephone cable.

**short wavelength**  The spectrum from 800 to 1000 nanometers.

**show cause**  In IBM's SNA, the reason code in the RECMS indicating to VTAM or NPDA the threshold that was exceeded and whether or not the threshold has been dynamically altered.

**shunt**  1. An alternate path in parallel with a part of a circuit. 2. A bypass or parallel path which diverts current from a circuit.

**SI**  1. Shift-In. 2. Suppress Index.

**SIAC**  Securities Industry Automation Corporation.

**SIC**  Service Initiation Charge.

**SID**  1. Sudden Ionospheric Disturbances. 2. System Identification Number.

**SID**  Systems Integration Division.

**sideband**  The frequency band on either the upper or lower side of the carrier frequency within which fall the frequencies produced by the process of modulation.

**sideband power**  The power contained in the sidebands. It is this power to which a receiver responds (not to the carrier power) when receiving a modulated wave.

**sidebands**  The additional frequencies, both above and below the carrier frequency, produced as a result of modulation of a carrier.

**sidelobe**  Off-axis response of an antenna.

**sidetone**  The effect of a telephone microphone that enables callers to hear their own voices or room noises through their own handset.

**SIF**  Signaling Information Field.

**SIG**  Signal.

**SIG PLT CP**  SIGnal PLaToon Command Post.

**SIGCEN**  SIGnal CENter.

**SIGINT**  SIGnal INTelligence.

**signal**  Aggregate of waves propagated along a transmission channel and intended to act on a receiving unit.

**signal, analog**  A nominally continuous electrical signal that varies in some direct correlation to an impressed signal.

**signal, digital**  A nominally discontinuous electrical signal that varies in some direct correlation to an impressed signal.

**signal conditioning**  The amplification or modification of electrical signals to make them more appropriate for conveying information.

**signal constellation**  An arrangement for graphically depicting the possible phase and amplitude combinations of a complex modem transmission scheme.

**signal converter**  Device that received input signal information and outputs it in another form. Also Modem or Data Set.

**signal distance**  Same as 'hamming distance'.

**signal element**  Each part of a digital signal, distinguished from others by its duration, position, or sense; can be a start, information, or stop element.

**signal ground** The common ground reference potential for all circuits except protective ground.

**SIGnal INTelligence (SIGINT)** The monitoring and analysis of communications traffic.

**signal latency** A synonym for propagation delay.

**signal level** The strength of a signal, generally expressed in either units of voltage or power.

**signal mode** An optical waveguide that is designed to propagate light of only a single wavelength. An optical fiber that allows the transmission of only one light beam, and is optimized for a particular lightwave frequency.

**signal reflection** A condition that results from an electrical signal passing through a transmission medium encountering a point where the impedance changes, causing a portion of the signal to return (reflect) back to its origin.

**Signal to Noise Ratio (S/N)** The ratio between the signal and the actual noise bed as illustrated. dBrn is a measurement of noise: $-90$ dBm = 0 dBrn.

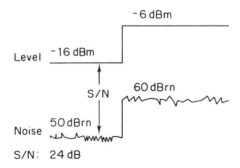

*Signal to Noise Ratio (S/N)*

**Signal Transfer Points (STP)** Tandem packet switches in a carrier network which route common channel signaling #7 (CCS#7) messages among service control points (SCPs) and service switching points (SSPs).

**signal unit** An integral number of 8-bit bytes contained in a CCITT (now ITU) Common Channel Signaling System 7 message.

**signaling** In telephony, the process of controlling calls, specifically the communication between stations of such information as whether a station is available (on-hook) or unavailable (off-hook). Digital telephony offers a broader range of services than analog telephony, and thus requires a more complex form of signaling.

**Signaling Connection and Control Party (SCCP)** In CCITT (now ITU) #7 terminology Recommendations Q.711–Q.714.

**Signaling Information Field (SIF)** In CCITT (now ITU) #7 terminology the body of a #7 MSU.

**Signaling Link Selection (SLS)** A portion of the #7 MTP label.

**signaling synchronization sequence** A logical sequence of six bit values carried in the framing bits of the 2nd, 4th, 6th, 8th, 10th, and 12th frames of a D4 superframe, or the 2nd, 6th, 10th, 14th, 18th, and 22nd frames of an ESF superframe. In combination with the frame synchronization sequence, this sequence indicates the locations in the bit stream of the 6th and 12th frames, which

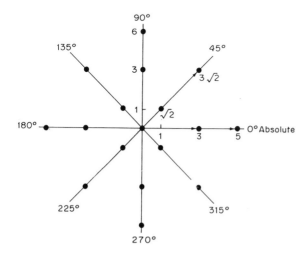

*signal constellation*

contain A and B signaling bits. The signaling synchronization sequence is the pattern 001110.

**Signaling System Multi-Frequency number 6 (SSMF6)** The UK version of CCITT (now ITU) R2 inter-register signaling. SSMF6 was never actually implemented.

**Signaling System number 6 (SS6)** The first common channel signaling system to be standardized under the auspices of the CCITT (now ITU). SS6 operates on analog telephone systems using 2.4 or 4.8 Kbps modems and is known as common channel interswitch signaling (CCIS) in North America.

**Signaling System No. 7** CCITT (now ITU) Common Channel Signaling System No. 7: the current international standard for out-of-band call control signaling on digital networks.

**Signaling Terminal Equipment (STE)** The CCITT (now ITU) term that defines an X.75 gateway switch.

**Signaling Transfer Point (STP)** The term used to describe a Signaling System 7 (SS7) packet switch.

**sign-on character** The first character sent on an ABR circuit; used to determine the data rate.

**SIGSEC** SIGnal SECurity.

**silicon** A dark gray, hard, crystalline solid. Next to oxygen, the second most abundant element in the earth's surface. It is the basic material for most integrated circuits and semiconductor devices.

**SIM 3278** A software program from Simware Inc. of Ottawa, Canada, which operates on an IBM mainframe and makes asynchronous terminal devices appear to the host as 3278 full-screen terminals.

**simple gateway** In IBM's SNA, a gateway consisting of one gateway NCP and one gateway SSCP.

**Simple Mail Transfer Protocol (SMTP)** The Internet standard protocol used for transferring electronic messages from one computer to another.

**simplex** Loosely, a communications system or equipment capable of transmission in one direction

only. Permits transmission in only one direction from "A" to "B" but not from "B" to "A."

**simplex circuit** 1. A circuit permitting the transmission of signals in either direction, but not in both directions simultaneously (CCITT). 2. A circuit permitting transmission of signals in one direction only (common usage).

**simplex mode** Operation of a communication channel in one direction only, with no capability for reversing.

**simplex transmission** Transmission in only one direction.

**Simplified Network Management Protocol (SNMP)** A network management protocol created by the Internet Activities Board responsible for TCP/IP network management standards.

**simulated logon** In IBM's VTAM, a session-initiation request generated when an ACF/VTAM application program issues a SIMLOGON macro instruction. The request specifies an LU with which the application program wants a session in which the requesting application program will act as the PLU.

**simulation** A technique through which a model of the working system is built in the form of a computer program. Special computer languages are available for producing this model, and the program is run on a separate computer. A complete system can be described by a succession of different models. These models can then be easily adjusted, and the system that is being designed or monitored can be used to test the effect of any proposed changes.

**simulator** A program in which a mathematical model represents an external system or process.

**SINCGARS** SINgle Channel Ground and Airborne Radio System.

**sine wave** A periodic wave form with amplitude proportional to the sine of a linear function of time, space, or both.

**SINET** Schlumberger Information Network.

**singing** An unintended whistle on the transmission circuit.

**single address message** A message that has only one address.

**Single Channel Ground and Airborne Radio System (SINCGARS)** A U.S. Army combat net radio that provides secure vehicular, manpack and airborne radio communications.

**Single Channel Per Carrier (SCPC)** A frequency division multiplexing technique in which each channel is modulated on its own carrier for transmission.

**single fiber cable** A plastic-coated fiber surrounded by an extruded layer of polyvinyl chloride, encased in a synthetic strengthening material and enclosed in an outer polyvinyl chloride sheath.

**Single Message-unit Rate Timing (SMRT)** A USA telephone company tariff under which local service is measured and calls timed in increments of 5 minutes or less—with a single message unit charge applied to each increment.

**single sideband** A system of radio communications in which either the upper or lower sideband is removed from amplitude modulation transmission to reduce the channel width and improve the signal-to-noise ratio.

**single threading** A program which completes the processing of one message before starting another message.

**single wire circuit** A single conductor that uses the ground return path to complete a transmission circuit.

**single-current transmission** A form of telegraph transmission effected by means of undirectional currents.

**single-domain network** In IBM's SNA, a network with one system services control point (SSCP).

**single-frequency signaling** A method of conveying dialing or supervisory signals, or both, with one or more specified single frequencies.

**single-mode fiber** A fiber lightguide with a slender core that confines light to a single path. It offers low dispersion and enormous information-carrying capacity. Single-mode fiber is generally used with laser diodes.

**single-phase** Pertaining to the simplest form of alternating current, where one outward and one return conductor are required for transmission.

**single-route broadcast** In a local area network the forwarding of specially designated broadcast frames only by bridges which have single-route broadcast enabled. Also called limited broadcast.

**single-sideband transmission** Transmission where one sideband of the carrier signal is transmitted while the other is suppressed.

**single-thread application program** In IBM's ACF/VTAM, an application program that processes requests for multiple sessions one at a time. Such a program usually requests synchronous operations from ACF/VTAM, waiting until each operation is completed before proceeding.

**sink** In data communications, a receiver.

**sink tree** For a given destination (sink) the paths taken by packets sent to it from all points of the network form a tree when fixed routing tables are used. This is the sink tree.

**S-interface** In ISDN, the 2-wire physical interface used for single customer termination at the S-reference point, between NT2 and Terminal Equipment. The S-interface is the standard, end-user interface for ISDN's Basic Rate Service. According to current ISDN standards, the S- and T-interfaces are physically and logically identical.

**SIO** 1. Scientific and Industrial Organization. 2. Serial Input/Output. 3. Service Information Octet.

**SIP** Societa Italiana per L'Esercizio Telefonico pA.

**SIR** Session Information Retrieval (IBM's SNA).

**SIT** Scanner Interface Trace (IBM's SNA).

**SITA** Société Internationale de Télécommunications Aeronautiques.

**Site** A node in a bulletin board system network.

**site number** A number assigned to a site in a bulletin board system network.

**skewing** The time delay or offset between any two signals.

**skirt** A descending edge on an adapter card behind its connector. The presence of a skirt may restrict the expansion slots into which the adapter card can be installed.

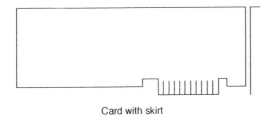

Card with skirt

Card without skirt

**Skylin X.25** A satellite-based shared X.25 network service offered by Scientific Atlanta, Inc., of Atlanta, GA. Skylin X.25 supports interactive communications between multiple VSAT locations and a shared network hub via a Ku-band satellite.

**Skylink.25** A third generation VSAT data network marketed by Scientific-Atlanta, Inc., which is based upon the CCITT X.25 architecture and supports a variety of commonly used protocols.

**Skynet** Digital transmission service offering from AT&T Communications, featuring on-site earth-station facilities for wideband satellite trans-

mission, with Accunet Reserved 1.5 circuits. Also called High-Capacity Satellite Digital Service (HCSDS) and Skynet 1.5.

**Skypaper** A trademark of National Satellite Paging of Washington, D.C., as well as the name for the first FCC licensed and operating nationwide paging system marketed by the vendor.

**Skyphone** A direct-dial telephone service for airline passengers offered by British Telecom International.

**SLAR** Side-Looking Airborne Radar.

**slave** A called unit under the control of commands and signals from a master (calling) unit.

**slave station** In point-to-point circuits, the unit controlled by the master station.

**SLC** Submarine Laser Communications.

**SLCSAT** Submarine Laser Communications SATellites.

**SLIC** Subscriber Line Interface Card.

**slicing level** A voltage or current level of a digital signal where a one or zero bit can be determined or not. Also threshold.

**slip** A defect in timing that caused a single bit or a sequence of bits to be omitted or read twice. Slips are primarily caused by improper synchronization resulting from wander and improper distribution of the network reference frequency.

**SLIP** Serial Line IP.

**SLM** System Logic Module.

**slot** 1. A unit of time in a TDM frame where a sub-channel bit or character is carried. 2. Basic information unit (an 18-bit pattern) sent on an IBM 3600 System Loop. Specifies a command, data, or a sync pattern.

**slot sharing** A procedure by which more than one terminal may contend for the same slot thereby increasing the number of terminals that may be attached an IBM System 3600 Loop.

**slot time**    A local network parameter that describes the contention behavior of a media access control. Nominally equal to twice the propagation delay.

**slotted ALOHA**    A packet broadcast system like ALOHA in which packets are timed to arrive at the center in regular time slots, synchronized for all stations.

**slotted frame**    A frame of information which is shared by two or more stations by dividing the frame into equal parts. This technology is used by many TDMs and on some LANs with ring topologies.

**SLS**    Signaling Link Selection.

**SLU**    Secondary Logical Unit.

**SL/1**    A PBX manufactured by Northern Telecom.

**SM**    1. Statistical Multiplexer. 2. Switching Module. 3. System Management.

**SMA**    Search Mode Acquisition.

**Small Computer System Interface (SCSI)**    An industry standard interface for connecting peripheral devices and their controllers to a microprocessor. The SCSI interface defines both hardware and software standards for communication between a host computer and a peripheral.

**Small-Scale Integration (SSI)**    A term used to describe a multifunction semiconductor device with a sparse density (10 circuits or less) of electronic circuitry contained on a single silicon chip.

**smart terminal**    A smart terminal has semiconductor-based memory. Its features include the ability to be polled, to store data in blocks, and to perform error checking. Functions such as editing can be performed on a smart terminal, but specific capabilities are built into the terminal and cannot be changed. A smart terminal is not user-programmable.

**SMARTCOM**    A series of communications programs marketed by Hayes Microcomputer Products, Inc., of Atlanta, GA, for use on personal computers. Various versions of SMARTCOM operate on Apple Macintosh and IBM personal computers.

**SmarTerm 240**    Terminal emulation software from Persoft, Inc., which makes a personal computer operate as a Digital Equipment Corporation (DEC) VT240 terminal.

**SmarTerm 470**    A terminal emulator program marketed by Persoft, Inc., of Madison, WI, which operates on IBM PC and compatible personal computers. SmarTerm 470 emulates Data General D470, D461, D460, D450 and Dasher terminals.

**SMARTMODEM**    A copyright of Hayes Microcomputer Products as well as a term used to reference a modem capable of recognizing commands.

**Smartnet**    A trademark of Telematics International of Calabasas, CA, as well as a family of wide area networking products that permit personal computers and workstations to access SNA, BSC and X.25 networks.

**SMATV**    Satellite Master Antenna TV.

**SMB**    Server Message Block.

**SMBASE**    Statistical Multiplexer BASE card.

**SMDR**    Station Message Detail Recording.

**SMDS**    Switched Multimegabit Data Service.

**SMEXP**    Statistical Multiplexer EXPansion feature.

**SMF**    System Management Facilities (IBM's SNA).

**SMN**    System Message Number.

**SMP**    System Modification Program (IBM's SNA).

**SMP/E**    System Modification Program Extended (IBM's SNA).

**SMRT**    Single Message-unit Rate Timing.

**SMS**    Service Management System.

**SMTP**    Simple Mail Transfer Protocol.

**S/N**    Signal-to-Noise ratio.

**SNA**  Systems Network Architecture.

**SNA Character String (SCS)**  In IBM's SNA, a data stream composed of EBCDIC controls, optionally intermixed with end-user data, that is carried within a request/response unit.

**SNA distribution services**  In IBM's Systems Network Architecture, the facilities permitting multiple SNA networks to be interconnected.

**SNA network**  In IBM's SNA, the part of a user-application network that conforms to the formats and protocols of Systems Network Architecture. It enables reliable transfer of data among end users and provides protocols for controlling the resources of various network configurations. The SNA network consists of network-addressable units, boundary function components, and the path control network.

**SNA Network Interconnection (SNI)**  An IBM product which permits multiple-domain SNA networks.

**SNA terminal**  A terminal that supports Systems Network Architecture protocols.

**SNADS**  SNA Delivery System.

**SNAP**  Standard Network Access Protocol.

**snapshot**  The capture of the state of a protocol analyzer during operation. This information content may include memory contents, register status, the flag status, etc.

**SNBU**  Switched Network BackUp (IBM's SNA).

**SNDCF**  SubNetwork Dependent Convergence Function.

**sneak currents**  Low-level currents that are insufficient to trigger electrical surge protectors and therefore able to pass them undetected. These currents may result from contact between communications lines and AC power circuits or from power induction, and may cause equipment damage from overheating.

**sneaker mail**  A term used to refer to delivery of messages by the postal service.

**SNEP**  Switching Network Evolution Program.

**SNET**  Southern New England Telephone.

**SNI**  1. SNA Network Integration. 2. SNA Network Interconnection.

**SNFA**  Shared Network Facilities Agreements.

**Sniffer**  A local area network protocol analyzer marketed by Network General that supports a variety of hardware to include Ethernet, Token-Ring, Arcnet and Starlan.

**SNMP**  Simple Network Management Protocol.

**SNR**  Signal-to-Noise Ratio.

**SNRM**  Set Normal Response Mode.

**SNUMB**  Submittal NUMBer—a job sequence number used to identify a particular job.

**soak**  A term referring to a means of uncovering problems in software and hardware by running them under operating conditions while they are closely supervised by their developers.

**SOC**  Service Oversight Center.

**Societa Italiana per l'Esercizio Telefonico pA (SIP)**  The Italian state telephone agency.

**Société Internationale de Télécommunication Aeronautique (SITA)**  Sponsor of an international data communications network used by many airlines. This network links most airlines of the world.

**socket**  A three-part identifier assigned to each Internet network consisting of a process port number, device address and network address.

**SofCram/Net**  Data compression software marketed by Dynatech Communications, Inc., of Woodbridge, VA, that employs artificial intelligence to compress data by up to 90 percent. SofCram/Net operates on IBM 4300 and 3090 mainframes and personal computers.

**soft error**  1. An intermittent error on a network that requires retransmission. The adapters are able to retransmit the data that had the difficulty and communication continues. 2. An error on a network that can impair the network's performance but does not, by itself, affect its reliability. If the number of soft errors reaches the ring error limit, reliability is affected.

**soft font**   Fonts that are stored on disk and loaded into a printer's memory for use.

**soft link**   An alternate name for an object or directory in a namespace. Soft links allow users to view names as forming an acyclic directed graph rather than a pure tree.

**soft turn-off**   A soft carrier frequency transmitted by a modem operating on the switched telephone network to prevent transients at the end of a message being misinterpreted as spurious space signals at the remote modem.

**softcopy**   An electronic document that is stored in a computer.

**Softerm PC**   A communications program marketed by Softronics, Inc., for use on the IBM PC and compatible personal computers. The program is known for its inclusion of over 50 exact terminal emulations and seven file transfer protocols.

**software**   A computer program or set of computer programs held in some kind of storage medium and loaded into read/write memory (RAM) for execution.

**software defined network**   A network constructed by a long distance carrier, in which a user is given a means to define a "virtual network" that operates in some respects as a private network, yet is assembled from conventional switching and trunking components of the carrier's public switched network.

**Software Defined Network Services (SDNS)**   An AT&T provided service that gives customers control of their own special call-routing programs stored in AT&T's network. This will provide fast and cost-effective network flexibility.

**software demultiplexing**   In multiplexer appli-cations, the connection of a computer directly to a multiplexer trunk without an intervening multi-plexer to permit the remotely multiplexed trans-mission to be demultiplexed by software running within the computer.

**software engineering**   A broadly defined discipline that integrates the many aspects of programming, from writing code to meeting budgets, in order to produce affordable software that works.

**software maintenance**   The continual improvements and changes required to keep programs up to date and working properly.

**SO**   Shift-Out.

**SOH**   Start Of Header.

**solar array**   A power generation method using solar cells.

**solar outage**   A condition in which the high radiated noise level of the sun can be many times stronger than a transmitted signal. In satellite transmission a solar outage occurs when the sun passes behind or near a satellite and within the field of view of an antenna. Satellite solar outages are predictable and occur twice a year.

**solar panel**   A device on satellites that converts solar energy into electrical energy.

**solicited message**   In IBM's ACF/VTAM, a response from VTAM to a command entered by a program operator.

**solid-state component**   A component whose oper-ation depends on control of electric or magnetic phenomena in solids such as a transistor, crystal diode or a ferrite core.

**SOM**   Start Of Message.

**SONET**   Synchronous Optical NETwork.

**SOP**   Standing Operating Procedure.

**sound-powered telephone**   A telephone that derives its operating power from the speech input.

**source**   1. In data communications, the originator of

407

a message. 2. In optical systems, a common term for the light source, such as a laser or light emitting diode.

**source code**  A file or listing of a high-level program in the original language in which it was coded.

**source of data, source**  A term used to describe any device capable of supplying data in a controlled manner.

**source rotary**  A subscription option for defining symbolic mail and a list of X.121 addresses associated with a name. The Packet Assembler/ Disassembler (PAD) attempts to connect to the first address on the list and continues with other addresses until connection is made or the list is exhausted.

**source routing**  A Routing Information (RI) field within the frame header of a token-ring frame which carries specific ring and bridge identifiers which indicate the path for the frame to follow to reach the destination station.

**source routing protocol**  A protocol which permits flexibility in designing network topologies to include moving workstations on a local area network without changing hardware or software and permitting redundant network components to be installed to maximize system availability.

**Source Service Access Point (SSAP)**  The one-byte field in a LLC frame on a LAN that identifies the sending Data Link client protocol.

**Southernnet**  An Atlanta-based long-distance communications carrier.

**SP**  1. SPace character. 2. SPace bar.

**space**  1. An ASCII or EBCDIC character that results in a 1-character-wide blank when printed. 2. In single-current telegraph communications, the open circuit or no-current-flowing condition. 3. In data communications, represents a binary 0.

**space division**  Any form of switching system which allots a single physical circuit to each connection.

**Space Physics Analysis Network (SPAN)**  A network operated by the U.S. National Aeronautics and Space Administration (NASA).

**space-division multiplexing**  A technique which uses a separate circuit or channel for each device. Essentially this means no multiplexing at all.

**space-division switching**  A switching technology where a separate physical path through the switch is maintained for each connection.

**space-hold**  The normal no-traffic line condition whereby a steady space is transmitted.

**space-to-mark transition**  The transition, or switching, from a spacing impulse to a marking impulse.

**span**  A T1 circuit connecting two endpoints with no intermediate switching, multiplexing, or demultiplexing; a T1 link.

**SPAN**  Space Physics Analysis Network.

**span line**  A repeated T1 line section between two central offices.

**spanned record**  A variable length record that is segmented and which spans one or more control intervals. Spanned records can occur on disk-resident sequential files.

**spanning tree**  A logical topology used for bridge forwarding in an extended LAN; it includes all bridges in the extended LAN but has no loops.

**SPARKLE**  Modem pooling software marketed by Concept Developments System of Kennesaw, GA.

**SPC**  Stored Program Control.

**SPCL**  A cellular telephone data transmission protocol developed by Spectrum Cellular Corp. of Dallas, TX. SPCL employs automatic retransmission and forward error control.

**speaker phone**  Telephone equipment which allows anyone in the room to hear the telephone conversation and to participate in the conversation through a microphone and speaker. Speaker phone options may be part of the telephone set itself, or provided as ancillary equipment. Speaker phones eliminate the need to repeat entire conversations to the rest of the staff.

**SPEARNET**  South Pacific Education And Research NETwork.

**Special Prefix Information Delivery Service (SPIDS)**  A service of Southwestern Bell Telephone Company which provides access to unique prefix telephone numbers on a subscription basis.

**Specialized Common Carrier (SCC)**  A company providing special or value-added communication services.

**specific-mode**  In IBM's ACF/VTAM: 1. The form of a Receive request that obtains input from one specific session. 2. The form of an Accept request that completes the establishment of a session by accepting a specific queued CINIT request.

**specifications**  Literature which clearly and accurately describe essential requirements for equipment, systems or services including the procedures which determine that the requirements have been met.

**spectral bandwidth**  The difference between wavelengths at which the radiant intensity of illumination is half its peak intensity.

**spectral density**  In T1, the specification concerning the space/mark ratio. AT&T specifies that no more than 15 consecutive zeros are allowed.

**spectrum**  1. A continuous range of frequencies, usually wide in extent, within which waves have some specific common characteristic. 2. A graphical representation of the distribution of the amplitude (and sometimes phase) of the components of a wave as a function of frequency. A spectrum may be continuous or, on the contrary, contain only points corresponding to certain discrete values.

**spectrum rolloff**  Applied to the frequency response of a transmission line, or a filter, it is the attenuation characteristic at the edge of the band.

**speech circuit**  A circuit designed for the trans-

mission of speech but which can also be used for data transmission or telegraphy.

**speech plus**  Technique used to combine voice and data on the same line by assigning the top part of the normal voice bandwidth to data.

**speech scrambler**  A device in which speech signals are converted into unintelligible forms before transmission, and are restored to intelligible forms at reception.

**speed**  Same as data rate.

**speed calling**  A telephone feature that permits caller to reach frequently called numbers by using abbreviated codes.

**speed conversion**  The changing of transmission speeds to enable two devices with unequal transmission speeds to communicate.

**speed dial**  A PBX or telephone set feature which allows users to preprogram phone numbers so that a pushbutton can be accessed to dial them. Individual speed dialing uses designated buttons on the specific phone. System speed dialing logs the number into a central memory buffer in the PBX.

**speed dialing**  Process of using short sequences of digits to represent complete telephone numbers.

**speed number**  A one-, three-, or four-digit number that replaces a seven- or ten-digit telephone number. These numbers are programmed into the switch in the carrier's office or in a PBX.

**SPIDS**  Special Prefix Information Delivery Service.

**spike**  A sudden surge of current in an electronic circuit. Spike is a transient characterized by a short rise-time.

**spill forward**  The transfer of full control on a call to the succeeding office by sending forward the complete telephone address of the called party.

**spin-off data set**  A data set that is deallocated (available for printing) when it is closed. Spin-off data set support is provided for output data sets just prior to the termination of the job that created the data set.

**spin stabilization**   A type of satellite attitude control achieved by spinning all or most of the satellite's exterior.

**spine network**   A local area network in which the user's computers and network servers connect to access networks, with the access networks in turn connected via gateways to the spine.

**Spiral Redundancy Checking (SRC)**   A validity-checking technique for transmission blocks where the information sent for receiver checking is accumulated in a spiral bit position fashion (IBM 2780).

**spiral-4 cable**   A cable used in many transmission line circuits which has two pairs of wires that are insulated with polyethylene and constructed in standard $\frac{1}{4}$-mile lengths.

**splice**   1. In lightwave systems, the joining of two fiber lightguides, end-to-end. Mechanical splices use coupling devices to connect lightguides; fusion splices use heat to melt and join them; and bonded splices "glue" lightguides together with an epoxy. 2. In wire systems, the joining of two ends of copper wire.

**split screen**   Two panels that appear together on a screen, one above the other or one to the right of the other in a condensed format.

**splitter**   An analog device for dividing one input signal into two output signals or combining two input signals into one output signal. Used to achieve tree topologies in CATV or broadband local area networks.

**spoofing**   1. The process of introducing spurious transmission into a communications system. 2. A term used to refer to the activity during which a perpetrator intercepts a valid user's signon request, fooling the user with a fake screen message that logoff is complete. The perpetrator then continues to use the active session. Also called piggybacking.

**SPOOL**   Simultaneous Peripheral Operation On Line.

**spooled data set**   A data set written on an auxiliary storage device.

**spooling**   A technique whereby output is stored on disk or tape for subsequent printing.

**SPOT**   Small Portable Operating Terminal.

**spot beam**   1. A satellite transmission, a narrow and focused down-link transmission. 2. A satellite antenna designed to direct all of the satellite's power to a relatively compact area on earth.

**spread spectrum communications**   The process of modulating a signal over a significantly larger bandwidth than is necessary for the given data rate for the purpose of lowering the bit error rate (BER) in the presence of strong interference signals.

**Spread Spectrum Multiple Access (SSMA)**   Refers to a frequency modulation technique.

**spread-spectrum signal**   A signal which occupies a bandwidth in excess of the minimum information bandwidth that would be required to transmit the information. Spread-spectrum communications can be used to reduce the probability of the intercept of communications, protect against communications jamming and minimize unintentional interference.

**Sprint**   A U.S. long distance communications carrier originally owned 80 percent by United Telecommunications and 20 percent by GTE Corporation. Sprint was one of the first long distance communications carriers to install a nationwide fiber optic network.

**SPST**   Single Pole, Single Throw Switch.

**spying**   In a communications environment, the process of line tapping.

**SQD**   Signal Quality Detector.

**SQL**   Structured Query Language.

**SQM**   Signal Quality Monitor.

**Squeeziplexer**   A trademark of Astrocom Corporation as well as a coaxial cable multiplexer designed to replace up to 32 individual cables that normally connect to ports on an IBM control unit.

**SRC**   Spiral Redundancy Checking.

**SRDM**   SubRate Data Multiplexer.

**SRHIT**   Small Radar Homing Intercept Technology.

**SRM**   Service Reference Model.

**SRPI**   Server–Requester Programming Interface.

**SRR**   Short Range Radio.

**SRS-300**   A dual port ISDN terminal adapter manufactured by Fujitsu. The SRS-300 provides access to ISDN B-channel network services through two RS-232 ports for asynchronous and synchronous transmission and includes a packet assembler/disassembler (PAD) which provides packet switched communications capability via an ISDN D-channel.

**SRV**   SeRVer.

**SS**   Surge Suppressor.

**SS No. 7**   CCITT Common Channel Signaling System No. 7.

**SSAP**   Source Service Access Point.

**SSB**   Single SideBand.

**SSBSC**   Single SideBand Suppressed Carrier.

**SSCP**   1. Service Switching and Control Point. 2. System Services Control Point (IBM's SNA).

**SSCP ID**   In IBM's SNA, a number uniquely identifying a system services control point (SSCP). The SSCP ID is used in session activation requests sent to physical units (PUs) and other SSCPs.

**SSCP rerouting**   In IBM's SNA network interconnection, the technique used by the gateway SSCP to send session-initiation RUs, by way of a series of SSCP–SSCP sessions, from one SSCP to another, until the owning SSCP is reached.

**SSCP–LU session**   In IBM's SNA, a session between a system services control point (SSCP) and a logical unit (LU). The session enables the LU to request the SSCP to help initiate LU–LU sessions.

**SSCP–PU session**   In IBM's SNA, a session between a system services control point (SSCP) and a physical unit (PU); SSCP–PU sessions allow SSCPs to send requests to and receive status information from individual nodes in order to control the network configuration.

**SSCP–SSCP SESSION**   In IBM's SNA, a session between the system services control point (SSCP) in one domain and the SSCP in another domain. An SSCP–SSCP session is used to initiate and terminate cross-domain LU–LU sessions.

**SSI**   Small-Scale Integration.

**SSMA**   Spread Spectrum Multiple Access.

**SSMF6**   Signaling System Multi-Frequency number 6.

**SSP**   1. Service Switching Point. 2. Advanced Communications Function for the System Support Programs.

**SSPA**   Solid-State Power Amplifier.

**SSU**   Subsequent Signal Unit. CCITT #6 terminology.

**SS6**   Signaling System number 6.

**stack**   A collection of items which can be thought of as arranged in sequence. One end of the sequence is the "top" of the stack. New items are added at the top, and items are also read and removed from the top of the stack. A stack can be stored as a list. Sometimes pointers in parts of the stack other than the top are retained.

**stand-alone workstation**   A workstation that establishes a display and/or printer sessions with the host computer, thereby providing access to host resources such as data or word processing applications and printer facilities.

**star**   1. In LAN technology, a network topology where the central control point is connected individually to all stations. 2. A multi-node network topology that consists of a central multiplexer node with multiple nodes feeding into and through the central node. The outlying nodes are usually interconnected with one another such that more than one path exists to any other node in the network.

**star network** A network topology in which each device is directly connected to a central node.

**star topology** A point-to-point wiring system in which all network devices are connected to a central server.

**STAR 3.4** A network design and capacity planning tool for private leased-line networks marketed by Network Synergies of Lafayette, IN.

**Starlan** A local area network specification of 1 Mbps baseband data transmission over two twisted pair (IEE 802.3).

**StarLan10** An Ethernet local area network from Hewlett-Packard that operates at 10 Mbps over unshielded twisted pair wiring.

**start bit** In asynchronous communication, the start bit is attached to the beginning of a character so that bit and character synchronization can occur at the receiver. Always a "0" or space condition.

**start element** The first element of a character in certain serial transmissions, used to permit synchronization. In Baudot teletypewriter operation, it is one space bit.

**Start Of Header (SOH)** A communication character used to identify the beginning of the header field in a message block.

**Start Of Message (SOM)** A control character or group of characters transmitted by a station to indi-

cate to other stations on the line that what follows are the addresses of stations to receive the message.

**Start of Text (STX)** A control character used to indicate the beginning of a message; it immediately follows the header in transmission blocks.

**start option** In IBM's ACF/VTAM, a user-specified or IBM-supplied option that determines certain conditions that are to exist during the time an ACF/VTAM system is operating. Start options can be predefined or specified when ACF/VTAM is started.

**start-of-routing signal** In U.S. military communications a signal consisting of two consecutive hyphens (- -) or dashes (— —) used to indicate the start of the addressee routing indicator(s).

**start/stop** A system in which each code combination is preceded by a start signal which serves to prepare the receiving mechanism for the reception of a character and is followed by a stop signal which serves to bring the receiving mechanism to rest. The start and stop signals (bits) are known as "machine information" or synchronizing bits.

**start–stop envelope** A form of digital signal which is anisochronous and comprises a short group of signal elements, fixed in number and length. For example, 11 elements at 300 elements per second is a common format. The envelope can begin whenever the line is clear of the previous envelope and the timing starts with the first signal element, the start element. The following elements carry the information and the envelope concludes with a stop element or elements.

**start–stop transmission** Asynchronous transmission such that a group of signals representing a character is preceded by a start bit and followed by a stop bit.

**state** In packet switching, information about the condition of a virtual call used to select appropriate processing routines.

**State-of-the-Art-Medium Terminals (SAMT)** SHF (X-band) satellite communications ground terminals used by the U.S. Army Communications

Electronics Command. SAMT is used to provide command, control and communications support to the U.S. Army defense satellite communications system network.

**static**  Any electrical disturbance caused by atmospheric conditions.

**static routing**  The use of manually entered routing information to determine routes.

**station**  A logically addressable end point in a communications network.

**Station Carrier System**  An analog carrier system used within a local loop to provide multiple communications paths over a single cable.

**station clock**  An internal oscillator and circuit that provide a clock source for an electronic device such as a multiplexer or transmission facilities management system.

**station controller**  Processor that controls communications to and from a terminal device. Provides control and error handling functions.

**station log**  A chronological record of station events, such as entries relating to message handling, equipment difficulties, personnel, etc.

**Station Message Detail Recording (SMDR)**  A PABX feature that provides magnetic tape records of calling information data which the end customers use to help analyze billing or calling patterns.

**station protector**  A device mounted on the customer's premise which prevents excessive voltage on the outside plant (lightning, power contacts, power induction, ground potential rise, etc.) from damaging equipment or harming personnel inside the building.

**station serial numbers for messages**  In manual teletypewriter operations a station serial numbering system permits consecutive numbering of messages by each station in the network. The net control station decides whether or not station serial numbers will be used. Generally, the decision is to use them either all the time or not at all.

When they are used, however, they are helpful in identifying messages and receipts. In U.S. military communications, a station serial number consists of three items: the originating station's call sign, the prosign NR, and the number of the transmission. Each station maintains a separate series of numbers, beginning with NR1 at 0001 hours each day, for messages it sends to each other station in the network.

**station set**  A telephone terminal used in voice communications. The station set can be a desktop set, wall telephone, mobile telephone, or public pay telephone.

**stationkeeping**  The process of making in-orbit adjustments with small rocket thrusters attached to the satellite, keeping the satellite in its correct geosynchronous position.

**station-to-station calling**  Direct dialing between stations without attendant intervention.

**statistic**  In IBM's Network Problem Determination Application (NPDA), a resource-generated data base record that contains recoverable error counts, traffic, and other significant data about a resource.

**statistical multiplexer**  A statistical multiplexer divides a data channel into a number of independent data channels greater than that which would be indicated by the sum of the data rates of those channels. It does this on the basis that not all the channels will want to transmit simultaneously, thus freeing up capacity to accommodate additional channels. The microprocessor in the statistical multiplexer scans the channels for input, formats the data into "blocks" and transmits it over the "composite" output channel. Buffers hold temporary bursts of data that exceed composite channel capacity. Statistical multiplexers reduce the high-speed channel idle time by "concentrating" more channels together using microprocessors, buffers and synchronous protocols.

**Status Activity Monitor (SAM)**  A testing device that provides full break-out box functions in

413

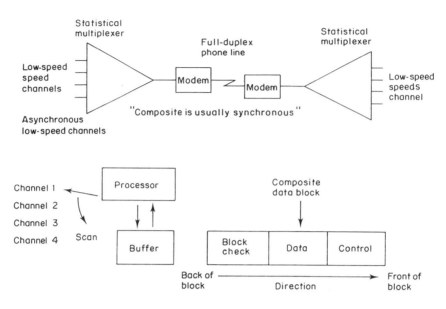

*statistical multiplexer*

addition to LED monitoring for indicating low or high lead activity.

**Status Alarm and Control System (SAC System)** Collects Telenet Processor (TP) cabinet and cage environmental information (e.g., power, temperature, fan, and door).

**status information** Information about the logical state of a piece of equipment. Status information is one kind of control signal. Examples are: (1) A peripheral device reporting its status to the computer. (2) A network terminating unit reporting status to a network switch.

**status maps** Tables which give the status of various programs, devices, input–output operations, or the status of the communication lines.

**STC** Serving Test Area.

**STD** Subscriber Trunk Dialing.

**STE** Signaling Terminal Equipment.

**steamer** A trademark of Datagram of East Greenwich, RI, as well as a device which performs data compression.

**step-by-step** 1. An automatic switching system in which a call is extended progressively step-by-step. 2. The common name given to the Strowger system of two-motion selector switching.

**step-index** A type of waveguide (optical fiber) preferred for single-mode operation and long distance transmission.

**step-index fiber** A fiber lightguide in which the refractive index changes in one big jump between core and cladding, and not gradually as in graded-index fibers. When the core is small, high transmission speeds over relatively long distances are possible. The core is of a uniform refractive index and has a sharp decrease in the index of refraction at the cladding.

**stop bit** In asynchronous transmission, the last bit used to indicate the end of a character; normally a mark condition which serves to return the line to its idle or rest state.

**stop element** The last element of a character in asynchronous serial transmissions, used to ensure recognition of the next start element. In Baudot teletypewriter operations, it is 1.42 mark bits.

**stop-and-wait ARQ**   A form of ARQ in which the sender will transmit one block of data and stop sending until an acknowledgment for that block is received from the receiver.

**storage**   Memory. A device into which data can be entered, held, and be retrieved at a later time. Any device that can store data.

**store**   To file a document.

**store and forward**   1. The operation of a data network in which the interruption of data flow from the originating terminal to the designated receiver is accomplished by storing the information *en route* and forwarding it at a later time. 2. A technique in which a message is received from the originator and held in storage until a circuit to the addressee becomes available. 3. The handling of messages or packets in a network by accepting them completely into storage before sending them forward to the next switch.

**stored program control**   Electronic switching equipment that can be programmed to perform a variety of functions.

**STP**   Signal Transfer Point.

**STR**   Synchronous Transmit Receive.

**straight-through pinning**   RS-232 and RS-422 configuration that matches DTE to DCE, pin for pin (e.g. pin 1 with pin 1, pin 2 with pin 2).

**straight-tip connector**   An optical fiber connector used to join single fibers together at interconnects or connect them to optical cross-connects.

**strain rate**   In a fiber optic system, the rate at which strain is increased during a fiber tensile test.

**strapping**   Options built into a modem and other devices that can be activated by the user for a variety of functions. Strapping options include switching from half- to full-duplex operation; going from two- to four-wire transmission; or converting a point-

to-point modem into a multipoint modem and vice versa.

**StrataView**   A network management system marketed by StrataCom for use with that firm's T1 multiplexers.

**stratum**   A classification of a clock in a synchronous network. The network master clock which is known as the Primary Reference Source (PRS) is a Stratum 1 clock. The following table lists the clock stratum of a synchronous network.

| Clock stratum | Typical location(s) |
|---|---|
| 1 | Primary Reference Source |
| 2 | Class 4 toll office |
| 3 | Class 5 end office |
| 4 | Channel bank, end-user multiplexer |

**stream encryption**   Encryption of a data stream (such as a stream of characters) with no hold-up of data in the system. Thus a single character can pass through the encryption, transmission and decryption processes without waiting for other characters. The encryption may depend on previous data in the stream.

**streaming**   A condition of a modem when it is sending a carrier signal on a multipoint communication line when it has not been polled.

**Streamline**   The name used by Infotron Systems Corp. of Cherry Hill, NJ, for a DS-3 multiplexer manufactured by its Licom, Inc., subsidiary.

**string code**   A technique for combining several sequential occurrences of the same character or bits.

**Strowger, Almon B.**   A Kansas City undertaker who invented the step-by-step switching system and the rotary dial.

**structured programming**   The practice of organizing a program into modules that can be designed, prepared, and maintained independently of each other. Makes programs easier to write, check, read, and modify.

**Structured Query Language (SQL)**   A formal language which is used for such data base operations as adding, deleting, changing and retrieving records.

**STS** Scandinavian Telecommunication Services.

**STU** Secure Telephone Unit.

**stub (tip) cable** The short (usually 25 feet or less) cable which terminates on or is terminated on a cable terminal, protector, block, etc., and extends the connection into the cable.

**stub-in/stub-out** Where both the input and the output sides of a cable terminal, protector, block, etc., are prewired, eliminating the need for infield splicing or punch-downs.

**stunt box** 1. A device to control the nonprinting functions of a teletypewriter terminal, such as carriage return and line feed. 2. A device to recognize line control characters and to cause specific action to occur when those characters are recognized.

**STX** 1. Start of TeXt. 2. System Network Architecture-to-X.25.

**ST3000** A portable teleconferencing system consisting of a round audio module containing three unidirectional microphones and a specially designed loudspeaker. The ST3000 is marketed by Shure Teleconferencing Systems of Evanston, IL.

**ST6000** A teleconference audio system marketed by Shure Teleconferencing Systems of Evanston, IL.

**SU** Signalling Unit. CCITT #6 and #7 terminology.

**SUB** SUBstitute character.

**subaddress** An additional piece of addressing information for use by a DTE, beyond that needed to identify a particular subscriber line or group of lines.

**subarea** In IBM's SNA, a portion of a network containing a sub-area node, any attached peripheral nodes, and their associated resources. Within a subarea node, all network-addressable units, links, and link stations that are addressable within the subarea share a common subarea address and have distinct element addresses.

**subarea address** In IBM's SNA, a value in the subarea field of the network address that identifies a particular subarea.

**subarea/element address split** In IBM's SNA, the division of a 16-bit network address into a subarea address and an element address.

**subarea link** In IBM's SNA, a link that connects two subarea nodes.

**subarea LU** In IBM's SNA, a logical unit in a subarea node.

**subarea node** In IBM's SNA, a node that uses network addresses for routing and whose routing tables are therefore affected by changes in the configuration of the network. Subarea nodes can provide boundary function support for peripheral nodes. Type 4 and type 5 nodes are subarea nodes.

**subarea PU** In IBM's SNA, a physical unit in a subarea node.

**subcarrier** A carrier that is applied as modulation on another carrier or on an intermediate subcarrier.

**subchannel** The result of subdividing a communications channel into narrower bandwidth channels.

**SUBCOR** SUBject to CORrection.

**Submarine Laser Communications Satellites (SLCSAT)** A proposed array of satellites that will use blue laser technology to relay high data rate messages to submerged U.S. Navy submarines.

**sub-multiframe** In European T1, a group of eight consecutive frames; half of a 16-frame multiframe.

**subnet** In local area networking, the portion of a network that is partitioned by a router or another device from the remainder of the network.

**subnet address** An extension of the Internet addressing scheme which enables a site to use a single Internet address for multiple physical networks.

**subnetwork** The communication subsystem of a

computer network. A collection of equipment and physical media which forms an autonomous whole and which can be used to interconnect systems for the purposes of communication. A public data communication service could function as the subnet. For example, an X.25 packet switched network or an HDLC datalink.

**Subnetwork Access Protocol (SNAP)** A form of LLC frame used on LANs where multiplexing is done using a 5-byte Protocol ID field. This allows higher layer protocols that are not national or international standards to be addressed.

**subnetwork dependent convergence function** The set of functions required to enhance the service provided by a particular subnetwork to that assumed by the Connectionless-mode Network Protocol (ISO 8473).

**subprogram** A program invoked by another program.

**subrate** On a digital circuit, a bit rate less than DS0 (64 Kbps). The subrates most commonly used are 9600 bps and 56 Kbps.

**subrate (DDS)** A data speed that is either 9600 bps, 4800 bps, or 2400 bps.

**subrate data** In the digital transmission hierarchy of the United States, subrate data are digital signals sent at speeds of less than the T1 rate of 1.544 Mbps, such as the popular rates of 300, 1200, 2400, 4800, 9600, 19 200 and 56 000 bps.

**Subrate Data Multiplexer (SRDM)** A unit that combines a number of data streams at or below some basic rate (2400, 4800, 9600 Kbps) into a single 64 Kbps time division multiplexed signal.

**subrate multiplexing** The multiplexing of subrate data (bit rates less than 64 Kbps) into DS0 channels or DDS circuits. With current technologies, subrate multiplexing imposes limits on the number and kinds of subrate channels that can be multiplexed.

**subroutine** 1. A procedure that alters data in an area which is common to both the subroutine and its caller. The procedure's instructions are set by a sequential group of statements that can be used in one or more computer programs at one or more points in a computer program. 2. A routine that can be part of another routine.

**subscriber** Anyone who pays for and/or uses the services of a communications system. A subscriber can also be called a party, station, customer, or user.

**subscriber line module** Circuitry that provides the interface between a PABX and the telephone line.

**Subscriber Trunk Dialing (STD)** The name given to the national subscriber dialing facility in the UK.

**subscriber's line** The telephone line connecting the exchange to the subscriber's station.

**subscriber's loop** Same as local loop.

**subscript** An integer or variable with a value that identifies a particular element or group of elements in a table or an array.

**subset** A subscriber set of equipment, such as a telephone. A modulation and demodulation device.

**subsplit** A method of allocating available bandwidth in a single cable broadband system.

**substitute character (SUB)** A control character used in the place of a character that has been found to be invalid or in error.

**subsystem** 1. A conceptual grouping of programs by function (X.400 specific). 2. In IBM's SNA, a secondary or subordinate system, usually capable of operating independently of, or asynchronously with, a controlling system. 3. A program, or a set of programs, that will accomplish a specific task between remotely connected computers and/or devices.

**subvoice-grade channel** A channel of bandwidth narrower than that of voice-grade channels. Such channels are usually subchannels of a voice-grade line.

**SUNET** Swedish University NETwork.

**SunOS** A version of UNIX that supports the MS-

417

DOS operating system. SunOS operates on Sun Microsystems Sun 386i workstations.

**Super High Frequency (SHF)**  The portion of the electromagnetic spectrum in the microwave region with frequencies ranging from approximately 2 to 20 GHz.

**Super T-1**  A 1.544 Mbps, full-duplex, point-to-point satellite based transmission service marketed by GTE Spacenet of McLean, VA. A 4.5 or 3.7 meter Very Small Aperture Terminal (VSAT) is required to be installed at each Super T-1 customer premises site.

**superframe**  A unit of consecutive frames on a multiplexed digital circuit. In D4 framing format, a superframe contains 12 frames; in ESF, a superframe contains 24 frames. In CCITT terminology, a superframe is called a multiframe. A European multiframe contains 16 frames.

**supergroup**  The assembly of five 12-channel groups (60 circuits), occupying adjacent bands in the spectrum, for the purpose of simultaneous modulation or demodulation.

**supermastergroup**  600 circuits processed as a unit in a carrier system.

**Super-Regional Hub**  In the RIME network a Super-Regional Hub receives messages from Hubs within a geographic area and calls the NetHub to exchange mail packets on a network basis.

**supervision**  The monitoring of the progress of a call.

**supervisory port**  An interface on communications equipment to an operator console that can monitor and control communications functions.

**supervisory programs**  Those computer programs designed to coordinate service and augment the machine components of the system, and coordinate and service application programs. They handle work scheduling, input–output operations, error actions, and other functions.

**supervisory signals**  1. Signals used to indicate the various operating states of circuit combinations.

2. A signal, such as "on-hook" or "off-hook," which indicates whether a circuit or line is in use.

**supervisory system**  The complete set of supervisory programs used on a given system.

**support programs**  A set of programs needed to install a system, including diagnostics, testing aids, data generator programs, terminal simulators, etc.

**suppressed carrier transmission**  That method of communication in which the carrier frequency is suppressed either partially or the maximum degree possible. One or both of the sidebands may be transmitted.

**SURAnet**  Southeastern Universities Research Association network.

**SURFNET**  Dutch University NETwork.

**surge**  A short-duration increase of current or voltage in a circuit.

**Survivable Adaptable Fiber Embedded Network (SAFENET)**  A U.S. Navy project which uses fiber optic technology aboard ships to obtain continued operations in the event a ship is hit during battle.

**SUSDUPE**  SUSpected DUPlication.

**SVC**  Switched Virtual Circuit.

**SVC 76**  A supervisor call instruction executed to record error incidents occurring under IBM's OS/VS.

**SVD**  Simultaneous Voice/Data.

**SVIP**  Secure Voice Improvement Program.

**SVS**  Switched Voice Service.

**swapping**  The process of temporarily removing an active job from memory, saving it on disk, and processing another job in the area of memory formerly occupied by the first job.

**SWBD**  SWitchBoarD.

**SWBT**  SouthWestern Bell Telephone.

**SWIFT**  Society for Worldwide Interbank Financial Telecommunications

**SWIFT II** A second generation network under development by the Society for Worldwide Interbank Financial Transmission.

**switch** 1. Informal for data PABX. 2. In packet switched networks, the device used to direct packets, usually located at one of the nodes on the network's backbone.

**switch hook** The switch on a telephone that requests service from the central office.

**switchboard** A device designed to provide an interconnection capability between telephones. Each telephone station is connected to a common switchboard and a switchboard operator either uses switches or cords with plugs for insertion into jacks connected to the ends of the lines from the switchboard to telephones to connect them to one another.

**switched backup** A mode for a telecommunications line within the network facility of IBM PC 3270 Emulation Program; switched mode allows you to establish communication to the host by dialing a number on a telephone.

**switched carrier** A modem-to-modem handshaking technique in which RTS generates CD at the other end. Can be used full-or half-duplex.

**switched circuits** Lines of transmission formed by the activation of switches along the transmission route which connects them with other lines to complete the transmission.

**Switched Data Service (SDS)** An FTS2000 service that provides synchronous, fully duplex, digital circuit switched data communications at 56 Kbps. SDS will be availabe at speeds up to 64 Kbps when clear channel capabilities become available.

**Switched Digital Integrated Service (SDIS)** An FTS2000 service which provides integrated access to on-net Switched Voice Service (SVS), Switched Data Service (SDS), Packet Switched Service (PSS), Compressed Video Transmission Service (CVTS), Dedicated Transmission Service (DTS),

excluding unchannelized T-1. The two digital interfaces supported under SDIS are ISDN and T-1.

**switched line** A communications link for which the physical path may vary with each usage, such as the public, switched telephone network.

**switched major node** Same as switched SNA major node (IBM's VTAM).

**Switched Multimegabit Data Service (SMDS)** An experimental transmission and switching technology developed by Bell Communications Research, Inc. (Bellcore). Under SMDS computers and local area network gateways are expected to be connected to telephone company central offices via dedicated interfaces operating at 45 Mbps.

**switched network** A network which is shared among many users, any one of whom can potentially establish communications with any other when required.

**switched network (public)** It includes all of the integrated network components required to provide a telecommunications service.

**Switched Network Backup (SNBU)** In IBM's SNA, an option where a switched, or dial-up line is used as an alternate path if the primary, typically a leased-line, is unavailable.

**switched SNA major node** In IBM's ACF/VTAM, a major node whose minor nodes are physical units and logical units attached by switched SDLC links.

**switched virtual call** In packet switching, a temporary association between end users in which Logical Channel Numbers (LCNs) at each X.25 interface are dynamically assigned.

**Switched Virtual Circuit (SVC)** In a packet switching environment, a type of connection established between two network devices via a CALL request command. The circuit is temporary for the duration of the call.

**Switched Voice Service (SVS)** Under FTS2000 a service that provides voice and low-speed data transmission capabilities up to 4.8 Kbps. FTS2000 to FTS2000 calls are completed by dialing a seven-digit number. Agency users can connect with non-FTS2000 locations by dialing the appropriate 10-digit public switched network numbers.

**Switched 57 Service** An equipment package marketed by Digital Access Corp. and Hubbell Inc.'s Pulsecom Telecommunications Division that supports 57.6 Kbps asynchronous transmission over existing dial-up, digital-voice circuits. The equipment package includes a service unit and an interface card.

**switching** The process of transferring a connection from one device to another by connecting the two circuits.

**switching center** An installation in a communications system in which switching equipment is used to provide exchange telephone service for a given geographical area. A switching center is also called a switching facility, switching exchange or central office.

**switching matrix** In LAN technology, the electronic equivalent of a cross-bar switch.

**Switching Module (SM)** The microprocessor-driven hardware on an AT&T 5ESS switch which provides the signaling interface for external lines and trunks.

**switching multiplexer** A multiplexer able to switch channels from one high-speed, multiplexed circuit to another, or among any of several low-speed circuits.

**Switching Network Evolution Program (SNEP)** An AT&T developed software program for area planning. This program is used by telephone companies to examine the economics of various alternatives and identify the best alternative.

**switching node** 1. In circuit switched systems, a location that terminates multiple circuits and is capable of physically interconnecting circuits for the transfer of traffic. 2. In message or packet switched systems, a location which terminates multiple circuits and which is capable of storing

and forwarding messages, or packets, that are in transit.

**switching office** Same as central office.

**Switching Module (SM)** An AT&T 5ESS switch unit composed of a module controller/time slot interface unit along with a number of peripheral units. It performs 95 percent of all switching done in the 5ESS switch.

**switching processor** Communications processor specialized for handling the function of switching messages/packets in a network.

**switchover** When a failure occurs in the equipment, a switch may occur to an alternative component. This may be an alternative communication line or an alternative computer. This switchover process may be automatic or manual.

**SWR** Standing Wave Radio.

**SX-200D** A PBX marketed by Mitel of Boca Raton, FL, which includes T1 and automatic call distribution features.

**symbol** In data transmission, a symbol is a discrete waveform, usually representing binary digits, modulated as appropriate to be understood by the receiver.

**symbolic name** A means used to identify a collection of stations (as in an access group) or computer ports (as in a resource class).

**symmetric cipher** A cipher function that requires the same key to decrypt as was used to encrypt.

**symmetric flow** A flow pattern with a symmetric traffic matrix and, on each line of the network, an (s,t) flow component equal and opposite to the (t,s) flow component for each pair s,t terminal points. Thus the overall flow is symmetric and so also is the routing of each flow component.

**symptom string** In IBM's SNA, a structured character string written to a file when VTAM detects certain error conditions.

**SYNAD exit routine**  In IBM's SNA, a synchronous EXLST exit routine that is entered when a physical error is detected.

**sync**  Short for synchronous or for synchronous transmission.

**sync bits**  Framing bits in synchronous transmission.

**synchronization, synchronizing**  The process of making the receiver be "in step" with the transmitter; usually achieved by having a constant time interval between successive bits, by having a predefined sequence of overhead bits and information bits, and by having a clock.

**synchronizing character (SYNC)**  A character used to synchronize the receiver so it can accept the transmitted data coherently.

**synchronous**  1. Having a constant time interval between successive bits or characters. The term implies that all equipment in the system is in step; also bit synchronous. 2. A digital network in which all network elements are referenced to a single frequency source.

**synchronous data channel**  A communications channel capable of transmitting timing information in addition to data. Sometimes called an isochronous data channel.

**Synchronous Data Link Control (SDLC)**  A communications line discipline, associated with the IBM system network architecture SNA: initiates, controls, checks, and terminates information exchanges or communications lines. Designed for full-duplex operations simultaneously sending and receiving data. It can also be used for half-duplex transmission. SDLC is a bit-oriented protocol. Instead of using a control character set as does bisynchronous, SDLC uses a variety of bit patterns to flag the beginning and end of a frame. Other bit patterns are used for the address, control and packet header fields which route the frame through a network to its destination. Typical SDLC transmission frames are illustrated.

**synchronous data network**  A data network in which the timing of all components of the network is controlled by a single timing source.

**synchronous idle**  In synchronous transmission, a control character used to maintain synchronization and as a time fill in the absence of data. The sequence of two SYN characters in succession is used to maintain synchronization following each line turnaround.

**synchronous modem**  A modem which can transmit timing information in addition to data. It must be synchronized with its associated terminal equipment by the exchange of timing signals. Sometimes called an isochronous modem.

**synchronous network**  A network in which all communications links are synchronized to a common clock.

**synchronous operation**  In IBM's ACF/VTAM, a communication, or other operation in which ACF/VTAM, after receiving the request for the operation, does not return control to the program until the operation is completed.

**Synchronous Optical NETwork (SONET)**  A Bellcore developed synchronous optical transmission protocol standard that accommodates the capability to add and drop lower bit rate signals from the higher bit rate signal without requiring electrical demultiplexing. The standard defines a set of

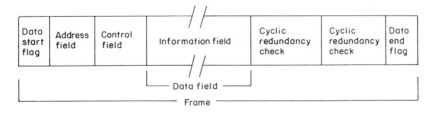

*Synchronous Data Link Control (SDLC)*

transmission rates, signals and interfaces for fiber optic transmission facilities. The base transmission rate of SONET, Optical Carrier 1 (OC-1) is approximately 51 times the bandwidth of a standard 1.544 Mbps T1 lines. The following table shows standardized SONET digital transmission rates. The basic rate, OC-1, sometimes referred to as STS-1 (Synchronous Transport Signal), builds on the existing DS-3 rate of 45 Mbps.

| OC level | Line rate |
| --- | --- |
| OC-1 (STS-1) | 51.840 megabits per second |
| OC-3 | 155.520 Mbps |
| OC-9 | 466.560 Mbps |
| OC-12 | 622.080 Mbps |
| OC-18 | 933.120 Mbps |
| OC-24 | 1.244 gigabits per second |
| OC-36 | 1.866 gbps |
| OC-48 | 2.488 gbps |

**synchronous request**   In IBM's ACF/VTAM, a request for a synchronous operation.

**synchronous routine**   A set of computer instructions that is called into service at periodic intervals.

**synchronous system**   A system in which sending and receiving equipment is operating continuously at the same frequency.

**synchronous terminal**   A data terminal that operates at a fixed rate with transmitter and receiver in synchronization.

**synchronous transmission**   Transmission in which the data characters and bits are transmitted at a fixed rate with the transmitter and receiver synchronized. This eliminates the need for individual start bits and stop bits surrounding each byte, thus providing greater efficiency.

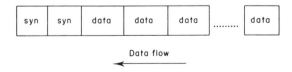

Data flow

**Sync-up**   A trademark of Universal Data Systems of Huntsville, AL, as well as a series of adapter cards designed for use in an IBM PC or compatible

personal computer which contain a popular modem and emulation hardware.

**syntax**   The rules of grammar in any language, including computer language.

**Syntran**   A restructured electrical DS3 (45-Mb/s) signal format for efficient synchronous transmission designed to add/drop DS1 (1.544-Mb/s) signals without demultiplexing the DS3 signal.

**SYSCON**   SYStems CONtrol.

**SYSCONCEN**   SYStems CONtrol CENter.

**SYSGEN**   SYStem GENeration.

**Sys/Master**   A component of the Net/Master network management system from Cincom Systems, Inc., of Cincinnati, OH, which provides console monitoring and operating system management.

**SYSOP**   SYStem OPerator. The owner or operator of a bulletin board system.

**system**   An implementation of a computer that supports the Network layer, the Transport layer, and the Session Control layer. Each system has a unique Network layer address.

**system administrator**   The person responsible for maintaining a computer system.

**System Application Architecture (SAA)**   A set of specifications developed by IBM which describes how application programs, communications programs, and users interface with one another. SAA represents an attempt by IBM to standardize the look and feel of applications.

**system clock**   The source designated as the reference for all clocking in a network of electronic devices, such as a multiplexer or transmission facilities management system.

**system configuration**   A process that specifies the devices and programs that form a particular data processing system.

**system control programming**   IBM-supplied programming that is fundamental to the operation and maintenance of the system. It serves as an interface with program products and user programs and is available without additional charge.

**System Generation (SYSGEN)** The process of loading an operating system in a computer.

**System Identification Number (SID)** In cellular telephone communications, a 15-bit binary code which when transmitted by a mobile station identifies its "home" network to which the unit subscribes and by which it is billed.

**System Management (SM)** An application process performing management functions on behalf of the entire OSI System. System Management contains all of the information relevant to the operation of a system (the Management Information Base) and may also include one or more managers providing various fuctionalities.

**System Management Facilities (SMF)** An optional control program feature of OS/360 and OS/VS that provides the means for gathering and recording information that can be used to evaluate system usage.

**System Message Number (SMN)** A unique number assigned by Telemail to each message.

**System Modification Program (SMP)** In IBM's SNA, a program that facilitates the process of installing and servicing an MVS system. For NPDA, either SMP or SMP/E is required for installation in an MVS system.

**System Modification Program Extended (SMP/E)** In IBM's SNA, a program that facilitates the process of installing and servicing an MVS system. For NPDA, either SMP or SMP/E is required for installation in an MVS system.

**system name** The name of a computer system. In IBM's PROFS and Office Vision a system name/user name combination is used when sending material to someone who uses a different computer system. Also called a location or nodeid.

**System One** The name of the airline reservation system owned by Texas Air Corp. of Houston, TX.

**system operator** The person responsible for operating a computer system.

**system restart** A restart that allows reuse of previously initialized input and output work queues. Synonymous with warm start.

**System Service Control Point (SSCP)** In IBM's SNA, a host-based network entity that manages the network configuration.

**System Support Program (SSP)** An IBM product program, which is made up of a collection of utilities and small programs, that supports and is required for the operation of the NCP.

**System 7** CCITT (now ITU) Common Channel Signaling System No. 7: the current international standard for out-of-band call control signaling on digital networks.

**System 12** A telephone switch marketed by Alcatel.

**System 75** A PBX manufactured by AT&T.

**systematic intrusion detection** A Nynex software program designed to help telephone companies track down computer users who program their PCs to make a series of calls using random number sequences until they identify a valid long distance calling card number.

**Systéme Interbancaire de Télécompensation (SIT)** A network in France designed for interbank transactions that uses Transpac, the nationwide X.25 packet switching network operated by France Telecom, the state-controlled network operator.

**Systems Network Architecture (SNA)** IBM's total description of the logical structure, formats, protocols, and operational sequences for transmitting information units between IBM software and hardware devices. Data communications system functions are separated into three discrete areas: the Application layer, the Function Management layer, and the Transmission Subsystem layer. The structure of SNA allows the ultimate origins and destinations of information—that is, the end users—to be independent of, and unaffected by,

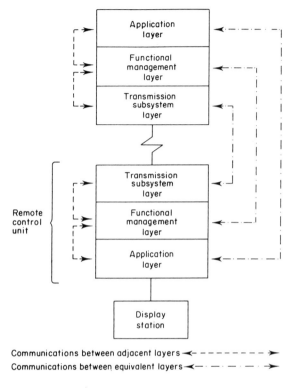

Communications between adjacent layers ◄ — — — — — — — ►
Communications between equivalent layers ◄ — · — · — · — · ►

Structure of SNA

the specific data communications system services and facilities used for information exchange.

**Systems Network Architecture-TO-X.25 (STX)** Software marketed by Telenet that can be installed in a host for 3270-to-asynchronous conversion.

**systems technology management**  A joint venture between Electronic Data Systems Corp., Dallas, TX, and the Korean Lucky-Goldstar Group.

**systems software**  Programs or routines that belong to the system and that usually perform a support function.

**SYSTIMAX PDS**  A trademark of AT&T as well as the new name for all of that vendor's premises distribution system (PDS) products. SYSTIMAX PDS products include unshielded twisted pair copper wiring for connecting products from multiple vendors; optical fiber and adapters for high-speed, high-capacity data communications networks; cross-connectors for wiring multiple local area networks and peripherals together; multiplexers; short-haul modems; support hardware; and protection devices.

**SYU**  SYnchronization signal Unit. CCITT #6 terminology.

# T

**TA** Technical Advisory.

**TA** Terminal Adapter.

**T/A** Traffic Analysis.

**table** A body of information (data structure) used to describe the connectivity, equipment characteristics, or operating parameters of a device in a network.

**table-driven** A logical computer process frequently used in network routing, access security, and modem operation. In a table-driven process, a user-entered variable is matched against an array of predefined values. If a match occurs, another variable in a table associated with the user-entered variable is selected, resulting in the process also being called a table-lookup.

**TABS** 1. Telemetry Asychronous Block Serial protocol. 2. Messages sent from a control center that access customer premises equipment registers that store performance data concerning errored seconds and failed seconds for a T1 line.

**TAC** 1. Technical Assistance Center. 2. Telenet Access Controller.

**TACL** Tandem Advanced Command Language.

**TACS** Theater Army Communications System.

**TACT** Terminal-Activated Channel Test.

**TADDS** Tactical Automatic Digital Switch.

**TAF** Terminal Access Facility (IBM's SNA).

**tag** A representation of the control that is placed in a document to control indexing, format, substitution, and the like.

**tail circuit** A feeder circuit or extension of an existing communication link to a network node; normally a leased line. In the example below, there is a point-to-point line from a corporate headquarters in Chicago to a Regional Office in Boston with three terminals. There are two modems with built-in multiplexers to accomplish data transfer. We now need a line that connects the computer in Chicago to a newly opened office in Providence, RI, having one terminal. It is possible to connect the modem in Boston to another modem in Providence, via a tail circuit. By doing saves the cost of another point-to-point line directly connecting Chicago to Providence. The data between Chicago and Providence will now go through the modem in Boston and be transmitted on to Providence.

**tailending** A condition that occurs in a dial-up communications system in which the system security is breached and one user is connected to the system under the account of another user.

**tailing** A feature on a multichannel modem that allows another modem link to be attached to one of the channels.

**TALC** Tactical Airborne Laser Communications.

**talker echo** The reflection of a modem's transmitted signal back into its receiver.

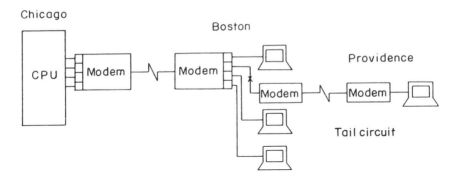

*tail circuit*

**talking battery**   The voltage sent by the central office to the subscriber which activates the carbon transmitter in the handset.

**talking path**   The connection in a telephone circuit between the tip and ring conductors.

**TAMS**   Telenet Access Management System.

**Tandem Ddvanced Command Language (TACL)**   A more powerful command interpreter for the Tandem non-stop systems.

**tandem data circuit**   A channel connecting two data circuit-terminating equipment (DCE) devices in series.

**tandem office**   A telephone company office with switching equipment which is used to interconnect central offices over tandem trunks in a densely settled exchange area where it is uneconomical to provide direct interconnection between all central offices. The tandem office completes all calls between the central offices but is not directly connected to subscribers.

**tandem operating system**   Known as Guardian, the operating system of the Tandem non-stop computer used by the Telemail system.

**tandem switch**   A special class of telephone company trunk-to-trunk switching system typically used in large metropolitan areas to interconnect central offices. Tandems are often classed as toll

switching systems although a large portion of the connection may be within the exchange area of a central office.

**tandem trunk**   A trunk circuit that connects a tandem, or intermediate, switch to the central office.

**tap**   In cable-based LANs, a connection to the main transmission medium.

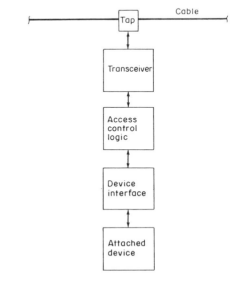

*Cable-based LAN tap*

**TAP**   Trace Analysis Program (IBM's SNA).

**tape copy**   A message in tape form, which is the result of a transmission.

**tapestry**   A local area network manager based network operating system marketed by Torus Systems of Redwood City, CA. Tapestry uses an icon-based user interface.

**TARA**   Threshold Analysis and Remote Access (IBM's SNA).

**tariff**   The published schedule of rates for specific equipments, facilities, or services offered by a common carrier; also, the vehicle by which regulatory agencies approve the rates. Thus, a contract between the customer and the common carrier.

**Tariff 9**   A tariff which enables AT&T to offer C and D conditioning on leased lines.

**Tariff 12**   A tariff which enables AT&T to offer custom voice and data networks to customers at a fixed price for a specific period of time.

**Tariff 15**   A tariff which allows AT&T to lower its switched network rates selectively (off-tariff service pricing). This tariff enables AT&T to negotiate rates with large customer accounts in response to off-tariff pricing of a rival.

**Tariff 16**   A tariff which allows AT&T to market services to certain U.S. Government customers under special rates and conditions.

**TAS**   Tactical Automatic Switch.

**TASI**   Time Assignment Speech Interpolation.

**task**   A sequence of instructions that has a starting point, an ending point, and performs some identifiable function.

**task group**   A named set of one or more tasks that has a common set of resources.

**TASS**   Tactical Automatic Switching Center.

**TAT**   TransAtlantic Telecommunications.

**TAT-8**   The eighth transatlantic telephone cable and the first to use single-mode fiber optic technology.

**TAT-9**   A transatlantic fiber cable scheduled to begin operations in 1991 that will link the United States and Canada to the United Kingdom, France and Spain.

**TA-TSY-000055**   A Bellcore technical advisory (TA) which covers intrusive testing of T1 links.

**TA-TSY-000194**   A Bellcore technical advisory (TA) that covers the T1 Extended Super Frame (ESF) format.

**TA-TSY-000783**   A Bellcore technical advisory (TA) that covers ISDN basic access line testing requirements.

**TA-TSY-000784**   A Bellcore technical advisory (TA) that covers ISDN basic access mediation function for Operations, Administration, Maintenance and Provisioning (OAM&P).

**TA-TSY-000892**   A Bellcore technical advisory (TA) that covers switching system operations general requirements for basic rate ISDN.

**TA-1/PT**   A U.S. military lightweight, weatherproof, sound-powered telephone intended for use on field wire lines in forward areas. This telephone can be used for communications with any local battery field telephone of local battery switchboard.

**TA-43/PT**   A U.S. military telephone set that is similar to the TA-312/PT but does not have a headset connector and associated items which the TA-312/PT has.

**TA-125/GT**   A U.S. military lightweight, weatherproof terminal box capable of terminating 12 telephone lines.

**TA-182/U**   A U.S. military telegraph–telephone signal converter. The TA-182/U converts 20 hertz ringing signals to either 1225 hertz or 1600 hertz ringing signals and 1225 or 1600 hertz to 20 hertz.

**TA-219/U**   A U.S. military telephone modem that consists of four channel modem subassemblies.

**TA-236/FT**   A U.S. military general purpose common battery telephone that has a range of approximately 8 km (5 miles).

**TA-264/PT**  A U.S. military local battery telephone set that has a built-in amplifier and is capable of operating over greater distances than can be obtained with ordinary local battery sets.

**TA-287/G**  A U.S. military two-way, single-channel, voice-frequency, unattended, battery-powered telephone repeater. The TA-287/G is used at intermediate or terminal locations to improve speech quality and to extend the talking range of a field wire (WD-1/TT) telephone circuit.

**TA-312/PT**  A U.S. military 2-wire, battery-operated field telephone. The TA-312/PT can be used in a simple point-to-point voice frequency wire communications link or in any 2-wire ring-down subscriber position of telephone communications system. The TA-312/PT has a range of approximately 22 km over WD-1/TT wire in dry conditions.

**TA-341/TT**  A U.S. military 4-wire, transistorized, local battery telephone set intended for use in sheltered areas.

**TA-838/TT**  A U.S. military ruggedized, solid-state field teleone designed for use with the SB-3614/TT switchboard or with the AN/TTC-25 or AN/TTC-38 tactical automatic switches.

**TA-938/G**  A U.S. military 2-wire common battery telephone intended for use in sheltered areas. The TA-938/G uses dual-tone multifrequency signaling and has a range of approximately 8 km (5 miles) from a central office.

**T-BERd**  Trademark of Telecommunications Techniques Corporation. A hand-held T1 BERT tester.

**TC**  1. Terminal Controller. 2. Transmission Control.

**TCAM**  Telecommunications Access Method.

**T-carrier**  A time division multiplexed digital transmission facility operating at an aggregate rate of 1.544 Mbps.

**TCAS**  Terminal Control Address Space (IBM's SNA).

**TCC**  1. Technical Control Center. 2. Telecommunications Center.

**TCF**  Technical Control Facility.

**TCM**  Time Compression Multiexing.

**TCO**  Telenet Central Office.

**T-connector**  A coaxial connector, shaped like a T, that connects two coaxial cables while supplying an additional connection for a network interface card of a local area network.

**TCP/IP**  Transmission Control Protocol/Internet Protocol.

**TCU**  Transmission Control Unit.

**TC97**  Technical Committee on Information Processing Systems of the ISO. The authors of the OSI model.

**TD**  1. Transmitted Data. 2. Transmitter–Distributor.

**TDB**  Terminal Descriptor Block.

**TDD**  Telecommunications Device for the Deaf.

**TDF-1**  The first French Direct Broadcast Satellite (DBS) launched by Arianespace on October 27, 1988.

**TDHS**  Time Domain Harmonic Scaling.

**TDM**  Time Division Multiplexer or Time Division Multiplexing.

**TDMA**  Time Division Multiple Access.

**TDR**  Time Domain Reflectometer.

**TDT2**  Telenet Diagnostic Tool 2.

**TD-202/U**  A U.S. military 12- or 24-channel TDM/PCM radio transmission interface unit. The transmit section of the TD-202/U accepts the output of one or two TD-352s, TD-660s, a TD-204 or TD-754, or another TD-202 and processes these outputs for radio transmission.

**TD-204/U**  A U.S. military multiplexer set that supports 12-, 24- or 48-channels. The transmit section of the TD-204/U accepts TDM/PCM signals from one or two TD-352s or TD-660s, another TD-204, or a TD-202 and processes these signals for transmission over coaxial cable.

**TD-206/G**  A U.S. military unattended repeater designed to restore the amplitude, waveshape, and timing of a TDM/PCM signal transmitted in either direction on a coaxial cable system.

**TD-352**  A U.S. military multiplexer set which converts 12 4-wire voice-frequency channels to a TDM/PCM signal for transmission over either a multichannel radio system or coaxial cable, or a combination of the two in tandem.

**TD-660/G**  A U.S. military multiplexer which supports 6 or 12 channels of voice-frequency communications over a TDM multichannel radio system, over a coaxial cable or over both in tandem.

**TD-754/G**  A U.S. military multiplexer that supports 6, 12, 24, or 48 voice-frequency channels.

**TE**  Terminal Equipment.

**Teammate**  A mobile U.S. military multistation ground-based communications intelligence system.

**Technical Advisory (TA)**  An AT&T vehicle for the release of network planning information and equipment compatibility specifications to independent Telephone Companies and the general trade. Technical Advisories have evolved into two basic types with two distinct purposes.

Type 1 are TAs that contain network planning information and equipment compatibility specifications for the benefit of the Independent Telephone Companies to ensure that the equipment and systems installed by the Independnts are compatible with the Bell System Network.

Type 2 are TAs that contain compatibility and interface information developed to enable the general trade to manufacture telecommunications equipment for possible purchase and use by the Bell System.

**Technical and Office Protocols (TOP)**  A Boeing Corporation version of the Manufacturing Auto-

mation Protocol (MAP) aimed at office and engineering applications.

**Technical Control Center (TCC)**  A centralized electronic system from which trained personnel monitor, test, and control the operation of a large data communication network.

**Technology Requirements Industry Forum (TRIF)**  Forums held by Bellcore, the research and engineering consortium jointly owned by seven regional Bell holding companies. TRIF's are designed to present proposed technical advisories as well as to obtain feedback from telephone companies and vendors.

**TeDe 400**  An electronic messaging system operated by TeleDelta in Sweden.

**TEI**  Terminal End-point Identifier (LAPD terminology).

**Tekelec, INC.**  A Calabasas, CA, manufacturer of data communications test equipment; manufacturer of the Chameleon 32 and the TE-820. A partly owned subsidiary of France's Tekelec-Airtronic.

**Tel Plus**  A series of key telephone systems manufactured by Siemens Information Systems which permits modem pooling, continuation of service during power failures, and the use of cartridge-type program modules for software enhancements.

**tel set**  Telephone set.

**Telaction Corporation**  A subsidiary of J. C. Penny which markets a videotex service that was co-developed with Cableshare of London, ON, Canada. Telaction combines telephone and cable in an interactive service that users control with ordinary tone-dial telephones.

**telco**  A general term for telephone common carrier or for a telephone company central office.

**Teldrin**  A paging network in Italy which enables page subscribers to be contacted in Italy, West Germany (Cityruf), France (Alphapage), and the United Kingdom (Europage). Each network operates in the UHF 466 frequency band.

**Telebit Corporation** A company located in Cupertino, CA, which specializes in the manufacture of high-speed modems under the Trailblazer trademark that are designed for operation on the switched telephone network.

**Telebox** The electronic mail service offered by Deutsche Bundespost in Germany.

**TeleBridge** A local area network bridge marketed by Shiva Corp. of Cambridge, MA.

**TeleCenter** A voice-mail system marketed by Votan of Fremont, CA.

**telecommunication line** Any physical medium such as a wire or microwave beam, that is used to transmit data. Synonymous with transmission line.

**telecommunications** A term encompassing the transmission or reception of signals, images, sounds, or information by wire, radio, optic, or infrared media. The term is derived from the Greek combining form "tele" meaning "far."

**Telecommunications Access Method (TCAM)** IBM teleprocessing access methods that controls the transfer of messages between the application program and the remote terminals and provides the high level message control language. TCAM macro instructions can be used to construct a message control program that controls messages between remote stations and application programs.

**Telecommunications Services Network Support (TSNS)** An IBM service which acts as a single point of service for its customers with mixed vendor equipment.

**telecommunity** Society in which the technologies of the information age allow anyone, anywhere, at any time, to send or receive any kind of information without technical barriers.

**Telecoms Integrated Management System (TIMS)** A Swedish Telecom-designed system which provides computerized assistance for handling subscriber requests for service, trouble handling, resource planning, line and telephone number assignment, and directory service.

**TELECON** TELEtypewriter CONference.

**teleconference (TELECON)** A conference between persons remote from one another but linked by a telecommunication system.

**teleconferencing** The process of conferring between persons in separate geographic areas by using telephonic means. Also refers to a type of communications (e.g., electronic mail) conducted by computers.

**telecopier** Facsimile machine.

**Teledis** An electronic data interchange service run by Milan-based Televas SpA. Teledis permits paperless trading of documents through the ITAPAC X.25 packet switching network and through the telephone network.

**Telefon Treff** A pilot program between Neumann Electronik in Mulheim and the German Bundespost that offers to set up conference calls involving as many as eight participants.

**Telefones de Losboa de Portugal (TLP)** One of two telecommunications service operating companies in Portugal.

**Telefonica** Spain's national communications carrier.

**telegram** A hard copy message delivered through the general telegraph service to an addressee.

**telegraph** A system employing the interruption of, or change in, the polarity of DC current signaling to convey coded informaton.

**telegraph channel** The apparatus and transmission path used for the sending and receiving of telegraph signals.

**telegraph distortion** Distortion which alters the duration of signal elements.

**telegraph terminal** A term used to reference teletypewriter modems in the U.S. Army inventory.

**telegraphy** Data transmission technique characterized by data rate of 75 bps where the direction, or polarity, of DC current flow is reversed to indicate bit states.

**Telemail**   An electronic mail service provided by Telenet Communictions Corp. of Reston, VA.

**Telemail Local Community (TM-LC)**   A group of subsystems responsible for accessing Telemail and delivering messages to local users.

**Telemail Message Transfer Agent (TM-MTA)**   A subsystem that handles connections with other Telemail systems. Also referred to as the Interconnect MTA.

**telemarketing**   A marketing system which combines telecommunications technology with management information systems for planned, controlled sales and service programs. Used effectively by small and large businesses to accomplish specific marketing goals.

**Telemessage**   The British Telecom Internation modern "Telegram" service that operates in the UK and the United States.

**telemetry**   The use of telecommunications for automatically indicating or recording measurements at a distance from the measuring instrument.

**Telemetry Asynchronous Block Serial Protocol (TABS)**   AT&T's proprietary protocol for the Link Data Channel (LDC) on Extended Superframe Format (ESF) circuits; used to convey statistical information on signal quality and hardware failure rates (retrieves ESF data from customer service units (CSUs)).

**Telenet**   Value Added Network service proved by GTE Telenet Corporation.

**Telenet Access Management System (TAMS)**   A network-level security system that screens virtual calls being placed in the network.

**Telenet Central Office (TCO)**   A location where Telenet network equipment is installed. There are four "classes" or levels of TCOs. Class I TCOs are hubs that are part of the network backbone. Class II TCOs are those TCOs that must, according to the rules of network architecture, be connected to Class I TCOs. Class III TCOs are large enough to contain several pieces of equipment; however, these TCOs in most cases must be connected to a Class II TCO to be connected to the network. Class IV TCOs are small, asynchronous-only locations having only a single equipment cabinet.

**Telenet Central Office, Class I (Class I TCO)**   The backbone hub site that connects to other Class I sites. This site also contains Class II and Class III types of equipment and services.

**Telenet Central Cffice, Class II (Class II TCO)**   An asynchronous/synchronous office. It connects to Class I and other Class II TCOs. It services Class III TCOs by providing access to the backbone.

**Telenet Central Office, Class III (Class III TCO)**   An asynchronous site only. It connects to a Class II TCO for access to the backbone.

**Telenet Central Office, Class IV (Class IV TCO)**   An asynchronous site only. This is a single-cabinet site that replaces the normal Foreign Exchange (FX) service when FX costs or services prohibit the use of the FX services.

**Telenet Diagnostic Tool 2 (TDT2)**   A program used in the Network Control Center (NCC) to remotely diagnose and correct problems in Telenet Processor 3000s (TP3s) and Telenet Processor 4000s (TP4s).

**Telenet Internal Network Protocol (TINP)**   A virtual circuit-based, proprietary, backbone, network protocol. TINP is a superset of the CCITT X.75 gateway protocol.

**Telenet Processor (TP)**   A TP is a data communications processor that interfaces terminals and host computers to the Telenet network. Generally, no changes to the customer's software or hardware are needed. The TP is available in different models designed to support various user requirements. The TP may be used as a terminal concentrator. Ports are provided by the TP to allow customer terminals access to the network. These access ports may be directly cable-connected to nearby customer or authorized user terminals, or they may be connected by leased or dial-in communications channels to distant terminals. TP4s are used as packet switches within the network to set up and manage virtual calls.

**Telenet Processor, 3000-series (TP3)**  A series of microprocessor-based concentrators in a Telenet Public Data Network (PDN). The TP3 is used for data transmission among hosts, terminals, and other TP devices in the network.

**Telenet Processor, 4000-series (TP4)**  A series of multi-microprocessor-based packet switches and concentrators used for data transmission among hosts, terminals, and other TP devices in the network.

**Telenet Processor, 5000-series (TP5)**  A Prime mini-computer-based Network Management System (NMS) located in the Network Control Center (NCC) and connected to a Telenet network as the X.25-interface host processor. The TP5 is used to monitor and control network facilities and usage.

**Telenet Processor Operating System (TPOS)**  TPOS controls process scheduling, buffer allocation, and intercard communications. The software also contains a set of subroutines for queue management and provides debug port and buffer management. Each card on a Telenet Processor 4000 (TP4) has its own TPOS.

**Telenet Processor Reporting Facility (TRPF)**  A software program in the Network Control Center (NCC) that receives messages regarding alarms or events sent by TPs. If the messages indicate problems, the Telenet Diagnostic Tool 2 (TDT2) program is used to rectify them.

**Telenet Users Association (TELUS)**  An association of Telenet value-added carrier users sponsored by Telenet of Reston, VA. TELUS provides Telenet customers a mechanism to exchange ideas as well as to provide feedback to the carrier's management.

**TELEOX**  Telecommunications Exchange.

**telephone**  A component of a voice transmission system which converts voice power into electrical power by the use of a diaphragm which makes the resistance of carbon granules vary at the same frequency as the sound wave. The change in resistance is used to vary the current flow on a transmission line.

**Telephone Administration**  An organization run by a government agency or a private business to operate a public switched network.

**telephone exchange**  A central office.

**telephone frequency**  The frequency range between 300 and 3300 Hz where audible voice quality is at a reasonable level for general telephony.

**telephone transmitter**  The portion of a telephone which converts sound waves into electric current which varies by waveform and frequency to changes in the sound waves.

**telephony**  A general term for voice telecommunications.

**Telephony User Part (TUP)**  The higher layer protocol in Common Channel Signalling System No. 7 that deals with end-user signalling for voice telephony.

**Teleplan**  An AT&T sponsored marketing agreement with hotels located outside of the United States which reduces the surcharge on calls to the US.

**Telepoint**  A mobile telephone service in the United Kingdom technically known as CT2.

**teleport**  A telecommunications service wholesaler, usually involved in local loop bypass, who serves a particular geographical region with specific classes of service, e.g., satellite access.

**teleprinter**  A terminal without a CRT that consists of a keyboard and a printer.

**Teleprinter Exchange Service (TELEX)**  A network of teleprinters connected over an international public switched network; uses Baudot code.

**teleprocessing** A form of information handling in which a data processing system utilizes communication facilities. (Originally, but no longer, an IBM trademark). Synonymous with data communications.

**Teleprocessing Access Method (TPAM)** Access method using a monitor specially made to interface teleprocessing devices. Examples include BTAM, TCAM, and VTAM.

**Teleset** A trademark of Aspect Telecommunications of San Jose, CA, as well as a proprietary digital telephone designed by that company.

**Telespazio** A subsidiary of IRI-STeT, the Italian state-owned holdings company under the control of the Ministry of State Participation. Telespazio is responsible for satellite ground stations in Italy.

**Teletel** The French videotext service.

**Teletex** 1. The new CCITT standard for text and message communications which is intended to replace ASCII Telex. Teletex operates at a high speed (2400 bps), can accommodate uppercase and lowercase characters, and has a well-defined format for transmission and presentation of text. 2. Two-way textual communication service.

**teletext** A system for the one-way transmission of graphics and text for display on subscriber television sets. More limited than two-way videotex.

**teletraining** Using the interactive audio and/or graphics and video capability of the telephone network to provide remote interactive training. Businesses use teletraining instead of sending employees to a training site; similarly, schools and universities use teletraining to allow handicapped or seriously ill students to participate in classes.

**Teletype** Trademark of Teletype Corporation. Commonly used to refer to one of their series of teleprinters or to compatible devices manufactured by other vendors.

**teletype grade** The lowest type of communications circuit, in terms of speed, cost, and accuracy. The term is used to establish a distinction between this type of service and voice-grade service. Teletype grade is also called teleprinter grade.

**teletypewriter** A generic term for a start–stop signalling device that consists of a keyboard transmitter and printing receiver.

**Teletypewriter Exchange Service (TWX)** An AT&T public switched teletypewriter service in which suitably arranged teletypewriter stations are provided with lines to a central office for access to other such stations throughout the United States and Canada. Both Baudot and ASCII coded machines are used. Business machines may also be used, with certain restrictions.

**Televerket** Sweden's national communications carrier (PT&T).

**Television Receive Only (TVRO)** A term used to refer to earth stations that have the capability to only receive satellite transmissions.

**Telex** The public switched low-speed (telegraph) data network which is used worldwide for the transmission of administrative messages. Uses Baudot code; however, numerous code conversion facilities are available for sending data on the Telex network.

**Telex network** Same as Telex.

**Telex Plus** A telex service offered by British Telecom which allows subscribers to send a message to up to 100 addresses from a single call from their machine.

**Telmex** The Mexican government majority-owned communications company responsible for providing voice telephone service to towns with a population of 2500 or more inhabitants as well as some data services on a non-exclusive basis.

**Telnet** An application level protocol in the TCP/IP protocol suite that allows a terminal on one computer to pass through to another computer on a local or remote network. Under Telnet the terminal appears as a local device to the other computer. The format of the Telnet command is: "telnet address.domain" or "telnet address.domain port#".

433

The term Telnet comes from teletype network and refers to the predominant type of terminal used in the early days of computing. There are two sides to the Telnet protocol: one for the user or client and one for the server or remote host. Telnet allows either the client or the server to request or respond to option settings that are related to the pass-through session—for example, a choice of character set (i.e. ASCII or EBCDIC). Most Telnet sessions operate in ASCII full-duplex mode.

**Telnet port** The port address on a computer which supports remote Telnet access. Normally port 23 is the default Telnet port.

**Telpak** The name given to a now obsolete pricing arrangement by AT&T in which many voice-grade telephone lines were leased as a group between two points.

**TEL-SENSE** A trademark of Tel Electronics, Inc., of American Fork, UT, as well as a series of telephone call accounting and management systems marketed by that vendor.

**TELUS** TELenet USers association.

**TEMA** Telecommunications Equipment Manufacturers Association, UK.

**template** In Digital Equipment Corporation Network Architecture (DECnet), a named collection of module-specific parameters which can be referenced by a client of the module without knowing their individual significance.

**temporary error** 1. A resource failure that can be resolved by error recovery programs. 2. IBM's NPDA, the resource failure that can be resolved by error recovery programs. Synonymous with performance error.

**Temporary Text Delay (TTD)** The TTD control sequence (STX ENQ) transmitted by a sending station when it wants to retain the line but is not ready to transmit.

**TER** Appended to a CCITT (now ITU) standard, it identifies a third version of the standard.

**terminal** A device for sending and/or receiving data on a communication channel. A wide variety of terminal devices have been built, including teleprinters, special keyboards, light displays, cathode tubes, personal computers, telephones, etc.

**terminal access facility** In IBM's NCCF, a facility that allows network operators to control a number of subsystems. In a full-screen or operator control session, operators can control any combination of such subsystems simultaneously.

**terminal adapter** A key element in Integrated Services Digital Network (ISDN) which permits non-ISDN terminals to be connected to an ISDN network.

**terminal cluster** A group of terminals, usually geographically colocated and controlled by a single unit (cluster controller).

**terminal component** A separately addressable part of a terminal that performs an input or output function, such as the display component of a keyboard–display device or a printer component of a keyboard–printer device.

**Terminal Control Address Space (TCAS)** In IBM's SNA, the part of TSO/VTAM that provides logon services for TSO/VTAM users.

**terminal control unit** Same as cluster control unit.

**Terminal Descriptor Block (TDB)** A list of the port parameters residing in software that describes the terminal with which the computer communicates.

**terminal emulation** In protocol testing, a technique in which the protocol analyzer terminates a circuit and performs the role of that circuit's normal terminal device.

**terminal emulator** A program enabling a personal computer to imitate another computer system and execute the programs written for that other system as though they were written for the personal computer.

**Terminal Equipment (TE)** 1. In ISDN the subscriber side of the S reference point. 2. CCITT terminology for ISDN access: TE1 type 1 = ISDN compliant; TE2 type 2 = ISDN non-compliant.

**Terminal Equipment Type 1 (TE1)** An ISDN terminal.

**Terminal Equipment Type 2 (TE2)** A terminal incompatible with ISDN recommendations. A TE2 requires the use of a terminal adapter (TA) located at the R reference point to operate on ISDN.

**terminal handler** A part of a data communication network which serves simple, character stream terminals. It has other names such as terminal processor, terminal interface processor (TIP) in the ARPA network and packet assembler and disassembler (PAD) in public packet networks.

**Terminal Interface Equipment (TIE)** An interface device that provides "auto-dial/auto-answer modem" functions and supports CSU, DSU and call setup capabilities.

**terminal multiplexer** A device such as the IBM 3299 Terminal Multiplexer which interleaves signals from many devices onto a single coaxial cable.

**Terminal Multiplexer Adapter (TMA)** An IBM adapter connected to the terminal adapter in the 3174 control unit and which provides control for a maximum of eight terminals.

**terminal node** In IBM's SNA, a peripheral node that is not user-programmable, having less intelligence and processing capability than a cluster controller node.

**terminal polling** Same as polling.

**terminal processor** In a packet switching network it is convenient to treat terminals like other processors needing communication. To this end, each terminal has a process looking after it. This can be regarded as residing in a terminal processor (a separate processor is not essential—it could be part of an interface computer, for example).

**terminal server** In LAN technology, a device that allows one or more terminals or other devices to connect to an Ethernet system.

**terminal station** A station within a tape relay network, which performs the same functions as a tributary station: receipt and processing of originating and terminating messages for transmission, delivery, or refile. The primary function of a terminal station is to service its associated relay.

**terminal/tributary** Synonymous terms to describe a communications facility in Automatic Digital Network that is capable of sending and receiving message, using communications equipment and cryptographic equipment.

**terminate** In IBM's SNA, a request unit that is sent by an LU to its SSCP to cause the SSCP to start a procedure to end one or more designated LU–LU sessions.

**terminated line** A circuit with resistance at the far end equal to the characteristic impedance of the line so no reflections or standing waves are present when a signal is entered at the near end.

**termination** 1. Placement of a connector on a cable. 2. An item that is connected to the terminal of a circuit of equipment. 3. An impedence connected to the end of a circuit being tested. 4. The points on a switching network to which a trunk or line may be attached.

**terminator** A resistor used at both ends of a local area network cable to ensure that signals do not reflect back and cause errors. A terminator is usually attached to an electrical ground at one end.

**ternary** Having three possible values. There are ternary number representations using, for example, the digits 0, 1, 2 and there are ternary signals which nominally take three possible values, for example +1 volt, 0, −1 volt.

**terrestrial** A term commonly used to define long-distance transmission that uses earth-bound transmission facilities as opposed to satellite transmission.

**terrestrial interference** Interruptions in a satellite signal caused by high-power land-based microwave links in the 4 GHz band.

**test center** A facility to detect and diagnose faults and problems with communications lines and equipment. Also network control center.

**Test Interface Module (TIM)** The component of Atlantic Research Corporation's Interview 7000 Series protocol analyzers that provides the physical interface to the circuit under test. Test interface modules are available for RS-232-C, V.24, and T1. The T1 TIM provides physical access to ISDN primary rate circuits.

**test mode** A condition of a modem or DSU in which its transmitter and receiver are inoperative due to a test in progress on the line.

**text** The part of a message between a start of text sequence and an end of text sequence; it includes the data to be processed.

**Text Direct** A service offered by British Telecom which enables customers to send and receive Telexes without a dedicated Telex terminal.

**text editor** Program to facilitate user preparation of text at a terminal.

**Text Search Service** A service of Dow Jones News/Retrieval Service which provides on-line full-text coverage of the *Wall Street Journal*, the *Washington Post*, and several other publications.

**TE1** Terminal Equipment Type 1.

**TE2** Terminal Equipment Type 2.

**TE-820** Trademark of Tekelec, Inc. A T1 framing tester, designed to test central office trunks.

**TFTP** Trivial File Transfer Protocol.

**TG** Transmission Group (IBM's SNA).

**TGID** Transmission Group IDentifier (IBM's SNA).

**TH** Transmission Header.

**The Real-time Operating Nucleus (TRON)** A Japanese project to develop an original computer operating system that will become a world standard.

**thermal noise** A type of electromagnetic noise in conductors or in electronic circuitry which is proportional to temperature. Also Gaussian noise.

**thermal printer** A printer that uses heated wires to melt ink on a ribbon and deposit it on paper to form characters and graphics.

**Thick Ethernet** 1. Standard Ethernet cabling 0.5 inches in diameter that is yellow-coated and considerably heavier than Thin Ethernet. 2. A cabling system using large-diameter, relatively stiff cable to connect transceivers. The transceivers connect to the nodes through flexible multiwire cable.

**Thin Ethernet** A lighter, black-coated variation of Ethernet cable that is 0.2 inches in diameter. This cable is specified under the IEE 802.3 standard and is used to save cable and installation costs but is restricted in effective distance. Also known informally as "Cheapernet."

**thin-film waveguide** A film of transparent material that provides a path about one ten-thousandth of an inch thick for light waves. Some of the uses for integrated optical circuits include transmission, switching, and filtering.

**thrashing** Unfortunate state in which overhead is consuming so much of a processor's attention that no useful work is being accomplished.

**thread** A chain of connected messages.

**three out 24** An industry-standard stress test for T1 lines where at least three characters out of a 24-character byte are set high (binary one). The sequence of seven zeros followed by a one stresses the circuit.

**three-way calling** A PBX or central office switching system feature permitting a user to add a third party line without the assistance of an attendant.

**threshold** 1. In IBM's Network Problem Determination Application (NPDA), refers to a percentage value set for a resource and compared to a calculated error-to-traffic ratio for the purpose of generating a performance event record. 2. In the IBM Token-Ring Bridge Program, refers to a value set for the number of frames per 10 000 that can be lost before an entry is made in the bridge performance statistics, and a notification is sent to any network manager program that has requested such reports.

**Threshold Analysis and Remote Access (TARA)** In IBM's SNA, the NPDA feature that can notify a central operator about network problems and errors. It provides remote control of IBM 3600 and 4700 controllers and can record, analyze, and display performance and status data on IBM 3600 and 4700 Finance Communications Systems.

**through channels** The term given to channels that are routed through a multiplexer, i.e., from an aggregate to another aggregate without being de-multiplexed.

**throughput** A measurement of processing or handling ability which measures the amount of data accepted as input and processed as output by a device, link, network, or a system.

**throughput delay** The length of time required to accept input and transmit it as output.

**thruster** A small rocket motor.

**THS** Tactical Hybrid Switch.

**TH-5/TG** A U.S. military telegraph terminal which converts neutral direct current signals from a tele-typewriter to frequency-shift telegraph signals, and frequency-shift telegraph signals to neutral direct current signals for the teletypewriter to print.

**TH-22/TG** A U.S. military lightweight, frequency-shift keying telegraph terminal.

**TI** Transmission Identification.

**TICC** Terminal-Initiated asynchronous Channel Configuration.

**TicketTaker** A BellSouth Corporation service which assigns a special local telephone number to a specific pay-per-view cable television program, permitting callers to have their orders transferred to a cable television computer.

**TIE** 1. Terminal Interface Equipment. 2. Time Internal Error.

**tie line/tie trunk** A private line communications channel provided to link two or more switchboards or PBX systems. Interconnection may be on a manual or dial access basis. A telephone private line directly connecting two Private Branch Exchanges (PBXs) through the telephone network. A configuration whereby two or more tie trunks are connected together is known as a Tandem Tie Trunk Network (TTTN).

**tightly coupled** A term used to describe the inter-relationship of processing units that share real storage and are controlled by the same control program.

**tilt** The difference between insertion losses. Signals at higher frequencies weaken more dramatically than signals at lower frequencies.

**TIM** Test Interface Module.

**timbre of sound** A term used to express the quality of a particular sound which helps to identify the object, instrument, or person that is its source.

**Time Assignment Speech Interpolation (TASI)** A technique for making trunk circuits more efficient by combining portions of conversations on the same circuit. This technique makes use of the fact that typical conversations have quiet periods during which the circuit can be used for other conversations.

**time bomb** A destructive program that remains until a pre defined event or time triggers its operation.

**time call** A call between two subscribers, where the called subscriber is within the caller's area code but outside the caller's free calling area.

**Time Compression Multiplexing (TCM)** A digital transmission technique that permits full-duplex data transmission by sending compressed bursts of data in an alternating or "ping-pong" fashion.

**Time Division Multiple Access (TDMA)** Refers to a form of multiple access where a single carrier is time shared by many users. Signals from earth stations reaching the satellite consecutively are processed in time segments without overlapping.

**time division multiplexers (TDMs)**  100 percent digital, dividing high-speed digital channels, such as a modem RS-232C output, into digital subchannels. TDMs either bit-interleave or character-interleave data.

**Bit-interleaving** is used primarily for synchronous multiplexing of protocols (Bisync, HDLC, X.25, SDLC, etc.) Bit-interleaving maintains the order and number of bits from input at one end of the channel to output at the other end. Synchronous protocols require maintaining the number and order of bits in each clock to ensure correct calculation of the block check characters at the receiving ends, and to ensure that the receiving end divides the receiving data into correct 8-bit bytes.

**Character-interleaving** is used primarily for asynchronous data. Asynchronous data is packaged in start bits (0s) and stop bits (1s). This makes it easy to distinguish each character and place it in a slot. Character-interleaved multiplexers strip start and stop bits, achieving about 20 percent increased efficiency. The extra space is used for more channels or to pass RS-232C control signals.

For instance, the RS-232C Request to Send control may be passed to the other end as Data Carrier Detect, this simulates the switched carrier modem function for one or more low-speed channels.

**time division switching**  Switching method for a TDM channel requiring the shifting of data from one slot to another in the TDM frame. The slot in question may carry a bit or byte (or, in principle, any other unit of data).

**Time Domain Reflectometer (TDR)**  A test device used to place a pulse of energy on a cable and connect the cable to a display where an electron beam moves from left to right as a function of time. By measuring the position of the beam deflection, one can determine the location of impedance discontinuity which indicates a cable fault.

**Time Interval Error (TIE)**  A measure of the phase variation of a given signal with respect to an ideal timing source over a defined observation interval. The TIE provides an indication of the magnitude and direction of a signal's drift over a defined time interval.

**Time of Availability (TOA)**  In U.S. Army terminology, that time when an incoming message has

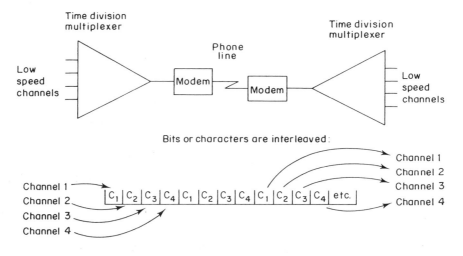

*time division multiplexers (TDMs)*

been completely processed and edited and is now available at the message center desk for the AG section.

**time of day reconfiguration** The ability of an unattended multiplexer to automatically install another configuration in place of an existing configuration at a predetermined time, either daily or on a specific date. This feature is often employed to optimize the multiplexer bandwidth use. For example, a daytime configuration that consists primarily of voice communication being replaced by a configuration that allows a high-speed CPU to CPU transfer at night.

**Time of Delivery (TOD)** The date and time at which a message is delivered to an addressee.

**Time of Receipt (TOR)** The date and time at which a communications center completes reception of a message transmitted to it by another communications center.

**Time Ready for Transmission (TRT)** In U.S. military communications, that time when the poking operator has finished preparing the message in tape form and has completed page copy of the message.

**time sharing** The sharing of an equipment between several processes by giving the processes access to the equipment in turn, i.e. sharing out its time. Usually applied to processor time, but TDM is also a form of time sharing.

**Time Sharing Option (TSO)** An IBM Host application.

**Time Sharing Option for VTAM (TSO/VTAM)** An optional configuration of the operating system that provides conversational time sharing from remote stations in a network using VTAM.

**time slot** 1. In LAN technology, an assigned period of time or an assigned position in a sequence. 2. In multiplexer technology, the reserved time in the data stream for a specific device connected to a time division multiplexer.

**time-derived channel** Any of the channels obtained from multiplexing a channel by time division.

**time-hopping** A spread spectrum communications technique in which a code sequence is used to control the transmission time and period. By making the time and duration of transmission as random as possible, access is denied to an unintended receiver.

**timeout** 1. The expiration of a predefined interval which then triggers some action—such as a disconnection that occurs following 30 seconds without any data activity (in a 30-second, no-activity timeout). 2. The length or existence of such an interval.

**TimePac** The name of a series of packet assembler/disassemblers (PADS) manufactured by Timeplex Inc.

**timesharing** A method of computer operation that allows several interactive terminals to use a computer and its facilities; although the terminals are actually served in sequence, the high speed of the computer makes it appear as if all terminals were being served simultaneously.

**TimeView** A network management system marketed by ASCOM Timeplex, Inc., for T1 multiplexer and packet switch networks.

**TIMS** Telecoms Integrated Management System.

**TIMS** Transmission Impairment Measuring Set.

**TINP** Telenet Internal Network Protocol.

**T-interface** In ISDN, the 4-wire physical interface at the T-reference point, between NT1 and NT2. In configurations where NT1 and NT2 are parts of the same physical device, the T-reference point resides inside the hardware, and there is no T-interface. According to current ISDN standards, the T- and S-interfaces are physically and logically identical. This interface can only be about 1 kilometer long.

**tip** The ground wire in a traditional telephone circuit.

**TIP** Terminal Interface Processor.

**tip and ring** Names for the two conductors in a conventional two-wire local loop.

**TIRKS** Trunks Integrated Record Keeping System.

**title page** The first page of a multipage display.

**titled display** A video display in which windows are drawn side-by-side and not overlapped.

**TLP** 1. Telefones de Losboa de Portugal. 2. Transmission Level Point.

**TL-1** Transaction Language-1.

**TMA** 1. Telecommunications Managers Association, UK. 2. Terminal Multiplexer Adapter.

**TM-LC** TeleMail Local Community.

**TMS** Transmission Message Unit.

**TMS380** A chip set designed and produced by Texas Instruments in conjunction with IBM which provides non-IBM Token-Ring adapters media-access level compatibility with IBM Token-Ring adapters.

**TNC** A threaded connector for miniature coax; TNC is said to be short for threaded–Neill–Concelman.

**tn3270** A version of the Telnet program which supports IBM 3270 terminal emulation.

**TNS** Transaction Network Service.

**TOA** Time Of Availability.

**TOD** Time Of Delivery.

**TOF** Time Of Filing.

**toggle** Activation or deactivation of a function or mode.

**token** In LAN technology, a packet (or part of a packet) used in explicit access LANs; the station that "owns" the token is the station that controls the transmission medium.

**token bus, token-passing bus** In LAN technology, a bus topology LAN that uses a token for explicit access.

Physical layout     Logical flow of control

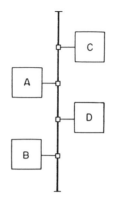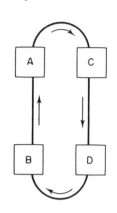

*token bus, token-passing bus*

**token passing** A method of controlling traffic on local area networks. A special message called a "token," or control packet is transmitted from node to node. When a node receives the token, it can transmit messages before transmitting the token on the next node.

**token ring** A local area network topology in which a control packet or token is passed from station to station in sequential order. Stations wishing access must wait for the token before transmitting data. In a token ring, the next logical station receiving the token is also the next physical station, as opposed to token bus.

**token-ring repeater** A device that extends the allowable distance between IBM Multistation Access Units on an IBM Token-Ring Network wired with the IBM Cabling System data grade media.

**TokenTalk NB** An Apple Computer Company token-ring card which enables Macintosh computers to be connected to a token-ring network. The NB stands for the NuBus the TokenTalk card is inserted into.

**toll cable** A cable whose outermost covering is made of lead. In general, a toll cable is used in permanent installations for long-distance transmission and may be either strung overhead on poles or installed underground.

**toll call**   A call outside the local exchange area.

**toll center**   A class 4 central office where channels and toll message circuits terminate. While this is usually one particular central office in a city, larger cities may have several offices where toll message circuits terminate. Also called "toll office" and "toll point."

**toll circuit**   (*American*) Same as trunk circuit (*British*). A circuit connecting two exchanges in different localities.

**toll restriction**   A PBX or central office switch feature that does not allow particular stations to make calls outside the local area.

**toll-free**   Any of a variety of services which use 800 in place of an area code. Calls made to 800 numbers are toll-free to the caller and are paid for by the receiver, usually at a lesser cost than normal long distance rates or collect calls. Most companies who wish to encourage customers to call them to place orders, to inquire for further information, or other customer services, use 800 services.

**toll-free directory assistance**   In the United States, you can dial 1-800-555-1212 to obtain the telephone number of organizations providing toll-free telephone service.

**toll office**   A switching office where trunks are interconnected to serve toll calls. Toll offices are in a hierarchical structure as follows:

| | |
|---|---|
| Regional center | Class 1 |
| Sectional center | Class 2 |
| Primary center | Class 3 |
| Toll center | Class 4 |

**toll service**   A telephone service across central office boundaries, or otherwise as provided in a relevant tariff, such that call duration and distance determine the cost billed to the subscriber.

**toll switching trunk**   (*American*) Same as trunk junction (*British*), which see.

**tone dialing**   Pushbutton dialing.

**tone modulation**   A type of trnsmission obtained by causing the radio frequency (RF) carrier amplitude to vary at a fixed audio frequency (AF) rate. When this type of transmission is keyed, it becomes modulated continuous wave (MCW).

**toner**   A black, powdered ink used in laser printers and cetain types of duplication equipment. Typically, an electrical charge is placed on a drum and the toner is attracted to the charge and transferred to the surface of the paper where it is melted or fused in place.

**TOP**   1. A network design tool of DMW Commercial Systems of Ann Arbor, MI. 2. Technical and Office Protocols.

**topology**   The shape of the arrangement of network components and of the interconnections between them. Common network topologies include linear bus, multiple bus, a circular ring, and a star.

**topology of networks**   The layout of the nodes (switches, concentrators) and lines of a network. The strict meaning refers to their pattern of connection but the word is used to include distance and geography.

**TOR**   Time Of Receipt.

**torn-tape switching center**   A location where operators tear off incoming punched paper tape and transfer it manually to a tape reader connected to the proper outgoing circuit.

**Tosser**   A program which takes incoming mail from another bulletin board system and tosses incoming messages into appropriate message areas in the BBS message bases.

**TOT**   Time Of Transmission.

**Touch-Call**   Proprietary term of GT&E used to refer to pushbutton dialing.

**Touch Tone**   A trademark of AT&T, referring to tone or pushbutton, rather than pulse or rotary, dialing.

**tower**   A protocol sequence along with associated address and protocol-specific information.

**TP**   Telenet Processor.

**TPAM**   TeleProcessing Access Method.

**T-Pause**   The data flow control mechanism used by Tandem Non-Stop Computer Systems designed to eliminate the chance of data loss due to buffer overflow.

**TPDU**   Transport layer Protocol Data Unit.

**TPOS**   Telenet Processor Operating System.

**TPRF**   Telenet Processor Reporting Facility.

**trace**   To record data that provides a history of events occurring in the system.

**Trace Analysis Program (TAP)**   In IBM's SNA, an SSP program service aid that assists in analyzing trace data produced by VTAM, TCAM, and NCP and provides network data traffic and network error reports.

**trace packet**   A special kind of packet in the ARPA network which functions as a normal packet but because its "trace" bit is set causes a report of each stage of its progress to be sent to the network control center.

**tracking**   Using earth-based equipment to follow a satellite's position.

**Tracking, Telemetry and Control (TT&C)**   Refers to satellite control stations used to monitor onboard satellite operations and to direct satellite electronics and rocketry equipment.

**Tracking and Data Relay Satellite (TDRS) Program**   A system of satellites used by NASA to communicate with other satellites in low-earth orbit. The TDRS program replaces a network of 23 ground stations.

**traffic**   The volume and intensity of transmitted and received messages over a communications facility.

**traffic analysis**   The statistical study of transmissions and the correlation of that study to provide information on volume, types, directional flow, length of messages, peak and low periods of trafffic.

**traffic capacity**   The maximum traffic per unit of time that a telecommunications system, subsystem, or device can carry under specified conditions.

**traffic control**   A means to stop and restart data transmission without losing data.

**traffic engineering**   The science of designing communications facilities to meet user requirements.

**traffic flow security**   The transmission of an uninterrupted flow of random text on a wire or radio circuit, with no indication to an interceptor as to when traffic is being passed.

**traffic matrix**   A matrix of which the $(i,j)$ element contains the amount of traffic originated at node $i$ and destined for node $j$. The unit of measurement could be calls, or packets per second, for example, depending on the kind of network.

**Trailblazer**   1. A modem manufactured by Telebit Corporation that can operate at data rates up to 19.2 Kbps over the switched telephone network. 2. A U.S. military ground-based communications intercept and direction finder primarily used for supporting army troops in a tactical environment.

**trailer or trace block**   Control information which is transmitted after the text of a message. It is used for tracing error events, timing, and recovering blocks after system failures.

**train**   The adjusting to the conditions of a particular line segment by a receiving modem. These conditions include amplitude response, delay distortion, and timing recovery.

**training**   The process in which a receiving modem achieves equalization with a transmitting modem.

**training pattern**   The sequence of signals used in training.

**training time**   The time that the modem with automatic equalizer uses to adjust its equalization parameters. Also learning time.

**transaction**   A message directed to an application program. In batch or remote job entry (RJE), a job or job step.

**Transaction Language-1 (TL-1)**  A Bellcore and ANSI T1-committee developed protocol that allows network management devices to talk with each other.

**Transaction Network Service (TNS)**  A common-user switched network service.

**transaction processing**  A real-time of data processing in which individual tasks or items of data (transactions) are processed as they occur—with no primary editing or sorting.

**transaction set**  A standardized electronic message format used in Electronic Data Interchange (EDI) communications. Tranaction sets specify formats for business documents to include purchase orders, invoices and bills of lading.

**Trans-Atlantic-Telecommunications (TAT)**  A series of cables ranging from copper to fiber medium that link North America to Europe. The following table lists current transatlantic cables.

| Cable | Year of operation | Medium |
|-------|-------------------|--------|
| TAT-1 | 1956 | copper |
| TAT-2 | 1959 | copper |
| TAT-3 | 1963 | copper |
| TAT-4 | 1965 | copper |
| TAT-5 | 1970 | copper |
| TAT-6 | 1976 | copper |
| TAT-7 | 1983 | copper |
| TAT-8 | 1989 | fiber |

**transceiver**  A generic term for a single device that combines the function of a TRANSmitter and a reCEIVER.

**Transcode**  An early six-bit transmission code used with an airline reservation system version of IBM's Binary Synchronous Communications protocol.

**Transcom**  The French switched 64 Kbps service.

**transducer**  A device that converts signals from one form to another.

**transfer orbit**  An intermediate elliptical orbit used to reach geosynchronous orbit, where the apogee is the same altitude as the final operating orbit.

**Transfix**  A service offered by France Telecom which provides digital leased lines operating at multiples of 64 Kbps to end users.

**transients**  Intermittent, short-duration signal impairments.

**Transistor–Transistor Logic (TTL)**  A common set of electrical characteristics used between various levels of integrated circuits, and, occasionally, as the interface between terminals and transmission equipment (e.g., modems).

**transit exchange**  European version of tandem exchange.

**transit network**  The highest level of a switched network. The transit network is well provided with links between its switching centers so that it rarely needs intermediate switching.

**Transit Network Identifier Code (TNIC)**  The code that identifies a transmitted country network in international calls. This code is used for accounting purposes.

**Transit switch**  In Telenet Processor 4000 (TP4), used for routing and switching functions over trunk lines using the Telenet Internal Network Protocol (TINP). A TP4 transit switch cannot be equipped with Packet Assembler/Disassembler (PAD) software and does not communicate with X.25/X.75 Data Terminal Equipment (DTEs).

**Translan**  A bridge product of Vitalink Communications Corp, of Mountain View, CA, that can be used to connect separate Ethernet local area networks via 56 Kbps or 1.544 Mbs communications facilities.

**translator**  A device that converts information from one system of representation into equivalent information in another system. In telephone equipment, it is the device that converts dialed digits into call routing information.

**transmission**  The passage of information through a communications medium.

**transmission block** A sequence of continuous data characters or bytes transmitted as a unit, over which a coding procedure is usually applied for synchronization or error control purposes.

**Transmission Control (TC)** Category of control characters intended to control or facilitate transmission of information over telecommunication networks. Samples of TC characters are: acknowledgment (ACK), data link escape (DLE), enquiry (ENQ), end of transmission block (ETB), negative acknowledgment (NAK), start of header (SOH), start of text (STX), and synchronization (SYN).

**transmission control character** Any control character used to control or facilitate transmission of data between data terminal equipment. Synonymous with communication control character.

**Transmission Control (TC) layer** In IBM's SNA, the layer within a half-session that synchronizes and paces session-level data traffic, checks session sequence numbers of requests, and enciphers and deciphers end-user data. Transmission control has two components: the connection point manager and session control.

**Transmission Control Protocol (TCP)** A data transport protocol developed by the Department of Defense (DOD) for the reliable delivery of datagrams from one computer to another. TCP uses checksums, duplicate detection and retransmission of lost segments.

**Transmission Control Protocol/Internet Protocol (TCP/IP)** A set of protocols used by the Internet to support such services as file transfer (FTP), electronic mail transfer (SMTP), and remote log-in (Telnet).

**Transmission Control Unit (TCU)** A control unit (such as an IBM 2703) whose operations are controlled solely by programmed instructions from the computing system to which the unit is attached; no program is stored or executed in the unit.

**transmission facilities** The equipment that a communications common carrier uses to provide a stated type of service. Some examples of the equipment are links, switching centers, and other devices.

**Transmission Group (TG)** In IBM's SNA, a group of links between adjacent subarea nodes, appearing as a single logical link for routing of messages. *Note.* A transmission group may consist of one or more SDLC links (parallel links) or of a single System/370 channel.

**Transmission Group Identifier (TGID)** In IBM's SNA, a set of three values, unique for each transmission group, consisting of the subarea addresses of the two adjacent nodes connected by the transmission group, and the transmission group number (1–255).

**Transmission Header (TH)** In IBM's SNA, control information, optionally followed by a basic information unit (BIU) or a BIU segment, that is created and used by path control to route message units and to control their flow within the network.

**transmission identification** A combination of letters and figures used to identify a transmission on a channel between two stations.

**Transmission Level Point (TLP)** The transmission level of any point in a transmission system is the ratio (in dB) of the power of a signal at that point to the power of the same signal at the reference point. In the direction of transmission, each end of the channel is said to be the 0 dB Transmission Level Point (0TLP). The 0TLP gives the maximum power applicable at this point. All other level points on the overall circuit are referenced to the 0TLP, they are identical in numbers throughout the system. For example, $-13$ dBm0 means that the test tone is 13 dB below the 0 reference when measured anywhere in the circuit. The 0TLP was at one time a point accessible to probes and measuring instruments, but is seldom so today. As a consequence of improving transmission, it is now normal to consider the outgoing side of the toll transmitting switch as $-3$ TLP. Signal magnitudes measured at this point are 3 dB lower than would be measured at the reference level point if such a measurement were possible.

**transmission line** 1. synonym for telecommunication line (IBM's SNA). 2. Any conductor or system of conductors used to carry electrical energy from its source to its load.

**transmission loss** 1. Total loss encountered in transmission through a system. 2. The gradual weakening of energy as a signal moves from a transmitter to a receiver. Loss is measured in decibels (dB) and is corrected when necessary by repeaters in analog systems or regenerative repeaters in digital systems.

**transmission media** Twisted-pair wire, fiber optic cable, microwaves, radio, and broadband or baseband coaxial cable.

**transmission medium** A physical carrier of electrical energy or electromagnetic radiation.

**transmission priority** In IBM's SNA, a rank assigned to a path information unit (PIU) that determines its precedence for being selected by the transmission group control component of path control for forwarding to the next subarea node of the route used by the PIU.

**transmission protocol** A set of rules for the exchange of data over a communications network.

**Transmission Rate of Information Bits (TRIB)** A measure of data communications efficiency or throughput

**Transmission Security** Measures developed to protect transmission, traffic analysis, and imitative deception.

**Transmission Services (TS) Profile** In IBM's SNA, a specification in a session activation request (and optionally, in the responses) of transmission control (TC) protocols (such as session-level pacing and the usage of session-level requests) to be supported by a particular session. Each defined transmission services profile is identified by a number.

**transmission speed** The number of information elements sent per unit time, usually expressed as bits, characters, or words. Preferred expression is bits per second (bps).

**Transmission Subsystem Component (TSC)** In IBM's SNA, the component of VTAM that comprises the transmission control, path control, and data link control layers of SNA.

**Transmission Systems Engineering Evaluation Facility (TSEEF)** A facility operated by the U.S. Army Information Systems Engineering Command used for the evaluation and development of computer and communications networks.

**transmission window** The wavelength at which a fiber lightguide is most transparent.

**transmissive star** A fiber optic transmission system that allows a single input light signal to be transmitted on multiple output fibers. Used primarily in fiber optic local networks.

**transmit flow control** A transmission procedure which controls the rate at which data may be transmitted from one terminal point so that it is equal to the rate at which it can be received by the remote terminal point.

**Transmitted Data (TD)** An RS-232 data signal (sent from DTE to DCE on pin 2).

**transmitter** A device which inserts data into a communications channel.

**Transmitter–Distributor (TD)** The device in a teletypewriter terminal which makes and breaks the line in timed sequence. Usage of the term can refer to a paper tape transmitter.

**Transpac** The French packet switched data network.

**transparency** The ability of a communications system to pass control signals or codes as data to the receiving unit, without affecting the communications system.

**transparent** A mode of transmission in which the transmission medium (receiving program or device) will not recognize control characters or initiate any control function.

**transparent mode** A transmission technique that places no restrictions on the format of user data.

**transponder** In satellite communications, a circuit that receives an up-link signal, translates it to another, higher frequency, amplifies it, and then retransmits it as the down-link signal in a broad wide-area coverage. Satellites generally have up to 24 transponders.

**transport connection** A virtual connection at the Transport layer between two transport service users.

**transport delay** The amount of time it takes to carry information through a network. Its value depends on the propagation delay and the switching method used in the network.

**Transport layer** The fourth layer in the OSI model; ensures error-free, end-to-end delivery.

**transport protocol** The basic level of protocol which is concerned with the transport of messages. The software which carried out this protocol was called a transport station in the Cyclades network and the term has been widely adopted.

**transport service user** A user of the service provided by the Transport layer. For example, session control.

**TransRING** A local area network bridge marketed by Vitalink Communications Corp. of Fremont, CA.

**TranstexT Universal Gateway** A videotext gateway service provided by BellSouth.

**transversal filter** In this filter the input is passed through a delay network and delayed versions of the signal, through suitable attenuators, are added to generate the output. The attenuators must be able to invert the signal.

**transverse parity check** A type of parity error checking performed on a group of bits in a transverse direction for each frame.

**Traveling Wave Tube (TWT)** Refers to the amplifier section of a transponder.

**Traveling Wave Tube Amplifier (TWTA)** Refers to an amplifier using a TWT and power supply.

**tree** A network topology, with only one route between any two network nodes. A network that resembles a branching tree, such as CATV networks.

**tree network** A complex form of bus network in which there are branches in the cable but only a single transmission path between any two stations.

**tree topology** A topology in which there is only one route between any two network nodes.

**trellis coding** A method of forward error correction used in some high-speed modems whereby each signal element (baud) is assigned a coded binary value to represent the element's phase and amplitude. Due to a convoluting coding scheme, the receiving modem can determine if the signal element was received in error and, if so, correct the error.

**trellis encoding** An advanced modulation technique which provides greater throughput and reliable transmission rates for speeds above 9600 bps. With trellis encoding, coding information is added to the traditional modulation scheme to provide a record of successive dependencies between transmitted signal points. By continuously looking backwards, comparing received data with newly presented information, trellis encoding provides a greater tolerance to noise for a given block error rate.

**TRIB** 1. Transmission Rate of Information Bits. 2. TRIButary.

**tributary** A circuit path connecting one or more stations, terminals, or devices to a network backbone, or to a centralized switching system.

**tributary station** A non-control station in a multipoint configuration.

**TRIF** Technology Requirements Industry Forum.

**TrimLink Express** A data compressor with a four-channel synchronous multiplexer marketed by RAD Data Communications Inc. of Rochelle Park, NJ.

**TRINTEX** A joint venture of Sears, Roebuck and Co. and IBM for a videotex service. The name was changed to Prodigy on June 1, 1988.

**TRI-TAC** TRI-service TACtical communications.

**Trivial File Transfer Protocol (TFTP)** The Internet standard protocol for file transfer. This protocol has both a minimal capability and minimal overhead.

**Trojan Horse** A destructive program attached to a file a person innocently loads. When loaded the program may erase disk storage or perform another harmful activity.

**TRON** The Real-time Operating Nucleus.

**tropospheric scatter radio systems** A radio system which provides directional, over-the-horizon radio links in the VHF, UHF and SHF bands (100 MHz to 10 GHz) by detecting signals reflected from the troposphere using large, sensitive antennae.

**Trouble Tracker** A PBX fault management system marketed by AT&T of Basking Ridge, NJ.

**TRT** Time Ready for Transmission.

**TR-TSY-000385** The Bellcore standard definition for billing transmission interface.

**true power** Power in a circuit when the load is purely resistive. This condition occurs when the phase angle between current and voltage is zero, resulting in the product of voltage and current (power) becoming zero.

**trunk** A telephone circuit connecting two or more telephone company (or PTT) central offices. A trunk may be either a high-speed digital circuit or a wideband analog circuit; however, all trunks now being installed are digital, using DS1 or higher rates.

**trunk circuit** (*British*) Same as toll circuit (*American*). A circuit connecting two exchanges in different localities. In Britain, a trunk circuit is approximately 15 miles long or more. A circuit connecting two exchanges less than 15 miles apart is called a junction circuit.

**trunk exchange** A telephone office primarily for switching trunks.

**trunk group** A group of telephone circuits treated as a unit and connecting PBXs, central offices, or other switching devices.

**trunk junction** (*British*), same as toll switching trunk (*American*). A line connecting a trunk exchange to a local exchange and permitting a trunk operator to call a subscriber to establish a trunk call.

**trunking protocol** The protocol, or rules of operation, that apply to the transmission, or trunking, of data across a digital facility.

**trustee** A user granted access to files on a private directory on a local area network using Novell's NetWare.

**TSAT** Very Small Aperture Terminals that handles data at T1 speeds.

**TSC** Transmission Subsystem Component (IBM's SNA).

**TSEEF** Transmission Systems Engineering Evaluation Facility.

**T-Series** A set of CCITT (now ITU) recommendations covering terminal equipment for use in telematic services to include facsimile, teletex and videotext.

T5    General aspects of group 4 facsimile apparatus.

T6    Facsimile coding schemes and coding control functions for group 4 facsimile apparatus.

T35    Procedure for the allocation of CCITT members' codes

T51    Coded character sets for telematic services.

T71    LAPB extended for half-duplex physical level facility.

T72    Terminal capabilities for mixed mode of operation.

T73    Document interchange protocol for the telematic services.

T90    Teletex requirements for interworking the telex service.

T91   Teletex requirements for real-time interworking with the telex service in a packet-switching network environment.

T101   International interworking for videotex services.

**TSNS**   Telecommunications Services Network Support.

**TSO**   Time Sharing Option.

**TSO/VTAM**   Time Sharing Option for VTAM.

**T-span**   A telephone channel through which a T-carrier operates.

**TSR**   Terminate and stay resident program.

**TST**   TeST.

**TS-712/TCC-11**   A U.S. military telephone test set used by repairmen servicing a cable system to test the operation of the unattended repeaters in the system.

**tt**   Teletypewriter.

**T-tap**   A passive line interface used for monitoring data flowing in a circuit.

**TTC&M**   Telemetry, Tracking, Control & Monitoring.

**TTD**   Temporary Text Delay.

**TTF**   Trunk Test Facility.

**TTL**   1. Transistor–Transistor Logic. 2. Transmission Test Line.

**TTNP**   Tactical Telephone Numbering Plan.

**TTY**   1. TeleTYpe. 2. TeleTYpewriter.

**TTY transmission**   Teletypewriter communications; usually asynchronous ASCII data communications.

**TUA**   Telephone Users Association, UK.

**TUCC flat cable**   AT&T trademark for Telephone Under Carpet Cable.

**TUG**   Transtext Universal Gateway.

**tunable laser**   A laser that can be made to vary the frequency of its light.

**tuning**   The process of adjusting a radio circuit so that it resonates at the desired frequency.

**TUP**   Telephony User Part.

**TurboCom**   A data compression program marketed by Datran Corporation of La Crescenta, CA. for use on IBM PC and compatible personal computers. TurboCom can be used to speed modem data transfers up to four times the modem data rate.

**turnaround time**   The actual time required to reverse the direction of transmission from sender to receiver or vice versa when using a half-duplex circuit. Time is required for line propagation effects, modem timing and computer reaction.

**turnkey**   A system sold by a vendor (often an OEM) which is self-contained and used intact by the customer. A computer system that is supplied, installed and sometimes managed by one vendor and can be used by an untrained person.

**turret**   A large multi-button telephone that can directly terminate private line and dial tone circuits. This allows the user almost instant communication and circuit status on a large number of buttons for trading or broker operations, telephone answering positions, command and control applications, and dispatch.

**TVRO**   TV Receive Only.

**TV-Sat-1**   A German Direct Broadcast Satellite launched in November 1987.

**TWA**   Two-Way Alternative.

**twinaxial cable**   A shielded coaxial cable with two center conducting leads.

**twisted pair**   Two insulated copper wires twisted

together. The twists, or "lays," are varied in length to reduce the potential for signal interference between pairs. In cable greater than 25 pairs, the twisted pairs are grouped and bound together in a common cable sheath. Twisted pair cable is the most common type of transmission media.

**TwoWay Access**  An MCI ISDN-equivalent service which allows customers to use the same dedicated access line for inbound and outbound calls.

**Two-Way Alternate (TWA)**  A transmission method in which data can send in both directions on a data link but never at the same time. A synonym for half-duplex operation.

**two-way loss**  The capability of measuring the loss of a circuit in both the transmit and receive directions.

**Two-Way Simultaneous (TWS)**  A transmission method in which data may be sent in both directions on a datalink at the same time. Synonym for full-duplex operation.

**two-wire circuit**  A circuit formed of two conductors, which are insulated from each other, that provide a "go" and "return" channel in the same frequency.

**TWS**  Two-Way Simultaneous.

**TWT**  Traveling Wave Tube.

**TWTA**  Traveling Wave Tube Amplifier.

**TWX**  TeletypeWriter eXchange service.

**TX**  Transmit.

**TxC**  Transmit clock (control signal), an interface timing signal that synchronizes the transfer of Transmit Data (TxD), provided by DCE.

**TXD**  Transmitted Data/Transmit Data (control signal).

**TXE 2**  A UK small local exchange system using electronic common control of a reed relay matrix ·switch.

**TXE 4**  A UK large local exchange system using stored program control of a reed relay matrix switch.

**TymDial X.25**  A synchronous dial-up service into the Tymnet packet switched network. TymDial X.25 provides users with the ability to connect X.25 devices to a central computer via the Tymnet network.

**Tymnet**  Value added network service provided by Tymnet Corporation (McAUTO). The Tymnet network has been acquired by British Telecom plc.

**Tymusa**  An enhanced gateway access service from the Tymnet public data network which permits fast and easy access of U.S.-based host computers from foreign locations, without the administrative burden of obtaining an account from the local PTTs.

**Tymview**  A software interface marketed by the Tymnet public data network which allows Tymnet network equipment to be managed by IBM's host-based Netview network management system.

**Tymvisa**  An enhanced gateway access service marketed by the Tymnet public data network which provides easy access to host computers in foreign countries without requiring the use of long number sequences.

**TYM-X.25 Access**  A Tymnet dial-up service that provides users with the benefits of end-to-end synchronous communication, without the cost of expensive leased lines. Error detection is performed throughout the entire session using the X.25 protocol, and 2400 bps data transmission speeds cut connect-time charges. Personal computers, minicomputers, and Packet Assemblers/Disassemblers (PADs) are provided switched access to applications based on host computers connected to the Tymnet network.

**Tymnet Asynchronous Outdial Service**  A service of Tymnet which provides the means to communicate with asynchronous devices not directly connected to the Tymnet network. Outdial can be used to send information from a central computer to remote asynchronous devices, from a PC to another

PC, or from a terminal or PC to an asynchronous central computer. No customer-dedicated facilities are required; the Outdial service is provided through public ports in selected Outdial nodes throughout the domestic United States.

**Type A Coax** In IBM 3270 systems, a serial transmission protocol operating at 2.35 Mbps which provides for the transfer of data between a 3274 Control Unit and attached display stations or printers.

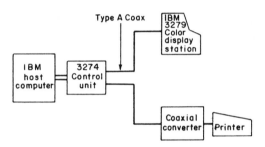

Printer coverter application

**Type 1 cable** In the IBM cabling system Type 1 cable consists of two twisted pairs of number 22 AWG wire, with each pair shielded with a foil wrapping. Both pairs are surrounded by an outer braided shield. Type 1 cable supports the longest transmission distance of all metallic cable types.

**Type 2 cable** In the IBM cabling system Type 2 cable is the same as Type 1 with the addition of four twisted pairs of copper conducted for telephone use. Type 2 cable is available for both non-plenum and plenum installations.

**Type 3 cable** An unshielded twisted-pair wire that meets IBM specifications for use in a token-ring network.

**Type 5 cable** In the IBM cabling system Type 5 cable consists of two 100/140 micron (micrometer) optic fibers. Type 5 cable is available for non-plenum and outdoor applications.

**Type 6 cable** In the IBM cabling system Type 6 cable is shielded, 26 AWG wire containing two twisted

pair (four conductors). Type 6 cable is smaller in diameter than Type 1 cable and is available in non-plenum only.

**Type 8 cable** In the IBM cabling system Type 8 cable is undercarpet cable. This cable contains two individually shielded, parallel cables of 26 AWG, solid, copper conductors.

**Type 9 cable** In the IBM cabling system Type 9 cable is a shielded, two-pair cable whose copper conductors are 26 AWG. Type 9 cable is smaller in diameter than Type 1 cable and is a lower cost alternative to the use of Type 1 cable. According to IBM Type 9 cable permits a transmission distance about two-thirds of that obtainable through the use of Type 1 cable.

**Type 105 test line** A test line in a telephone central office with the ability to allow access to a responding type unit.

**typeface** The printed design of characters. Commonly used typefaces include Courier, Helv, and Letter Gothic.

**T1 (time division multiplexing, level 1)** The original, and still more common, name for DS1 Service. T1 describes a digital, time division multiplexed bit stream of 1.544 Mbps (in North America) or 2.048 Mbps (in Europe).

**T1 Committee** A committee of the American National Standards Institute (ANSI) organized to set U.S. standards for digital telephony, especially ISDN. *Note*. Despite its name, Committee T1 is not chartered to set standards for T1 circuits; this is a point of frequent confusion.

ISDN Activity Working Groups include:

| | |
|---|---|
| T1S1.1 (formerly T1D1.1) | ISDN Architecture and Services |
| T1S1.2 (formerly T1D1.2) | Signaling and Switching Protocols |
| T1S1.3 (formerly T1X1.1) | Common Channel Signalling |
| T1S1.4 (formerly T1X1.2) | Individual Channel Signalling |

**T1 line**   A digital transmission line that carries data at a rate of 1.544 Mbps in North America (DS1 level). In a digital service it is used for short haul links (Intracity Megaroute).

**T1 timer**   In packet switched networks, used to measure timeout intervals in link initialization and data exchanges.

**T-1C**   An American Telephone & Telegraph Company digital facility using the DS-1C format and operating at a speed of 3.152 Mbps.

**T1DM (T1 Data Multiplexer)**   A multiplexer used for time division multiplexing for up to twenty-four 64 Kbps channels with synchronizing information into a DS1 line.

**T2**   An American Telephone & Telegraph Company digital facility using the DS2 format and operating at a speed of 6.312 Mbps.

**T3**   An American Telephone & Telegraph Company digital facility used to transmit a DS-3 formatted digital signal at 44.736 Mbps and above.

**T.4**   The CCITT standard recommendation which specifies a modified Huffman data compression technique for use in facsimile transmission.

**T30BIS**   A new specification drafted by the CCITT to add error-correction techniques to FAX transmission. Under the proposed standard images to be faxed are broken down into data blocks of 64 or 256 bytes. Each block is further divided into smaller frames with a cyclic redundancy check (CRC) character added to each frame.

**T45**   An AT&T service which permits data to be transferred at 45 Mbps (actually 44.736 Mbps).

# U

U   Unprotected.

UA   1. Unnumbered Acknowledgment (response). 2. User Agent.

UART   Universal Asynchronous Receiver/Transmitter.

UC   Up Converter.

UDI   Unrestricted Digital Information.

UDLC   Sperry Data Link Control.

UDLC   Universal Digital Loop Carrier.

UDP   User Datagram Protocol.

UDR   User Destination Routing.

UDVM   Universal Data Voice Multiplexing.

UEJ   Unattended Expendable Jammer.

UFD   User File Directory.

UFGATE   A program which enables a FIDO compatible bulletin board system to exchange UUCP mail with UUCP sites.

UFGATE site   A FidoNet node which runs UFGATE or equivalent software which enables a Fido BBS to exchange UUCP mail with other UUCP sites.

UHF   Ultra High Frequency; ranging from 300 MHz to about 3 GHz, includes television channels 14 through 83, and cellular radio frequencies.

UI frame   An unnumbered information frame in HDLC used to carry data which is not subject to flow control or error recovery.

U-interface (ISDN)   In the United States, the U-Interface is the 2-wire physical interface between the NT1 and the provider's local loop. The U-Interface exists only to satisfy legal requirements that the provider's network be managed separately from customer premises equipment.

UIS   Universal Information Services.

UL   Underwriters Laboratory.

u-law   The coding law used by the U.S. standard 24-channel PCM systems.

ULCS   Unit Level Circuit Switch.

ULSI   Ultra Large-Scale Integration.

Ultimate
A series of high-speed modems manufactured by Paxdata of the United Kingdom.

Ultra Large-Scale Integration (ULSI)   A term used to describe a multifunction semiconductor device with an ultra-high density (over 10 000 circuits) of electronic circuitry contained on a single silicon chip.

Ultra Small Aperture Terminal (USAT)   A satellite earth station that has an antenna less than 1.2 meters in diameter.

**Ultra 96**  A Hayes Microcomputer Products CCITT V.32 full-duplex, 9600 bps dial-up modem which incorporates V.42 bis data compression to obtain data throughput up to 38.4 Kbps.

**ULTRASONICS**  Frequencies above the audible range, normally 20 000 Hz or above.

**UNA**  Universitats-Netz Austria.

**unattended file transfer**  A feature of a communications program which permits the transmission and reception of messages on an unattended basis.

**unattended messaging**  The ability to preprogram transmissions of electronic messages, voice-mail, or file transfers.

**unattended mode**  A term that describes the operation of a device, such as an auto-answer modem, designed to operate without the manual intervention of an operator.

**unattended operations**  The automatic features of a station's operation permit the transmission and reception of messages on an unattended basis.

**unbalanced line**  A transmission line in which the magnitudes of the voltages on the two conductors are not equal with respect to ground; e.g., a coaxial line.

**unbalanced to ground**  (Two-wire); the impedance-to-ground on one wire is measurably different from that of the other.

**unbillables**  Telephone calls for which revenue is not collectable as the "bill to" number cannot be charged for collect or third party calls.

**unbind**  In IBM's SNA, a request to deactivate a session between two logical units (LUs).

**unbundling**  Separation of vendor provided services.

**uncontrolled slip**  The overflow or underflow of a T1 unframed buffer resulting from excessive jitter or wander.

**uncontrolled terminal**  A user terminal that is on line all the time and does not contain control logic for polling or calling.

**undefined characters**  Refers to those bytes in a code which have no character assignments.

**undel message**  A message sent to the originator of an electronic message indicating that the message could not be delivered.

**Underwriters Laboratories (UL)**  A private testing laboratory concerned with electrical and fire hazards of equipment.

**unformatted**  In IBM's VTAM, pertaining to commands (such as LOGON or LOGOFF) entered by an end user and sent by a logical unit in character form. The character-coded command must be in the syntax defined in the user's unformatted system services definition table. Synonymous with character-coded.

**unformatted display**  A screen display in which no attribute character (and, therefore, no display field) has been defined.

**Unformatted System Services (USS)**  In IBM's SNA products, a system services control point (SSCP) facility that translates a character-coded request, such as a logon or logoff request into a field-formatted request for processing by formatted system services and translates field-formatted replies and responses into character-coded requests for processing by a logical unit.

**unframed BERT**  A reference pattern for Bit Error Rate Testing that uses the full bandwidth of the circuit under test without following framing conventions.

**unidirectional**  In one direction only.

**Unified Network Management Architecture (UNMA)**  An AT&T integrated network management system based upon the International Standards Organization's (ISO) evolving Open Systems Interconnection network management standards. UNMA is a three-tiered architecture that serves as a blueprint for future network management products and services.

**Uniform Call Distribution (UCD)**  A PBX or central office switch feature which allows calls coming in on a group of lines to be assigned stations as smoothly as possible so that all stations can handle similar loads. UCD is a form of load balancing.

**uniform numbering plan** Permits station users at a PBX to place calls over tie trunks using a uniform dialing plan. For example, the prefix 8 might access a tie line to Chicago.

**Uninet** A common carrier offering a X.25 PDN that merged with Telenet.

**Uninett** Nordic University Network.

**uninterpreted name** In IBM's SNA, a character string that an SSCP is able to convert into the network name of an LU. *Note*. Typically, an uninterpreted name is used in a logon or initiate request from an SLU to identify the PLU with which the session is requested. The SSCP interprets the name into the network name of the PLU in order to set the session. When the PLU eventually sends a BIND to the SLU, the BIND contains the original uninterpreted name.

**Uninterruptible Power System (UPS)** Consists of a battery package and the necessary power conversion and inversion equipment, switch, gear, etc. The system carries the design load at all times. This system will switch to battery power upon commercial power loss. System will run for 10 to 15 minutes on battery power before system shutdown. If an Emergency Power System (EPS) is present, it will take over before UPS shuts down. The UPS creates a battery-powered sine wave of electricity as a substitute for the alternating current sine wave generated by an electric utility.

**unipolar non-return to zero signalling** A simple type of signal ling used for early key telegraphy. Currently used with private line teletypewriter systems as well as representing the signal pattern used by RS-232 and V.24 interfaces. In this signal scheme, a DC current or voltage represents a mark, while the absence of current or voltage represents a space. When used with transmission systems, line sampling determines the presence or absence of

current, which is then translated into an equivalent mark or space.

**unipolar return to zero signalling** A variation of unipolar non-return to zero signalling is unipolar return to zero. Here the signal always returns to zero after every "1" bit. While this signal is easier to sample, it requires more circuitry to implement and is not commonly used.

**unipolar/bipolar signals** A digital signal technique that uses a positive or negative excursion and ground as the two binary signal states for unipolar signals where bipolar has amplitude in both directions from the reference axis.

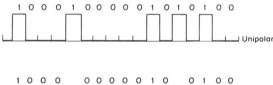

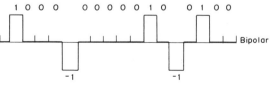

**unit element** A signal element of one unit element duration.

**unit interval** A unit interval is the duration of the shortest nominal signal element. The duration of the unit interval in seconds is the reciprocal of the telegraph speed expressed in baud.

**United States Independent Telephone Association** The association of telephone companies, excluding AT&T and its former operating company subsidiaries.

**Universal Character Set (UCS)** A printer feature that permits the use of a variety of character arrays.

**universal data channel**  A telephone company service delivery mechanism by which local loop facilities are used to provide digital access to end users.

**Universal Data Voice Multiplexing (UDVM)**  A multiplexing technique used to combine voice and digital signals over a single telephone line to the local telephone company central office.

**Universal Digital Loop Carrier (UDLC)**  A digital transmission system in which the usual analog service interface signals are converted to digital at the telephone company central office terminal and remote terminal locations.

**Universal Information Services (UIS)**  AT&T's vision for the future that will permit users of telecommunications networks to receive any kind of voice, data, or image service in any combination, with a maximum of convenience and economy at any time, in any place.

**Universal Product Code (UPC)**  Bar code used to identify products and their manufacturers in inventory systems.

**Universal Protocol Platform (UPP)**  Software from Excelan, Inc., of San Jose, CA, which separates hardware specific components from the underlying transport and network functions, permitting multiple protocol suites to simultaneously operate on the same hardware. This eases the migration from one network protocl set to another.

**universal service**  The policy of providing basic telephone service at a cost which is affordable by the entire U.S. population.

**Universal Synchronous Receiver/Transmitter (USRT)**  An integrated circuit designed to interface with synchronous circuits on behalf of a central processing unit, such as a microprocessor, and to perform basic functions such as character assembly/disassembly, and the conversion of parallel data into serial form for transmission over a synchronous data channel.

**Universal Synchronous Asynchronous Receiver/Transmitter (USART)**  An integrated circuit fabricated as a chip which is designed to interface with asynchronous circuits on behalf of a central processing unit, such as a microprocessor. The USART performs basic functions such as character assembly/disassembly operations and the conversion of data in parallel form from the CPU into several forms for transmission. The USART combines the functions of a UART and a USRT.

**universally administrated address**  A permanently encoded address in a token-ring adapter card which is unique. This address is administered by the IEEE.

**UNIX**  A computer operating system developed at AT&T Bell Laboratories in the late 1960s which provides a multi-user, multi-tasking capability. Also a trademark of Unix Systems Laboratories, Inc., which is owned by Novell, Inc.

**unloaded line**  A pair of wires running from one building to another, using no loading coils or switching equipment; a DC continuity line.

**UNMA**  Unified Network Management Architecture.

**unprintable characters**  Refers to those characters in a code which control printer or data link operation and hence, are not printed (such as LF, CR, NU, or SP).

**unprotected field**  On a display device, a display field in which the user can modify, or erase data.

**unrouted message**  On a bulletin board system network, a message sent to every board in the network. Also known as plain mail.

**unsolicited message**  In IBM's SNA, a message, from VTAM to a program operator, that is unrelated to any command entered by the program operator.

**unsuccessful call**  A call attempt that does not result in the establishment of a connection.

**UP**  User Part. CCITT #7 terminology.

**UPC**  Universal Product Code.

**uplink** The earth-to-satellite transmission channel used to transmit information to a geosynchronous satellite; complement of downlink.

**up-loop** A fixed pattern (repeating 10 000) which forces a receiving CSU to disconnect a signal loop-back.

**UPP** Universal Protocol Platform.

**UPS** 1. Uninterruptible Power Supply. 2. Uninterruptible Power Source.

**upstream** 1. In the direction of data flow from the end user to the host. 2. The direction opposite to message flow in a token-ring network.

**upstream device** For the IBM 3710 Network Controller, a device located in a network such that the device is positioned between the 3710 and a host. A communication controller upstream from the 3710 is an example of an upstream device.

**upstream line** A telecommunication line attaching an IBM 3710 Network Controller to an upstream device.

**uptime** Uninterrupted period of time that networks or computer resources are accessible and available to a user.

**upward compatible** Application that can be configured to function in some vendor or protocol enhanced mode. As opposed to downward compatible.

**US** The Unit Separator character.

**U.S.** United States.

**US access** A subsidiary of Telephone Electronics Corporation which operates its own fiber optic network between Denver and Colorado Springs and is constructing digital microwave links in Mississippi and Tennessee.

**USA DIRECT** A service mark of AT&T, this service permits persons outside the United States to rapidly and economically make calls to the U.S. from certain foreign locations. Callers can use their AT&T card or call collect. USA DIRECT can be used from any telephone in certain foreign locations or from designated USA DIRECT phones which places you in contact with an AT&T operator in the U.S. who places your call. Dial access countries and their telephone numbers are listed in the following table.

| Australia | 0014-881-011 |
|---|---|
| Belgium | 11-0010 |
| Br. Virgin Is. | 1 800 872-2881 |
| Denmark | 0430-0010 |
| France | 19-0011 |
| Germany | 0130-0010 |
| Hong Kong | 008-1111 |
| Japan | 0039-111 |
| Netherlands | 06-022-9111 |
| Sweden | 020-795-611 |
| United Kingdom | 0800-89-0011 |

**USACC** United States Army Communications Command.

**USACEEIA** United States Army Communications Electronics Engineering Installation Agency.

**usage-sensitive service** A "pay-for-what-you-use" method of charging for local calls fostered during the deregulation of the telephone industry.

**USART** Universal Synchronous/Asynchronous Receiver/Transmitter.

**USASC** United States Army Signal Center.

**USASCII** USA Standard Code for Information Interexchange; same as ASCII.

**USAT** Ultra Small Aperture Terminal.

**USDCFO** United States Defense Communications Field Office.

**$\mu$sec** Microsecond.

**USENET** USEr's NETwork. A network message sharing system that exchanges messages using a standard format. Messages are grouped into categories known as newsgroups.

**user**  Anyone who requires the services of a computing system.

**User Agent (UA)**  A set of processes that are used to do message-related functions, such as composing, editing, and displaying messages. It provides the commands to submit messages to the Message Transfer System (MTS) and retrieves messages that have been delivered to the user's mailbox.

**user correlator**  In IBM's SNA, a 4-byte value supplied to VTAM by an application program when certain macro instructions such as REQSESS are issued. It is returned to the application program when subsequent events occur (such as entry to a SCIP exit routine upon receipt of BIND) that result from the procedure started by the original macro instruction.

**User Datagram Protocol (UDP)**  The Internet standard protocol that allows an application program on one machine to send a datagram to an application program on another machine. Under UDP messages include a protocol port number, allowing the sender to distinguish among multiple destinations (application programs) on the remote machine.

**User Destination Routing (UDR)**  A term originally used by Codex Corporation for the addition of asynchronous port switching with contention and queuing to some of its statistical multiplexers.

**user dictionary**  A directory that contains unique information pertaining to each registered entity (user, node, bulletin board, etc.) within the Telemail system.

**user exit**  A point in an IBM-supplied program at which a user exit routine may be given control.

**user exit queue**  In IBM's SNA, a structure built by VTAM that is used to serialize the execution of application program exit routines. Only one exit routine on each user exit queue can run at a time.

**User File Directory (UFD)**  A specific work area on the Telenet Processor 5000 (TP5) system that contains files and from which software programs are operated.

**user interface**  The means by which a computer and a user communicate. It is the collection of everything the user sees, hears and does while activating the computer to perform some task.

**user name**  A name assigned by a system administrator which enables a computer system user to be uniquely recognized and identified as a user assigned to a particular computer system.

**user proprietary channel**  Allocated to the user for input of information for use in maintenance activities and remote alarms external to the span equipment.

**user-application network**  In IBM's SNA, a configuration of data processing products, such as processors, controllers, and terminals, established and operated by users for the purpose of data processing or information exchange, which may use services offered by communication common carriers or telecommunication administrations.

**USITA**  United States Independent Telephone Association.

**USOC**  Universal Service Ordering Code.

**USRT**  Universal Synchronous Receiver/Transmitter.

**USS**  Unformatted System Services (IBM's SNA).

**USTS**  UHF Satellite Terminal System.

**utilization**  The fraction of the available capacity that is being used—see processing, memory, and communications facilities.

**UTL**  Universal Test Line.

**UTP**  Unshielded Twisted Pair.

**UTS**  Universal Telephone Service.

**UUCP**  Unix-to-Unix Copy Program.

**UUCP network**  A loose association of computer systems communicating with each other at specific intervals using the UUCP protocol. UUCP routing is on a store and forward basis based upon the addresses in the bang path.

**Uudecode** A utility program which converts an ASCII file which represents binary data back onto binary.

**Uuencode** A program which encodes a binary file into ASCII code, usually for transmission between two computer systems.

**UULink** Software from Vortex Technology of Topanga, CA, which turns a DOS-based personal computer into a UNIX-based UUCP networking environment node.

# V

**V** CCITT (now ITU) code designation for standards dealing mainly with modem operations.

**V** Volt.

**V interface** (ISDN) The 2-wire physical interface used for single-customer termination from a remote terminal.

**VAC** 1. Value-Added Carrier. 2. Volts Alternating Current.

**VADS** Value-Added Data Service.

**validation** Checking data for correctness, or compliance with applicable standards, rules, and conventions.

**validity check** The method used to check the accuracy of data after it has been received.

**validity of data** The status of information being processed by hardware component or communications equipment.

**value-added carrier** A carrier that supplies specialized services, such as computer-oriented services or facsimile transmission.

**Value-Added Network Service (VANS)** A data transmission network which routes messages according to available paths, assures that the message will be received as it was sent, provides for user security, high-speed transmission and conferencing among terminals.

**Value-Added Network (VAN)** A network source that "adds value" to the common carrier's network services by adding computer control of communications.

**VAN** Value-Added Network.

**VANS** Value-Added Network Services.

**VAR** Value-Added Reseller.

**variable** 1. A name used to represent a data item with a value that can change while the program is running. 2. In IBM's NCCF, a character string beginning with "&" that is coded in a command list and is assigned a value during execution of the command list.

**Variable Quantizing Level (VQL)** A method of speech encoding that quantizes and encodes an analog voice conversation for transmission on a digital network. VQL usually encodes a voice conversation at 32 Kbps.

**variable tag** In IBM's PROFS, a set of special commands that substitute information in a Revisable-Form Text (RFT) document in place of variable controls.

**varistor** A circuit component designed to suppress voltage surges. A varistor is a type of diode that conducts current only when voltage above a certain threshold value is present.

**VAX** The trademarked name for a family of computers manufactured by Digital Equipment Corporation.

**VBI** Vertical Blanking Interval.

**VC** Virtual Call.

**VDC** Volts Direct Current.

**VDM** Voice Data Multiplexer.

**VDT** Video Display Terminal.

**VDU** Video Display Unit. Same as VDT.

**vector graphic image** An image formed by lines or curves drawn from point to point.

**vendor** A seller of telecommunications services or equipment.

**vendor code** A term used to indicate that software was written by the same company that manufactured the computer system.

**vent safe** Unique design which continues to provide surge protection even in the remote circumstance of gas venting from the device to the atmosphere.

**verify** 1. To determine whether a transcription of data or other operation has been accurately accomplished. 2. To check the results of keypunching.

**VERONICA** (Very East Rodent-Oriented Net-wide Index to Computerized Archives.) An Internet accessable service which maintains an index of titles of Gopher items and provides a keyword search of those titles. VERONICA is accessed through a Gopher client which provides a connection to a VERONICA server.

**VersaTerm Pro** A full-featured asynchronous communications program from Peripherals, Computers and Suppliers, Inc., of Mt Penn, PA, that operates on the Apple Macintosh. VersaTerm Pro includes VT-100 and Tektronix graphics terminal emulation capability.

**Vertical Blanking Interval (VBI)** A stripe of lines transmitted between each frame of a picture which was originally incorporated to provide vacuum-tube TV sets with time to move the picture-painting electron gun from the bottom right to the top left of the screen for each frame of a picture. Today in European countries, several lines are used for public teletext transmission while some additional lines are being considered for data transmission. In the United States portions of the VBI are used as closed captions for the hearing-impaired.

**vertical parity** Vertical parity checks character integrity. It is decided that the one bits in each character will add up to an odd or even number. As the character is transmitted, the one bits are counted and if they do add up to the predetermined odd or even number, the data is considered to be error-free. This is not a foolproof system, however. If an even number of bits are destroyed, the character will be seen as correct even though it is not. The illustration is an example of even vertical parity.

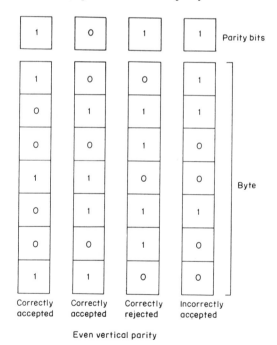

Even vertical parity

**vertical parity (redundancy) check** See parity check, vertical.

**vertical redundancy check (VRC)** In ASCII-coded

blocks, the parity test (even or odd) performed on each character in the block.

**Vertical Tabulation (VT)**   A format effector which advances the active position to the same character position on the next predetermined line.

**Very High Frequency (VHF)**   The portion of the electromagnetic spectrum with frequencies between approximately 30 and 300 MHz. This is the operating band for TV channels 2 through 13 and most FM radio.

**Very Large-Scale Integration (VLSI)**   A term used to describe a multifunction semiconductor device with a very high density (up to 10 000 circuits) of electronic circuitry contained on a single silicon chip.

**Very Low Frequency (VLF)**   That portion of the electromagnetic spectrum having continuous frequencies from approximately 3 to 30 kHz.

**Very Severe Burst (VSB)**   An error condition on high-speed digital circuits; defined as a period longer than 2.5 continuous seconds during which the bit error rate exceeds 1 percent. A very severe burst causes the receiving hardware to enter a yellow alarm condition.

**Very Small Aperture Terminal (VSAT)**   In satellite communications, a small diameter receiver station that normally operates in the Ku frequency band.

**VF**   Voice Frequency.

**VFCT**   Voice Frequency Carrier Telegraph.

**VG**   Voice Grade.

**VGA**   Video Graphics Array.

**VHF**   Very High Frequency.

**VHSD**   Very High Speed Data.

**Via net loss**   The net losses of trunks in the North American long-distance switched network.

**video**   1. The portion of the frequency spectrum used for TV signals. 2. A signal with a bandwidth of approximately 5 MHz generated from TV scanning.

**video attribute**   A description of the way video appears on a cathode ray tube display, such as reverse, blink, underline, low intensity, or a combination of the preceding.

**video conferencing**   Two-way transmission of video between two or more locations. Requires a wideband transmission facility, usually a satellite. Bandwidth ranges from 56 Kbps (freeze frame) to T1 (1.544 Mbps).

**Video Display Terminal (VDT)**   Terminal providing user temporary display (not hard copy) of input and output.

**Video Graphics Array (VGA)**   A personal computer video standard that provides a resolution of 640 by 480 pixels and 16-color simultaneous display.

**video resolution**   The number of dots (pixels) that can be displayed on a video screen at one time. It is presented as the number of horizontal dots times the vertical dots (H * V).

**video stream**   British Telecom's video conferencing service. Both full-motion monochrome and color video conferencing between customer's premises are offered.

**Video Transmission Service (VTS)**   An FTS2000 video conferencing service available in compressed and wideband modes.

**Videotel**   A videotext service offered by SIP, Italy's network operator.

**Videotel Inc.**   A Houston-based company that has exclusive rights to the French Minitel technology in the United States. Videotel markets a modified version of the Minitel terminal manufactured by Alcatel.

**videotex**   An interactive communications system designed to allow users to access data bases using television sets or low-cost terminals.

**viewdata**   The use of equipment based on teletext techniques to access data bases, through the telephone network, has been given the generic name of viewdata.

**Viewtron**  An experimental videotex service marketed by Knight-Ridder.

**VINES**  VIrtual NEtworking System.

**VINES Mac Mail Gateway**  A product from Banyan Systems, Inc., of Westborough, MA, which allows Macintosh and MS-DOS users to exchange electronic mail and files over that vendors VIrtual NEtworking System (VINES) network.

**VIP**  Visual Information Projection.

**virtual**  A term generally used by vendors, implying infinite capacity, but permitting effective use.

**virtual call**  Same as a virtual circuit.

**virtual circuit**  A communication path which is not dedicated to a single conversation for the duration of the call, e.g. a packet switching network facility gives the appearance to the user of an actual end-to-end circuit established for the duration of a call; in contrast to a physical circuit. Also a variable network connection which enables transmission facilities to be shared by many virtual users. Also called a logical circuit.

**virtual connection**  An apparent communications channel between two stations, whereby information or data transmitted by one station is automatically routed through the network via the most expeditious path to the other station. No long-haul circuit capacity is preassigned to a virtual connection; rather, capacity is made available only as data is transmitted by the stations.

**virtual disk server**  In a local area network, a type of disk server which gives each user access to a section of the disk, called a volume. Sometimes referred to as disk server.

**virtual file server**  In a local area network, a type of disk server which allows users to share files. Also referred to as file server.

**Virtual Machine (VM)**  1. An IBM program product whose full name is the Virtual Machine/System Product. It is a software operating system controlling the execution of programs. 2. A composite module containing both hardware and software which appears to the remainder of the software as a single functional entity.

**Virtual Machine Facility**  An IBM system control program, essentially an operating system that controls the concurrent execution of multiple virtual machines on a single mainframe.

**Virtual Machine/System Product (VM/SP)**  An IBM interactive, multiple-access operating system consisting of a control program and the conversational monitor system. It can be used with separate licensed programs to provide additional services.

**virtual memory**  The use of disk memory to hold portions of a computer's internal memory, permitting large programs to be segmented to operate in internal memory.

**virtual network**  An economical and flexible network made up of packet switched virtual circuits, either in a local area network (LAN) or a wide-area network (WAN) made up of several LANs.

**Virtual Node Control Application (VNCA)**  An IBM software package which provides host-node control.

**virtual private network**  A communications carrier provided service in which the public switched telephone network provides capabilities similar to those of private lines to include conditioning, error testing, and full-duplex, four-wire transmission.

**Virtual Route (VR)**  In IBM's SNA, a logical connection (1) between two subarea nodes that is physically realized as a particular explicit route, or (2) that is contained wholly within a subarea node for intra-node sessions. A virtual route between distinct subarea nodes imposes a transmission priority on the underlying explicit route, provides flow control through virtual-route pacing, and provides data integrity through sequence numbering of path information units (PIUs).

**Virtual Route Identifier (VRID)**  In IBM's SNA, a virtual route number and a transmission priority number that, when combined with the subarea addresses for the subareas at each end of a route, identify the virtual route.

**virtual route pacing**  In IBM's SNA, a flow control technique used by the virtual route control component of path control at each end of a virtual route to control the rate at which path information units (PIUs) flow over the virtual route. VR pacing can be adjusted according to traffic congestion in any of the nodes along the route.

**virtual route selection exit routine**  In IBM's VTAM, an optional installation exit routine that modifies the list of virtual routes associated with a particular class of service before a route is selected for a requested LU–LU session.

**virtual route sequence number**  In IBM's SNA, a sequential identifier assigned by the virtual route control component of path control to each path information unit (PIU) that flows over a virtual route. It is stored in the transmission header of the PIU.

**virtual storage**  An internal memory system which allows the computer to make efficient use of the available memory by drawing into the main memory only that portion of a program from a disk which is needed at a specific point of time. Also called virtual memory.

**Virtual Storage Access Method (VSAM)**  In IBM's SNA, an access method for direct or sequential processing of fixed and variable-length records on direct access devices. The records in a VSAM data set or file can be organized in logical sequence by a key field (key sequence), in the physical sequence in which they are written on the data set or file (entry-sequence), or by relative-record number.

**Virtual Storage Extended (VSE)**  An IBM program product whose full name is the Virtual Storage Extended/Advanced Function. It is a software operating system controlling the execution of programs.

**Virtual Telecommunications Access Method (VTAM)**  IBM teleprocessing access method that

gives users at remote terminals access to applications programs. It also provides resource sharing, a technique for efficiently using a network to reduce transmission costs.

**Virtual Telecommunications Access Method Entry (VTAME)**  An IBM program product that provides single-domain and multiple-domain network capability for 4300 systems using VSE. The set of programs control communication between terminals and application programs running under DOS/VS, OS/VS1 and OS/VS2.

**virtual terminal**  A technique that allows a variety of terminals with different characteristics to be accommodated by the same network by converting to a standard network format.

**virtual terminal protocol**  In the ISO Model, an application layer protocol intended to facilitate the interconnection of specific terminal types with the more general constructs of the Model.

**VIRUS**  A program which attacks memory or data storage, usually by replication to choke system resources.

**VisaNet**  The communications network used by Visa U.S.A. which links approximately 19 000 financial institutions to data processing services at VisaNet Interchange Centers.

**Visitel**  A videophone manufactured by Mitsubishi Electric Company aimed at residential users. The Visitel can be cabled to a standard handset similar to a telephone answering machine.

**Vista-Mail**  An electronic mail system marketed by Datapoint Corp. of San Antonio, TX, for use on that vendor's 7XXX and 8XXX/RMS computer systems.

**VIT**  VTAM Internal Trace.

**VLF**  Very Low Frequency.

**VLSI**  Very Large-Scale Integration.

**VM**  Virtual Machine.

**VM**  Virtual Machine Operating System (IBM's SNA).

**V-Mail**  The microfilming process used during World War II whereby letters to or from servicemen, to be sent overseas, were written on a standard form, reduced to microfilm on reels containing hundreds of letters, then sent to destinations to be printed in a readable size. The printed letters were then mailed in free official window envelopes.

**VMS**  Voice Mail System.

**VM/SP**  Virtual Machine/System Product—an IBM Operating System. Synonym for VM.

**VNCA**  Virtual Node Control Application.

**Vocoder**  An early type of voice coding device composed of a speech analyzer and a speech synthesizer that was used to reduce the bandwidth requirement of speech signals.

**Vodafone**  A cellular payphone marketed in the United Kingdom by Racal.

**VOGAD**  Voice Operated Gain Adjusting Device

**voice annotated text**  A system in which a message can be delivered to a user's terminal and displayed on a screen so that certain portions of the message can be selected which then result in voice commentary being delivered through a speaker of telephone handset connected to the terminal.

**voice compression**  A technique for reducing the bandwidth occupied by a voice telephone call. ADPCM and CVSD are voice-compression techniques.

**voice digitization**  The conversion of analog voice signals into a digital form.

**voice encoding**  The process of encoding an analog conversation into a digital data stream.

**voice grade**  An access line suitable for voice, low-speed data, facsimile, or telegraph service. Generally, it has a frequency range of about 300–000 Hz.

**voice jack**  Most data modems in today's environment are installed using a voice jack. The common nomenclature for such a jack is RJ-11. The deter-

mination for the type of jack to be used rests with the manufacturer of the modem. Specifications for the jack will be determined by how the modem is registered with the Federal Communications Commission (FCC). Information concerning the installation and jack requirements will be found in the manufacturer's brochure which came with the modem. With a voice jack requirement, it is "required" that the modem transmits at a fixed level of −9.0 dBm.

**Voice Link Plus**  A service of Metropolitan Fiber Systems which is designed as a replacement for analog four-wire leased lines used to support data transmission at rates up to 9.6 Kbps. The vendor digitizes customer analog signals and delivers the data to either another location on the network or a long haul carrier's point of presence.

**voice mail**  A general term used to describe a system that has the capability to answer calls, store messages, switch calls to another party and playback previously recorded messages. Voice-mail systems accept commands entered from touch-tone telephones and are also known as phone-mail and voice-processing systems. The following table describes popular voice-mail features.

| Feature | Description |
|---|---|
| Access code | A password entered into a voice-mail system through a touch-tone telephone or computer keyboard. The access code is used to control access to subscriber mailboxes. |
| Auto-Attendant | A feature which enables the transfer of calls to any telephone extension. An auto-attendant feature is normally obtained by connecting a voice-mail system to a PBX or another in-house telephone switching system. |
| Call-handling | A feature which enables voice-mail users to specify how incoming calls are handled. |
| Call-holding | A feature in which a voice-mail system takes messages without ringing the telephone extension of a mailbox. |

Call-screening | A feature which prompts a caller for their name, calls the desired extension, plays back the name and gives the mailbox owner the option of accepting or rejecting the call.

Distribution | The ability to send a voice-mail message to all persons on a given distribution list.

Individual greeting | A feature which allows each mailbox owner to record their own message which then greets callers routed to that mailbox.

Local phone | A standard telephone that plugs into a phone connector on a voice-mail system, permitting you to listen to incoming messages and record greetings or other messages.

Mailbox | A location where voice messages are stored in a voice-mail system before being claimed.

Mailbox owner | The person authorized to retrieve messages from a given mailbox in a multi-user voice-mail system. Also called subscriber or owner in some systems.

Message forwarding | A feature which enables a mailbox owner to have recorded messages routed to a different telephone extension for playback.

Message notification | A feature of a voice-mail system in which the mailbox owner is notified that a message is waiting. Notification can include turning on a message-waiting light on a person's telephone, calling them at another specified number or by the use of a beeper.

Personal message | The ability of a mailbox owner to record different messages for different persons.

System administrator | The person responsible for setting up a voice-mail system, initially distributing access codes and performing mailbox maintenance and other functions.

System greeting | The first greeting message a caller hears when a multi-mailbox voice-mail system answers an incoming telephone call.

System owner | The recipient of all messages on a single-user voice-mail system or an answering machine.

**Voice Manager** A voice-mail system marketed by Fujitsu Business Communications Systems of Tempe, AZ.

**voice message service** A service that permits a caller to send a one-way spoken message to a service user; the message is digitally stored in message box.

**voice PABX, voice-only PABX** A PABX for voice circuits; a telephone exchange.

**voice recognition** The process by which a machine, such as a computer, accepts, decodes, and acts on human speech.

**Voice Relay** A voice-mail system marketed by Lanier Voice Products of Atlanta, GA.

**voice response** The conversion of computer output into spoken words and phrases.

**Voice Saver** A solid-state voice greeting device manufactured by ComDev of Sarasota, FL. The Voice Saver assures that incoming callers receive a consistently clear, pleasant, and correct greeting.

**voice sounds** Complex sounds which contain different sets of harmonics.

**voice store-and-forward systems** A system that enables a computer to accept a message and store it until a transmission path or receiver is available.

**voice synthesis** The process by which a machine creates human-sounding speech.

**voiceband** A bandwidth for audio transmission, generally 300 to 3300 Hz. Also, voice grade (VG), voice frequency (VF).

**Voicecard** A telephone calling card marketed by U.S. Sprint which first routes the caller to equipment that checks the caller's voice against a previously recorded sample. Only if a match occurs will the call be routed to its destination.

**voice/data PABX**  A device which combines the functions of a voice PABX and a data PABX, often with emphasis on the voice facilities.

**voice-frequency**  Any frequency within that part of the audio-frequency range essential for the transmission of speech of commercial quality, i.e., 300–3400 Hz.

**voice-frequency carrier telegraphy**  That form of carrier telegraphy in which the carrier currents have frequencies such that the modulated currents may be transmitted over a voice-frequency telephone channel.

**voice-frequency multichannel telegraphy**  Telegraphy using two or more carrier currents the frequencies of which are within the voice-frequency range. Voice-frequency telegraph systems permit the transmission of up to 24 channels over a single circuit by use of frequency division multiplexing.

**voice-grade channel, voice-grade line**  A channel or line that offers the minimum bandwidth suitable for voice frequencies, usually 300 to 3400 bps.

**VoiceMark**  An AT&T public messaging service which allows users to record and transmit voice messages to telephones in the U.S. and 150 countries. VoiceMark users dial (800)562-6275 to record and send a one-minute message.

**voice-operated device**  A device used on a telephone circuit to permit the presence of telephone currents to effect a desired control. Such a device is used in most echo suppressors.

**Voice-Operated Gain-Adjusting Device (VOGAD)**  A device somewhat similar to a compandor and used on some radio systems; a voice-operated device which removes fluctuation from input speech and sends it out at a constant level. No restoring device is needed at the receiving end.

**volatile**  A term used to describe a data storage device (memory) that loses its contents when power is lost. Contrast with nonvolatile.

**volatile memory**  A storage medium that loses all data when power is removed.

**volatile storage**  A storage device whose contents are lost when power fails.

**voltage**  A measure of electrical potential expressed in units of volts and named after Count Alessandro Volta.

**Volt-Ohm-Meter (VOM)**  A test device that is used to measure and display voltage, resistance and current.

**volume**  A logical portion of a disk due to its partition. Each volume is treated like a separate hard disk by anyone using it.

**VOM**  Volt-Ohm-Meter or Volt-Ohmmeter.

**VOX**  Voice-Operated Device.

**VOW**  Voice Order Wire.

**VPN56**  A switch 56 Kbps digital transmission service marketed by U.S. Sprint.

**VQL**  Variable Quantizing Level.

**VR**  Virtual Route.

**VRC**  Vertical Redundancy Check.

**VRID**  Virtual Route IDentifier.

**VRU**  Voice Response Unit.

**VS**  Virtual Storage.

**VSAM**  Virtual Storage Access Method.

**VSAT**  Very Small Aperture Terminal.

**VSB**  Very Severe Burst.

**VSE**  Virtual Storage Extended operating system.

**VSE/AF**  Virtual Storage Extended/Advanced Function operating system. Synonym for VSE.

**V-Series**  A set of CCITT (now ITU) recommendations for the transmission of data over the public switched telephone network.

## GENERAL

V.1 Equivalence between binary notation symbols and the significant conditions of a two-condition code.

V.2 Power levels for data transmission over telephone lines.

V.3 International Alphabet No. 5

V.4 General structure of signals of International Alphabet No. 5 code for data transmission over public telephone networks.

V.5 Standardization of data signalling rates for synchronous data transmission in the general switched telephone network.

V.6 Standardization of data signalling rates for synchronous data transmission on leased telephone-type circuits.

V.7 Definitions of terms concerning data communication over the telephone network.

## INTERFACE AND VOICE-BAND MODEMS

V.10 Electrical characteristics for unbalanced double-current interchange circuits for general use with integrated circuit equipment in the field of data communications. Electrically similar to RS-423.

V.11 Electrical characteristics of balanced double-current interchange circuits for general use with integrated circuit equipment in the field of data communications. Electrically similar to RS-422.

V.15 Use of acoustic coupling for data transmission.

V.16 Medical analogue data transmission modems.

V.19 Modems for parallel data transmission using telephone signalling frequencies.

V.20 Parallel data transmission modems standardized for universal use in the general switched telephone network.

V.21 300 bps duplex modem standardized for use in the general switched telephone network. Similar to the Bell 103.

V.22 1200 bps duplex modem standardized for use on the general switched telephone network and on leased circuits. Similar to the Bell 212.

V.22 bis 2400 bps full-duplex 2-wire.

V.23 600/1200 baud modem standardized for use in the general switched telephone network. Similar to the Bell 202.

V.24 List of definitions for interchange circuits between data terminal equipment and data circuit-terminating equipment. Similar to and operationally compatible with RS-232.

V.25 Automatic calling and/or answering equipment on the general switched telephone network, including disabling of echo suppressors on manually established calls. RS 366 parallel interface.

V.25 bis Serial RS 232 interface.

V.26 2400 bps modem standardized for use on 4-wire leased telephone-type circuits. Similar to the Bell 201 B.

V.26 bis 2400/1200 bps modem standardized for use in the general switched telephone network. Similar to the Bell 201 C.

V.26 ter 2400 bps modem that uses echo cancellation techniques suitable for application in the general switched telephone network.

V.27 4800 bps modem with manual equalizer standardized for use on leased telephone-type circuits. Similar to the Bell 208 A.

V.27 bis 4800/2400 bps modem with automatic equalizer standardized for use on leased telephone-type circuits.

V.27 ter 4800/2400 bps modem standardized for use in general switched telephone telephone network. Similar to the Bell 208 B.

V.28 Electrical characteristics for unbalanced double-current interchange circuits (defined by V.24; similar to and operational with RS-232).

V.29 9600 bps modem standardized for use on point-to-point 4-wire leased telephone-type circuits. Similar to the Bell 209.

V.31 Electrical characteristics for single-current interchange circuits controlled by contact closure.

V.32 Family of 4800/9600 bps modems operating full-duplex over two-wire facilities.

V.32 bis Family of 14400 bps modems operating full duplex over two-wire facilities.

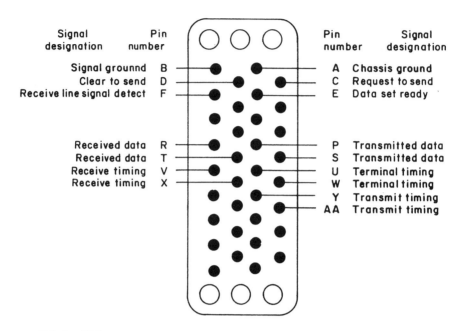

| Signal designation | Pin number | | | | | Pin number | Signal designation |
|---|---|---|---|---|---|---|---|

*V-Series: V.35*

**V.33**    14.4 Kbps modem standardized for use on point-to-point 4-wire leased telephone-type circuits.

**V.34**    28800 bps modem that operates full duplex over two-wire facilities.

**V.35**    Data transmission at 48 kilobits per second using 60–108 kHz group band circuits. CCITT balanced interface specification for data transmission at 48 kbps, using 60–108 kHz group band circuits. Usually implemented on a 34 pin M block type connector (M 34) used to interface to a high-speed digital carrier such as DDS.

**V.36**    Modems for synchronous data transmission using 60–108 kHz group band circuits.

**V.37**    Synchronous data transmission at a data signalling rate higher than 72 kbps using 60-108 kHz group band circuits.

## ERROR CONTROL

**V.40**    Error indication with electromechanical equipment.

**V.41**    Code-independent error control system.

**V.42**    The recommendation for error-control techniques for asynchronous modems that support public switched network communications at up to 9.6 Kbps. V.42 specified two error-control techniques—LAP M and the Microcom Network Protocol (MNP).

**V.42 bis**    A data compression standard for modems based upon British Telecom's Lempel-Ziv string encoding algorithm.

## TRANSMISSION QUALITY AND MAINTENANCE

**V.50**    Standard limits for transmission quality of data transmission.

**V.51**    Organization of the maintenance of international telephone-type circuits used for data transmission.

**V.52**    Characteristics of distortion and error-rate measuring apparatus for data transmission.

**V.53**    Limits for the maintenance of telephone-type circuits used for data transmission.

**V.54**    Loop test devices for modems.

**V.55** Specification for an impulse noise measuring instrument for telephone-type circuits.

**V.56** Comparative tests of modems for use over telephone-type circuits.

**VSPC** Visual Storage Personal Computing.

**VT** Vertical Tabulation.

**VTAM** Virtual Telecommunications Access Method.

**VTAM application program** In IBM's SNA, a program that has opened an ACB to identify itself to VTAM and can now issue VTAM macro instructions.

**VTAM definition** In IBM's SNA, the process of defining the user application network to VTAM and modifying IBM-defined characteristics to suit the needs of the user.

**VTAM definition library** In IBM's SNA, the operating system files or data sets that contain the definition statements and start options filed during VTAM definition.

**VTAM Internal Trace (VIT)** A trace used in VTAM to collect data on channel I/O, use of locks, and storage management services.

**VTAM operator** In IBM's SNA, a person or program authorized to issue VTAM operator commands.

**VTAM operator command** In IBM's SNA, a command used to monitor or control a VTAM domain.

**VTAM terminal I/O Coordinator (VTIOC)** In IBM's SNA, the part of TSO/VTAM that converts TSO TGET, TPUT, TPG, and terminal control macro instructions into SNA request units.

**VTAME** Advanced communications function for the Virtual Telecommunications Access Method Entry (IBM's SNA).

**VTEK** A trademark of Scientific Endeavors Corporation of Kingston, TN, as well as a software terminal emulator program that operates on IBM PC and compatible computers. VTEK emulates DEC VT102, VT100 and VT52 terminals as well as Tektronix 4105, 4010 and 4014 terminals.

**VTERM** The name used by Coefficient Systems Corporation of New York City, NY, for a series of terminal emulator programs that operate on the IBM PC and compatible personal computers.

**VTERM/100** A DEC VT100 emulation program from Coefficient Systems Corporation of New York City, NY, which supports four file transfer protocols. VTERM/100 emulates DEC VT52, VT100, VT101 and VT102 terminals when operated on IBM PC and compatible personal computers.

**VTERM/220** A DEC VT220 emulation program from Coefficient Systems Corporation of New York City, NY, which supports four file transfer protocols. VTERM/220 emulates DEC VT52, VT100, VT101, VT102 and VT220 terminals for IBM PC and compatible computers.

**VTERM/4010** A monochrome Tektronix graphics emulation program from Coefficient Systems Corporation of New York City, NY. VTERM/4010 emulates the Tektronix 4010 and 4014 terminals as well as DEC VT220, VT100 and VT52 devices for IBM PC and compatible computers.

**VTERM/4105** A Tektronix color graphics emulation program from Coefficient Systems Corporation of New York City, NY. VTERM/4105 provides a complete emulation of the Tektronix 4105, 4010 and 4014 terminals as well as the DEC VT100 for IBM PC and compatible computers.

**VTERM/4208** A Tektronix 4208 terminal emulator from Coefficient Systems Corporation of New York, NY. VTERM/4208 operates on IBM PC and compatible personal computers and provides a complete emulation of the Tektronix 4208 terminal as well as Tektronix 4105, 4107, 4205, 4207 and DEC VT220 functions.

**VTIOC** VTAM Terminal I/O Coordinator (IBM's SNA).

**VTP** Virtual Terminal Protocol.

**VTS**   Video Transmission Service.

**VT102**   A Digital Equipment Corporation video display terminal which is similar to the firm's VT 100 product but also supports a printer.

**VT240**   A Digital Equipment Corporation 132-column terminal which many third-party vendors market emulation software for to enable personal computers to operate as this type of device. The VT240 has a 800 by 240 pixel resolution.

**V&H**   Vertical and Horizontal grid coordinates are used by AT&T and other communications carriers in mathematical formulas to determine the airline distance between any two rate centers.

**V+TU**   Voice plus Teleprinter Unit.

# W

**WACK (wait before transmit)** The WACK character sequence allows a receiving station to indicate a "temporarily not ready to receive" condition to the transmitting station.

**wafer** A thin disk of a purified crystalline semiconductor, typically silicon, that is divided into chips after processing. Typically, a wafer is about one-fiftieth of an inch thick and four or five inches in diameter.

**WAIS** Wide Area Information Service.

**wait state** A period of time when a processor does nothing. It is employed to synchronize circuitry or devices operating at different speeds.

**WAN** Wide Area Network.

**wander** Jitter which occurs at a rate less than 10 Hz is the long-term variation in the phase of a digital signal. The primary causes of wander are clock drift and propagation delay changes.

**Wang Office** An electronic mail system marketed by Wang Laboratories, Inc., of Lowell, MA, for use on that vendor's Wang VS Series of computers.

**warm boot** The complete reload of a "warm" computer's operating system. The computer is "warm" at the start because the electricity is on.

**warm start** In IBM's SNA, a synonym for system restart.

**Washington Interagency Telecommunications System (WITS)** A network of central office equipment in the Washington, D.C., area dedicated to U.S. Government agency users.

**watchdog timer** A timer set by a program to generate an interrupt to the processor after a given period of time, ensuring that a system does not lose track of buffers and communications lines because of a hardware error.

**WATS** Wide Area Telephone Service.

**Watson** A voice mail system marketed by Natural MicroSystems Corporation of Natick, MA.

**watt** A basic unit of absolute optical power.

**WATTC** World Administrative Telegraph and Telephone Conference.

**wave propagation** The transfer of electromagnetic or acoustical energy from one place to another through a suitable transmission medium.

**waveform** The pattern resulting from plotting the voltage of a signal as a function of time.

**waveguide** A hollow conductor used to efficiently transmit high-energy, high-frequency waves in the centimeter to micrometer range. Fiber optic cables are one form of solid conductors for micrometric and smaller frequencies.

**wavelength**  The distance between successive peaks of a sinusoidal wave. It is the interval of one complete cycle and is normally expressed in meters. The wavelength is equal to velocity divided by the frequency.

**Wavelength Division Multiplexing (WDM)**  The simultaneous transmission of more than one information-carrying channel on a single fiber lightguide using two or more light sources of different wavelengths.

**Wayfarer**  A system marketed by Sony Corporation of America which supported mobile truck to a dispatch center communications via satellite. This system has been superseded by the vendor's 2-Wayfarer Mobile Communications System.

**WBT**  Wait Before Transmit.

**WCCS**  Wireless Crew Communication System.

**WDM**  Wavelength Division Multiplexing.

**WD/1TT**  U.S. military field wire which consists of two twisted, individually insulated, American Wire Gage (AWG) No. 23 conductors. Each conductor has four tinned-copper strands and three galvanized steel strands.

**WD-36/TT**  U.S. military telephone cable which uses solid aluminum, American Wire Gage (AWG) No. 23 conductors.

**WECO 310**  A large tip-and-ring jack connector used in digital telephony.

**WEFAX**  WEather FAcsimile.

**Weibull plot**  A statistical graph that is used to show data plotted as cumulative failure probability versus a load, stress, or time at failure. Used with cable testing.

**WESTAR**  The name of the family of satellites providing domestic communications services owned and operated by Western Union.

**WET T1**  A T1 circuit with a Bell Operating Company (BOC) powered interface.

**Whisper**  A communications software program and plug-in encryption board marketed by Mika Limited Partnership of Roswell, GA, which includes real-time data encryption and other security features for DOS computers.

**white line skipping**  A facsimile compression scheme in which scanned lines containing no information are encoded, rather than transmitted as a large number of blank bits.

**white noise**  Background noise caused by thermal agitation of electrons. Also referred to as Gaussian noise.

**white pages**  A term used to refer to a data base that contains basic information about subscribers on a network, such as their name, E-mail address, telephone number and postal address.

**WHOIS**  A data base produced and maintained by the Internet Network Information Center (NIC) which lists both users and sites as well as separate data bases maintained by organizations connected to the Internet.

**Wide Area Information Service (WAIS)**  A distributed text searching system which enables you to search the contents of archives on the Internet for occurrences of those words. WAIS requires the use of a client program on the local workstation or host to access the system.

**Wide Area Network (WAN)**  A network which connects geographically separated areas.

**Wide Area Telecommunications Service (WATS)**  Same as Wide Area Telephone Service (WATS).

**Wide Area Telephone Service (WATS)**  A service provided by telephone companies in the United States which permits a customer by use of an access line to make calls to telephones in a specific zone for a flat charge per hour of call duration. Under the WATS arrangement, the U.S. is divided into six zones for interstate usage as well as 50 zones for intrastate usage.

**wideband**  In general, having a large bandwidth. While "wideband" and "broadband" mean basically the same thing, "wideband" is the preferred term in telephony. "Broadband" is usually applied to communication at radio frequencies such as broadband local area networks. Generally, wideband circuits operate at data rates in excess of 9.6 Kbps and often at rates of 56 Kbps to 1.544 Mbps.

**wideband channel**   A telecommunications channel with a bandwidth greater than the bandwidth of a voice-grade channel.

**wideband packet technology**   A method of combining all communication traffic to include voice, data, and certain kinds of images for high-speed transport through a wideband packet system capable of switching millions of packets per second.

**Wideband Video Transmission Service (WVTS)**   An FTS2000 video conferencing service provisioned over satellite facilities for multiple signal distribution.

**WIN**   Worldwide Intelligent Network.

**Winchester disk**   A high-density magnetic storage system originally developed by IBM in which the media is non-removable. Also called a fixed disk and hard disk.

**window**   1. A flow control mechanism whose size governs the number of units of data that may be transmitted before an acknowledgment is required. 2. In IBM's SNA, (a) The path information units (PIUs) that can be transmitted on a virtual route before a virtual-route pacing response is received, indicating that the virtual route receiver is ready to accept more PIUs on the route. (b) The requests that can be transmitted on the normal flow in one direction in a session before a session-level pacing response is received, indicating that the receiver is ready to accept the next group of requests. A synonym for pacing group. 3. A box containing information or a menu that appears on your screen over the document you are working with.

**window protocol**   A protocol that permits a number of blocks of windows to be outstanding prior to requiring an acknowledgment.

**window size**   1. In a packet switching environment, refers to a traffic flow mechanism which limits the number of elements which can be sent through the network before an acknowledgment is received. Window size is defined on both packet level and link level. 2. In IBM's SNA: (a) The number of path information units (PIUs) in a virtual route window. The window size varies according to traffic congestion along the virtual route. (b) The number of requests in a session-level window. The window size is set at session activation. A synonym for pacing group size.

**windows**   A method of displaying information on a screen in which viewers see what appear to be several sheets of paper similar in appearance to a desktop. The viewer can shift the positions of the windows on the screen, resize different windows, and select a window to work with the data it contains.

**wink**   The momentary interruption in a single frequency tone which indicates that a distant central office is ready to receive the digits you just dialed.

**wink pulsing**   A sequence of recurring pulses where the off-pulse is very short with respect to the on-pulse.

**wire**   The twisted wire pair which provides telephone service. By extension, any telephone circuit which uses twisted wire pairs or coaxial cable for transmission.

**wire center**   A building in which one or more local central office switching systems are installed and where the outside cable plant is connected to central office equipment.

**wire center serving area**   The area of an exchange served by a single wire center.

**wire fault**   An error condition caused by a break or short between the wires in a segment of a cable.

**wire pair** Two conductors, isolated from each other, that associate to form a communications channel.

**wire pairs** Transmission medium in which a pair of conducting wires form a circuit.

**wireline** A term used to reference a cellular telephone network licensed to the indigenous telephone company.

**wiring blocks** Molded plastic blocks designed in various pair configurations which terminate cable pairs and establish pair location.

**wiring closet** A closet within which a number of wires or cables connected to individual telephones or stations terminate. A wiring closet is normally used to facilitate the rearrangement of telephone sets to telephone lines.

**wiring concentrator** A lobe concentrator that allows multiple attaching devices access to the ring at a central point such as a wiring closet or in an open work area.

**WIS** WWMCCS Information System.

**WITS** Washington Interagency Telecommunications System.

**Wizard Mail** A trademark of H&W Computer Systems of Boise, ID, for an electronic mail program that operates on IBM DOS and OS computer systems.

**WM** Work Manager.

**word** In telegraphy, six operations or characters (five characters plus one space). ("Group" is also used in place of "word.") In computing, a sequence of bits or characters treated as a unit and capable of being stored in one computer location.

**word length** The number of bits in a word, determined as an optimal size for processing, or transmission; often based on register size and internal operation of a computer (8 bit, 16 bit, 32 bit) (e.g., IBM System/370 uses 32-bit words, IBM PC employs 16-bit).

**word processor** Computer software used to create, edit, and format text. The availability of word processing software and personal computers has made a significant impact on the way businesses handle administrative functions.

**Words Per Minute (WPM)** A common measure of speed in telegraph systems.

**work area** An area in which terminals to include display stations, keyboards, and printers are located.

**work location wiring subsystem** That part of a premises' distribution system that includes the equipment and extension cords from the information outlet up to the terminal device connection.

**workgroup** A corporate department or division in which computer users require data access and sharing capabilities to carry out their departmental objectives.

**working diskette** A computer diskette onto which programs and files are copied from an original diskette for use in everyday operation.

**working draft** In ISO, a working draft is the initial stage of a standards document which describes the standard as envisioned by a working group of standards committee or subcommittees.

**workstation** Input/output equipment at which an operator works; a station at which a user can send data to or receive from a computer for the purpose of performing a job. For example, a terminal or microcomputer connected to a mainframe or to a local or wide area network.

**World Administrative Telegraph and Telephone Conference (WATTC)** A conference organized by the International Telecommunications Union (ITU) to seek agreement on new regulations governing the flow of international telecommunications.

**World Communications, Inc. (WORLDCOM)** A former part of Western Union Corporation involved in international private-line business that was sold to Tele-Columbus AG of Baden, Switzerland.

**world numbering plan** A CCITT (now ITU) numbering plan that divides the world into nine zones. Each zone is allocated a number that forms the first digit of the country code for every country in that zone.

**WORLDCOM** WORLD COMmunications Inc.

**Worldnet** A global satellite television network operated by the United States Information Agency.

**Worldwide Airborne Command Post (WWAB-NCP)** Specially modified U.S. Air Force planes whose mission is to provide reliable, survivable and endurable communications between the National Command Authorities (NCA), the individual Commanders-in-Chief (CINCs) and the U.S. allied forces.

**Worldwide Intelligent Network (WIN)** An international voice and data network operated by the Wall Street brokerage firm Shearson Lehman Hutton, Inc.

**Worm** A computer program which replicates itself and may destroy files on a computer.

**WORM** Write Once Read Many.

**WPM** Words Per Minute.

**wrap count** In IBM's Network Problem Determination Application (NPDA), the number of events

that can be retained on the data base for a specific resource.

**wrap test** A test that checks circuitry without checking a mechanism itself by returning the output of the mechanism as input. A wrap test can be used to transmit a specific character pattern to or through a modem in a loop and then compare the character pattern received with the pattern transmitted.

**wraparound** The movement of the cursor as it reaches the right edge of screen, disappears, and "wraps around" to the beginning of the next line.

**write** To make a permanent or transient recording of data in a storage device or on a data medium.

**Write Once Read Many (WORM)** An optical storage technology in which data can be recorded to disk once but thereafter read as many times as desired.

**WRU** The who-are-you character.

**WVTS** Wideband Video Transmission Service.

**WWABNCP** WorldWide AirBorNe Command Post.

**WWMCCS** WorldWide Military Command and Control System.

**W/XMODEM** A windowed version of the X-MODEM protocol in which up to four packets can be transmitted and outstanding prior to requiring an acknowledgment.

# X

**X Windows** The graphical user interface used by many UNIX systems.

**XA** Extended Architecture.

**XC** Cross-Connect.

**XDOS** A software utility developed by Motorola and Hunter Systems, Inc., of Mountain View, CA, which translates DOS applications into binary code that can be executed on Unix Systems using the Motorola 68000 and 88000 microprocessors.

**XDR** External Data Representation.

**XEDIT** An IBM full-screen editor that can be used to create or change programs or files.

**XENIX** Microsoft Corporation trade name for a 16-bit microcomputer operating system that was derived from AT&T's UNIX operating system. The operating system offers features required for use in a business environment.

**Xerox** The company that developed Ethernet.

**Xerox CIN** Corporate Internet.

**Xerox Network Services (XNS)** Xerox Corporation's layered data communications protocols.

**Xerox Network Systems' Internet Transport Protocol (XNS/ITP)** In LAN technology, a special communications protocol used between networks. XNS/ITP functions at the third and four layer of the OSI model. Similar to TCP/IP.

**Xerox RIN** Research Internet.

**XID** EXchange station IDentification.

**XID** In IBM's SNA, a data link control command and response passed between adjacent nodes that allows the two nodes to exchange identification and other information necessary for operation over the data link.

**XID frame** A frame in HDLC used to exchange operational parameters between the participating stations.

**XMIT** Transmit.

**XMODEM** A popular public-domain personal computer link control protocol. This protocol blocks groups of asynchronous characters together for transmission and computes a checksum which is appended to the end of the block. The checksum is obtained by first summing the ASCII value of each data character in the block and dividing that sum by 255. Then, the quotient is discarded and the remainder is appended to the block as the checksum. The figure illustrates the XMODEM protocol block format. The Start of Header is the ASCII SOH character whose bit composition is 00000001, while the one's complement of the block number is obtained by subtracting the block number from 255. The block number and its complement are contained at the beginning of each block to reduce the possibility of a line hit at the beginning of the transmission of a block causing the block to be retransmitted.

| Start of Header | Block number | One's complement block number | 128 data characters | Checksum |
|---|---|---|---|---|

XMODEM protocol block format

**XMODEM 1K** A modified version of the X-MODEM protocol which transmits 1K blocks to obtain faster data transfers.

**XMTR** Transmitter.

**XNS** Xerox Network Services.

**XNS/ITP** Xerox Network Systems' Internet Transport Protocol.

**XOFF** Transmitter off. The communication control character that instructs a terminal to suspend transmission.

**XON** Transmitter on. The communication control character that instructs a terminal to start or to resume transmission.

**X-ON/X-OFF** A handshaking protocol. When the terminal's buffer is nearly full, it transmits an X-OFF to the computer to stop transmission when the buffer is almost empty, an X-ON is transmitted to the host to resume transmission.

**X/Open Co.** A consortium of computer companies promoting open systems.

**X/OS** Olivetti's implementation of a Unix operating system.

**X.PC** An error-connection protocol designed and sponsored by Tymnet which provides the functions typical of the OSI Network Layer 3.

**Xpedite** An electronic mail service for facsimile and telex users provided by Xpedite Systems, Inc., of Eatown, NJ.

**X-Press Information Services** A videotext service offered by McGraw-Hill and Telecommunications, Inc., that offers a news service called X-CHANGE which allows subscribers to access a variety of international newswire services.

**Xpress Transfer Protocol (XTP)** A local area network (LAN) protocol designed to allow workstations to transmit data at the rates emerging LAN's support to include the 16 Mbps token ring, broadband ISDN and the 100 Mbps Fiber Distributed Data Interface (FDDI).

**X-Series** A set of CCITT (now ITU) recommendations for data transmission in public data networks.

## CCITT STANDARDS

### SERVICES AND FACILITIES

X.1 International user classes of service in public data networks.

X.2 International user services and facilities in public data networks.

X.3 Packet assembly/disassembly facility (PAD) in a public data network.

*X.3 Parameters/X.25 Level 1*

| Parameter number | Parameter function | Values |
|---|---|---|
| 1 | PAD Recall Defined Character | 0 or 1<br>2–127 |
| 2 | Local Echo | 0 or 1 |
| 3 | Data Forwarding Characters | 0,2,4,8,<br>16,32,<br>64,126,<br>127,128+$n$,<br>254 or 255 |
| 4 | Data Forwarding Timeout | 0–255 |
| 5 | PAD to Terminal Flow Control | 0–4,10,<br>11 |
| 6 | Control of PAD Service Signals | 0,1,4,5 |
| 7 | PAD Action on Receipt of Break from Term | 0,1,2,8,<br>21 |
| 8 | Discard Output | 0 or 1 |
| 9 | Padding after Carriage Return | 0–7 |
| 10 | Line Folding | 0–255 |
| 11 | Async. Speed (Read-only parameter) | 0–14 |

| Parameter number | Parameter function | Values |
|---|---|---|
| 12 | Terminal to PAD Flow Control | 0,1,4 |
| 13 | Line Feed Instruction | 0–15 |
| 14 | Padding After Line Feed | 0–7 |
| 15 | Editing | 0 or 1 |
| 16 | Character Delete Defined Character | 1–127 |
| 17 | Buffer Delete Defined Character | 1–127 |
| 18 | Buffer Display Defined Character | 1–127 |
| 19 | Editing Service Signals | 0,2,8, 32–126 |
| 20 | Echo Mask | 0–128 |
| 21 | Parity Treatment | 0,1,2,3 |
| 22 | Page Wait | 0–255 |

X.4  General structure of signals of International Alphabet No. 5 code for data transmission over public data networks.

X.10  Categories of access for data terminal equipment to public data transmission services provided by PDNs and/or ISDN through terminal adapters.

X.15  Definitions of terms concerning public data networks.

*Interfaces*

X.20  Interface between data terminal equipment (DTE) and data circuit-terminating equipment (DCE) for start–stop transmission services on public data networks.

X.20 bis  Use on public data networks of data terminal equipment (DTE) which is designed for interfacing to asynchronous duplex V-Series modems.

X.21  Interface between data terminal equipment (DTE) and data circuit-terminating equipment (DCE) for synchronous operation on public data networks. This recommendation defines the physical characteristics and control procedures to set-up calls, transfer data and terminate calls through the network. Switched or leased circuit communications is supported at data rates up to and beyond 64 Kbps. X.21 is a mix of three protocols: physical, link and network. The physical link is a 15-pin connector. The electrical specification (V.11) is capable of data rates of 64 Kbps and higher.

| Interchange circuit | Name | Direction | |
|---|---|---|---|
| | | To DCE | From DCE |
| G | Signal ground or common return | | |
| Ga | DTE common | X return | |
| T | Transmit | X | |
| R | Receive | | X |
| C | Control | X | |
| I | Indication | | X |
| S | Signal element timing | | X |
| B | Byte timing | | X |

X.21 bis  Use on public data networks of data terminal equipment (DTE) which is designed for interfacing to synchronous V-Series modems.

X.22  Multiplex DTE/DCE interface for user classes 3–6.

X.24  List of definitions for interchange circuits between data terminal equipment (DTE) and data circuit-terminating equipment (DCE) on public data networks.

X.25  Interface between data terminal equipment (DTE) and data circuit-terminating equipment (DCE) for terminals operating in the packet mode on public data networks.

X.26  Electrical characteristics for unbalanced double-current interchange circuits for general use with integrated circuit equipment in the field of data communications.

X.27  Electrical characteristics for balanced double-current interchange circuits for general use with integrated circuit equipment in the field of data communications.

X.28  DTE/DCE interface for a start–stop mode data terminal equipment accessing the packet assembly/disassembly facility (PAD) in a public data network situated in the same country.

X.29    Procedures for the exchange of control information and user data between a packet assembly/disassembly (PAD) and a packet mode DTE or another PAD.

X.30    Support of X.21 and X.21 bis based DTEs by an ISDN.

X.31    Support of packet-mode terminal equipment by an ISDN.

X.32    Interface between DTE and DCE for terminals operating in the packet mode and accessing a PSPDN through a PSTN or a circuit switched Public Data Network (CSPDN). Dial X.25 connection.

X.33    Procedures and arrangements for DTEs accessing circuit-switched digital data services through analog telephone networks. X.310 (1984) was renumbered as X.33 (1988).

## TRANSMISSION, SIGNALLING AND SWITCHING

X.40    Standardization of frequency-shift modulated transmission systems for the provision of telegraph and data channels by frequency division of a group.

X.50    Fundamental parameters of a multiplexing scheme for the international interface between synchronous data networks.

X.50 bis  Fundamental parameters of a 46 Kbps user data signalling rate transmission scheme for the international interface between synchronous data networks.

X.51    Fundamental parameters of a multiplexing scheme for the international interface between synchronous data networks using 10-bit envelope structure.

X.51 bis  Fundamental parameters of a 48 Kbps user data signalling rate transmission scheme for the international interface between synchronous data networks using 10-bit envelope structure.

X.52    Method of encoding anisochronous signals into a synchronous user bearer.

X.53    Numbering of channels on international multiplex links at 64 Kbps.

X.54    Allocation of channels on international multiplex links at 64 Kbps.

X.55    Interface between synchronous data networks using 6+2 envelope structure and SCPC-satellite channels.

X.56    Interface between synchronous data networks using an 8+2 envelope structure and SCPC-satellite channels.

X.57    Method of transmitting a single lower-speed data channel on a 64 Kbps data stream.

X.58    Fundamental parameters of multiplexing scheme for international interface between synchronous non-switched data networks using no envelope structure.

X.60    Common channel signalling for circuit switched data applications.

X.61    Signaling System No. 7—Data user part.

X.70    Terminal and transit control signalling system for start–stop services on international circuits between anisochronous data networks.

X.71    Decentralized terminal and transit control signalling system on international circuits between synchronous data networks.

X.75    Terminal and transit call control procedures and data transfer system on international circuits between packet switched data networks.

X.80    Internetworking of interexchange signalling systems for circuit-switched data services.

X.81    Interworking between an ISDN and a CSPDN.

X.82    Detailed arrangements interworking between CSPDNs and PSPDNs T.70.

X.87    Principles and procedures for realization of international user facilities and network utilities in public data networks.

## NETWORK ASPECTS

X.92    Hypothetical reference connections for public synchronous data networks.

X.96    Call progress signals in public data networks.

X.110   Routing principles for international public data services through switched public data networks of the same type.

## SYSTEM DESCRIPTION TECHNIQUES

X.250    Formal description techniques for data communications protocols and services.

X.290    OSI conformance testing methodology and framework for protocol recommendations for CCITT applications.

## INTERNETWORKING BETWEEN NETWORKS

X.300    General principles and arrangements for internetworking between public data networks, and between public data networks and other public data networks.

X.301    Description of general arrangements for call control within a subnetwork and between subnetworks for provision of data transmission services.

X.302    Description of general arrangements for internal network utilities within a real subnetwork and intermediate utilities between real subnetworks for provision of data transmission services.

X.305    Functionalities of subnetworks relating to support of OSI connection-mode network service.

X.310    Procedures and arrangements for DTEs accessing circuit switched digital data services through analog telephone networks. Renumbered as X.33 (1988).

X.316    Accuracy and dependability performance values for PDNs when providing international packet switched service.

X.320    General arrangements for interworking between ISDNs for provision of data transmission services.

X.321    General arrangements for interworking between CSPDNs and ISDNs for provision of data transmission services.

X.322    General arrangements for interworking between PSPDNs and CSPDNs for provision of data transmission services.

X.323    General arrangements for interworking between PSPDNs.

X.324    General arrangements for interworking between PSPDNs and mobile systems for provision of data transmission services.

X.325    General arrangements for interworking between PSPDNs and ISDNs for provision of data transmission services.

X.326    General arrangements for interworking between PSPDNs and CCSN.

X.327    General arrangements for interworking between PSPDNs and private data networks for provision of data transmission services.

X.340    Principles for networking involving both communication and transmission between a PSPDN and the telex network.

X.350    General requirements to be met for data transmission in international mobile satellite systems.

X.351    Special requirements to be met for packet assembly/disassembly facilities (PADs) located at or in association with coast earth stations in the maritime satellite service.

X.352    Internetworking between public packet switched data networks and the maritime satellite data transmission system.

X.353    Routing principles for interconnecting the maritime satellite data transmission system with public data networks.

X.370    Arrangements for transfer of internetwork management information.

## MESSAGE HANDLING SYSTEMS (MHS)

X.400    High-level protocol recommendation for message handling providing protocol and technical specs for interconnection of separate computer-based message handling system. MHS system and service overview.

X.401    Basic service elements and options user facilities.

X.402    MHS—Overall architecture.

X.403    MHS—Conformance testing.

X.407    MHS—Abstract service definition conventions.

X.408    MHS—Encoded information type conversion rules.

X.409    Presentation transfer syntax and notation.

X.410    Remote operations and reliable transfer server.

X.411 MHS—Message transfer system: Abstract service definition.

X.412 Merged with X.419.

X.413 Message store: Abstract service definition.

X.414 Merged with X.419.

X.419 MHS—Protocol specifications.

X.420 MHS—Interpersonal messaging system.

X.430 Access protocol for teletex terminals.

X.500 The Directory—Overview of concepts, models and services.

X.501 The Directory—Models.

X.509 The Directory—Authentication framework.

X.511 The Directory—Abstract service definition.

X.518 The Directory—Procedures for distributed operation.

X.519 The Directory—Protocol specifications.

X.520 The Directory—Selected attribute types.

X.521 The Directory—Selected object classes.

**XTC**   EXternal Transmit Clock.

**XTP**   Xpress Transfer Protocol.

**XWIN Graphical Windowing System**   An AT&T software application which provides an interface between application programs and a windowing environment so that one or several simultaneous applications can be displayed and operated from a graphically organized desktop. The AT&T XWIN System features the AT&T OPEN LOOK Window Manager which creates an easy to use environment of mouse-driven screen windows, icons, and graphical layouts. XWIN supports a hierarchy of overlapping windows, which may be created by the end user to any depth of windows within windows.

**X.21 Communication Adapter**   An IBM 3710 Network Controller Communication adapter that can combine and send information on one line at speeds up to 64 Kbps, and conforms to CCITT X.21 standards.

**X.25 Cause & Diagnostic Codes**   X.25 Cause Codes are received from data terminating equipment (DTE) and Diagnostic Codes are received from data circuit-terminating equipment (DCE). The following tables indicate the X.25 cause and diagnostic code values and their meanings.

*X.25 Clearing Cause Codes*

X'00'   DTE originated

X'01'   Number busy

X'03'   Invalid facility request

X'05'   Network congestion

X'09'   Out of order

X'0B'   Access barred

X'0D'   Not obtainable

X'11'   Remote procedure error

X'13'   Local procedure error

X'15'   RPOA out of order

X'19'   Reverse charging acceptance not subscribed

X'21'   Incompatible destination

X'29'   Fast select acceptance not subscribed

X'39'   Ship absent (used in mobile maritime service)

X.25 Clearing Cause Codes greater than X'80' represent the clearing or restarting cause field which is generated by the remote DTE.

*X.25 Resetting Cause Codes*

X'00'   DTE originated

X'01'   Out of order (PVC only)

X'03'   Remote procedure error

X'05'   Local procedure error

X'07'   Network congestion

X'09'   Remote DTE operational (PVC only)

X'0F'   Network operational (PVC only)

X'11'   Incompatible destination

X'1D'   Network out of order (PVC only)

X.25 Resetting Cause Codes greater than X'80' represent the resetting or restarting cause field which is generated by the remote DTE.

*X.25 Diagnostic Codes from the DCE*

A second field displayed along with the Call Rejected, Call Cleared and X.25 Reset error messages contains the X.25 Diagnostic Code. The contents of this field are generated by the X.25 network and provide additional information on the reason for an error message. *Note.* When a Cause Code is DTE originated (i.e. X'00' or greater than X'80') the information provided by the Diagnostic Code may not be always correct; some hosts generate their own Diagnostic Codes which do not follow CCITT Recommendation X.25.

| | |
|---|---|
| X'00' | *No additional information* |
| X'01' | Invalid P(S) |
| X'02' | Invalid P(R) |
| X'10' | *Packet type invalid—general* |
| X'11' | For state r1 |
| X'12' | For state r2 |
| X'13' | For state r3 |
| X'14' | For state p1 |
| X'15' | For state p2 |
| X'16' | For state p3 |
| X'17' | For state p4 |
| X'18' | For state p5 |
| X'19' | For state p6 |
| X'1A' | For state p7 |
| X'1B' | For state d1 |
| X'1C' | For state d2 |
| X'1D' | For state d3 |
| X'20' | *Packet not allowed—general* |
| X'21' | Unidentifiable packet |
| X'22' | Call on one-way logical channel |
| X'23' | Invalid packet type on a PVC |
| X'24' | Packet on unassigned logical channel |
| X'25' | Reject not subscribed to |
| X'26' | Packet too short |
| X'27' | Packet too long |

| | |
|---|---|
| X'28' | Invalid general format identifier |
| X'29' | Restart with nonzero general format identifier |
| X'2A' | Packet type incompatible with facility |
| X'2B' | Unauthorized interrupt confirmation |
| X'2C' | Unauthorized interrupt |
| X'2D' | Unauthorized reject |
| X'30' | *Time expired—general* |
| X'31' | For incoming call |
| X'32' | For clear indication |
| X'33' | For reset indication |
| X'34' | For restart indication |
| X'40' | *Call setup or call clearing problem—general* |
| X'41' | Facility/registration code not allowed |
| X'42' | Facility code not allowed |
| X'43' | Invalid called address |
| X'44' | Invlaid calling address |
| X'45' | Invalid facility/registration length |
| X'46' | Incoming call barred |
| X'47' | No logical channel available |
| X'48' | Call collision |
| X'49' | Duplicate facility requested |
| X'4A' | Non-zero address length |
| X'4B' | Non-zero facility length |
| X'4C' | Facility not provided when expected |
| X'4D' | Invalid CCITT-specified DTE facility |
| X'50' | *Miscellaneous—general* |
| X'51' | Improper cause code from DTE |
| X'52' | Non-aligned byte (octet) |
| X'53' | Inconsistent Q-bit setting |
| X'70' | *International problem—general* |
| X'71' | Remote network problem |
| X'72' | International protocol problem |
| X'73' | International link out of order |

486

X'74' International link busy

X'75' Transit network facility problem

X'76' Remote network facility problem

X'77' International routing problem

X'78' Temporary routing problem

X'79' Unknown called DNIC

X'7A' Maintenance action

X.25 Diagnostic Codes greater than X'80' are reserved for network-specific diagnostic information, for example, Diagnostic Codes applicable to national networks. Refer to the documentation from your network administration for applicable specifications.

**X.25 Dial Service**  A service marketed by Telenet Communications Corporation of Reston, VA, which enables concurrent communications with up to 25 host applications. X.25 Dial Service provides public data access as well as private dial ports and supports asynchronous, SDLC, BSC and 2780/3780 protocols.

**X.25 frame structure**  The X.25 frame structure is illustrated. Each X.25 frame begins and ends with a single 8-bit octet called a FLAG whose bit structure is 01111110. The information field begins with a General Format Identifier (GFI) which is half an octet (4 bits) in length and contains the Q (Qualifier) bit, D (Delivery Confirmation) bit, and SN (Packet Sequence Numbering) bits. The second half of the octet contains the Logical Group Number (LGN). This is followed by the Packet Type Identifier (PTI) and the actual data.

X.25 FRAME STRUCTURE

| FLAG | ADDRESS FIELD | CONTROL FIELD | INFORMATION FIELD | | | | | FCS |
|---|---|---|---|---|---|---|---|---|
| | | | GFI | LGN | LCN | PTI | DATA | |
| 1 Octets | 1 | 1 | 1/2 | 1/2 | 1 | 1 | 1–28 | 2 |

**X.25 PAD**  A device that permits communication between non-X.25 devices and the devices in an X.25 network.

**X.25 NCP Packet Switching Interface (NPSI)**  The X.25 Network Control Program Packet Switching Interface, which is an IBM program product that allows SNA users to communicate over packet switched data networks that have interfaces complying with Recommendation X.25 (Geneva 1980) of the International Telegraph and Telephone Consultative Committee (CCITT). It allows SNA programs to communicate with SNA equipment or with non-SNA equipment over such networks. In addition, this product may be used to attach native X.25 equipment to SNA host systems without a packet network.

**X3S3**  The Data Communications Technical Committee of the American National Standards Institute (ANSI). X3S3 Data Communications Technical Committee task groups are listed in the following table:

X3S31  Planning

X3S32  Glossary

X3S33  Transmission Format

X3S34  Control Procedures

X3S35  System Performance

X3S36  Signaling Speeds

X3S37  Public Data Networks

**X3T5**  Information Processing Technical Committee. Working Groups of the X3T5 committee are organized as follows:

X3T5.1  OSI Architecture and the Reference Model.

X3T5.4  OSI Management Protocols and the Directory.

X3T5.5  OSI Session, Presentation, and Application Layers.

**X.400**  A Consultative Committee for International Telephone and Telegraph standard which facilitates communications between messaging systems.

**X.500**  A Consultative Committee for International Telephone and Telegraph standard which defines a directory system used for permitting E-mail users to obtain an E-mail address.

# Y

**YCP**   (yellow) crossover patch cable.

**Yellow Alarm**   An alarm condition on a T1 circuit; the hardware at one end of a T1 circuit enters the Yellow Alarm condition when it receives either a Red Alarm signal or a Very Severe Burst over the circuit. In ESF, a device signals a Yellow Alarm by transmitting a continuous pattern of Hex FF Hex 00 over the Facility Data Link.

**Yellow Book**   The CCITT edition of colored books in which all of its recommendations for the 1977 through 1980 period are gathered.

**YMODEM**   A version of the XMODEM protocol that sends data in 1024 byte blocks with a 2-byte CRC. Also known as XMODEM 1K. The YMODEM protocol supports the transfer of multiple files.

**YMODEM G**   A streaming protocol that sends an entire file prior to waiting for an acknowledgment. This protocol is designed for use with modems that have built-in error-correction capability, such as Hayes Microcomputer Products V-Series modems.

# Z

**Z**  Impedance.

**ZAP**  To eradicate all or part of a program or data base.

**ZBTSI**  1. Zero Bit Time-Slot Insertion. 2. Zero Byte Time Slot Interchange.

**zero beat**  The condition where two frequencies are exactly the same and therefore produce no beat (audible) note.

**zero-bit insertion**  A technique in bit-oriented protocols such as HDLC/SDLC to achieve transparency. A zero is inserted into sequences of one bits that would cause false flag detection. Then, the inserted zero is removed at the receiving end of the data link. Also called bit stuffing.

**Zero Bit Time-Slot Insertion (ZBTSI)**  The most complex technique for maintaining ones' density on framed T1 circuits in which an area in the ESF frame carries information about the location of all-zero bytes consisting of eight consecutive zeros in the data stream. This technique requires intelligent rearrangement of the bit stream at both ends of the circuit and is rarely used. Also referred to as Zero Byte Time Slot Interchange (ZBTSI).

**zero code suppression**  The insertion of a "one" bit to prevent the transmission of eight or more consecutive "zero" bits; used with digital T1 and related facilities.

**zero insertion**  In SDLC, the process of including a binary 0 in a transmitted data stream to avoid confusing data and SYN characters; the inserted 0 is removed at the receiving end.

**zero slot lan**  A local area network that uses the built-in serial port of a personal computer instead of a separate network card. This reduces the cost of the LAN, although it operates slower then conventional LANs.

**Zero Transmission Level Point (0 TLP)**  A reference point for measuring the signal power gain and loss of a circuit.

**ZIF**  Zero Insertion Force.

**ZIP**  The extension of a file compressed using PKZIP, a sharware program from PKWare Inc. of Brown Deer, WI. Also a term used to refer to compressing a file.

**Zmodem**  A file transfer protocol which permits simultaneous transmission in both directions and the resumption of transmission from where a previous transmission was suspended due to line noise or another type of interruption.

**ZMODEM**  A streaming version of the XMODEM protocol. The ZMODEM protocol keeps sending data unless it receives a negative acknowledgment (NAK). When a NAK is received the protocol starts retransmission from the failed block. This protocol was designed for use with transmission via packet switched networks.

**zone**  In a bulletin board system based network, a

zone is a large geographic area that can cover one or more countries or continents.

**zone coordinator**   A person responsible for maintaining the nodelist of a zone for a bulletin board based network and sharing that list with other zone coordinators.

**Zone Mail Hour**   (ZMH) A common hour in a BBS FidoNet network zone when all boards are available for transmitting and receiving NetMail. During the ZMH a BBS does not accept human calls, file requests of EchoMail transfers.

**ZSTEM**   The name for a family of DEC VT terminal emulation products marketed by KEA Systems Ltd of Vancouver, Canada.

@   When used in an Internet address, separates a user name or computer name from a domain address, e.g. user@host.

%   When used in an Internet UUCP address, acts as a forwarding request, resulting in mail being sent ot the address between the % and @ after it is received by the primary address, e.g., user%host1@host2.

!   When used in an Internet address, defines the path data flows between addresses on each side of the exclamation point, e.g., host1!host2!user.

::   When used in an Internet address, separates the host and user name. The double colon is primarily used in Digital Equipment Corporation computer networks, e.g., host::user.

# 0–9

**029**  An IBM card punch.

**059**  An IBM card verifier.

**083**  An IBM punch card sorter.

**1AESS**  An AT&T stored-program-controlled switch that supports analog facilities.

**1A2 multi-line phone**  A type of telephone with lighted buttons to select up to 5, 9 or 19 phone lines, plus a hold button. It's easily identified by its thick 25-pair cable that ends in a $3\frac{1}{2}$ inch long connector to attach the phone to the line.

**1A2**  An electromechanical key system marketed by AT&T which can be used to obtain station features from a telephone company's central office.

**1-persistent**  In LAN technology, same as persistent.

**2B+D**  A term which denotes two bearer (2B) and one signaling (D) ISDN channel and which is synonymous with the term basic rate interface (BRI).

**2B1Q**  Two binary, one quaternary. The transmission line code for the ISDN basic rate interface (BRI). This means that two binary bits of data are compressed and transmitted in one time state as a four-level code. It is the T-1 committee's recommended draft standard for the line code for echo cancellation devices used in North America. This ISDN standard for digital line code sends a 160-Kb/s duplex signal over a copper pair using a quaternary pulse and echo cancellation. The signal can be transmitted over a pair of wires up to three miles without the use of repeaters.

**2W E&M**  Two-wire Ear and Mouth signaling.

**2-Wayfarer Mobile Communications System**  A two-way truck communications and tracking system marketed by Sony Corporation of America. This system uses radio signals to pinpoint the location of trucks and enables dispatchers and drivers to communicate with one another via satellite.

**3B4000**  A minicomputer system marketed by AT&T which supports up to 400 simultaneous users.

**4B3T**  A line code for echo cancellers advocated by the West German standards group, in which four binary data bits are compressed to transmit in three time states.

**4B/5B**  A digital signal encoding technique in which four bits of data are mapped into a 5-bit code. This code is used in Fiber Distributed Data Interface (FDDI) transmission where each element of the 4B/5B code is encoded using Non-Return to Zero Inverted (NRZI) coding. The following table lists the 32 5-bit codes used in FDDI.

| Decimal | Code | 4B/5B Code Symbol | Assignment |
|---------|------|--------|------------|
| *Line state symbols:* | | | |
| 00 | 00000 | Q | QUIET |
| 31 | 11111 | I | IDLE |
| 04 | 00100 | H | HALT |

## 4B/5B Code

| Decimal | Code | Symbol | Assignment |
|---------|------|--------|------------|

*Starting delimiter:*

| Decimal | Code | Symbol | Assignment |
|---------|------|--------|------------|
| 24 | 11000 | J | First of sequential SD pair |
| 17 | 10001 | K | Second of sequential SD pair |

*Data symbols:*

| Decimal | Code | Symbol | Hex | Binary |
|---------|------|--------|-----|--------|
| 30 | 11110 | 0 | 0 | 0000 |
| 09 | 01001 | 1 | 1 | 0001 |
| 20 | 10100 | 2 | 2 | 0010 |
| 21 | 10101 | 3 | 3 | 0011 |
| 10 | 01010 | 4 | 4 | 0100 |
| 11 | 01011 | 5 | 5 | 0101 |
| 14 | 01110 | 6 | 6 | 0110 |
| 15 | 01111 | 7 | 7 | 0111 |
| 18 | 10010 | 8 | 8 | 1000 |
| 19 | 10011 | 9 | 9 | 1001 |
| 22 | 10110 | A | A | 1010 |
| 23 | 10111 | B | B | 1011 |
| 26 | 11010 | C | C | 1100 |
| 27 | 11011 | D | D | 1101 |
| 28 | 11100 | E | E | 1110 |
| 29 | 11101 | F | F | 1111 |

*Ending Delimiter:*

| Decimal | Code | Symbol | Assignment |
|---------|------|--------|------------|
| 13 | 01101 | T | Used to terminate the data stream |

*Control indicators:*

| Decimal | Code | Symbol | Assignment |
|---------|------|--------|------------|
| 07 | 00111 | R | Denoting logical ZERO (reset) |
| 25 | 11001 | S | Denoting logical ONE (set) |

*Invalid code assignments:*

| Decimal | Code | Symbol | Assignment |
|---------|------|--------|------------|
| 01 | 00001 | V or H | These code patterns |
| 02 | 00010 | V or H | shall not be transmitted |
| 03 | 00011 | V | because they violate consecutive code-bit zeros or duty |
| Decimal | Code | Symbol | Assignment |
| 05 | 00101 | V | cycle requirements. |
| 06 | 00110 | V | Codes 01, 02, 08 |
| 08 | 01000 | V or H | and 16 shall, however, |
| 12 | 01100 | V | be interpreted as |
| 18 | 10000 | V or H | HALT when received. |

(12345) = sequential order of code-bit transmission.

**4GL**  Fourth-Generation Language.

**4W E&M**  Four-wire Ear and Mouth signaling.

**5E4.2**  AT&T's ISDN basic rate interface specification.

**10BASE5**  An IEEE 802.3 media standard abbreviation for 10 Mbps, baseband, 500 meters.

**10NET**  A registered trademark of Digital Communications Associates as well as a local area network marketed by that vendor's 10NET Communications Division in Dayton, OH.

**15 Pitch**  A pitch with fifteen characters per inch.

**23B+D**  A term which denotes twenty-three bearer (23B) and one signaling (D) ISDN channel and which is synonymous with the term primary rate interface (PRI) in North America.

**31-bit storage addressing**  In IBM's SNA, the storage address structure available in an MVS/XA operating system.

**42A**  This is the most common connector used with 4-wire private line circuits. This is the type of connection that was found on all telephones prior to the advent of the RJ11C modular connector. The four wires found under the cover of the 42A block are terminated with screw-down terminals. Red and green wires are the transmit pair and black and yellow wires are the receive pair.

**50-pin connector**  A connector used to terminate a telephone company 25-wire pair. The 50-pin connector is normally used to connect 25 lines to a PBX.

**353M**  An AT&T premises distribution system adapter which lets businesses use twisted-pair wiring rather than more expensive coaxial cable to connect IBM 3170- and 3270-type data equipment.

**357**  An IBM Badge and/or Serial Card Reader Input Unit for a 357 Data Collection System.

**369A**  An AT&T premises distribution system adapter which allows businesses to use unshielded twisted-pair wiring in an IBM AS/400 computer

| Line | Wire colour | | Wire colour |
|------|------------|---|-------------|
| Line 1 | white/blue | | white/blue |
| | orange/white | | white/orange |
| Line 2 | green/white | | white/green |
| | brown/white | | white/brown |
| Line 3 | slate/white | | white/slate |
| | blue/red | | red/blue |
| Line 4 | orange/red | | red/orange |
| | green/red | | red/green |
| Line 5 | brown/red | | red/brown |
| | slate/red | | red/slate |
| Line 6 | blue/black | | black/blue |
| | orange/black | | black/orange |
| Line 7 | green/black | | black/green |
| | brown/black | | black/brown |
| Line 8 | slate/black | | black/slate |
| | blue/yellow | | yellow/blue |
| Line 9 | orange/yellow | | yellow/orange |
| | green/yellow | | yellow/green |
| Line 10 | brown/yellow | | yellow/brown |
| | slate/yellow | | yellow/slate |
| Line 11 | blue/violet | | violet/blue |
| | orange/violet | | violet/orange |
| Line 12 | green/violet | | violet/green |
| | brown/violet | | violet/brown |
| | slate/violet | | violet/slate |

*50-pin connector*

environment, eliminating the need for more expensive, shielded twinaxial cable.

**560 series** The model designation for the basic series of key telephone manufactured by AT&T's Western Electric Company and other vendors.

**564 series** The model designation for a desk telephone equipped with six keys that funtions as a key telephone. The left key is non-locking and operates a HOLD feature while the other five keys are locking (LINE-PICK-UP) that connect the telephone to up to five lines.

**565 series** An AT&T key telephone set model similar to the 564 series but which has an exclusion key associated with the first pick-up button.

**630 series** The model designation for an AT&T call director telephone equipped with 12 key buttons which can be expanded to 18 key buttons with an adapter.

**631 series** The model designation for an AT&T call director equipped with 18 key buttons which can be expanded to 30 keys.

**636 series** The model designation for an AT&T call director telephone set similar to the 630 series except that it is equipped with a jack for plugging in an operator's handset.

**637 series** The model designation for an AT&T call director telephone set similar to the 636 series except that it is equipped with 30 key buttons.

**802.1** The IEEE committee specification which provides an overview of the IEEE 802 standards.

**802.2** The IEEE committee specification for logical link control (LLC).

**802.3** The IEEE committee specification that describes a physical bus which uses the Carrier Sense Multiple Access/Collision Detection (CSMA/CD) access method and which is almost identical to Ethernet.

**802.4** The IEEE committee specification which describes a physical bus utilizing token-passing as the access method. This standard was adopted by General Motors as the data link layer of their Manufacturing Automation Protocol (MAP).

**802.5** The IEEE committee specification which describes a physical ring utilizing token-passing as the access method.

**802.6** The IEEE committee specification for a metropolitan area network.

**900** The three-digit telephone number prefix normally associated with "information and entertainment" telephone services, such as 976-XXXX.

**911** A three-digit telephone number which provides access to emergency service in most areas of the United States.

**1287** An IBM optical reader.

**1288** An IBM optical page reader.

**2501** An IBM card reader.

**2701** IBM Data Adapter Unit.

**2702** IBM Transmission Control Unit.

**2730B** An AT&T premises distribution system enhanced fiber optic multiplexer which can be used to provide IBM data communications equipment users with the higher speed and capacity of fiber optics at lower cost than coaxial cable. The 2730B multiplexer accepts the IBM 3299 protocol and connects IBM 3174/3270 controllers with up to 32 IBM type A terminals.

**3088** An IBM Multisystem Channel Communications Unit used for interprocessor communications over block multiplexer channels.

**3101** An IBM stand-alone CRT display terminal that uses ASCII code and has an asynchronous communications interface.

**3104** An IBM CRT display terminal that can be attached to that vendor's 8100 Information System or 4331 Processor.

**3151** A series of IBM ASCII displays that provide emulation capabilities compatible with Digital Equipment Company, Wyse, Data General, and TeleVideo display stations.

**3174** The most recent addition to IBM's series of cluster controllers. Members of the 3174 control unit series can support 8, 16 or 32 display stations and have either a token-ring network adapter or an asynchronous emulation adapter installed in the device.

**3178** An IBM compact, lightweight, monochrome display station.

**3179** An IBM compact, lightweight, color display station.

**3180** An IBM low-riced, multiple screen format, monochrome display station.

**3191** An IBM monochrome display station which has a built-in printer port on some models, is manufactured with 12-, 14-, or 15-inch screens, supports 102-, 104-, and 122-key keyboards, and contains a record/play/pause function which permits up to 1500 characters to be stored for later recall.

**3192** An IBM color display station which has a built-in parallel printer port on some models, is manufactured with 14- or 15-inch screens, supports 102-, 104-, and 122-key keyboards and contains a record/play/pause function which permits data to be stored in RAM for later recall.

**3193** An IBM high-resolution, portrait-like, monochrome display station. The Model 1 has a 122-key typewriter keyboard while the Model 2 has a 102-key IBM enhanced keyboard.

**3194** An IBM series of color and monochrome display stations. The members of the 3194 series support four host windows, two notepads and screen management functions. The 3194 display stations have a built-in parallel printer port, support the use of 102-, 104-, and 122-key keyboards, and have a record/play/pause feature which permits up to 30 000 characters to be stored in RAM for recall.

**3210** An IBM selectric console typewriter that is used as an I/O unit for certain S/370 computers.

**3251** An IBM interactive computer graphics display station that can be used with a S370, 30XX or 43XX processor.

**3258** An IBM channel-attached control unit which supports up to 16 3251 display stations.

**3262** A series of IBM line printers, used with S/34 and S/38 minicomputers, which can be attached to 3274 and 3276 control units and other IBM products.

**3270 Data Module** An AT&T product which enables IBM 3270 type terminals to communicate through a 5ESS switch using standard twisted-pair building wiring.

**3270 data stream**  A coded character data stream.

**3270, 3270 Information Display System**  A very popular IBM data entry and display system which consists of control units, display stations, printers, and other equipment.

**3274**  An IBM stand-alone control unit. Locally attached models 21A, 21B, 21D, 31A 31D, 41A and 41D can control clusters of up to 32 display stations and printers. Remotely attached models 21C, 31C and 41C can control clusters of up to 32 display stations and printers. Remotely attached models 51C and 61C control mid-sized clusters, with the model 51C supporting up to 12 display stations and printers while the model 61C can control up to 16.

**3276**  An IBM table-top control unit which is integrated into a display station and controls up to eight display stations and printers. Models 1, 2, 3, 4 operate at 1200, 2400, 4800, 7200 bps using the BSC protocol. Models 11, 12, 13, 14 operate at 1200, 2400, 4800, 7200 and 9600 bps using the SNA/SDLC protocol.

**3278**  An older monochrome display station no longer manufactured by IBM.

**3279**  An older color display station no longer manufactured by IBM.

**3290**  An IBM display station that features a large, flat plasma panel as its visual display medium.

**3299**  An IBM terminal multiplexer which permits up to eight terminals to be connected to an IBM 3174 or 3274 control unit via a single coaxial cable.

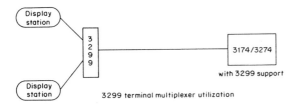

3299 terminal multiplexer utilization

**3601**  An IBM programmable communication controller which is used to attach 3600 Finance Com-

munication System terminals to several types of IBM processors.

**3663**  An IBM supermarket terminal.

**3667**  An IBM checkout scanner.

**3683**  An IBM point of sale terminal.

**3684**  An IBM point of sale control unit.

**3704**  A low-cost programmable communications processor announced by IBM in 1974.

**3705**  IBM's first programmable communications processor that was released in 1972.

**3710**  An IBM network controller capable of handling up to 31 lines using mixed protocols to include X.25.

**3725**  An IBM programmable communications processor, announced in March 1983, which can also be used as a node processor or remote concentrator.

**3726**  The expansion chassis for the IBM 3725.

**3737**  IBM's Remote Channel-to-Channel Unit. This device is a stand-alone control unit that provides host-to-host connectivity through a high-capacity communications facility at data rates up to 1.544 Mbps.

**3745**  An IBM programmable communications processor which supports up to eight T1 circuits.

**3747**  An IBM data converter used to convert batched data from diskette to one-half inch magnetic tape.

**3780**  An IBM data transmission terminal that uses the bisynchronous transmission protocol.

**3800 compatibility mode**  Operating the IBM 3800 model 3 as a 3800 model 1.

**3800 model 3 startup**  That process part of system initialization when the IBM 3800 model 3 is initializing.

**3812**  An IBM multifunction, nonimpact page printer of table-top design that can be connected to an IBM 3270 Information Display System.

**3814** An IBM Switching Management System that is used to switch processor channels among I/O control units in a data processing center.

**3833** An IBM 2.4 Kbps modem designed for use on leased lines.

**3834** An IBM 4.8 Kbps modem designed for use on leased lines.

**3845** An IBM data encryption device which supports asynchronous, bisynchronous and SDLC communications at data rates from 110 to 19 200 bps.

**3846** An IBM rack-mounted data encryption device similar to the IBM 3845.

**3863 model 1** An IBM 2400 bps modem designed for use on leased lines.

**3864 model 1** An IBM 4800 bps modem designed for use on leased lines.

**3865** An IBM 9600 bps modem designed for use on leased lines.

**3866** An IBM multimodem enclosure that provides housing, ventilation and power for certain IBM rack-mounted modems.

**3868 model 1** An IBM 2400 bps modem designed for use on leased lines.

**3868 model 2** An IBM 4800 bps modem designed for use on leased line.

**3868 model 3** An IBM 9600 bps modem designed for use on leased line.

**3868 model 4** An IBM 9600 bps modem designed for use on leased line.

**4224** A series of IBM serial dot-matrix impact printers that attach to an IBM 3270 Information Display System through a 3174, 3274 or 3276 control unit.

**4234** A heavy-duty IBM intermediate-speed impact-matrix printer.

**4245** An IBM high-speed printer which includes an optical character recognition (OCR) feature.

**4250** A high-resolution IBM printer which produces camera-ready print masters with text and line-out graphics intermixed.

**4745** An Amdahl Corporation front end processor designed to compete against IBM's 3745 communications controller. The front end processor is compatible with IBM 37XX communications controllers.

**4829** An IBM 2.4 Kbps modem developed on a half-size adapter card that is designed for use inside an IBM PC or compatible computer.

**4860** The IBM PC jr home computer.

**5150** The original IBM personal computer.

**5151** An IBM monochrome display designed for use with the IBM PC series of personal computers.

**5152** An IBM graphics printer marketed for use with the IBM PC series of personal computers.

**5155** The IBM portable personal computer.

**5160** The IBM PC XT.

**5161** An IBM expansion unit that can be connected to the IBM PC or IBM PC XT.

**5210** A desktop IBM impact printer that uses a bi-directional printwheel to produce letter-quality output.

**5540** An IBM multiworkstation that can function as a Japanese-language personal computer for business applications, a Japanese-language word processor, and a Japanese-language on-line communications terminal.

**5550** An IBM multiworkstation similar to the 5540.

**5560** An IBM multiworkstation similar to the 5540.

**5660** An NCR Comten front end processor that supports up to 1024 communications ports, eight mainframes and 16 Mbytes of memory. The 5660 is designed to work in an IBM System Network Architecture environment.

**5811** An IBM baseband modem capable of operating at 2400, 4800, 9600 or 19 200 bps.

**5811** An IBM limited-distance modem.

**5812** An IBM limited-distance modem.

**5822** An IBM Data Service Unit/Channel Service Unit (DSU/CSU).

**5852** An IBM stand-alone 2.4 Kbps modem designed for use with personal computers.

**5865** An IBM 9600 bps modem which operates with the vendor's Communications Network Management Facility, which is part of NetView.

**5866** An IBM 14.4 Kbps modem which operates with the vendor's Communications Network Management Facility, which is part of NetView.

**5868** An IBM 9600/14,400 bps modem.

**5979 MODEL L41** An IBM baseband modem capable of operating at 2400, 4800, 9600 or 19 200 bps.

**6150** The IBM RT personal computer.

**6151** The IBM RT personal computer which features a 32-bit reduced instruction set microprocessor.

**6504** An AT&T ISDN key telephone. This telephone has ten user-option buttons that can be programmed by the user with call access features or call appearances in any combination.

**6505** A Codex Corporation Packet Assembler/Disassembler (PAD).

**6525** A Codex Corporation X.25 packet switch.

**7102** An AT&T analog voice terminal which provides single-line capability for Definity 75/85 Communications System, existing System 75, and System 85 customers, as well as for System 25, MERLIN ® communications system, and Dimension ® system customers.

**7170** An IBM protocol converter.

**7172** An IBM protocol converter.

**7410** An AT&T digital voice terminal which supports up to 10 call appearances for customers requiring access to multiple calls. The 7410 will support a headset or speakerphone adjunct. The 7410 is supported by AT&T Definity 75/85 Communications System (Generic 1 and Generic 2) as well as System 75 and System 85.

**7434** An AT&T digital voice terminal which can support up to 34 call appearances on the Definity 75/85 Communications System (Generic 2) and up to 10 call appearances and 24 customized features (identical to the 7405), on System 75 and System 85 and the initial software release of Generic 1. The 7434 will support a headset or speakerphone adjunct.

**7495** An IBM counter-top order entry workstation.

**7505** An AT&T ISDN modular terminal that has a voice/data capability. This terminal has ten user-option buttons that can be programmed by the user with call access features or call appearances in any combination.

**7507** An AT&T 40-button ISDN telephone set that includes calling number identification capability.

**7820** An IBM terminal adapter which attaches synchronous host computer channels, controllers or terminals to networks with ISDN Basic Rate services. Up to two data terminal equipment cards supporting V.24, V.35 or X.21 interfaces can be installed in the 7820.

**7860** A series of IBM modems that operate at data rates from 4.8 to 19.2 Kbps and which are designed to work with NetView.

**7866** An IBM rack system which houses both IBM 7868 analog modems and IBM 5822 DSU/CSU units.

**8086** An Intel Corporation 16-bit microprocessor that uses a 16-bit data path for input/output operations.

**8088** An Intel Corporation 16-bit microprocessor that uses an 8-bit data path for input/output operations.

**8209**  An IBM LAN bridge which connects an IBM Token-Ring Network with an Ethernet IEEE 802.3 LAN.

**8218**  An IBM repeater designed for use with copper wire type 3 media on a token-ring network.

**8219**  An IBM optical fiber repeater that can be used with type 5 cable on a token-ring network.

**8220**  An IBM dual rate (4 or 16 Mbps) optical fiber converter used for transmission between wiring closets on a token-ring local area network.

**8228**  An IBM multistation access unit which permits up to eight token-ring devices to be connected to a token-ring network.

**8514/A**  An IBM video graphics board which generates a 1024 by 768 pixel resolution with 16 colors or gray levels displayable out of a palette of 256 000 colors.

**8525**  The IBM Personal System/2 Model 25 with a 3278/3279 Emulation Adapter.

**8530**  The IBM Personal System/2 Model 30 with a 3278/3279 Emulation Adapter.

**8550**  The IBM Personal System/2 Model 50 with a 3270 connection.

**8560**  The IBM Personal System/2 Model 60 with a 3270 connection.

**8570**  The IBM Personal System/2 Model 70 with a 3270 connection.

**8580**  The IBM Personal System/2 Model 80 with a 3270 connection.

**9270**  An IBM series of voice response units. The Model 40 supports four telephone lines and 4 Mbytes of main memory and 40 Mbytes of disk storage. The earlier Model 1 also supports four telephone lines but only half the storage of the Model 40.

**9274**  An IBM series of voice response units. The 9274 Model 40 supports four telephone lines while the Model 80 supports 12 telephone lines.

**9370**  An IBM series of computer systems designed for departmental usage.

**9722**  An IBM PBX designed specifically for small businesses and branch offices of large companies.

**9800**  A Codex Corporation Integrated Network Management System which can be used with that vendor's modems and multiplexers and which interfaces IBM's NetView.

**10XXX code**  The five digits that are entered prior to dialing 1-700-555-4141 to determine the long distance carriers that service your area. If you simply dial the 1-700-555-4141 number you will find out who is your primary long distance carrier. 10XXX codes and the likely carrier are listed in the following table.

| | |
|---|---|
| 10000-1-700-555-4141 | AT&T |
| 10011-1-700-555-4141 | Metromedia |
| 10220-1-700-555-4141 | MCI |
| 10223-1-700-555-4141 | Cables & Wireless |
| 10288-1-700-555-4141 | AT&T |
| 10333-1-700-555-4141 | U.S. Sprint |
| 10444-1-700-555-4141 | Allnet |
| 10488-1-700-555-4141 | ITT |
| 10777-1-700-555-4141 | U.S. Sprint |
| 10888-1-700-555-4141 | MCI |

**68030**  A 32-bit microprocessor manufactured by Motorola Corporation.

**68882**  A mathematical coprocessor manufactured by Motorola Corporation. The 68882 is designed for use in conjunction with Motorola's 68030 32-bit microprocessor.

**80386**  An Intel Corporation 32-bit microprocessor. Different versions of the 80386 operate at 16, 20 and 25 MHz.

**80386SX**  An Intel Corporation microprocessor that has a 32-bit architecture but uses a 16-bit input/output (I/O) bus.

**80486**  An Intel Corporation microprocessor that is binary-compatible with the firm's 80386 microprocessor. The 80486 integrates a floating-point unit, provides 8 Kbytes of cache memory and includes memory management with paging.

**88000**  A Motorola Reduced Instruction Set Computer (RISC) microprocessor.

# LOCAL AREA NETWORKING

## PROTECTING LAN RESOURCES
### A Comprehensive Guide to Securing, Protecting and Rebuilding a Network

With the evolution of distributed computing, security is now a key issue for network users. This comprehensive guide will provide network managers and users with a detailed knowledge of the techniques and tools they can use to secure their data against unauthorised users. Gil Held also provides guidance on how to prevent disasters such as self-corruption of data and computer viruses.

**1995   0  471  95407  1**

## LAN PERFORMANCE
### Issues and Answers

The performance of LANs depends upon a large number of variables, including the access method, the media and cable length, the bridging and the gateway methods. This text covers all these variables to enable the reader to select and design equipment for reliability and high performance.

**1994   0  471  94223  5**

## TOKEN- RING NETWORKS
### Characteristics, Operation, Construction and Management

This  timely book provides the reader with a comprehensive understanding of how Token-Ring networks operate, the constraints and performance issues that affect their implementation, and how their growth and use can be managed both locally and as part of an Enterprise network.

**1993  0  471  94041  0**

## ETHERNET NETWORKS
### Design, Implementation, Operation,  and Management

**1994  0  471  59717  1**

# REFERENCE

## DICTIONARY OF COMMUNICATIONS TECHNOLOGY
### Terms, Definitions and Abbreviations

**1995  0  471  95126  9 (Paper)**
**       0  471  95542  6 (Cloth)**

## THE COMPLETE MODEM REFERENCE
### 2nd Edition

**1994  0  471  00852  4**

## THE COMPLETE PC AT AND COMPATIBLES REFERENCE MANUAL

**1991  0  471  53315  7**

# ABOUT THE AUTHOR

Having gained a B.S.E.E. from Pennsylvania Military College, Gilbert Held majored in computer science for his M.S.E.E. from New York University. He also holds an M.B.A. and the M.S.T.M. degree from the American University.

Gibert Held is Chief of Data Communications for the United States Office of Personnel Management. He serves as a consultant to a number of companies and teaches several college courses in Computers and Decision Theory, Management Information Systems and Data Communications. He is the author of a large number of books and articles.

The only person to win the Interface Karp award for excellence in technical writing twice, Gilbert Held is also the winner of the Association of American Publishers Professional and Scholarly Publishing Division award. He has been selected as one of the Federal Computer Week top 100 professionals in Government, Industry and Academia who have made an outstanding contribution in the field of computer science. He has also received several Government awards for exceptional performance.